Earth Structures

Earth Structures

In Transport, Water and Environmental Engineering

Ivan Vaníček
Faculty of Civil Engineering, Czech Technical University,
Prague, Czech Republic

and

Martin Vaníček
Faculty of Civil Engineering, Czech Technical University,
Prague, Czech Republic

Ivan Vaníček
Faculty of Civil Engineering
Czech Technical University
Prague
Czech Republic

Martin Vaníček
Faculty of Civil Engineering
Czech Technical University
Prague
Czech Republic

ISBN: 978-1-4020-3963-8   e-ISBN: 978-1-4020-3964-5

Library of Congress Control Number: 2007941935

© 2008 Springer Science+Business Media B.V.
No part of this work may be reproduced, stored in a retrieval system, or transmitted
in any form or by any means, electronic, mechanical, photocopying, microfilming, recording
or otherwise, without written permission from the Publisher, with the exception
of any material supplied specifically for the purpose of being entered
and executed on a computer system, for exclusive use by the purchaser of the work.

Printed on acid-free paper

9 8 7 6 5 4 3 2 1

springer.com

Contents

1 **Introduction – Design Specification for Earth Structures** 1

2 **Soil as a Construction Material** 7
   2.1 Soil Classification – Specification for Utilization in Different Parts of Earth Structures 7
      2.1.1 Suitability of Soils for Transport Engineering Earth Structures 15
      2.1.2 Suitability of Soils for Hydro Engineering Earth Structures 17
   2.2 Borrow Pit – Soil in Natural State 20
      2.2.1 Soil Genesis 20
      2.2.2 Geotechnical Investigation 26
      2.2.3 Workability 31
      2.2.4 Soil Treatment Before the Utilization 32
   2.3 Soil Compaction 33
      2.3.1 Compaction Mechanism 34
      2.3.2 Laboratory Compaction Methods 36
      2.3.3 Compacting Machinery 43
      2.3.4 Compaction Quality Control 53
      2.3.5 Soil Properties as a Function of Compaction Effort 66

3 **Geosynthetics in Earth Structures** 71
   3.1 Basic Types 72
   3.2 Properties 75
      3.2.1 Index Properties 76
      3.2.2 Hydraulic Properties 77
      3.2.3 Strength Characteristics 81
      3.2.4 Friction Resistance of Soil – Geosynthetic Interface 83
      3.2.5 Mechanical Resistance 85
      3.2.6 Chemical Resistance 86
   3.3 Different Functions in Earth Structures 87
      3.3.1 Filtration Function 87
      3.3.2 Drainage Function 92

		3.3.3	Reinforcing Function 94
		3.3.4	Separation Function 97
		3.3.5	Protective Function................................... 97
		3.3.6	Surface Erosion Protection Function 98
		3.3.7	Sealing Function................................... 101
		3.3.8	Summary of Applications and Properties................ 104
	3.4	Geosynthetic Handling on Construction Site 105
4	**Soil Improvement** 109		
	4.1	Speeding Up of Consolidation 109	
		4.1.1	Preloading 115
		4.1.2	Vertical Drains 118
	4.2	Soil Stabilization.. 123	
		4.2.1	Mechanically Stabilized Soil........................ 123
		4.2.2	Stabilization with the Help of Cement, Lime and Other Hydraulic Bonding Agents 125
	4.3	Soil Reinforcement...................................... 133	
		4.3.1	Geosynthetic Soil Reinforcement 133
		4.3.2	Soil Nailing..................................... 145
5	**Limit States for Earth Structures** 153		
	5.1	Principles of Design Procedure for Earth Structures.............. 153	
		5.1.1	Risk Principle in the Design of Geotechnical Structures – Geotechnical Categories.................... 153
		5.1.2	Ultimate Limit States and Serviceability Limit States 155
		5.1.3	Actions, Design Situations 155
		5.1.4	Ultimate Limit States 157
		5.1.5	Basic Logic of Design Approach Based on Calculation of Limit States 165
		5.1.6	Other Methods of Limit States Verification 167
	5.2	Basic Limit States for Earth Structures 169	
		5.2.1	Limit State of Stability............................. 169
		5.2.2	Limit State of Deformation 187
		5.2.3	Limit State of Internal Erosion 204
		5.2.4	Limit State of Surface Erosion 217
	5.3	Modelling of Limit States 223	
		5.3.1	Numerical Models 224
		5.3.2	Laboratory Models................................ 234
		5.3.3	Modelling of In Situ – Monitoring 246
6	**Earth Structures in Transport Engineering** 259		
	6.1	Basic Assumptions – Situation of Traffic Network, Longitudinal Section..................................... 260	
		6.1.1	Application of Waste and Recycled Materials............ 261
		6.1.2	Landslide Prone Areas 262

	6.2	Transport Embankment – Fill 266
		6.2.1 Classical Case – Embankment on Good Quality Subsoil ... 266
		6.2.2 Embankment on Soft Subsoil 275
		6.2.3 Embankment Widening 299
	6.3	Cutting 303
		6.3.1 Classical Case – Limit States 303
		6.3.2 Ground Water Lowering 308
		6.3.3 Cuts Widening 310
	6.4	Specific Cases of Interaction 317
		6.4.1 Retaining Walls...................................... 317
		6.4.2 Bridge Abutments.................................... 327
		6.4.3 Interaction of Fill with Different Underpasses 330
	6.5	Specificity for Roadways 334
		6.5.1 Unpaved Roads...................................... 339
		6.5.2 Interaction of Construction Layer with Subsoil........... 343
		6.5.3 Parking Place 345
	6.6	Specificity for Railways 346
		6.6.1 Track Reconstruction for Higher Speed 348
		6.6.2 Construction of New Tracks for High Speed Trains 355
	6.7	Specificity for Airfields 361
		6.7.1 Practical Examples 362
		6.7.2 Experience Gained from the Construction of Airfields in Challenging Geotechnical Conditions 365
	6.8	Relation to Environments.................................... 370
		6.8.1 Impact of Waste Material Used in Earth Structures on Environment 370
		6.8.2 Collection of Surface Run-off and Subsoil Sealing 377
		6.8.3 Noise Protection Barriers............................. 378
7	**Earth Structures in Water Engineering** 381	
	7.1	Basic Cases of Earth Structures in Water Engineering and Their Features 382
	7.2	Earth and Rock-Fill Dams, Typical Cross-Section 383
	7.3	An Analysis of the Most Frequent Failures of Earth and Rockfill Dams 385
	7.4	Conditions and Reasons of the Tensile Cracks Development 387
		7.4.1 Conditions of the Tensile Cracks Development........... 392
		7.4.2 Reasons of the Tensile Cracks Development............. 397
	7.5	Stress-Strain Behaviour of Soils in Tension..................... 410
		7.5.1 Tensile Tests Under Undrained Conditions 412
		7.5.2 Results of Undrained Tensile Tests..................... 415
		7.5.3 Results of Drained Tensile Tests....................... 426
	7.6	Soil Behaviour During Seepage Through Crack 432
		7.6.1 Model of Seepage Cracks for Cohesive and Frictional Soils 433

	7.6.2	Swelling Potential 435
	7.6.3	Forces Between Individual Particles 437
	7.6.4	Dispersive Soils 441
	7.6.5	Flocculated Soils 447
7.7	Filtration Contact Stability for Protected Soil with a Crack 451	
	7.7.1	Protected Cohesive Soil with Crack 452
	7.7.2	Applicability of Geotextile Filters in Dam Engineering 455
7.8	Practical Examples from Earth and Rock-Fill Dam Construction ... 456	
	7.8.1	Dalešice Dam 456
	7.8.2	Hriňová Dam 459
	7.8.3	Jirkov Dam 461
	7.8.4	Low Dams in Southtown in Prague 462
	7.8.5	Teton Dam 463
	7.8.6	Failures Remediation 467
7.9	Summary of Earth and Rockfill Dam Design 471	
	7.9.1	Recommendation for Elimination of Tensile Cracks Development 471
	7.9.2	Recommended Logical Way for Fill Dams Design 475
7.10	Specificity of Small Fish Pond Dams 479	
	7.10.1	Rehabilitation of Old Earthfill Dams Failed During Heavy Floods 479
	7.10.2	Limit States of Failure 480
	7.10.3	Principles of the Reconstruction Design 484
	7.10.4	Some Recommendation Obtained from the Reconstruction 488
	7.10.5	Conclusions 489
7.11	Specificity of Flood Protection Dams (Dikes) 491	
	7.11.1	Details for Individual Countries 491
	7.11.2	Fundamental Specificity of the Protection Dams 493

8 Earth Structures in Environmental Engineering 497
- 8.1 Spoil Heaps 499
  - 8.1.1 Behaviour of Stored Clayey Material with Large Voids – Macropores 500
  - 8.1.2 Stability of Spoil Heaps 506
  - 8.1.3 Deformation of Spoil Heaps 511
  - 8.1.4 Experiences with New Construction on Spoil Heap Surface 515
- 8.2 Tailings Dams 526
  - 8.2.1 Types and Construction of Tailings Dams 527
  - 8.2.2 Tailings Dams with Sediments Resulting from Burning of Solid Fuel 530
  - 8.2.3 Tailing Dams with Sediments from Concentrator Factories, Uranium Tailings 545
  - 8.2.4 Tailings Dams with Other Sediments 555
- 8.3 Landfills 556
  - 8.3.1 Landfill Classification 557

	8.3.2	Approaches to the Landfill Design	561
	8.3.3	Demands on Sealing Systems for Landfills	567
	8.3.4	Drainage Systems for Landfills	579
	8.3.5	Properties of Stored Waste in Municipal Landfill	582
	8.3.6	Limit States of Landfills	585
	8.3.7	Application of Geosynthetics in Landfill Construction	588
	8.3.8	Landfill Monitoring	589
	8.3.9	Remediation of Old Landfills	591
	8.3.10	Utilization of Landfill Surface for New Construction	594

References 607

Index 629

Preface

One of the plausible perceptions about safe and generally optimal geotechnical structure, leads to the conclusion that its success is supported by four columns. The first column relies on the understanding of natural sciences such as Geology, Engineering Geology and Hydrogeology on one side, and on the understanding of Mechanics, Theory of Elasticity on the other side. The second column relies on the application of existing findings on the behaviour of soils and rocks under different stress-strain states – we are speaking about support from Soil Mechanics and Rock Mechanics. The third column relies on the combination of the theoretical findings with practical technologies during execution of Foundation Engineering and Underground Structures (Tunnelling). Finally the fourth column relies on a certain feeling of geological environment which Terzaghi (1959) denotes as "capacity for judgment" and he specifies that "this capacity can be gained only by years of contact with field conditions".

The authors are confident that the third column relevant to practical application should be strengthened about "Earth Structures", about application on structures, which belong to the oldest engineering structures utilizing the fundamental structural material – soil. At the same time they believe that Earth Structures will establish their position in the near future, as it is gaining another interesting field which Earth Structures are also part of, namely the field of Environmental Geotechnics.

That was the main reason why at the Faculty of Civil Engineering of the Czech Technical University, the subject "Earth Structures" was included in the postgraduate study 15 years ago. However up to date students have to their disposal only different author's publications from a wide branch of geotechnical engineering such as textbooks including Soil Mechanics, Foundation Engineering, Remediation of landfills, old ecological burdens, Civil engineering and environment protection, and monographs such as Development and behaviour of cracks in fill dams or Reinforced soils – limit states applied on slopes, retaining walls and bridge abutments.

During the discussions with many colleagues at home and abroad, the authors were challenged number of times to prepare this new comprehensive publication addressing the fundamental principles of Earth Structures. Once this idea gained real shape, during these discussions, the authors obtained significant support, including at least that from Prof. H. Brandl and colleagues from the ELGIP platform.

This publication is the result of these discussions and the authors believe that the book will first be accepted as a stimulative step in the area, where the design and execution of earth structures can, with the good knowledge of all aspects and with the support from the remaining columns, introduce many improvements, risk reductions, and last but not least financial savings.

The preparation of the publication in its final version was not only possible with the support from the family of authors, but also from the team of colleagues and group of research students from the Geotechnical department of the CTU, in particular during assembling of figures, where Ing. D. Jirásko had the main role. Some research activities and publication preparation was also supported from the Czech Ministry of Education and from the Czech Grant Agency. The authors herewith thank all of the above mentioned for their support.

Prague

Ivan Vaníček
Martin Vaníček

Acknowledgement

The authors have prepared practically all figures in this book in electronic form themselves. Where appropriate the original authors of the figures are being cited under the figure description and individual authors or publishers have been asked for permission to use the figures. The authors would like to express their gratitude to all of them.

The book also includes so-called generally known schematic figures, e.g. about compaction of soils, function of geosynthetics, principles of calculating limit states of stability and deformation, etc, which have been used by the authors in their academic activities for many years. If anyone identifies a general schematic figure for which he/she is the first author and no proper acknowledgement has been made, we as authors would like to apologize in advance.

Anyone who has not been properly credited is requested to contact the publishers, so that due acknowledgement may be made in subsequent editions.

Chapter 1
Introduction – Design Specification for Earth Structures

From the beginning it is necessary to mention a certain paradox. Although the earth structures belong together with wooden (timber) structures to the oldest structures, this term is a relatively new one and considerably less used as a typical term for other structures like steel structures, concrete structures or reinforced-concrete structures. A similar situation is at the Universities, where in the faculties of civil engineering, subjects like Concrete and Steel Structures are the most important ones. However, in the branch of geotechnical engineering, the most important subjects are Soil Mechanics, where principles of soil behaviour are covered, and Foundation Engineering, where basic principles for shallow and deep foundation are treated. Earth structures are usually only partly covered in subjects where whole problems of Transport Engineering, Water Engineering and Environmental Engineering are discussed, but very often using different approaches to the design of earth structures, and not giving an overall view of them.

Sometimes the term Earth structures is substituted by the term Earthworks, but it is not correct because this term is connected mostly with technological problems of soil excavation, transport and back-deposition.

Nevertheless with time this situation is improving, but sometimes rather slowly. For the first time the International Society for Soil Mechanics and Geotechnical (Foundation) Engineering recognized this situation; Session 8 during the 11th International Conference ISSMFE in San Francisco 1985 had the title: "Comparison of prediction and performance of earth structures".

One of the main aims of this publication is to help to correct this paradox and contribute to the knowledge enhancement of earth structures not only among civil engineers, but also among the general population.

From the view of specification of earth structures, we will mostly discuss those where the soil is the main construction material. The amount of soil used for the earth structures reaches many millions of cubic meters each year and soil is the far most used construction material. Among earth structures we can distinguish different types of earth fills for both transport engineering – road and railway embankments, parking areas, airports – and for hydro engineering, where we are first of all speaking about different types of dams, from small dams up to large ones or about embankments of small fishponds, protection embankments like flood protection system (dikes), floating channel etc. And finally ending with earth structures in environmental engineering, such as different spoil heaps from waste material after mineral resources extraction, tailings dams composed from fine waste material that is transported to the place of deposition by hydraulic means, as it is in the case of ash, sludge from the chemical concentrator factory or different landfills starting with municipal waste materials and ending with different dangerous chemical materials. But to the category of earth structures belong also those, which are realized in earth environment as are in particular communication trenches and cuts.

The convenience of such a compendious publication is also given by the complexity of the problem and also the relatively great dissimilarity from the design logic of the other types of structures. Again it is paradoxically in contrast with general opinion, because not only the common citizen but also many civil engineers believe that the construction of any fill is not a great problem compared to the design and realization of steel and concrete structures.

In the case of these structures the basic problem is connected with design and construction of beam, column and slab. On the basis of structural analysis the structural designer recommends for steel structures a certain structural element with required properties, and has great confidence that the steel factory will supply this structural element with high confidence in properties, first of all from the tensile and compression strength and from the deformation modulus point of views. For concrete structures the basic logic steps are similar, quality of concrete (strength, modulus) are checked not only in the concrete mixing plant but also at the construction site, where control samples are collected and the quality of concrete is tested in the laboratory as a function of time. Roughly the same is valid for wooden (timber) structures, where for different sorts of wood, basic properties (again strength and deformation modulus) are known in relatively narrow scatter.

On the other hand the logic of the design of earth structures is significantly more complex and indirect. The designer of earth structures needs to ensure that these structures will not fail from the limit state of total stability – that the part of the structure will not suffer from shearing and that slippage will not occur. Furthermore the designer has to ensure that the deformations will not exceed limit values and therefore the functionality of structures themselves or other adjacent structures would not be endangered, or that the permeability of earth structure will fulfil all requirements from the seepage point of view for different types of dams and protection barriers of different types of landfills etc.

The designer of the earth structure is in the first step limited by the possibility of obtaining appropriate soil for construction. Therefore he is looking for an appropriate borrow pit or is checking the possibility to use the soil obtained from the excavation, e.g. from cuts for transport enginecring. Nevertheless soil in natural deposition in the required volume does not exactly have the same properties. They are varying in a certain scatter not only from the point of view of particle size distribution (size of particles), but also from the view of content and type of clay minerals, from moisture content and from many others properties. The earth structure designer can only select a representative samples from the whole borrow pit. The material obtained from samples can be compacted in a laboratory by conventional energy and then on these compacted samples laboratory tests could be performed as shear, tensile, compression and permeability tests. Again the scatter of results can be significant, even some tens of percent. A challenge for the designer (geotechnical engineer) is to recommend so called characteristic values of shear strength, deformation modulus and coefficient of permeability. After that the recommended values are used for the further steps of the design of the proposed earth structure. Subsequently the earth structure designer is recommending how this soil has to be compacted on the construction site and how the quality of compaction has to be measured and checked. In most cases the recommendation is not defining basic properties as strength, compression or permeability but index properties like dry density and moisture content. Only in limited cases are the basic properties checked on site with the help of field tests, as it is for example a field test of permeability on clay liner for sanitary landfills or deformation characteristics determined with the help of a small field loading plate. In most cases the control is performed via dry density and moisture content of compacted soil, in order to prove how individual particles are close to each other for a given moisture content.

When we look back on the whole process through which the geotechnical engineer has to go, we arrive at the conclusion that the degree of accuracy is significantly lower than for steel or concrete structures, where the differences in the design can be in the order of a few percent, while for earth structures these differences can be in the order of tens of percents. That is why on one side the excellent knowledge of soil behaviour and treatment of soil as construction material can bring significant savings against conventional design but on the other one disregarding this can lead to failures of earth structures.

In the case of earth structures constructed by wet way, by hydraulic sluicing, as is the case for tailing dams or just by free fall, as it is the case for most of the spoil heaps, the estimation of soil properties is even more difficult.

The main aim of this book is to show basic principles of earth structure design, to show basic philosophy and the most sensitive places, which will be shown on the limit state design concept according to the Eurocode 7 – Geotechnical design. This concept combines different unfavourable situations, evaluates their probability and recommends partial factors of safety for material properties, for loads and resistances. The whole procedure has the main aim to design the earth structure in accordance with potential risk of failure, which can be accepted for the proposed structure.

To utilize the most appropriate way of construction, the knowledge of soil properties and characteristics are needed. Therefore basic knowledge of soil mechanics is assumed. To understand the expected variability of soil properties, brief information about soil development is incorporated, but the main attention is devoted to the compaction as to the main method by which the soil is treated as the construction material. To be able to give to the earth structures new possibilities, the description of different geosynthetic materials are described together with the role they can fulfil in these structures. All above mentioned roles are discussed in Chapter 2 – Soil as a construction material, Chapter 3 – Geosynthetics in earth structures, and in

Chapter 4 – Soil improvement. Detailed description of basic limit states is covered in Chapter 5 – Limit states for earth structures – where also different ways of soil structure modelling are summarized. Altogether these first five chapters generate common introduction, principles, which are valid for all earth structures.

Practical applications for transport, water and environmental engineering are described in Chapters 6–8, because each of them has many specialities. For transport engineering the limit states of stability and deformation play the most important role. For water engineering it is the limit state of internal erosion, and finally for environmental engineering we add to these limit states, also as the main aspect, the protection of environment against contaminant spreading from the deposited waste material.

Nevertheless there are some differences between individual chapters – e.g. Chapters 6 and 8 show typical problems for different situations which can arise, whereas Chapter 7 concentrates on the most problematic case of tensile cracks in the earth and rock fill dam body – as a preferential pathway for seepage and consequently for internal erosion. Therefore the stress-strain behaviour of soil in tension is also described in more detail than it is common in classical soil mechanics.

In a nutshell we can state that the main attention is devoted to the principles, to the sensitivity of individual cases on changes, which can occur with time and how the structure will react onto them. Slightly less attention is devoted to the calculation methods, which are presented only in the basic versions because with time, all computing methods are improving in order to be able to simulate the earth structures behaviour in a more precise way.

This is in agreement with the recommendation of Eurocode 7, where it is stated that "It should be considered that knowledge of the ground conditions depends on the quality and extent of the geotechnical investigations. Such knowledge and the control of workmanship are usually more significant to fulfilling the fundamental requirements than is precision in the calculation models and partial factors".

The last aspect, which the authors would like to mention, relates to the up to day tendency to consider the technical solution as a necessary condition but unsatisfactory as it is necessary to add to it new additional conditions, above all the environmental one. Therefore special attention is devoted not only to the earth structures for environmental engineering but also to the other types of structures where instead of natural soil, some recycled or secondary materials or soil excavated from a pit for the foundation of a new building, or soil excavated during the realization of underground structures (as is for example construction of metro), can be used. We feel the need to give preference to this solution protecting the natural soil and thus limiting the areas used as borrow pits. This principle is slightly complicated by EU legislative tendency to declare any excavated soil as waste material. Therefore it is useful to find very quick and cheap methods to declare used soil or recycled or secondary material as a material appropriate for earth structure. In the case of potential leachate from deposited material the solution has to prove that the influence of this leachate on the surrounding environment is within acceptable limits. That is why the contaminant spreading from the deposited material (not only from such structures as landfills) is described in more detail.

The authors have mostly utilized their own experience based on extensive consulting, research and teaching activities. However, to publish a comprehensive complexity of problems connected with earth structures, the authors have drawn from the experience of many other authors to give to the readers the overview of literature for deeper study.

Chapter 2
Soil as a Construction Material

As was mentioned in the introduction, soil is the oldest and also the most often used building material. Due to the variability of soils that can be found in nature, not every soil type is usable for a given type of earth structure. For example, gravely soils can be used in drainage layers for removal of water from earth structures. Only clayey soils are able to fulfil demanding criteria for permeability for example in sealing systems of different landfills. The demands for filters, transition layers between protected soil and drainage layers, are sometimes very strict and are located in a very narrow band in relation to the particle size distribution. Therefore, during the selection of soil suitable for a particular earth structure, it is necessary to consider:

- If the respective soil, based on existing experience, is suitable for the given function;
- If the selected soil is in the borrow pit in a required amount and its variability is within acceptable limits, so its suitability for a given purpose will not be limited.

Once the above conditions are satisfied, which can be declared by using relatively simple index tests, the soil will be tested in more detail. These detail tests will concentrate on the possibilities of modifying the soil prior to placement into the earth structure and finally to the soil treatment – on soil compaction. All these phases are discussed in this chapter.

2.1 Soil Classification – Specification for Utilization in Different Parts of Earth Structures

Soil classification systems have several purposes. At first, there is a tendency to classify soil into defined groups based on simple so-called index tests. These groups are given uniform names and symbols. The first reason is that engineers will be able to imagine the same type of soil based on a name and symbol. The second is the possibility of assigning certain common properties to each group. These properties are determined from experience, and from elaborated specific tests called mechanical-physical tests, which represent tests of strength, deformation and permeability. This experience collected mostly within the last century is very valuable and can be used as a certain guide during quick assessment or in the case when a designed structure does not represent any risks for its environment. Design based on experience is one of the basic principles of Eurocode 7 – Design of Geotechnical Structures. Here, the detail of design is related to the potential hazard that the collapse of a structure can cause.

The basis of most classification systems is particle size distribution of soil. The size of individual particles, described as an ideal sphere of a certain diameter, is most often divided into the following ranges:

- Clay particles, smaller than 0.002 mm,
- Silt particles, in between 0.002 and 0.06 mm,
- Sand particles, in between 0.06 and 2.0 mm,
- Gravel particles, in between 2.0 and 60 mm,
- Cobbles, in between 60 and 200 mm,
- Boulders, greater than 200 mm.

This basic division needs the following comments:

- The borders using the numerals of 2 and 6 are very useful, as they allow a more detailed division, for example for sand particles into fine, medium or coarse sand. These borders proposed already in the MIT classification in 1931 and used in many other cases like BS 5930:1999 or ČSN 731001:1987 and are also preferred by Terzaghi and Peck (1948) although they mention that some classifications use a slightly different border between silt and sand particles as this is given by the sizes of sieves that are used for testing – most often these borders are 0.063 mm or 0.074 mm.
- For detailed specification, many classification systems use particles smaller than 60 mm after removal and estimation of cobble and boulder contents.
- Silt and clay particles are together described as a fine fraction (f), sand (s) and gravel (g) particles are commonly called a coarse fraction and cobbles with boulders are called a very coarse fraction. Lately, a lot of classifications have been based on the above mentioned three fractions – f, s and g.
- The shape of soil grains has a big influence on soil properties although their size is described by an ideal sphere of the same size. All properties like permeability, stiffness and strength can show significant differences.
- During the evaluation of soil as a whole it is useful to account for the proportions of particle size – shown as a particle size distribution curve – see Fig. 2.1. From the character of the particle size distribution curve, some additional information can be obtained. A continuous particle size distribution curve, see b), shows a uniform distribution of grains in a reasonably wide range of grain sizes. On the other hand, the curves in a) or in c) represent an example when we deal with soil with just a selected fraction or the soil misses a certain fraction. For the determination of the particle size distribution curve character, the following characteristics are used:

2.1 Soil Classification – Specification for Utilization in Different Parts of Earth Structures

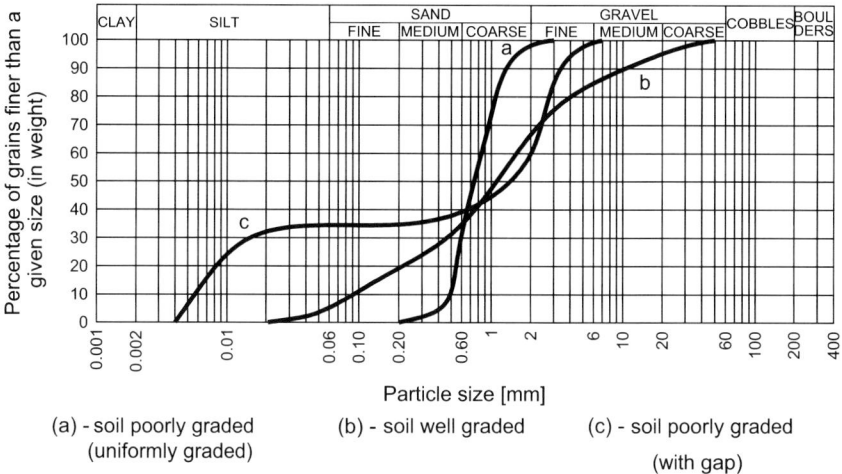

Fig. 2.1 Particle size distribution curve

– Coefficient of uniformity

$$C_u = d_{60}/d_{10} \qquad (2.1)$$

– Curvature number

$$C_c = (d_{30})^2/(d_{10} \cdot d_{60}) \qquad (2.2)$$

These characteristics are used for detailed specification of coarse grained soil into:

- Well graded – W, if $C_u > 6$ for sands or $C_u > 4$ for gravels and at the same time C_c is between 1 and 3
- Poorly graded – P, if the conditions for W are not fulfilled
- Poorly graded uniform – P_u
- Poorly graded with a gap – P_g

• The properties of fine grained soils, with the majority of fine fraction, are highly influenced by the type of clayey mineral. Soils with an identical particle size distribution curve can have completely different properties. Generally, there are three main groups of clay minerals – montmorillonite, illite and kaolinite groups.

Therefore, for more precise specification of fine grained soils or coarse grained soils with a higher percentage of fines, it is necessary to define in more detail the behaviour of this fine fraction, mainly clay particles. Clay minerals have strong coupling with a water phase in soil pores and yield plasticity to the soil, which strongly depends on the activity of clay minerals. The most active are the clay minerals from the montmorillonite group followed by the illite group and the least activity has the kaolin group. It is useful to mention here that nonclayey minerals as quartz and feldspar, even if milled to the size of clay minerals, do not have plastic behaviour.

For identification of plastic behaviour of soils Atterberg limits, the liquid limit w_L and the plastic limit w_p are generally used. Soil moisture at which the soil is starting to have some shear strength determines the limit between liquid state and plastic state and is called the liquid limit. On the other hand, the limit at which the soil changes state from plastic to stiff state is called the plastic limit. The difference between the liquid and plastic limits is defined as the plasticity index I_p, where:

$$I_p = w_L - w_p \tag{2.3}$$

Plasticity index is an important characteristic of fine grained (cohesive) soils. It shows the capability of soil to absorb water without changing its state. This happens because clay minerals are not electrically neutral and therefore they can attach to their surface molecules of water that create a dipole. The range of plasticity index is influenced not only by clay mineral groups but also by the character of exchangeable cations in the crystalline structure of clay minerals. The ones most commonly occurring in soils are calcium Ca^{2+}, magnesium Mg^{2+}, sodium Na^+ and potassium K^+. That is the reason why we distinguish, for example Na-montmorillonite, Ca-montmorillonite etc., while the highest activity, ability to attract water, is that of the sodium cation Na^+. For indirect identification of clay minerals, or at least for its main groups, the index of colloidal activity of clays I_A may be used, see Skempton (1953):

$$I_A = I_p / \text{proportion of grains smaller than 0.002 mm} \tag{2.4}$$

Generally, it is a relationship between an increase of the plasticity index and a proportion of the clay fraction. Typical values of this index are:

- For nonclay minerals: quartz – 0, limestone – 0.18, mica – 0.23,
- For clay minerals: kaolin – 0.33–0.46, illite – about 0.9, Na-montmorillonite – up to 7.2.

Skempton classifies clays as:

- Nonactive clays: $I_A \leq 0.75$
- Normal clays: $I_A = 0.75 - 1.25$
- Active clays: $I_A > 1.25$

General classification systems for engineering purposes use in principle the particle size distribution curve and Atterberg limits as well as the plasticity index. The most commonly used system worldwide is USCS – the United Soil Classification System. This classification is originally based on the classification of Casagrande, which he established for quick assessment of soil suitability for construction of field airfields during World War II.

2.1 Soil Classification – Specification for Utilization in Different Parts of Earth Structures

This classification is widely spread mainly due to the rich source of information on individual soil groups. The most important is the information about soil as a construction material, and about the properties of soils once compacted using a standard compaction method (Proctor standard). It gives typical values for shear strength, permeability and compressibility. US Bureau of Reclamation uses this classification for the purpose of small rockfill/earth dams, as will be specified in more detail later.

The USCS classification system divides soils into:

- Gravels (G), if the soil contains more than 50% of coarse grains ($s + g > f$) and $g > s$,
- Sands (S), if the soil contains more than 50% of coarse grains ($s + g > f$) and $s > g$,
- Silts (M) – M – soils, if the soil contains more than 50% of fine particles ($f > s + g$) and in Casagrande plasticity chart – see Fig. 2.2 – it is located below line "A",
- Clays (C), if the soil contains more than 50% of fine particles ($f > s + g$) and in Casagrande plasticity chart it is located above line "A",
- Organic soils (O),
- Peats (Pt)

For coarse soils, an additional symbol W or P can be added to specify the shape of the particle size distribution curve, or additional symbols M or C if the portion of fine particles is significant. In fine soils, such as silts and clays, the letter L and H, standing for low and high, gives an indication of the liquid limit of the soil, whether it is higher or lower than 50%. A summary of general soil types and their symbols within the USCS classification system is presented in Table 2.1. The USCS classification system is used with many other modifications. These modifications take into account:

- Soils with a higher proportion of a very coarse fraction.
- More detailed distinction of plasticity into low (L) when $w_L < 35\%$, intermediate (I) when $w_L = 35 - 50\%$, high (H) when $w_L = 50 - 70\%$, very high (V) when $w_L = 70 - 90\%$ and extremely high (E) when $w_L > 90\%$, see Fig. 2.2b.
- More precise determination of coarse grained soils – they have symbols G or S if the proportion of fine grains is lower than 35%, which means that coarse grains are in contact with each other and define the soil properties. While when the proportion of fine grains is greater than 35% it is possible that coarse grains can "swim" in the fine fraction and that the properties are mostly dependent on the properties of the fine grains fraction.

One of these classification systems is ČSN 731001:1987 "Foundation of structures. Subsoil under shallow foundations", or ČSN 721002:1993 "Soil classification for road constructions". The basis of this classification is a triangular diagram for the classification of soils with grains smaller than 60 mm (see Fig. 2.3) with additional specification of fines, clays C and silts M as based on the plasticity diagram

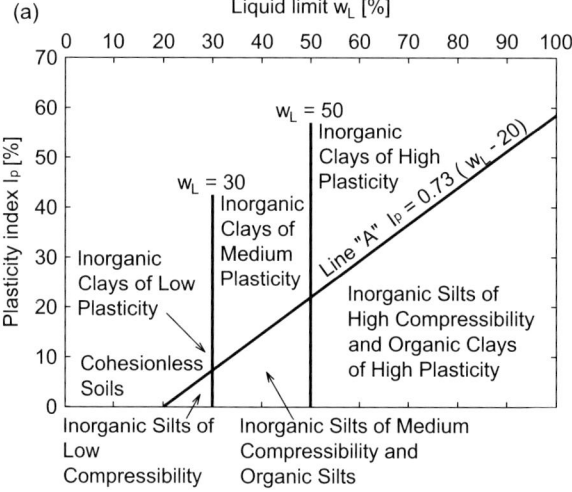

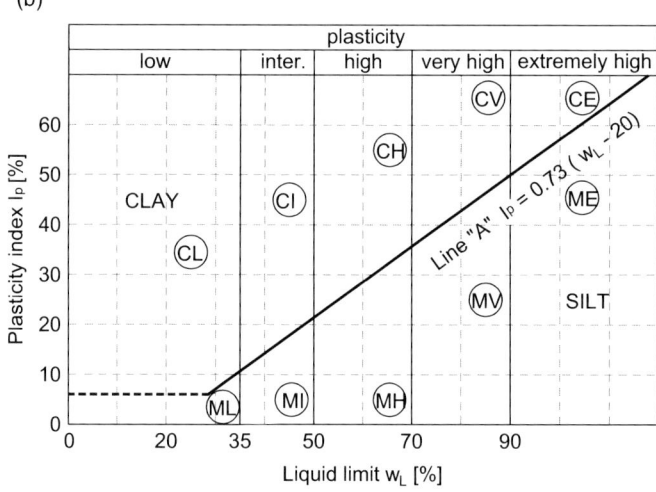

Fig. 2.2 Casagrande plasticity chart. (**a**) US Bureau of reclamation. (**b**) Czech classification of soils

as in Fig. 2.2b. A more detailed description of names and symbols is presented in Fig. 2.4a,b,c (Šimek et al., 1990).

As was already mentioned, apart from particle size distribution and plasticity there are other characteristic criteria of soils that are used in their classification (Vaníček, 1982a). A significant aspect is mainly determination of the proportion of *organic matter*, as it is susceptible to decay and therefore to changes in mechanical-physical properties. As a quick check, the amount of organic matter can be determined from the colour or characteristic odour of the soil and, apart from this, organic remains (decomposing remains of plants etc.) are described visually. In the

2.1 Soil Classification – Specification for Utilization in Different Parts of Earth Structures

Table 2.1 USCS soil classification system

Symbol	Soil type
GW	Well graded gravels, gravel sand mixtures, little or no fines
GP	Poorly graded gravels, gravel sand mixtures, little or no fines
GM	Silty gravels, gravel sand silts mixtures
GC	Clayey gravels, gravel sand clay mixtures
SW	Well graded sands, gravely sands, little or no fines
SP	Poorly graded sands, gravely sands, little or no fines
SM	Silty sands, sand silt mixtures
SC	Clayey sands and clay mixtures
ML	Inorganic silts and very fine sands
CL	Inorganic clays of low to medium plasticity
OL	Organic silts and organic silty clays of low plasticity
MH	Inorganic silts, micaceous fine sandy silty soils
CH	Inorganic clays of high plasticity
OH	Organic clays of medium to high plasticity, organic silts
Pt	Peat or other highly organic soils

laboratory, the amount of organic content is defined as the percentage of the original weight of dry soil. Generally, three methods can be used for the determination of organic content in soils:

- Annealing at temperatures between 700 and 800°C,
- Oxidation by hydrogen peroxide,
- Oxidation by dichromate potassium.

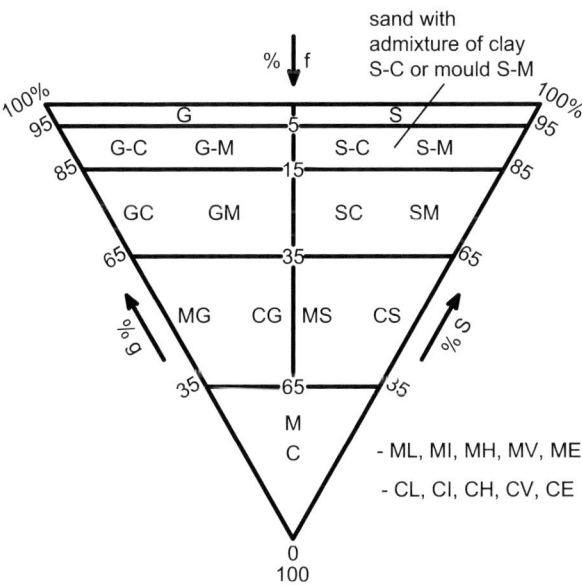

Fig. 2.3 Classification of soil with grains smaller than 60 mm

Amount of carbonates – the presence of lime as a technically important element in soil manifests itself in the form of calcium carbonate. Based on the amount of $CaCO_3$ it is possible to distinguish soils:

- Slightly calcareous (5–25%),
- Calcareous (25–50%),
- Highly calcareous (> 50%).

For an approximate in situ determination of carbonates in soil, the 5% solution of hydrochloric acid is used. The amount of carbonates can be assumed from the

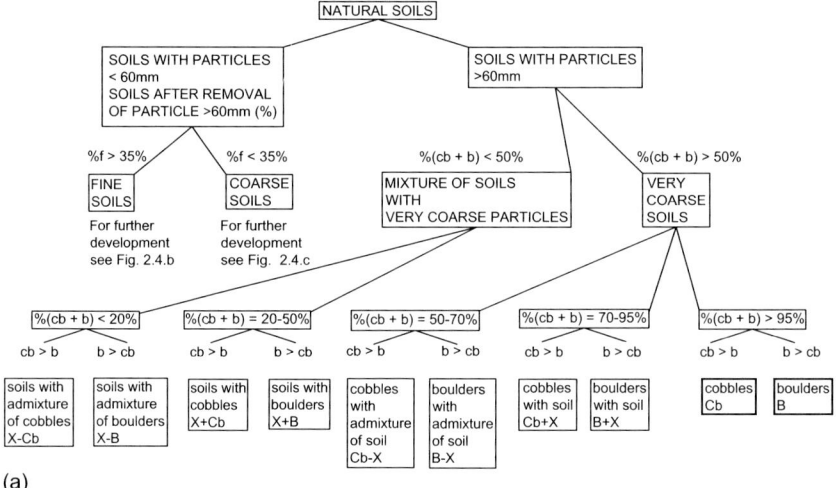

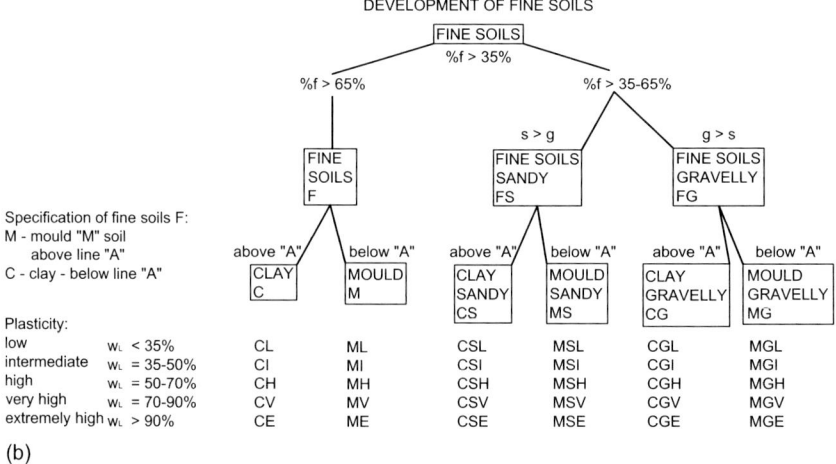

Fig. 2.4 Soil classification system – detail description of soils. (**a**) Natural soils. (**b**) Fine soils. (**c**) Coarse soils

2.1 Soil Classification – Specification for Utilization in Different Parts of Earth Structures 15

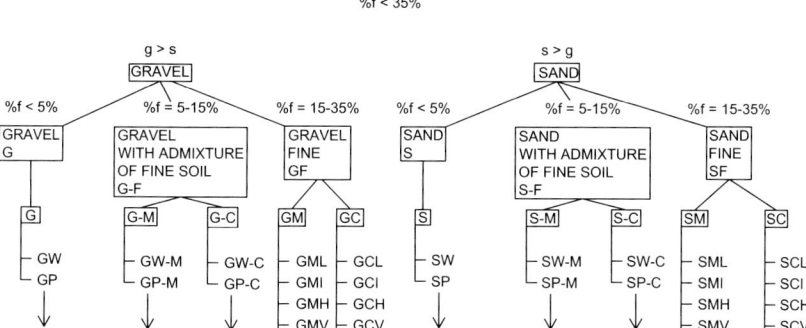

(c)

Fig. 2.4 (continued)

chemical reaction process. In the description it is necessary to mention if the calcium carbonate is evenly distributed, if it creates clusters, if it is extruded in layers, or if it creates efflorescence.

Amount of mica – in the description of soil, the admixture of certain minerals is also mentioned. Special attention is devoted to the amount of mica minerals, as they can significantly influence permeability and shear strength. Then the soil is described as:

– Marginally micaceous – contains only individual mica leaves,
– Medium micaceous – contains plenty of mica leaves or mica flakes,
– Highly micaceous – contains so much mica that the soil has a shiny savour.

Soil colour – is its basis characteristic, but it is essential to mention under which state the colour was determined (i.e. for dry or wet samples). Mixed colours are described using combinations of basic colours (black, grey, white, brown, yellow, red, blue, green, purple) and, if required, with a light or dark attribute.

2.1.1 Suitability of Soils for Transport Engineering Earth Structures

The basis for this use is ČSN 721002:1993, which implements more detailed expansion of the original USCS classification. This code uses only one dissimilarity as compared to the specification shown in Fig. 2.4, which is that soil types MS and CS have two subtypes MS_1 & MS_2 and CS_1 & CS_2. In order to specify the soil with index 1, the proportion of fines should be within 35–50% and the liquid limit

$w_L < 60\%$. Index 2 is assigned to MS or CS soils that do not fulfil conditions for index 1.

This code divides soils based on the suitability for use in embankments (for which it defines four groups) and for use in embankment subsoil (where it distinguishes ten groups – from the most suitable to those unsuitable), see Table 2.2a,b for more details. At the same time, expected limits of results from the Proctor standard compaction test and CBR (California Bearing Ratio) test are mentioned there. The following notes should be added to this table:

- The classification serves only for rough assessment; it should be used for the early determination of necessary measures to take or for the proposal of soil treatment and for the early determination of the road construction type.
- The extent of usability is given mostly by the current state of soil, for fine grained soils it will be consistency, for subsoil the level of ground water, the possibility of capillary action, the topography of the area, possibilities for surface water drainage etc. An approximate determination of frost susceptibility is presented in Fig. 2.5 (amended Schleible criteria).

Table 2.2a Brief classification of suitability of soils for road construction – for fill and subsoil (according ČSN 721002:1993)

No.	Symbol	Suitability for fill				Suitability for subsoil									
		Unsuitable	Less suitable	Suitable	Very suitable	I	II	III	IV	V	VI	VII	VIII	IX	X
1	F1 MG		X	X							X	X	X		
2	F2 CG		X	X							X	X	X		
3	F3 MS1			X	X			X	X	X					
4	F3 MS2	X										X	X	X	
5	F4 CS1			X					X	X					
6	F4 CS2	X										X	X	X	
7	F5 ML	X	X									X	X	X	
8	F5 MI	X	X									X	X	X	
9	F6 CL	X	X										X	X	X
10	F6 CI	X	X										X	X	X
11	F7 MH	X	X									X	X	X	
12	F7 MV	X											X	X	X
13	F7 ME	X												X	X
14	F8 CH	X	X										X	X	X
15	F8 CV	X											X	X	X
16	F8 CE	X												X	X
17	S1 SW				X	X	X								
18	S2 SP				X	X	X								
19	S3 S-F				X		X	X	X						
20	S4 SM			X	X		X	X	X						
21	S5 SC			X	X		X	X	X						
22	G1 GW				X	X	X								
23	G2 GP				X	X	X	X							
24	G3 G-F			X	X	X	X	X							
25	G4 GM				X	X	X	X							
26	G5 GC			X	X		X	X	X						

2.1 Soil Classification – Specification for Utilization in Different Parts of Earth Structures

Table 2.2b Brief classification of suitability of soils for road construction – ranges of Proctor standard test and CBR test (according ČSN 721002:1993)

No.	Symbol	Proctor standard test		CBR (%)	
		Max dry density ($kg \cdot m^{-3}$)	Optimum moisture content (%)	For optimum moisture content	For degree of saturation $S_r = 0.95$
1	F1 MG	1550–1900	10–25	8–18	5–10
2	F2 CG	1550–2000	12–30	5–10	3–7
3	F3 MS1	1750–2000	10–25	5–25	4–15
4	F3 MS2	1600–1950	12–30	3–15	2–5
5	F4 CS1	1650–2000	12–30	5–30	5–20
6	F4 CS2	1550–1850	15–35	2–20	0–4
7	F5 ML	1600–1800	12–20	3–20	2–7
8	F5 MI	1500–1750	15–25	2–15	1–6
9	F6 CL	1600–1950	10–30	3–20	1–8
10	F6 CI	1550–1900	15–35	2–20	0–6
11	F7 MH	1400–1700	15–33	3–7	0–4
12	F7 MV	1380–1650	20–35	2–6	0–3
13	F7 ME	1350–1550	22–38	2–5	0–2
14	F8 CH	1380–1700	17–37	2–7	0–3
15	F8 CV	1360–1650	19–39	1–7	0–3
16	F8 CE	1330–1500	20–40	1–6	0–3
17	S1 SW	–	–	–	–
18	S2 SP	–	–	–	–
19	S3 S-F	1700–2100	8–16	8–70	6–25
20	S4 SM	1730–2050	8–18	6–50	4–15
21	S5 SC	1760–2000	8–20	4–30	2–12
22	G1 GW	–	–	–	–
23	G2 GP	–	–	–	–
24	G3 G-F	1800–2150	6–16	20–90	6–60
25	G4 GM	1750–2100	8–19	10–60	4–40
26	G5 GC	1700–2000	10–23	5–30	3–20

- The use of soils that are less suitable or even unsuitable is not rejected, but they can be used only if additional appropriate measures are taken, such as improvement of soil properties, soil reinforcement, and the use of this soil in sandwich types of embankment constructions or placement in locations where it would have a limited impact on overall safety.
- The use of special types of soils, such as organic soils, volume unstable soils like loess, swelling clays and clay shale or the use of carbonate soils and evaporate soils or made ground should be judged individually. For organic soils, the upper limit for usability is usually defined as 5%.

2.1.2 Suitability of Soils for Hydro Engineering Earth Structures

As a basis for the evaluation of soil suitability for the construction of earth and rockfill dams, ČSN 736850:1975 "Earth dams" was selected. This code is based on

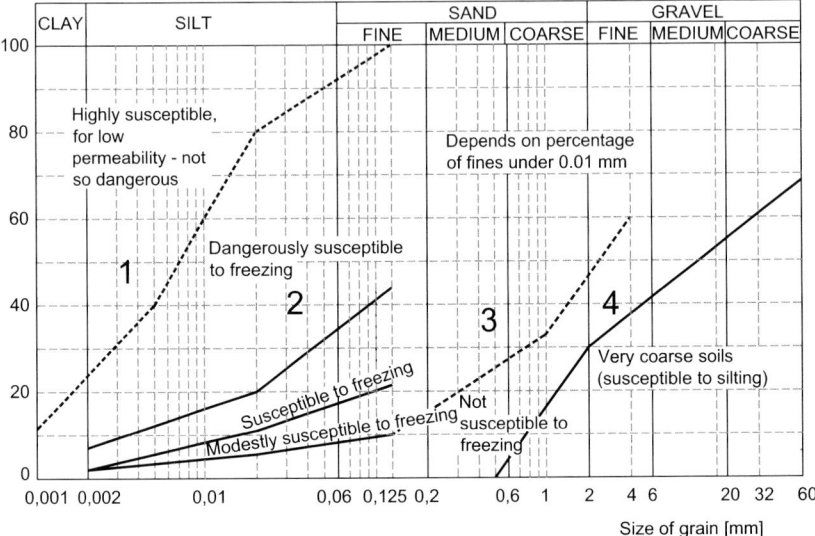

Fig. 2.5 Approximate determination of frost susceptibility (amended Schleible's criteria)

the USCS classification and on the recommendation of U.S. Corps of Engineers – see e.g. Sherard et al. (1963, 1967).

The suitability of the use of different soil types within selected zones of compacted earth dams may be judged using Table 2.3, which should be used as a guide for soils with moisture contents very close to the optimum.

Table 2.3 Suitability of different types of soils for different parts of earth dams

Soil type symbol	Homogeneous dam	Sealing core	Stabilization upstream and downstream zones
GW	unsuitable	unsuitable	excellent
GP	unsuitable	unsuitable	excellent
GM	excellent	very good	less suitable
GC	excellent	excellent**	less suitable
SW	unsuitable	unsuitable	suitable
SP	unsuitable	unsuitable	suitable*
SM	suitable	suitable	suitable*
SC	very good	excellent	unsuitable
ML	less suitable	less suitable	unsuitable
CL	suitable	very good	unsuitable
OL	less suitable	less suitable	unsuitable
MH	less suitable	less suitable	unsuitable
CH	less suitable	less suitable	unsuitable
OH	unsuitable	unsuitable	unsuitable

Note: Suitable are also transient types – for homogeneous dam and core it is GW–GC, SW–SC.
* if contains also gravel
** be careful of degree of weathering

2.1 Soil Classification – Specification for Utilization in Different Parts of Earth Structures

The information from Table 2.4 can be used for the determination of typical values of soil properties and the workability of individual soil groups. The values presented are valid for soils very well compacted with moisture-density control.

Roughly, it can be stated that for sealing sections of zoned and homogeneous dams, sealing trenches and upstream sealing layers (carpets) sealing soil can be used under the following conditions:

- The coefficient of permeability k is lower than 1×10^{-8} m·s^{-1},
- The grain size distribution curve is roughly in range 2, or potentially in 1 according to Fig. 2.5,
- The content of organic matter is lower than 5%,
- The liquid limit is lower than 50%,
- The maximum diameter of particles is 100 mm,
- For soil types ML, CL, the plasticity index is higher than 8.

If soils that do not satisfy the above conditions are to be used then it is required that their suitability is proven individually by laboratory testing.

For the stabilization of upstream and downstream zones, other types of soils are permitted if they are permeable once compacted, if they are not influenced by volume changes due to weather conditions and/or permeated water and if they have high enough shear strength. These soils should not contain organic matter and also substances, which once dissolved in water, could be aggressive to the body of the

Table 2.4 Approximate correlation between embankment properties and types of soils

Soil type symbol	Relative permeability	Probable range of k (m.s^{-1})	Relative shear strength	Compressibility after wetting	Relative workability
GW	permeable	$10^{-5} - 10^{-3}$	very high	negligible	very good
GP	permeable to very permeable	$5 \times 10^{-5} - 10^{-3}$	high	negligible	very good
GM	semipermeable	$10^{-9} - 10^{-7}$	high	negligible	very good
GC	impermeable	$10^{-10} - 10^{-7}$	high		very good
SW	permeable	$5 \times 10^{-6} - 5 \times 10^{-4}$	very high	negligible	very good
SP	permeable	$5 \times 10^{-7} - 5 \times 10^{-3}$	high	very low	good to fair
SM	semipermeable to impermeable	$10^{-9} - 5 \times 10^{-6}$	high	low	good to fair
SC	impermeable	$10^{-10} - 5 \times 10^{-7}$	high to medium	low	good to fair
ML	impermeable	$10^{-10} - 5 \times 10^{-7}$	medium to low	medium	fair to very poor
CL	impermeable	$10^{-10} - 10^{-7}$	medium	medium	good to fair
OL	impermeable	$10^{-10} - 10^{-7}$	low	medium	fair to poor
ML	very impermeable	$10^{-11} - 10^{-9}$	low	high	poor to very poor
CH	very impermeable	$10^{-12} - 10^{-10}$	low to medium	high	very poor

dam and its structures. It is suggested that the particle size distribution curve would be within zone 4 or potentially in zone 3 as per Fig. 2.5.

For the construction of rockfill dams it is possible to use very coarse soil particles, and their suitability will be checked based on deformation properties, strength in compression and shear strength determined under dry condition and after saturation by water. Resistance against climatic conditions, the height of the dam and the proposed location of very coarse particles within the dam body will also influence the suitability. Note that direct testing of compressibility and shear strength of very coarse particles requires non standard testing equipment of big dimensions, which is used only in rare situations, Marsal (1973). Most often indirect methods are chosen or the evaluation of properties is performed based on experience.

2.2 Borrow Pit – Soil in Natural State

Using engineering aspects, a wide, practically unlimited range of soils can be found in nature, from hard pieces of rock through gravel, sand, silt and clay to organic sediments of soft compressible peat. Therefore, the soils in borrow pits can have highly variable properties. The estimation of the properties is highly dependent on the soil genesis. Very briefly said, the following can be found:

- Residual soils that originated from weathering in place and have not been transported.
- Sediments – soils that have been transported from the place of their origin (where they weathered from the original rock) via transport media to the place, where the energy of the transport media diminished and could not carry the soil particles further. More detailed description of this type of soil depends on the transport media.
- Man-made sediments – sediments that were created by the activity of man, mainly during the last century. Made ground is a result of deposition of different extracted soils or waste materials on the Earth surface. As they are creating a new earth structure, their specification is described in Chapter 8. Made ground can be used as a source of material for the new construction, and today this is preferred, it only depends on the volume, which we are able – with respect to its properties – to embed into the earth structures.

2.2.1 Soil Genesis

In principle, soils make up the top layer of the Earth's crust, from orders of meters towards tens of meters and exceptionally even hundreds of meters. The remaining part of the Earth's crust is made up of rocks with a thickness of about 25–50 km. The Earth's surface is not constant, huge changes are going on in time. Geology describes these changes and classifies them with respect to time in the form of names of geological groups or eras. Atkinson (1993) states, in brief, that materials of the Cenozoic age are generally regarded as soils for engineering purposes; materials of the Mesozoic age are generally regarded as soft rocks and materials of the Paleozoic age are regarded as hard rocks. Geological evolution on the Earth is still going on and it is possible to describe it using a closed geological cycle: denudation – deposition – sediment formation – crustal movements. Individual processes can be described as follows, Vaníček (1982a).

Denudation covers all processes that contribute to the removal of the top layer of the Earth's crust. The most important process is the weathering process, which includes all destructive mechanical, chemical and biological processes that disturb the existing composition of the Earth's surface. The weathering process is connected with erosion and transport of weathered products by different means (gravity, water, wind) from one area to the other – mostly from higher areas to lower ones.

Deposition describes the process of accumulation of transported mass.

Sediment formation describes the processes by the influence of which accumulated sediments are hardened, they are changing their composition towards rock etc.

Crustal movement includes slow (epirogenetic) movements, movements generated by unloading of areas (uplift) or by loading from new sediments (downthrow) as well as rapid movements (tectonic movements).

From the above mentioned it is obvious that the character of soils and their behaviour will be influenced mainly by weathering, the type of transport and sedimentation. These processes will be described in more detail hereafter.

2.2.1.1 Weathering

Weathering of soils and rocks is a destructive process the result of which is a new composition of higher stability. Destruction of rocks is influenced by several processes of physical, chemical and biological nature. The physical process reduces the size of particles, increases the surface area of particles and their porosity. Both chemical and biological processes can cause complex changes of both chemical and physical properties.

Physical weathering is caused by several effects. Temperature changes (in both daily and yearly cycles) cause deformations of the rock massive and in this way they create weakened zones. Springing is caused by the effects of unloading, creation of ice crystals in the fissures causes cracking. Shrinkage of softened surface layers can cause striking out of crystals, roots of plants and trees cause destruction, etc.

Virtually all processes of chemical weathering depend on the presence of water. Hydration, i.e. surface water absorption, is the required prerequisite for other chemical processes, such as:

- Hydrolysis – reaction between water ions of H^+ and OH^-, the condition for hydrolysis is flowing water that transports off the products of weathering like silica acid and aluminium hydroxide,
- Exchange of cations,
- Oxidation or reduction,
- Carbonization,
- Dissolution.

The end product of the weathering process does not in many ways depend only on the original composition of the respective rock but also on the weathering conditions i.e. climate, topography, time and biological factors.

Within intact rocks that are the main source of soil on the Earth, the most common mineral is feldspar (about 60%), pyroxene and amphibole (17%). The amount of quartzite is about 12% and mica 4%, therefore for other minerals there remains only 8%, although about 2 000 of them have been identified and from those about 200 are considered as rock-forming ones (Zeman, 1979).

These rock-forming minerals are also called primary. Some primary minerals are transformed by weathering into secondary minerals, part of those are clayey minerals. Soils are therefore composed of both rock-forming minerals and secondary minerals including clayey ones. The properties of soils can be influenced by clayey minerals much more than it would correspond to their proportion.

In soils from primary minerals the most common is quartzite with a small amount of feldspar and mica. Pyroxenes and amphiboles are very rare. From carbonates the most common ones are calcite and dolomite, mainly as sediments in deep seas. Oxides of iron and aluminium are frequent in residual soils of tropical areas. From the proportion of feldspar it is obvious that it is the mineral most influenced by weathering, and the final product of such process are clay minerals.

2.2.1.2 Transport

The weathering zone is eroded by the effect of a water stream, rain, wind, waves, and glaciers and afterwards transported. We distinguish surface erosion by rain, erosion by water stream, wind erosion, etc. During the transport of eroded particles, additional physical changes of transported material can be going on, predominantly separation of particles by their size and abrasion.

The maximum size of transported particles highly depends on the type of transport. Ground water carries only colloids, running water, wind and waves particles of sand size, in the mountains the water stream is capable of transporting even bigger than gravel particles. During gravitational transport from steep slopes boulders are transported, the same applies to transport by glacier. In front of a glacier a glacial moraine is formed, which contains a wide range of particles from the smallest ones to the biggest ones, the particle size distribution curve is very smooth, and the remaining material is usually called boulder clay.

The transport distance depends on the type of transport media. If eluvium with uniform particle size distribution is washed off from headwaters, after the velocity of water is decreased first coarser particles and later finer ones are deposited.

2.2.1.3 Sedimentation

Based on transport media sediments are differentiated into:

– Colluvial soils – transported downhill by gravity and partly by rain water,
– Alluvial soils – river sediments,

- Glacial sediments,
- Sea sediments,
- Eolic soils – wind blown.

The most difficult deposition conditions are generally connected with alluvial sediments, which in terms of particle size vary from gravely particles to clayey particles, with a potential organic content. The variability in their plan and section is given mainly by the changes of the watercourse forming river meanders. Special problems are associated with deltas of big rivers.

Changes in the character of sediments may be very well associated with the place of deposition, e.g. for glacial sediments we can distinguish:

- Basal till – moraine soil just underneath the glacier,
- Boulder clay – soil of the moraine itself as already mentioned,
- Layered character of sediments in tarns (glacial lakes), which is given by sedimentation of coarser particles during summer thaw and finer particles in winter periods.

A similar situation applies to sea sediments, where very fine particles are deposited on the sea (ocean) bed, while coarser particles are deposited on the continental shelf.

A typical type of aeolian deposits are loess deposits that are composed of silt size particles, they have a special collapsible structure and, for engineering purposes, represent a very sensitive soil group. Aeolian sands occur mostly in desert areas.

Sedimentation conditions are of fundamental importance for the final arrangement (structure). During sedimentation in lakes and seas the type of water (fresh, salty or estuarial/brackish) is significant. Another factor is the periodicity of sedimentation, mostly the yearly cycle given by temperature changes. A feature typical of the already mentioned example of tarns is layering of thicker coarser material and thinner finer material. This structure is called varved clay.

2.2.1.4 Changes After Deposition

Many changes can happen within the time frame between sedimentation and the present state, and these changes may affect the final, current properties. On the one hand, there are ongoing processes of hardening and on the other processes weakening the bindings. Individual soil particles are loaded by gradual thickening of the sedimentation layer and thus the particles are influenced by each other directly on their contacts or via the liquid phase.

Another significant process that leads to hardening is cementation. This happens as a result of the presence of a significant amount of dissolved chemicals in soils – carbonates, sulphates, substances containing iron and aluminium, organic matter and others. Under certain conditions, these chemicals can develop chemically stable cements, crystal forms, gels that develop significant strength.

Under extreme conditions, these processes can lead to the development of sedimentary rocks, such as mudstones, sandstones and siltstones. The development of such strong bindings is due to high temperature, pressure and time. In geology, the processes leading to changes in particle fabric, transformation of one mineral into another and creation of new bonds between particles are called diagenetic processes.

Near the surface these changes are not as significant and also here the most important processes leading to strength changes, mostly lowering, take place. Among them we may include:

- Drying out – this leads to the strengthening of the surface layer due to shrinkage, but at the same time shrinkage can cause tensile cracks,
- Weathering,
- Wash out – leads to strength lowering – typical examples are sea clay sediments that are now above sea level, e.g. in Norway and Canada. Percolation of fresh water causes removal of soluble salts and by that conditions for the collapse of the internal structure are developed – this collapse is then called liquefaction of quick clays. Calcareous sediments are prone to wash out and void creation,
- Development of cracks and fissures – cracks can develop as a result of cyclic drying and wetting of alluvium sediments. This factor plays a more important role in unloaded sediments, as the cracks are slowly opening. For example, Most clay from the northern part of Bohemia is described as fissured tertiary clay, and these fissures play a significant role during the excavation of these clays which overlay brown coal seams.

Since soils have at least a partial stress history, this information is therefore useful. Normally consolidated soils are generally those that have not experienced a higher surcharge in the past than at present. This is mainly true of young sediments. Overconsolidated soils are those, whose surcharge in the past (due to self weight or glacier) was higher than the present one. To the group of overconsolidated clays belongs also Most clay, which is overconsolidated and fissured.

Figure 2.6 shows general soil profiles. For residual soils, the size of particles of original, not yet weathered rock increases with the depth. Sea sediments are characterised by much higher homogeneity in comparison to river sediments, within which we can expect much more rapid changes between coarse layers and fine soil layers up to extremely muddy horizons.

Most nowadays soil profiles in the topmost parts comprise topsoil in a thickness from a few decimetres to few meters. This layer is not suitable for earth structures as it includes a high percentage of organic content. It is only suitable for use as the topmost layer on structures where it is placed to support the growth of vegetation, e.g. on motorway embankments. Most countries have introduced strict regulatory requirements for this soil layer, it is necessary to store it before digging out the cutting or borrow pit and then reuse it.

This part may be concluded by stating that the knowledge of the soil origin and all the processes involved in its history is very important for the assessment of the

2.2 Borrow Pit – Soil in Natural State

Fig. 2.6 Photo of soil profiles. (**a**) Deluvial sediments with visible individual boulders. (**b**) Deluvial sediments with irregular interbedded layers of coarse material. (**c**) Man made sediments – spoil heap composed of clay clods

(a)

(b)

(c)

behaviour of soils for engineering purposes. This is relevant not only for soils that will be used as a structural material, but also for soils that will be used as subsoil for new earth structures. Good cooperation with a geologist or an engineering geologist is therefore of high importance for obtaining as much information on the soil as possible.

2.2.2 Geotechnical Investigation

The task of geotechnical investigation is not only to determine the properties of soils in a particular location, but also to determine the best place for the extraction of a suitable material. Generally, the following may be used for the construction of earth structures:

- Classical borrow pit,
- Material from the cutting of a linear construction – road or railway, when the material from cuttings can be used for embankment construction,
- Material from other excavations outside the given construction – e.g. excavations for new buildings, underground structures like metro (underground),
- Waste material, e.g. mine spoil, ash from coal burning after improvement, in the form of recycled material.

A classical borrow pit has the main advantage in that its selection is predetermined by the search for the most suitable material for the earth structure. Another advantage of it is that it is well investigated and the volume is sufficient for the whole earth structure. However, each new borrow pit causes an interference with the environment, and its influence on the environment must be assessed (for very large borrow pits even as part of EIA process). These days, it is usually required to define the way of reclamation after the borrow pit extraction even before it is opened. Property ownership relations also participate in the process where digging on somebody else's property means paying compensations per excavated cubic meter. For specialized types of earth structures like drains and filters, when the requirements are rather strictly prescribed, this usually includes transport from distant purpose run quarries. The distance of imports usually plays an important role not only for financial reasons, but also in terms of making an agreement on transport routes as these routes can have a negative impact on the neighbouring public, especially when the volumes are huge. When constructing small or large dams, an intriguing solution appears to be to create the borrow pit within the future flooded area by which way it is possible to overcome the above mentioned problems. But the sealing effect should not be affected by this borrow pit as this could cause leakage of the reservoir via its bed or sides.

The material from the cuttings of infrastructure projects is, in a way, explicitly given and the earth structure must adjust to it. The common question is if the future linear construction route can be utilized as a transport route or whether to use the

2.2 Borrow Pit – Soil in Natural State

existing route system. Another problem is the time relation between the excavation of the cutting and the filling of the embankment in different parts of the route.

The properties of materials from excavations of foundation pits or underground structures are also already given. The disproportion between the volume of excavations and the requirements for fills can as well be problematic. The environmental point of view is becoming an issue as some regulations declare any excavation as waste and require special handling of such excavated materials.

Linear infrastructure projects are presently the most common place where it is possible to safely use large volumes of waste materials. In this situation, the properties and the volumes of the material are known in advance, but in this case it is also necessary to review the environmental risks of storing such materials within the earth structure, mainly with regard to leakage. This problem will be discussed in more detail in chapters devoted to different earth structures.

2.2.2.1 Phases of Investigation

The process of gathering of information, data and detailing of the conditions in the borrow pit is continuous, when the next step follows the previous one and utilizes knowledge gathered previously. Nevertheless the whole process can be divided into three general phases, i.e. Vaníček (2002d):

- Preliminary phase – desk study,
- First phase of site investigation, which should in general verify the information from desk study and propose the next stage of site investigation,
- Second phase of site investigation is used to determine in detail the properties of the borrow pit material from both areal and depth point of view.

During the preliminary phase the information is gathered mainly from already available sources also in wider area, mainly from maps. For example in the Czech Republic it is "Set of geological and purpose maps of natural resources" prepared in the scale of 1 in 50 000. This set incorporates more than ten different sheets, from which the most important for earth structures are:

- Geological map,
- Engineering geology map,
- Hydrogeological map.

Of course it is possible to use other sheets if necessary, as chemistry of surface waters map or map of geophysical indicators and interpretations or map of geofactors – important topological features or conflicts of interest. For certain areas like the capital Prague there are also maps in bigger scale – 1 in 5 000 available. From the other sources it is possible to use the database of past site investigations, which in the Czech Republic are collected, compiled, evaluated and archived by Geofond Praha. Also useful is to use information from aerial survey. All of this background information is possible to use together with site walkover and can serve not only to a rough assessment of the subsoil stratigraphy with groundwater levels but also for locations of suggested ground investigation exploratory holes during the first phase of site investigation.

The first and second phase of site investigation are generally utilizing the same methods, the difference is in their scale. It is possible to distinguish among boreholes with sample collection, trial pits, static or dynamic penetration tests and geophysical methods. For borehole excavations it is possible to sort them with respect of possibility to collect undisturbed sample (on which tests for compressibility, shear strength and permeability can be performed) or disturbed samples on which index tests are performed in order to classify the soil and determine the soil state (consistency, relative density). When a bigger disturbed sample is collected than in laboratory, it is recompacted by standard compaction energy and tests are performed on the compacted sample for physical-mechanical properties – compressibility, shear strength, permeability.

In the first group of investigative techniques that can collect undisturbed samples, we can generally include borehole techniques when the soil is mainly collected by means of rotation or percussion or their combination without additional water. Here we can include:

– Hand window sampling – usually used for small depths of about 2 m, diameter about 30 mm, while the tube is hammered into the soil by wooden mall;
– Mechanical window sampling – where the sampling tube is vibrated into the soil by engine powered hammer;
– Hand rotary drilling – where the core of about 50 mm of max. depth of about 10 m based on soil type is drilled into the soil by hand;
– Mechanical rotary drilling – where the core is drilled into the soil using light-weight engines located on off-road car or trailer;
– Cable percussion drilling – where the sampler is hammered into the soil by mechanically driven heavy hammer located on trailer towed by off-road car;
– Mobile rotary drilling – these are used for bigger diameter core drilling and bigger depth of about 50 m, these drills are located on heavy trucks.

The classical hand dug trial pits are used nowadays fairly rarely, as the mechanically dug ones can be performed much more easily and quickly with bigger extent and therefore give a better picture about the investigated location. Under both types of trial pits it is advised to document it also using photos and/or video.

From geophysical methods there is available a huge variety of options that can be used to investigate the subsoil using physical fields, either natural or induced. Most of them are cheap and flexible methods. Generally they are trying to collect a continuous picture in between probes. During the investigation of borrow pits the investigation starts using surface probes, continues using combination of surface and hole probes and ends with utilising cross-hole techniques. Physical representation of nonhomogeneity of subsoil is described by geophysical anomaly. Optimal situation during the investigation is when the anomaly corresponds to the physical parameter of soil that is investigated. With respect to the physical background of investigated fields it is possible to divide the geophysical methods on methods gravimetric, radiometric, atmogeochemic, geothermometric, geoelectric, seismic and their borehole alternatives.

Penetration investigative methods are measuring the resistance against penetration of cone into subsoil either by continuous penetration or by dynamic using hammer. For the dynamic one it is measured by the amount of blows per certain penetration (e.g. 200 mm). Diameter of cone is ranging between about 25 and 100 mm. The advantage is rapid evaluation of the soil profile from the cone resistance point of view. This is useful for investigation of the quality of subsoil of large areas, where we can limit the amount of classical boreholes with soil sampling and testing.

2.2.2.2 Evaluation of In Situ and Laboratory Tests

While evaluating the results of in situ and laboratory tests for potential borrow pit, two main aims are coming to mind:

- Determine how homogeneous is the borrow pit, if it does not comprise of more than one set of material, which is not something exceptional especially for eluvial soils in vertical direction or river sediments, where the change can be also in horizontal direction;
- Determine how for an assumed quasi-homogeneous set of materials the investigated soil properties are distributed around average value or around the direction of investigated relationship.

For the evaluation it is possible to use principles of statistical methods and assess generally any investigated property. However most often the evaluation is done based on index properties of soils on disturbed soil samples. Here it is possible to independently evaluate single property or relationship of two or more investigated properties. Credibility of the set of data depends on the amount of tests. Generally it can be concluded that the higher amount of data the higher is the accuracy of the evaluation. As the minimum for credible evaluation of a given set is generally considered five tests, especially when a relationship of two properties is in question.

While assessing one variable, the frequency expression is applied where the horizontal axis displays the monitored variable and the vertical axis the frequency of individual values or a selected value range. Individual variables may e.g. stand for a percentage of a certain soil fraction, e.g. a proportion of clay particles, sandy particles etc., typical particle size, e.g. d_{50}, moisture content, liquid limit, plasticity limit, porosity, void ratio, degree of saturation, total unit weight, unit weight, dry unit weight etc. The results may indicate uniform distribution within a narrow band, Fig. 2.7a, within a wider band, Fig. 2.7b or they show that most probably two sets are in question, which must be separated from each other – distinguished in a more detailed way, Fig. 2.7c.

When assessing a dependency of two variables, the simplest case is given by the direct proportion of the monitored relation, i.e. by linear relationship. The distribution of the results obtained around this line depends on the homogeneity of the set – Fig. 2.8a, where the plasticity index – clay fraction proportion relationship is

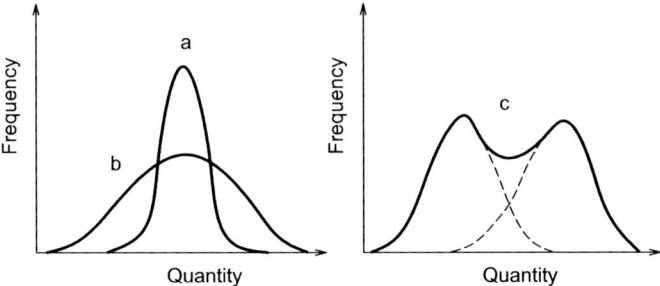

Fig. 2.7 Different distributions of obtained results for one variable. a – narrow band distribution; b – wide band distribution; c – two sets of distributions

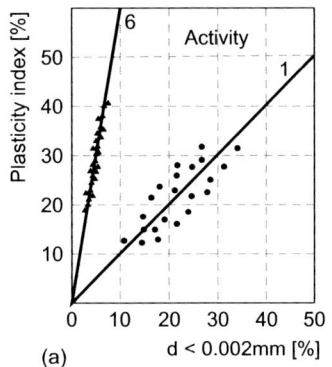

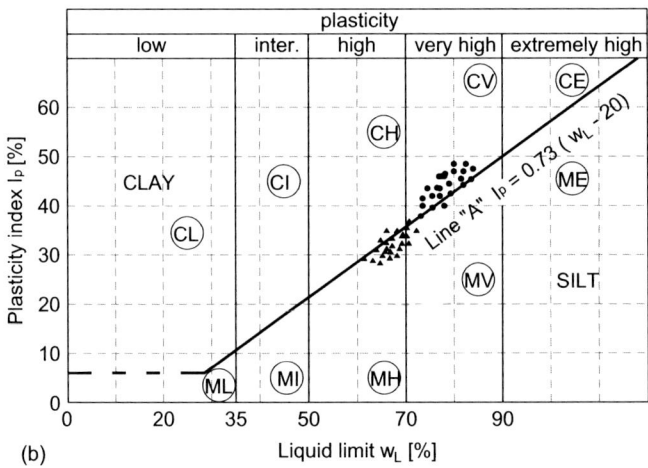

Fig. 2.8 Distributions of obtained results for two variables. (**a**) Determining clay activity index I_A. (**b**) Determining soil plasticity for soil classification

2.2 Borrow Pit – Soil in Natural State

shown, serving for the clay colloidal activity index determination. Very often, the homogeneity of a set is assessed by the relationship between plasticity index and liquid limit – a so-called plasticity diagram – Fig. 2.8b. It is generally expected that the relationship obtained will follow line "A" already defined in Fig. 2.2.

In order to assess a relation of three variables, the already mentioned triangular diagram may be used into which the results of the percentage share of three fractions, sand s, gravel g and fines f, are plotted, Fig. 2.9. This dependence is very useful for distinguishing vertical stratigraphy of residual soils in the borrow pit, where we may expect that the size of coarser particles will increase with depth.

2.2.3 Workability

Only brief notes will be made on the issues concerning workability as each case can be very individual and the approach of different companies to it may vary depending also on the size and extent of their mechanical equipment. On the other hand, demands for workability will have a significant impact on price. In the Czech Republic, the code of practice ČSN 733050 Earthwork is used, classifying soils and rocks into seven workability classes where the main aspect is the size of extracted particles, the subsoil weathering rate or adhesiveness for clayey soils. By rough classification, the designer provides very useful information for the tendering procedure. If it happens that in reality the workability class turns out to be different, the contractor is mostly entitled to negotiate with the investor potential bidding price modifications.

The first step starts with stripping the topsoil layer and its further direct use or placement on a temporary stockpile for successive reuse in surface reclamation. The so-called sub-topsoil layer cannot be neglected either, firstly for the reason that it might contain a higher percentage of organic matter than allowed for the respective earth structure, and secondly also with a view to future reclamation as the most suitable method applied in surface reclamation (e.g. of landfill surfaces) is placing this sub-topsoil layer under the humus layer. This procedure, however, is usually taken

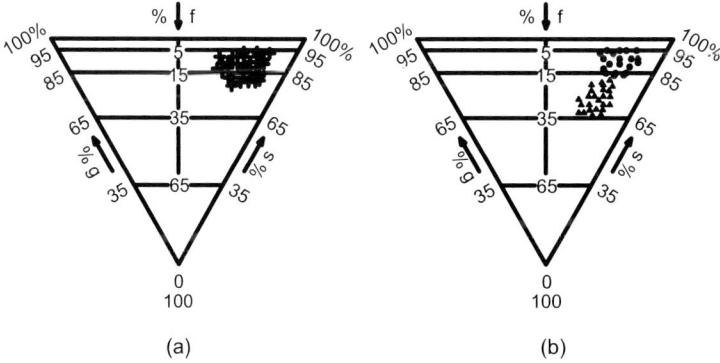

Fig. 2.9 Distribution of obtained results for three variables. (**a**) One homogeneous set. (**b**) Two nearby sets

step by step, as stripping the protective layer entirely along the whole borrow pit area at the same time can lead to changes in moisture content, both to increased moisture content after heavy rainfall, and drying of the exposed surface layer during sunny weather. Greater susceptibility of soil to surface erosion must not be forgotten either.

The groundwater level depth also significantly affects workability. For coarse soils, mainly gravels, this fact does not have to be crucial as extraction under the water level does not present any problem and is quite common. For fine-grained soils, prior to extraction it is desirable to make drain trenches for lowering the groundwater level below the presumed extraction depth. The reason is not only problems with extraction, but mainly an effort to reduce the soil moisture content, because when fully saturated the soil has greater moisture than is the optimum one for compaction.

The extraction procedure either at full height or in individual layers is given by the homogeneity of the borrow pit. For a homogeneous borrow pit, extraction in full height is an advantage. It applies extraction methods using dredgers or bucket chain excavators with direct loading onto lorries or via conveyer belts and hoppers. In variable subsoil, extraction is split vertically, and various scrapers may also be used, mainly if the transport distance towards the dumping place is economical – scrapers can place soil in required horizontal layers at the dumping place thus eliminating a need for additional intermediate machinery as is the case of transport by lorries.

The biggest problems of workability are connected with residual soils containing large boulders, sometimes up to m^3, which may float in fine soil. These boulders must be separated and individually blasted to decrease their size. An unfavourable situation may also occur for highly micaceous soils, which are easily washed during heavy rainfall.

The case where soil is washed to its destination would deserve a separate chapter. Apart from using this transport method in mud-settling ponds and practically also in dam construction (even though presently only exceptionally), it may be also used for specific cases. One of them is using a water stream for on-slope material disintegration and successive transport of the resulting sludge of a mix of soil and water to its destination by gravity chutes. After particle sedimentation, the water is collected and reused in the cycle. In the 1930s, new shoe making factories of Bat'a Company in the vicinity of the town of Zlín were situated on plots that had undergone such reclamation where deposited soil reached up to several meters. The territory, up to then often affected by floods, was unattractive, and land prices very low. After sedimentation and consolidation, the quality of the new subsoil was sufficient for production halls, whose level made them safe from flooding. In today's view, such procedure would likely meet with environmental concerns because of potential leakage of very fine particles into existing water ways.

2.2.4 Soil Treatment Before the Utilization

Provided that extracted soil is to be used into an earth structure, in fact the single treatment available is its moisture content regulation.

Soils with too much moisture extracted from the borrow pit may be spread out on temporary dumping grounds after extraction where the soil may be turned over with a cutter allowing for natural drying. The reliability highly depends on the weather even though temporary stockpiles may be protected by sheet cover. Soil drying in rotary drums was exceptionally applied in the past but because of high energy demands this method is presently practically unrealistic. The method more preferred is mixing moister soils with lime for better compaction, this procedure will be described later in Section 4.1.

For drier soil, on the contrary, a greater moisture content is achieved on the site where the soil is spread out into layers with a required thickness and road sprinklers are used here. Sometimes even in the case when the soil moisture content in the borrow pit is close to the optimum moisture content for compaction the road sprinklers are used, mainly due to warm sunny weather when the moisture content may lower very fast after spreading.

An untraditional treatment procedure may be used e.g. for tertiary over-consolidated clays. Their consistency is mostly stiff to hard, and the size of extracted clods substantially bigger than is commonly required for compaction. Extracted clays are placed on temporary stockpiles and kept there for up to several months. During this time, the clods take in moisture, start to weather and disintegrate into smaller and moister clods. These, in turn, may be used for embedding into earth structures, e.g. for sealing municipal waste disposal sites.

2.3 Soil Compaction

Soil compaction is the most common way of soil treatment as a construction material. Detailed knowledge of compaction technology, on one hand, allows using a wider scale of soils as construction material, and on the other hand, reducing energy demands.

The aim of compaction is transformation of an extracted and loosely spread soil layer into a new condition where compacted soil shows improved mechanical and physical properties, i.e. above all shear strength, compressibility and permeability. This improvement is achieved by applying loading with a compacting means, which results in tighter placement of individual particles (or clods) of soil – i.e. reducing porosity. In other words, considering soil as a three-phase system, then, in general, compaction is a change in the proportion of the gaseous, liquid and solid phase in favour of the solid phase. To make the application of the compaction maximally efficient, we need to know the compaction mechanism. The knowledge of this mechanism will enhance decision-making on the appropriate compaction method, which is affected in particular by the following factors:

- What type of soil is to be compacted,
- In what state (consistency, moisture content) it will be placed into the compacted body,
- How thick layers will be compacted,

- What compacting mechanism will be used for compaction,
- What degree of compaction should be achieved.

As early as at this stage, it is desirable to distinguish between non-cohesive, granular soils and cohesive, fine-grained soils. With a view to the content of fines and a capability of clay minerals to bind water, their compactibility greatly depends on the moisture content of compacted soil, while for granular soils this dependence is insignificant, the compactibility of granular soils depends more on the compaction energy and vibration effect.

2.3.1 Compaction Mechanism

This chapter refers to cohesive, fine-grained soils. Using a given compaction method and compaction effort, soil is compacted to reach a certain density in relation to moisture content. Figure 2.10 shows that the dry density ρ_d of compacted material grows with the moisture content until the maximum dry density $\rho_{d,max}$ is reached under the so-called optimum moisture content w_{opt}. After exceeding this point, the value of ρ_d starts falling despite the continuous moisture content growth.

For each compaction method and compaction effort (compaction energy) there exists only one maximum value of dry density under optimum moisture content. On the wet side of the compaction curve, starting with the optimum this side runs roughly parallel to the curve of fully saturated soil ($S_r = 1$), which does not contain any air filled voids. The distance of the wet side curve from the line of fully saturated soil represents the amount of air contained in compacted soil.

If the soil is compacted using the same method but under different compaction efforts (compaction energy), we will get a set of compaction curves with a similar shape – Fig. 2.11. With growing compaction energy the maximum value of dry density grows with the falling optimum moisture content. In this set of compaction curves, the line connecting the peaks for individual curves is roughly parallel to the curve of 100% saturation. On the wet side, starting from the optimum the curves obtained for various compaction energy values nearly overlap, which means that

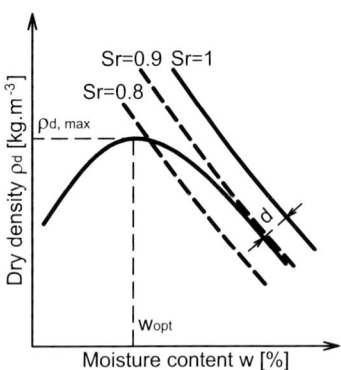

Fig. 2.10 Typical compaction curve for cohesive soils

2.3 Soil Compaction

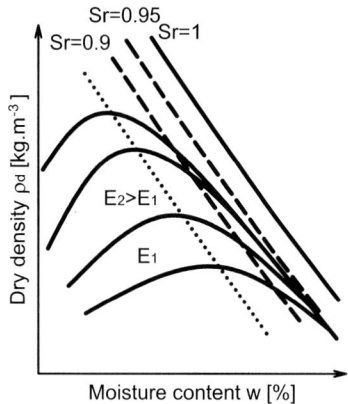

Fig. 2.11 Set of compaction curves for different compaction efforts

for specific moisture content there exists a certain reachable maximum of saturation ratio independent of the compaction energy. Dry sides of compaction curves run roughly parallel to each other.

These results make it clear that an increase in compaction energy is much more efficient for drier soil than for soil moister than the optimum. Figure 2.12 displays a qualitative relation between the growing compaction effect and the growing weight of dry soil ρ_d for soils with identical initial values of ρ_d but different moisture contents (points a, b). For higher compaction effects, the dry density ρ_d grows much faster at point a than at point b. This difference in the relevance of the compaction effect growth in dry density is given by the fact that moister material is softer and the shear stress acting on soil is greater than its shear strength. Thus the compaction energy is used for shearing of the compacted material and not for reducing porosity. On the dry side of the compaction curve, the material is tougher with more energy being converted into soil compaction to more settled state. Compression, of course, requires side restrain preventing soil extrusion into the sides.

For each soil, a certain compaction method provides a different set of compaction curves. In the laboratory where compaction tests may be carried out under controlled conditions, compaction curves may be obtained with relative ease and credibility.

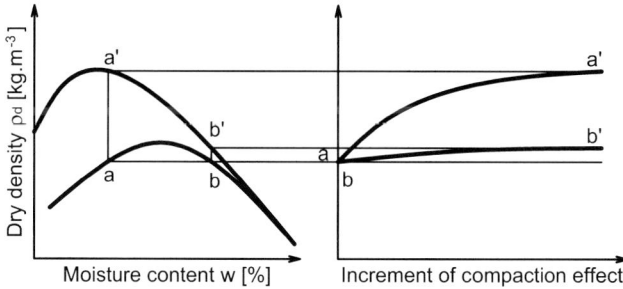

Fig. 2.12 Effects of compaction energy increase for wet and dry sides of compaction curves

However, it is much more difficult to obtain compaction curves for material compaction in embankments for various types of compaction mechanisms, e.g. rollers, as the variability of all factors affecting the test results is difficult to control.

2.3.2 Laboratory Compaction Methods

Laboratory compaction methods serve several purposes. The major one was principally discussed in the introduction. A laboratory-compacted sample is used for individual tests to obtain the parameters of shear strength, compressibility, permeability or other required values, and these results are successively used for calculations in the design of earth structures, in assessing their stability, deformation, permeability or other limit states. On-site activity, in turn, is restricted to checking, mostly by indirect means, via prescribed moisture content and dry density, if the required values were achieved.

Another important variable is determination of the optimum moisture content, which indicates if fine-grained soil presumed for application in an earth structure possesses a higher or, on the contrary, lower moisture content than required, and if potential treatment changing it to fit in the range required after placement into the earth structure is necessary.

The compaction curve shape, however, also indicates material sensitivity to moisture content changes. For example, slightly loamy gravely sand is a little sensitive to moisture content changes as by its properties it is close to granular soil. Apparently similar is the behaviour of clayey soils with high plasticity where a smaller change in moisture content from the optimum will result in a small change in the value of dry density. The soils showing the highest sensitivity to moisture content changes are fine-grained soils with low plasticity, such as sands with small amount of fine-grained soil, like sandy loam where a relatively small change in moisture content from the optimum, e.g. 2–3%, can cause a considerable drop in the value of ρ_d – see Fig. 2.13. Numerous methods of laboratory testing have gradually been developed differing mainly by the magnitude of compaction energy in relation to the subsequent on-site application. There was, on one hand, a requirement for soil to show – after compaction and placement into an earth structure – sufficiently good properties, e.g. its successive deformation should not exceed limit values, while, on the other hand, a requirement for minimizing the energy needed to achieve this state. The laboratory test that gradually became most widespread was the Proctor test, Proctor (1933). This test has two basic versions – the Proctor standard test and the Proctor modified test differing by the compaction energy applied. In the laboratory test, soil is compacted in a cylinder of a given volume with a rammer of prescribed weight falling freely from a prescribed height. The differences are evident from Table 2.5 where for the Proctor standard test, the cylinder is filled up in three layers (after the compaction of the last layer it slightly exceeds the height of the mould without an extension) while for the Proctor modified test there are five layers. The rammer as well is heavier for the modified test (4.5 kg) as compared to 2.5 kg used in the standard test and falling from a greater height (0.45 m as compared

2.3 Soil Compaction

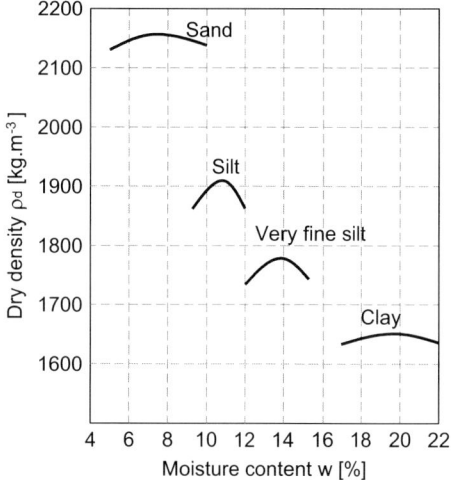

Fig. 2.13 Different compaction curves for different soils

to 0.30 m). The surface of the last compacted layer slightly exceeds the surface of the mould and after the removal of the extension it is removed. The compacted sample is pushed out of the cylinder, weighed and samples (mostly from its centre) are taken to determine moisture content.

Detailed test implementation is specified by national codes, in the Czech Republic e.g. the Methodologies of laboratory tests in soil and rock mechanics, Zavoral et al. (1987). The weight of a prepared technological sample (taken from one representative place of the borrow pit) must be at least 15 kg for the standard test and 25 kg for the modified test. The soil sample is crushed into small clods (clayey soils should be cut into small slices) and spread out on trays. Soil is left to dry so that it is easily separated. Then the soil is disintegrated to fall through a sieve with 16 mm or 5 mm mesh. The undersize product is used for selecting a representative soil sample while the oversize sieve residue is not used for further testing but its share is carefully recorded. The screened sample is newly spread out and left to dry to reach constant weight. The air temperature during drying must not exceed 60 °C. Once the moisture content is determined the sample is divided into at least five parts adding a previously calculated amount of water to each so that the driest sample has a moisture content of about 4–5% for coarse soils or about 8–10%

Table 2.5 Specification of Proctor tests

	Compaction test	
	Standard	Modified
Weight of rammer [kg]	2.5	4.5
Stroke of rammer [m]	0.3	0.45
Diameter of rammer [mm]	51	51
Diameter of container [mm]	101.5	101.5
Height of container [mm]	117	117
Number of layers	3	5
Number of strokes	25	25

Fig. 2.14 Rough estimation of optimum moisture content and dry density based on liquid limit

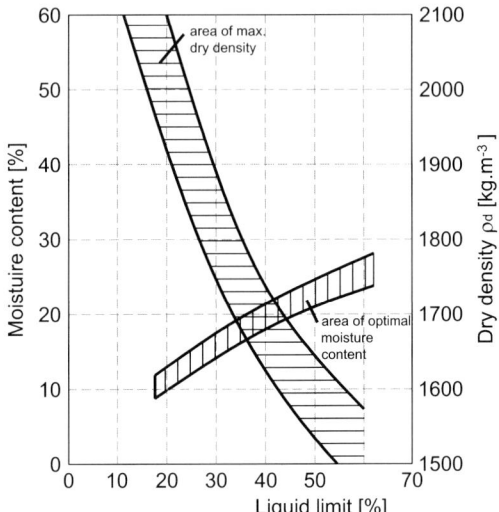

for clayey soils. Each consequent sample will be prepared with moisture content by 1–2% higher for coarse soils and by 2–3% for clayey soils. The recommended moisture content distinction is such that at least three samples should have moisture content lower than the optimum moisture content and at least one higher than the optimum. For rough determination of the optimum moisture content the relation of the optimum moisture content to Atterberg limits may be used – see e.g. Zavoral et al. (1987), Fig. 2.14. To reach proper moisture content distribution, after adding the calculated amount of water, the sample is thoroughly mixed and left to rest under constant temperature for at least 15 hours.

The two basic compaction tests – the Proctor standard and the Proctor modified – show a relatively great difference in compaction energy – $59.44\,\text{J/cm}^3$ ($0.6\,\text{N.mm.mm}^{-3}$) for the standard test, or $267.5\,\text{J/cm}^3$ ($2.7\,\text{N.mm.mm}^{-3}$) for the modified test, which is also reflected in the results – Fig. 2.15. In practice, for common embankments it turned out that compaction to a value of the maximum dry density under the Proctor standard test is sufficient to eliminate substantial settlement. For smaller embankments or backfills lower compaction usually suffices being expressed in the percentage of the maximum value – e.g. 95% Proctor standard. Compaction corresponding to the energy of Proctor modified was selected to comply with strict requirements for airport subsoil. Exceptionally, it is also used for very high rockfill dams where the energy recommended is slightly higher than that corresponding to the Proctor standard test.

The Proctor test is generally referred to as a dynamic compaction test. This type of test is presently preferred to the so-called static compaction test where soil is loaded with static load of a recommended magnitude based on the expected vertical loading of the respective earth layer in the earth structure. The volume density reached during this static test was subsequently recommended for a specific site. During the dynamic test, shear deformations occur as well allowing easier mutual

2.3 Soil Compaction

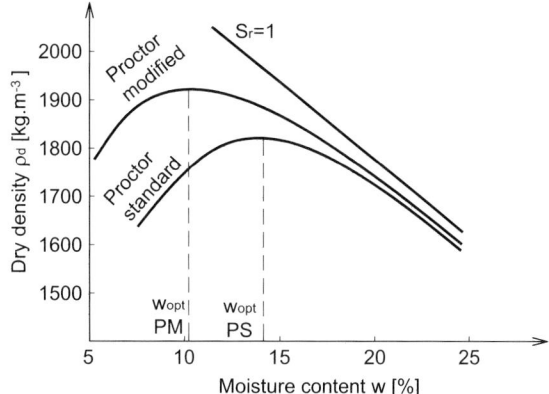

Fig. 2.15 Differences for compaction curves for standard and modified Proctor tests

connection of individual clods of compacted soil. During the static test, mainly for lower expected loads, the macropores between individual clods are prominent, and subsequent additional in situ deformation due to moisture content changes at contacts may occur, generally due to creep strain. Therefore, this static compaction test procedure has been mostly abandoned (it is used when some pre-compaction by means of a consolidation fill is needed, which is removed after a certain time).

Thus the present-day basis for defining the required degree of compaction is the Proctor test. The prescription is based on two values:

- Minimum value of the dry density ρ_d required (most often defined in relation to the maximum value reached under the optimum moisture content $\rho_{d,max}$) – the term compaction to 95% PS means the required ratio $\rho_d/\rho_{d,max} \geq 0.95$
- Required range of moisture content under which soil can be compacted, e.g. $w = w_{opt} \pm 1.5\%$.

This fact is displayed in Fig. 2.16. In the laboratory, samples from this defined range are tested, their bands are determined, and finally a so-called characteristic value is selected from them to be used by the designer of the earth structure in calculations.

Note 1: The upper limit of ρ_d is defined only exceptionally for fear of so-called over-compaction, e.g. at the contact of two different soil types where differences in deformation modules in certain intervals are required to avoid significant stress transfer from one area to the other. EN 1997-1 (EuroCode 7) states that over-compaction can cause the following undesirable effects: the development of slickenside and high soil stiffness in slope, high earth pressures on buried and earth retaining structures or crushing of materials such as soft rocks, slag and volcanic sands used as light weight fills.

Note 2: An exception to this rule is the requirement for the moisture content of compacted soil for the sealing layers of e.g. landfills where the moisture content required will be rather higher than the optimum – this question will be further discussed in the following chapters.

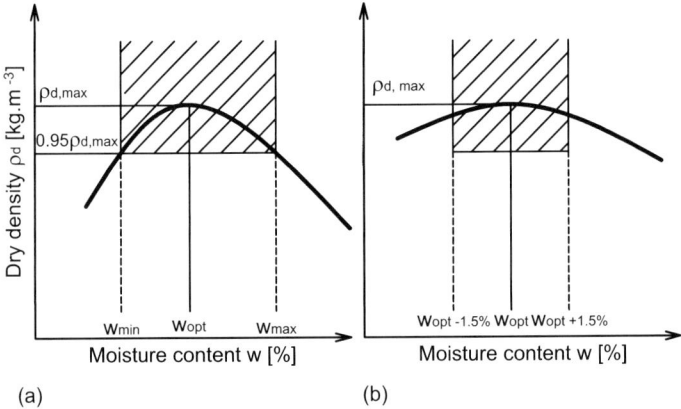

Fig. 2.16 Different approaches to compaction specification. (**a**) Based on dry density; boundaries of moisture content are given by intersection with Proctor curve. (**b**) Based on range of acceptable moisture content and recommended value of dry density

A certain problem arising between the results of laboratory tests and the in situ compaction rate required is the fact that the maximum grain size used in the laboratory during the Proctor test is 5 mm or 16 mm. The size of 16 mm can be already too big with a view to the cylinder diameter of 101.5 mm. That is why a cylinder with a larger diameter – 152 mm can be used increasing the number of blows per one layer to 56 to achieve identical compaction energy. However, it generally holds true that if the laboratory compaction test is performed using a finer material than in reality, the optimum moisture content and the maximum dry density must be recalculated as larger gravel particles do not bind water. If the same moisture content was kept on the site as in the laboratory, coarse particles would be surrounded by the remaining fraction with higher moisture, and the properties would be substantially negatively affected.

$$w = (w_f(100 - m_g) + w_g \times m_g)/100 \tag{2.5}$$

where

w – moisture content for whole sample
w_f – moisture content of finer fraction (lower than 5–16 mm)
w_g – moisture content of the coarser fraction (above 5–16 mm)
m_g – weight of coarser fraction as percentage of the whole sample

Note: valid for $m_g < 30\%$

$$\rho_d = \frac{100 \rho_{df} \times \rho_s}{\rho_s (100 - m_g) + \rho_{df} \times m_g} \tag{2.6}$$

2.3 Soil Compaction

where

ρ_d – dry density for whole sample
ρ_{df} – dry density for finer fraction
ρ_s – unit weight of coarser solid particles

At this point, another problem must be mentioned. The samples compacted in the laboratory, which successively undergo other tests, are mostly samples with a maximum grain size of 5 mm (e.g. with a view to the sample diameter in the triaxial test, the sample height in the oedometer). The properties obtained in laboratory, at which the designer of the earth structure can rely on, differ from the properties of in situ compaction if the particles are larger than those used in the laboratory test. This presents another principal complication distinguishing the design of earth structures from other types of building structures. Sometimes, however, this aspect is neglected as it is mostly on the safe side, soil with larger grains tends to show lower compressibility or higher shear strength.

For completely non-cohesive soils – sands and gravels – the Proctor test is not commonly used – as was already said – the value of the maximum dry density shows only slight dependence on the moisture content. An additional factor is the fact that granular soils are not so sensitive to the type of stress applied in the Proctor test but are more sensitive to vibrations. That is why their definition of required compaction is not based on this test but the requirement for their compaction rate is based on the relative density index I_D.

$$I_D = (e_{max} - e)/(e_{max} - e_{min}) \tag{2.7}$$

Here, the values e_{max} and e_{min} (maximum and minimum porosity numbers) represent the minimum and the maximum density states defined by a laboratory method and thus also the boundary values of dry density. The laboratory methods recommended in different countries may differ but this drawback can be eliminated by the relative density index as in principle it recommends the dry density which should be achieved on the site. The test of determining the minimum density usually involves pouring oven-dried soil into a container. The test of determining the maximum density, on the other hand, usually involves some form of vibration under vertical loading, see Fig. 2.17. Under ČSN 721018:1970 the maximum density is determined in cylinders similar to those used for the Proctor test. For grains greater than 32 mm a cylinder with a volume of 20 litres is used. The cylinders are loosely filled with dry material (in the same way as in the minimum density test), loaded with a weight exerting surplus loading by 15 kPa, mounted onto a vibrating table with an electric vibrator with a frequency of 3000 ± 300 cycles per minute and a deviation amplitude under full loading of 0.35 ± 0.07 mm, and undergo vibrations for 8 minutes. As an alternative, the test may be performed in wet state. The results of both states do not exclude that sometimes the density of soil can be lower (e.g. if fine sandy particles sediment in water) or, on the contrary, can be higher (under higher loading, longer vibration time etc.).

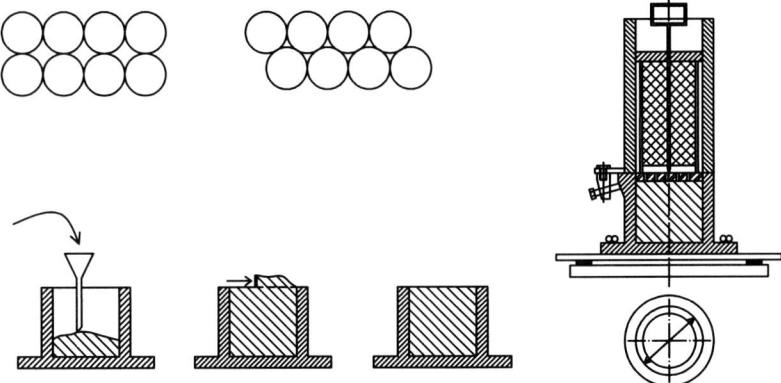

Fig. 2.17 Scheme of laboratory determination of maximum and minimum density of coarse soils

In practice, the most commonly prescribed compaction to I_D ranges within 0.8–0.95 as necessary. This recommended condition (of dry density) serves for performing laboratory tests envisaged in the design of an earth structure.

Note: The limitation concerning the grain size applies here, too. For substantially greater particles than may be tested in the laboratory on the above-mentioned device, we must either take into account the results of laboratory tests on significantly larger samples – e.g. compressibility tests or potentially shear strength tests, or base our assumptions on a back analysis of actual monitored earth structures.

The results of the minimum and maximum dry density vary in relation to the shape of the particle size distribution curve, particle size and their angularity. The highest values are reached for soils with a smooth particle size distribution curve. There in the structure of coarser particles smaller particles are infilling the pores while in evenly graded soils the pores between roughly identical particles are remaining free. Thus more internal contacts between particles are achieved for a smooth curve via which load is transferred.

This is the reason why the porosity number e is determined on these soils in the laboratory. The porosity number e is determined by laboratory procedure and defines the relative soil density from the loosest to the most settled.

To sum up laboratory tests and experience of them, we may conclude that compaction involves several basic principles:

- Exerting sufficient contact pressure on soil (mainly to reduce the macropores between individual clods),
- The moisture content at which soil is compacted determines the efficiency of the contact pressure used,
- Higher values of dry density are reached for non-uniformly-grained soils with a smooth particle size distribution curve,
- To achieve good compaction, soil must be limited at sides with a view to existing shear stress.

The last point is directly related to poor soil compactibility at the edges of embankments.

2.3.3 Compacting Machinery

Compacting machines exert various effects and compaction is done by pressure, kneading, impact, vibration or a combination of these effects. The most commonly applied compacting mechanisms are rollers, which apply in particular:

– Static effect,
– Dynamic effect.

Apart from rollers, the dynamic effect is also applied by compacting plates and rammers.

Rollers, in turn, can be subdivided by the runner type into rollers with a smooth surface, sheep-foot rollers and rubber-tyre rollers. They may also be classified by specific pressure on the runner, by total weight, by travel capability into self-travelling and towed etc. Some other classification methods will be more closely specified in the text below.

2.3.3.1 Static Rollers

Static machines used to be the basic type of compacting machines up to the 1950s. Since then vibration technology is gradually replacing static rollers. The efficiency of static rollers is presently exploited mainly at jobsites where, apart from compaction, there are high demands for the evenness of the compacted surface, such as road pavement wearing courses, or compaction of asphalt-concrete carpets.

As the static effect is of prominent importance in compacting machines, next to the vibration effect, some aspects will be considered in this chapter. In a static roller, the major parameter is specific pressure at the runner-ground contact – the heavier the rollers, the more efficient they are. Apart from the static effect, they, to a lower extent, exert kneading, which is more important in sheep-foot rollers.

The efficiency depth of static rollers is given by the above-mentioned specific pressure and the contact surface. A roller of the same weight but with a larger runner diameter exerts, on the one hand, lower specific pressure, but, on the other hand, its loading surface is larger. During the first passes of roller the contact surface is the largest and so the depth at which the pressure increases due to loading from the roller is the greatest. During the following passes, the contact surface is reduced and the area on which the pressure increase acts is smaller, but the pressure is higher see Fig. 2.18. Therefore, the compacted layer thickness must respect this fact; the aim is to reach a maximally homogeneous compaction rate within the compacted layer height. The most common compacted layer thickness is 0.2–0.3 m.

Sheep-foot rollers, which are fitted with protruding stumps along their perimeter, were and still are very popular in the construction of earth dams. They derive their name from historical embankment compaction using sheep feet. Today's stump shape is governed by the requirement to avoid soil chipping at the moment when the

Fig. 2.18 Contact area and pressure combined with effective depth of compaction for smooth static roller

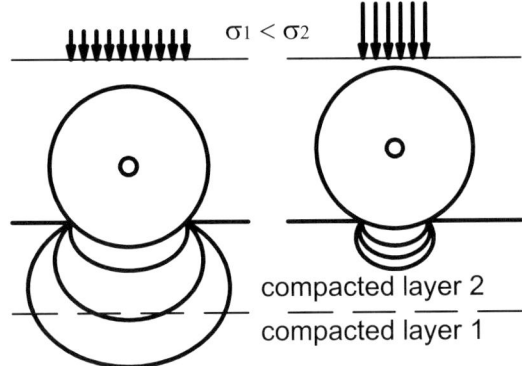

point leaves the soil. The stump height is about 0.15–0.32 m being of truncated pyramidal shape with rounded edges. During the first passes, the roller weight is transferred through the soil under the roller's drum, which exerts a relatively low pressure, while the soil under the stumps' surfaces, which exert a relatively high pressure onto the bottom layer, is properly compacted. As compaction proceeds, the load is transferred only via the stumps. In this case, the roller weight is transferred via a small surface with a high specific pressure, see Fig. 2.19. That is why in some recommendations there is a requirement to complete compaction only after the pressure is transferred only via the stumps and the drum itself no longer touches compacted soil. By their kneading and pressure, the stumps crush even larger, tougher clods or chips of soft rock very well. Therefore, in such cases, a greater number of passes is mostly required, about 10–12. The advantage of sheep-foot rollers is good interconnection of successively compacted layers.

Rubber-tyred rollers exert their compaction effect by both static loading and kneading. Although their contact pressure is slightly lower during the first passes,

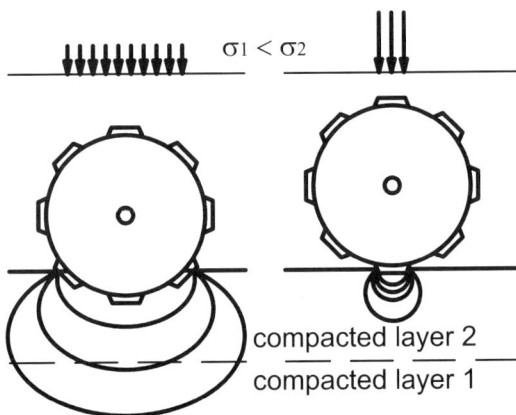

Fig. 2.19 Contact area and pressure combined with effective depth of compaction for sheep-foot static roller

it remains at about the same level for all passes. It roughly equals to the pressure in the tyres so that this pressure directly affects the compaction effect. Therefore, some rollers allow fast pressure changes in the tyres to raise the pressure with the growing number of passes.

2.3.3.2 Compacting Machines Applying Dynamic Effect

Impact Compaction

Impact compaction, e.g. by the fall of steel plates from a great height or by explosion rammers, causes such a heavy pressure wave at the surface of compacted material, which propagates into a great depth. Thus the static effect of the plate itself at the end of its fall is many times multiplied due to kinetic energy. The development of this method is connected with Ménard's name and so the method is often also called Ménard's method. Other common terms for it are heavy tamping or dynamic consolidation. Sometimes this method is also referred to as shock compaction as high energy fully disrupts internal bonds in the soil and before the soil can be extruded into the sides, it is compacted. Recently the weight of the falling plate has been growing, from the standard of 10–40 tons falling from heights of 10–20 m to extremes of weights up to 200 tons and falling heights of 40 m. The method is suitable when it is necessary to improve the subsoil quality consisting of soils with a low bearing capacity, fluvial deposits and peat and so it is not very common in soil treatment as a construction material in earth structures. An exception is compaction of landfill material, which will be further specified in the chapter on landfills. It may naturally be used if the subsoil quality under the routes of transport structures must be improved. In this way, subsoil with a thickness of up to several meters, i.e. thicknesses which cannot be reached with conventional vibratory rollers, can be improved.

Vibration Compaction

Vibration used to be applied mainly to compact granular soils – sand, gravel. Nowadays it is equally common for compacting cohesive soils. Vibration of non-cohesive soils results in a considerable decrease of their shear strength, which accounts only for a small percentage of their strength in still state. In this state, under vibration and loading, the particles get closer together. A certain problem for evenly graded granular materials, mainly sands, is lower compaction of the topmost layer. Brandl (2001a,b) refers to this phenomenon as "surface-near re-loosening" generally not recommending compaction of too thin layers. The range of layer thickness stated is 25–50 cm using a vibratory roller with a weight of 10–25 tons.

For cohesive soils, mainly with lower moisture contents, the shear strength falls less steeply being time-dependent. Vibrations first cause a shear strength decline and successively a gradual, relatively small decrease of the internal strength of clay (internal structure disruption), which rises to higher values only after a certain time after completing the vibration (e.g. thixotropic consolidation). This effect can be observed during loading plate compaction controls where the results received immediately after completing the compaction may vary (show lower values) from those obtained if the checks take place several days later. This drop in internal strength caused by a mechanical effect – vibration may also be achieved using other types of mechanical effects, such as kneading, shaking. For example, the compressibility of clayey soil with the same dry density ρ_d will vary depending on the preparation procedure whether this volume density was reached by applying static loading or kneading. The above-mentioned effect, however, refers mainly to soils with higher contents of active clay minerals, which is not a common case in earth structures. For sealing systems of landfills where the main criterion is permeability this effect is of minor importance.

From vibration components – frequency and amplitude – the one with a greater effect on particle transport in cohesive soils is the amplitude. This, in turn, is manifested by means of the kinetic energy of vibrating mass. At the contact with soil, it changes into static pressure thus increasing the static pressure due to the self-weight of the compacting means acting on the compacted element, Forssblad (1965). Therefore, a combination of a high weight with large amplitude is of utmost benefit for cohesive soils. In actual compacting machines, rollers, plates, rammers, however, high values cannot be reached considering the workers operating these machines and also their high wear and tear rate. Nevertheless, vibrating plates and rammers in general apply lower frequencies putting a greater stress on the amplitude. In vibratory rollers with a circular exciter system, both combinations can be chosen, the former implying selection of smaller amplitude and a higher frequency, but in some, on the contrary, greater amplitude and a lower frequency. The latter option is common for compacting thicker layers of both granular and cohesive materials, mainly in embankment and dam compaction.

An ideal frequency should correspond to the resonance frequency of the compacted soil and the compacting machine system. This resonance frequency changes with the soil type, its compaction rate and characteristics of the compacting machine used, ranging within 800–1 700 oscillations per minute (about 28–13 Hz). The design of a compacting mechanism with a variable frequency in relation to the resonance frequency is feasible, though constant frequencies are mostly selected for rollers, ranging, as a rule, in the upper band of the above-stated values. When scanning the pressures in the compacted subsoil layer, both the basic static pressure of the machine and the amplitude and frequency effects are evident, see Fig. 2.20.

The basic classification of compacting mechanisms applying vibrating effects, therefore, distinguishes vibratory rollers, vibratory plates or vibratory rammers.

Vibratory Rollers

Vibratory rollers take up the largest share and may be further subdivided according to various aspects, such as total weight, drum surface – into smooth, sheep-foot, rubber-tyred, the number of axles into one-axle and two-axle types, if they are hand-guided or driven etc., see Fig. 2.21.

2.3 Soil Compaction

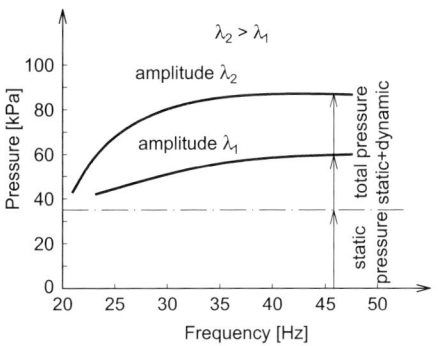

Fig. 2.20 Combined effect of static and dynamic compaction

The classification according to the vibrating effect application – and its effect on the resulting compaction rate mostly splits rollers into vibratory rollers, oscillatory rollers, Vario rollers and Vario Control rollers where the vibratory roller exerts mainly vertical surcharge by a rotating mass excentre, while the drum of an oscillatory roller oscillates torsionally, Brandl (2001b). Vibratory rollers represent the most commonly used type. In the case of sand compaction, the abovementioned "surface-near re-loosening" effect may be encountered. In the given case, the result depends on sand uniformity, selection of an optimum roller type, or it is recommended to turn off vibration for the last roller pass. In the case of an oscillatory roller, the soil is loaded horizontally in addition to the vertical static line load. The cyclic and dynamic horizontal forces cause shear deformations of the surface-near sand connected with shear peelings and cracks development. Oscillatory rollers are therefore not suited for the compaction of sand or other evenly graded granular materials. They are in a much greater scale used for compacting asphalt layers. In a Vario roller two counter-rotating exciter masses, which are concentrically distributed on the axis of the drum, cause a directed vibration. The direction of excitation can be adjusted to local conditions by turning the complete exciter unit in order to optimize the compaction effect. If the exciter direction is almost vertical or inclined, the compaction effect can be compared with that of a vibratory roller. The exclusively vertical direction allows total prevention of the development of shear peelings and cracks. A Vario roller can be employed universally, the individual optimum direction of excitation can be found on a test field. The Vario Control roller is an automatically controlled roller. In this roller type the direction of excitation is turned automatically by using defined control criteria, like the maximum soil contact force or the respective operating condition of the drum.

The travel speed of rollers – and vibratory rollers is small, like that of static rollers, ranging mostly from 1.5 to 6.0 km/hour. The compaction rate grows with the falling speed as the loading effect on the compacted element is longer. Nevertheless, it also depends on the required compaction rate and overall capacity, compaction efficiency – generally it is possible to reach the same compaction rate per constant time with a greater number of passes at a greater speed. It generally applies that for higher compaction rates required the optimum speeds are lower than for

(a)

(b)

(c)

Fig. 2.21 Photos of different rollers

2.3 Soil Compaction

lower compaction rates. To verify this, a compaction field trial can be carried out. Analogically, low speeds are applied for sheep-foot and rubber-tyred rollers. The exception are tamping rollers, which have stumps similar to sheep-foot rollers – stumps shaped like perpendicular rectangular pyramids with flat seating faces. They are designed mounted on undercarriages of heavy loaders working usually at high speeds of up to 28 km/hour.

Surface Shape Apart from the above-mentioned roller types – smooth, sheep-foot, rubber-tyred, tamping rollers, relatively unusual shapes are becoming available. One of them is so-called Polygon-Drum – e.g. Adam and Markiewicz (2001), which is composed of five octagonal elements, see Fig. 2.22. The second and fourth elements are wider than the other elements in order to achieve equivalent total contact widths. The authors indicated that the main achievement for this Polygon-Drum is

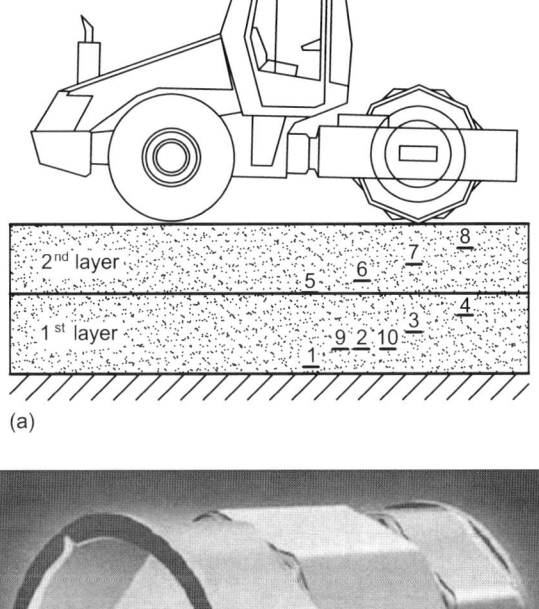

Fig. 2.22 Roller with octagonal elements – polygon drum (according to Adam and Markiewicz) **a**) schematic cross-section, **b**) detail of polygon drum

the prevention of bow and back waves usually caused by pore pressures in cohesive soils. Another case is described by Pinard (2001). Putting a greater stress on the importance of impact compaction, he says that the laboratory compaction test (Proctor test) is also more based on "a high amplitude and low frequency which, in practice, relates more to the process of impact compaction rather than to either static or vibratory compaction, in which energy is delivered at a low amplitude and high frequency". Rollers of this type are termed as High Energy Impact Compactors. The distinguishing feature of all impact compactors is their non-circular compacting masses which have a series of points alternating with flat compacting faces. The faces vary in relation to their shape (three, four or five-sided), mass and configuration – see Fig. 2.23.

Vibratory Plates

Vibratory plates are suitable mainly for the compaction of backfills of trenches and embankments and fills in the vicinity of structures (Fig. 2.24). Heavier plates may also be applied for levelling roller-compacted surfaces or their additional compaction. It generally holds true, however, that the contact stress applied is lower and the depth efficiency small, which is reflected in the recommended height of compacted layers, ranging mostly within 0.05–0.15 m. The contact surface is even, sometimes, mainly at sides, slightly rounded. With a view to the constant contact surface, they constantly exert the same pressure, independent of the ground compaction rate, contrary to rollers. They move using the horizontal component of the vibrator's dynamic force.

Vibratory Rammers

Like vibratory plates, vibratory rammers work at lower frequencies with the major parameter being the amplitude (Fig. 2.25). Vibratory rammers are easy to operate, hand-guided, moving again due to the effect of the horizontal component of the vibrator's centrifugal force. Like vibratory plates, they are suited for backfills and packing.

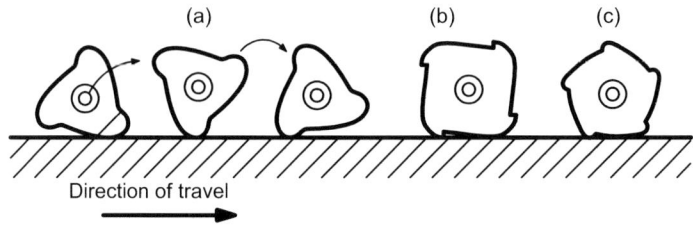

Fig. 2.23 High energy impact compactor drums (according to Pinard)

2.3 Soil Compaction

Fig. 2.24 Photo of compaction vibratory plate

2.3.3.3 Suitability of Compacting Machines

The selection of the most suitable compacting machine is a very complex issue depending on multiple factors, among which the basic ones are:

- Character and state of soil to be compacted,
- Type of site,
- Required construction speed/schedule.

A natural target in selecting the most suitable means will be not only the fulfilment of prescribed qualities of the compacted earth body in the required time, but the fulfilment of these objectives with minimum energy demands, or at a minimum price. The selection below will be focused on compaction rollers, because

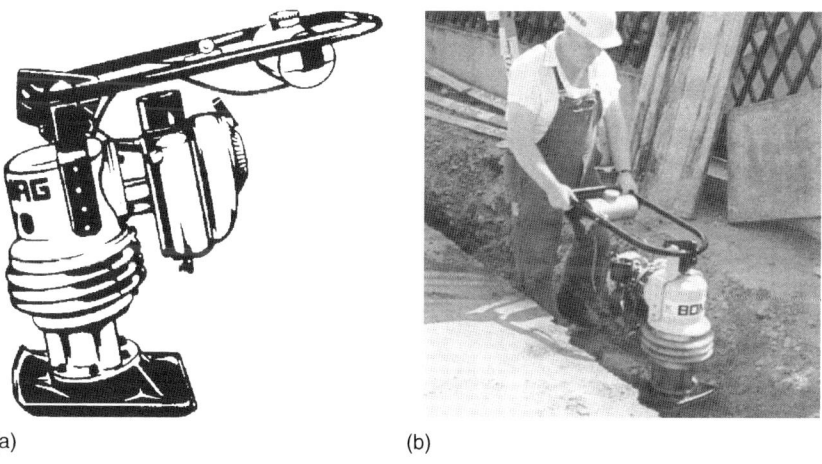

Fig. 2.25 Scheme and photo of compaction vibratory rammer

impact compaction – dynamic consolidation or application of vibratory rammers and plates has its own specificities fitting the strictly defined above-mentioned purposes.

Considering the first point, there will be great differences between granular soils (which are generally very well compactable – despite certain problems connected with grain size uniformity) and cohesive soils. The fact commonly stated is that soil is well compactable if its plasticity index I_p is lower than 50%. This, however, need not present a limitation, mainly in sealing layers of landfills. In relation to moisture content, drier soils are better compactable, needing, however, more energy. The only potential hazard is successive settlement after watering if larger pores were left among dried clods. Significant problems are connected with moister soils where high moisture content may lead up to the sinking of the machine into the compacted layer, and sometimes there are some problems with soil sticking to the drum perimeter. Here, preliminary measures must be taken leading to reducing the moisture content or improving the compaction potential by adding smaller amounts of lime. The above-mentioned facts concerning moisture content hold true in general, though this problem should not affect soils embedded into earth structures whose moisture content is in the band recommended on the basis of the Proctor test.

When considering the type of site or required construction pace, the greatest differences are between the sites involving large volumes of placed soil, and smaller-scale sites with many details. In the former case, the choice is in favour of heavy vibratory rollers with high efficiency given by the capability of compacting layers of greater thicknesses, ca. 0.4–1.2 m applying a smaller number of passes, while in the latter case the thickness of layers is smaller, in the order of 0.15–0.3 m. Smaller contracting companies with a smaller number of compacting rollers will put more effort on using their machine for a wider scope of applications even with a lower efficiency, while a large company can exploit more efficient machines in a narrow scope. The question, however, is the uptime of individual machines. All these problems lead to a situation when individual firms (large ones in particular) define the optimum way of their machinery fleet application. On the other hand, manufacturers specify for individual roller types the presumed capability of proper soil compaction at different layer thicknesses, at different speeds and number of passes. This kind of information may serve for calculating the overall efficiency, or efficiency per volume unit.

2.3.3.4 Compaction Field Trial

The recommendations above referring to the selection of a suitable compaction mechanism are a very good guide, nevertheless for a specific site and a specific soil type it is desirable to carry out a compaction field trial for large-scale compaction work in which the following variables are monitored for a selected compaction roller:

– Compacted layer thickness,
– Required compaction degree,
– Number of roller passes,
– Compacting machine speed.

2.3 Soil Compaction

In principle, the maximum thickness of a compacted layer is sought which is easily compactable to a required compaction degree using the standard number of passes (4–12). For this purpose, a trial section is made on subsoil corresponding to real site conditions as subsoil stiffness can strongly affect the result achieved. Several strips with a mild slope are chosen to reach variable thicknesses of filled and spread soil, see Fig. 2.26. Different speeds are used for different strips and after a specified number of passes samples from the bottom of compacted soil layer are taken. The results are checked by comparison with the required compaction degree. The output is the recommended optimum with a view to the layer thickness, the number and speed of passes.

2.3.4 Compaction Quality Control

The compaction quality control is the basic control measure of the earth structure implementation. In this view, the relations of the three principal parties in the construction process – the designer, contracting company and investor (owner) are important as it really deals with the determination of the relation between the recommended compaction degree and the actual, implemented degree.

Bulletin No. 56 Quality control for fill dams ICOLD 1986 states that there are a large number of possibilities in the sharing of the varied aspects of the quality control program between owner, engineer and contractor. The extreme situations are where either the owner has total responsibility for all aspects of design and

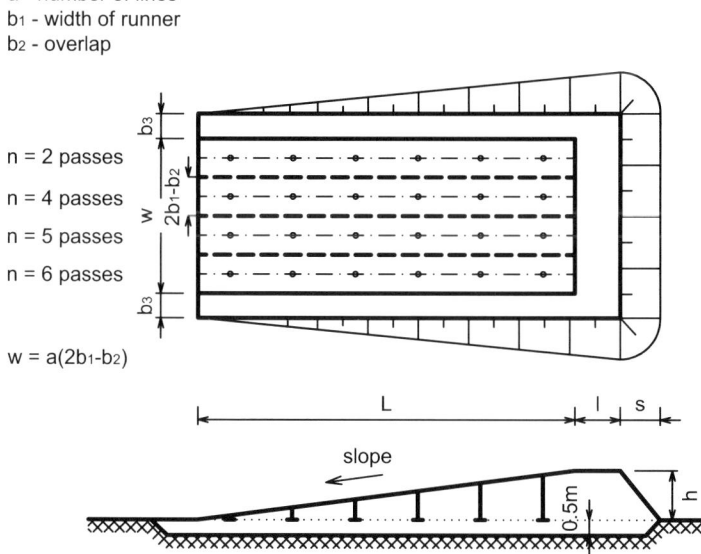

Fig. 2.26 Scheme of compaction field trial

construction with its own staff and own means, or where the contractor has total responsibility for all aspects of design and construction.

When considering a classic relation where the designer – mostly geotechnical engineer, recommended a compaction rate on the basis of dry density and moisture content, simultaneous on-site checks of both of these variables have been preferred to-date. This classic approach brings about the following problems:

- The quality control is carried out at spots on samples of limited volume – within the conventional control scope, in fact, not the entire earth structure, but only its very limited part is inspected, in the order of 1:1 000 000. A highly sensitive issue here is to define how many samples must be taken, and what their scattering range may be. The results, however, have official relevance proving compliance with the demands required, serving as a basis for the earthwork acceptance by the investor. Noncompliance with the demands, on the contrary, leads to additional modifications, additional compaction or removal of a partly embedded earth structure and its replacement with new material. Here, it must be said that the control cannot be aimed only at the compaction degree but at verification if the soil embedded into the earth structure was presumed for this purpose, and this fact must be proved by means of a series of index tests which commonly serve for soil classification – in particular the particle size distribution and consistency limits.
- The control using the above-mentioned tests is not instantly completed, but with the classical verification of an oven-dried soil sample it usually requires 1 day. The contracting company, however, cannot stop work and wait for the final verdict – therefore it performs continuous control by itself, often using indirect methods to ensure that the conditions were fulfilled but that they were not unnecessarily over-fulfilled at higher cost.

The contradiction above presently leads to a situation to make continuous control carried out by the contracting company recognizable on the investor's part as well. The thing is to eliminate in this way both "over-compaction and re-loosening of layers" and "non-uniform, heterogeneous compaction degree". Therefore, the main goal of cooperation between geotechnical and mechanical engineering is the development of "intelligent" compaction equipment which itself reacts to locally varying soil/granular material properties by automatically changing its relevant machine parameters – Brandl (2001b). Description of these new quality control procedures will be part of this chapter.

2.3.4.1 Fast, Indirect Check Procedures

As was said above, fast compaction quality control tests are used mainly by contracting companies for operative management of compaction works. From fast, indirect methods the simplest one is the penetration test applying different types of penetrometers – e.g. a pocket penetrometer, Proctor needle etc. It determines the magnitude of cone insertion under standard load compared with the magnitude of insertion into identical material with a required compaction rate. Another group includes methods using gamma radiation – "nuclear methods" – nuclear density

2.3 Soil Compaction

Fig. 2.27 Compaction control by nuclear density gauge

gauge, see Fig. 2.27. The penetration of radiation rays between the radiation source and the receiver depends not only on dry density, but also on moisture content. Then a calibrated device can provide results very fast and with reasonable credibility. More advanced machines e.g. Troxler do not require additional calibration being able to determine the compaction rate achieved directly, and the result for dry density is determined with a greater precision of about ±0.2% while for moisture content with a lower precision of ±2%. In general, this group also includes other geophysical, non-destructive methods, such as the continuous surface wave technique (CSW) or the spectral analysis surface wave method (SASW) Brandl (2001b). CSW utilizes an electromagnetic vibrator placed on a level ground surface to generate surface waves and a row of geophones to detect these waves. A significant advantage of the CSW technique is its deep-reaching capacity which makes a post-control of entire soil structures possible.

In many cases, the contracting company relies on the precise observation at the start of works, which is defined for large-scale volumes on the basis of the compaction field trial. There are, however, additional empirical rules, e.g. Sherard et al. (1963) for the USA state that compaction using a sheep-foot roller is considered as sufficient if only the stumps are penetrated into the ground, and the drum body remains above the surface.

2.3.4.2 Classical Compaction Quality Check

As was already mentioned, the method is based on checking dry density and for cohesive soils also actual moisture content. The dry density determination requires determination of the volume of the sample taken. For cohesive soils, this usually does not present a problem and various sampling tubes are used, see Fig. 2.28. For granular soils, an indirect method of volume determination of the sample taken must

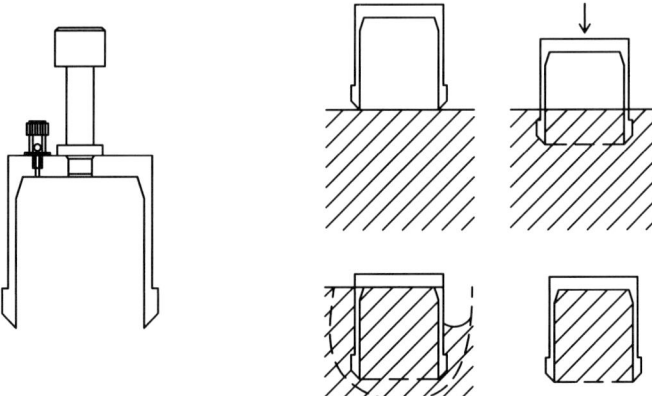

Fig. 2.28 Classical compaction control by sampling

be used. Here, the standard sand method is described; sand is filled in a prescribed way into the pit after soil sampling. The condition of this test method is the knowledge of the volume density of loosely poured standard sand. The total weight of standard sand filled in the pit serves for the calculation of its volume. Other possibilities are offered by membrane volumemeters, of air or liquid type, see Fig. 2.29. For rock fill or fill containing a large amount of coarse particles the measured volume must be proportional to the size of individual particles. Therefore, for very coarse particles, the volume of extracted material (gravel, stones, boulders) is determined by a geometric method using a grid placed on the surface with a selected size of square fields for whose centres the average depth to the pit bottom is measured.

The extent of tests is defined in national codes, requiring mostly one check sample from $500\,\text{m}^3$ of placed material or one sample taken from an area of $2000\,\text{m}^2$. For example, ČSN 736850:1975 Earth dams recommends, apart from the above-mentioned requirement for uncritical spots, that proof tests have to be carried out at least once per working day, from each treated layer, and under weather changes when the properties of the fill are substantially affected. An increased number of

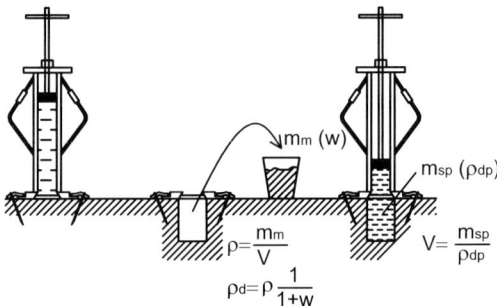

Fig. 2.29 Compaction control with the help of membrane volume-meter

2.3 Soil Compaction

samples must be taken at extremely critical spots, such as filters, layers that are in contact with foundation soil on hillsides and with dam structures etc. The selection of sampling spots must follow a certain system (uniform distribution, selected profiles, using a random number system etc.). Samples are also taken at spots where there are doubts concerning a sufficient compaction rate. In total, the number of proof tests and samples taken depends on local conditions, compaction technology, the variability of the fill and the extent of work. It is desirable to modify it during construction according to gathered experience and the results of previous tests.

The assessment of the results may be done in two ways:

- By comparing the results achieved with the requirements, fulfilment of absolute values,
- By evaluating a set of results in terms of its uniformity – homogeneity requirement.

In the first case, we may apply:

- *degree of compaction* D:

$$D = \rho_d / \rho_{d,max} \tag{2.8}$$

where

ρ_d is dry density measured on the tested sample,
$\rho_{d,max}$ – dry density from Proctor test for optimum moisture content

- *coefficient of efficiency of used compaction machine* C:

$$C = \rho_d / \rho_{od} \tag{2.9}$$

Where ρ_{od} is dry density for Proctor test for the same moisture content as obtained from the tested sample,

- *moisture content difference* Δw:

$$\Delta w = w - w_{opt} \tag{2.10}$$

This gives the difference in moisture content for the tested sample and optimum moisture content from Proctor test in the laboratory.

The degree of compaction D and moisture content difference Δw are decisive values for compaction quality checking.

Very often more than 90% of samples have to fulfil these conditions; none of them should be out of the range of 10% from the demanded value.

The coefficient of efficiency C provides very useful information about the used compaction machine, if it is really able to fulfil the demanded values with the given compaction effort.

In the second case, the homogeneity of the set obtained is determined by means of the coefficient of variation v, where

$$v = s_x/x \qquad (2.11)$$

where s_x = standard deviation, x = mean value.

For road and motorway earth structures, Brandl (2001b,c) recommends the following values of the coefficient of variation v for the replacement method (one test for $\geq 300\,\text{m}^2$), which are part of the Austrian Road Construction code:

- $v \leq 5\%$ for subgrade,
- $v \leq 4\%$ for sub-base,
- $v \leq 3\%$ for base

which are valid for a set of 20 tests. For more tests, the variation coefficient can be a little bit higher. This condition will guarantee uniform compaction rates along the whole area, which will avoid non-uniform deformations of the earth body after the application of external load.

A critical point to add to this classical quality control form is that while providing relatively accurate information, these inspection approaches have several disadvantages:

- Require continuous observation for method/process control,
- Offer measurements only for a small percentage of the fill volume for spot tests – typically 1:1 000 000,
- Require construction delays to allow time for testing,
- Result in downtime for data analysis,
- Cause safety issues due to personnel in the vicinity of equipment.

Note: Due to complications and problems with checking the quality of coarse particle compaction, an indirect method is preferred in the majority of cases based on the measurement of the deformation of the fill surface using established survey points (levelling method), which is, in some countries, regarded as a classical and generally evidential method. The fill is considered to be sufficiently compacted if the deformation during the last pass does not exceed a defined value. In most cases, sufficient compaction is reached in the case where the settlement difference after the last two passes is smaller than 0.5% of the layer thickness. At the same time, no visible elastic deformation under the roller runner occurs. Brandl (2001c) states that for the compaction of high motorway embankments of rock fill and coarse soil the following compaction criterion proved suitable:

$$\Delta s_n \leq a \times \sum_{i=1}^{n-1} \Delta s_i \qquad (2.12)$$

where

Δs_n = settlement increase of the layer during the last pass of the compacting roller,
$a = 0.05 - 0.1$, depending on the material (determined by trial compaction),
n = number of compaction passes of the roller,
Δs_i = average settlement increase of the particular fill layer during the roller pass i.

2.3.4.3 Direct Control Methods

As direct methods of compaction quality control we may consider the measurement of basic mechanical and physical properties – permeability, compressibility and shear strength. Shear strength is not used for checking purposes in practice, while the prevailing tests are those of permeability and deformation of the compacted layer surface. Permeability is verified for the compaction of individual landfill sealing layers. Due to its complexity, this test will be more closely specified in the chapter on landfills. The compressibility of compacted layers is most often required for transport structures or for large halls when defining the demands for the subsoil quality.

Plate Test

The static load test of subsoil and road base courses using a small loading plate – is presently the most common type of testing the deformation characteristics of compacted layers (Fig. 2.30). It is based on elasticity theory of elastic half-space deformation. The settlement of the centre of the circular flexible plate s under the loading of elastic half-space is calculated from the equation:

$$E_{def} = \pi \cdot \sigma_{ol} \cdot r \cdot (1 - v^2)/(2s) \qquad (2.13)$$

where

σ_{ol} – uniform surcharge at the plate-subsoil contact,
r – plate radius,
v – soil Poisson's number,
E_{def} – modulus of subsoil deformation determined for given surcharge.

Prior to the start of the measurement the surface of the tested soil must be levelled and the soil must be in intact state. The maximum subsoil particle size is 100 mm. For levelling the surface coarse-grained silica sand is used in as minimum amount as necessary so that the plate is in a horizontal position and the load acts exactly at its centre. The load is applied by using counterweight and hydraulics. To ensure appropriate contact, a short application of the load is recommended not exceeding a contact pressure of 0.025 MPa. Successively, step by step, the surcharge at the

Fig. 2.30 Photo of static load test by small plate

contact of the plate with the subsoil is increased and the vertical deformation y is read after setting thus proceeding up to the required maximum surcharge value, ranging mostly around 0.2 MPa, where the maximum vertical deformation for the first loading cycle $y_{max,1}$ is read. The unloading phase follows down to zero load where the vertical deformation $y_{0,1}$ is read. Immediately afterwards, the second loading phase starts proceeding up to the maximum value, where the vertical deformation $y_{max,2}$ is measured.

The test serves for the determination of the modulus of deformation for the first loading cycle $E_{def,1}$ and analogically the modulus of deformation for the second loading cycle $E_{def,2}$, Fig. 2.31. A simplified relation is mostly applied:

$$E_{def,1} = \frac{1.5 \times r \times \Delta\sigma}{y_{max,1}} \text{[MPa]} \quad (2.14a)$$

$$E_{def,2} = \frac{1.5 \times r \times \Delta\sigma}{y_{max,2} - y_{0,1}} \text{[MPa]} \quad (2.14b)$$

Where r – plate radius and $\Delta\sigma$ – maximum pressure (plate loading).

In order to determine the subsoil quality, compaction quality, both the modulus of deformation for the second loading cycle $E_{def,2}$ and the ratio of the determined moduli $E_{def,2}/E_{def,1}$ are used. For example, the code ČSN 736133 Road earthworks – design and execution requires $E_{def,2}$ of at least 45 MPa for the active zone (base layer) surface. Additional criteria are a 100% compaction degree achieved by Proctor standard test and a requirement that none of the measured values of the modulus of deformation of the road base layer is by more than 10% lower, or that only 10% of the results fail to fulfil the required criterion $E_{def,2}$. The Czech Railways define the required modulus of deformation $E_{def,2}$ in relation to the design speed of the railway track, i.e. higher values are demanded for higher speeds. The ratio of the moduli, in most cases, should not exceed a value of 2.2. For a higher ratio the modulus of deformation for the first cycle is small, and therefore we may conclude

2.3 Soil Compaction

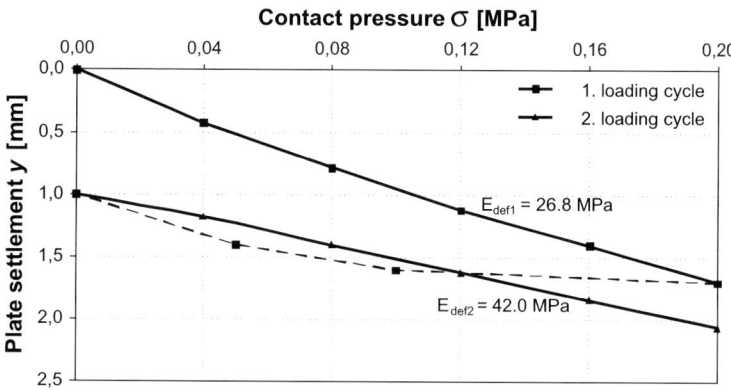

Fig. 2.31 Evaluation of modulus of deformation from static load test by small plate

that the surface was not sufficiently compacted, the particles got close together only after the application of loading.

A few notes can be made concerning the evaluation of the static load test required in several codes. The modulus of deformation does not need to be determined only at the start and at the end of the loading cycle, but also for individual surcharge stages. The second requirement for the moduli ratio in many cases proves to be not generally valid. For example, for checking the compaction of high road embankments, Brandl (2001c) mentions that in many cases this ratio exceeded – even considerably – the most commonly used limit of 2.2 with no negative impact on the quality of compaction achieved. "These examples, just as thousands of others, underline that the ratio $E_{def,2}/E_{def,1}$ is definitely not a generally valid, reliable criterion to assess the compaction quality". Practical experiences are showing that coarser the material is the higher the ratio $E_{def,2}/E_{def,1}$ is even for good compaction. This practical experience made it into the ČSN 721006:1998 "Soil and fill compaction control". This code for coarse grained soils require this ratio to be lower then 2.3, 2.5, 2.6 in relation to the degree of compaction D (>100, >98, >97). Maximum value of this ratio (4.0) is allowed for rockfill.

This test, which is becoming increasingly more common, is the source of many reservations. They include e.g.:

- Problems concerning the loading plate size (diameter). The diameter is mostly small, 0.3 m or 0.37 m, which corresponds to an area of 0.1 m². A standard depth that is affected by load increase from plate is about 1–2 times the plate diameter, i.e. about 0.3–0.7 m. In assessing the subsoil quality, e.g. the subsoil of large halls, this value is insufficient as the surcharge at the underside of the footing acts over larger area and hence to substantially greater depths. If the soil has poorer properties at such depths the result measured by the loading plate with a small diameter may substantially underestimate the total settlement. The effort to carry out the loading test with a larger plate diameter exists but encounters additional problems, e.g. providing the necessary counterweight for the application

of loading within the required limits. For small plates, the counterweight of a small off-road vehicle will suffice.
- Problems concerning the time required for the load plate test – its implementation, even though there is waiting time until the "setting" of deformation for individual loading steps. For clayey soils, this time need not suffice for obtaining the modulus of deformation under drained conditions, as the result is much closer to the modulus of deformation under undrained conditions. Another aspect concerns the time that passed between the compaction and the loading test. Load plate tests performed one or more days after compaction frequently yield higher results than immediately afterwards. Such self-strengthening could be observed mainly with carbonate soils or grain mixtures and with cohesive soils of some thixotropic behaviour – Brandl (2001b).

Nevertheless, the static load plate test is becoming a kind of index test, fast to carry out, easy to define and has considerably high credibility allowing comparison of individual results. In order to apply the moduli determined in practice, in the subsequent monitoring of the earth body deformations, the geotechnical engineer must show a great deal of foresight.

The same applies, even to a greater extent, to "dynamic load plate tests", e.g. ČSN 736192 (Fig. 2.32).

2.3.4.4 Continuous Compaction Control

We now witness a boom of continuous control methods, implemented already during compaction, which principally apply the measurement of the response of the compacted layer to the passes of the vibratory roller, Adam (1996).

The development basically follows two directions, in the first case there is a special vibration detector, mounted onto the non-rotary vibrating part of the vibratory runner whose values are assessed and successively shown on a display in the roller crew cabin. The compaction rate value representing the minimum requirement for the compaction rate is saved in the system memory. The magnitude of this value is determined within the compaction field trial using identical material and under identical conditions as those of a specific site. The output provides information on the compaction rate achieved in relation to the required one, evaluating the compaction growth between successive passes. With the help of this data, a state when the compaction rate is achieved may be determined, or we may evaluate a case when the required compaction rate may not be achieved even after subsequent passes. Additional measuring equipment provides information on the roller travel speed, vibration frequency and it may display the plan with plotted compaction rates for individual areas.

In the second case, a similar device is already a component of the manufactured vibratory roller, and the advantage of the device is that having evaluated the compaction rate achieved, its growth with the number of passes, it can automatically adjust the machine's vibration and speed to improve its efficiency conditions. Moreover, the information on the compacted area can be combined with the GPS

2.3 Soil Compaction

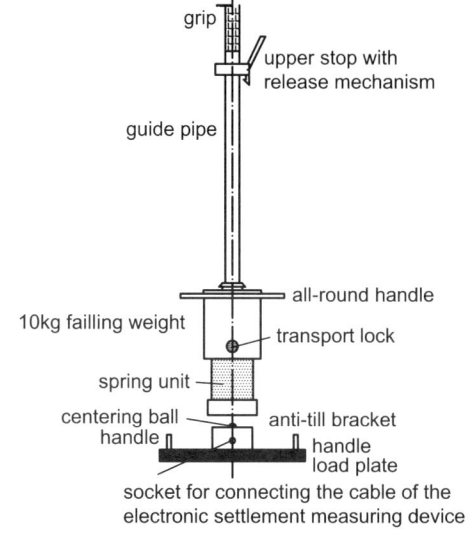

(a)

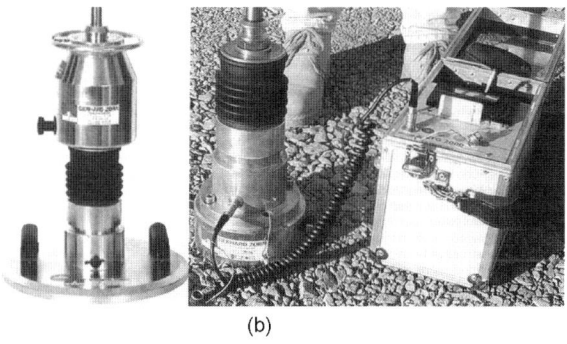

(b)

Fig. 2.32 Scheme and photo of dynamic load plate test

navigation system so that the output about the compacted area is not relative, but specified in the form of position coordinates. This new control method is often referred to as "intelligent compaction". Intelligent compaction helps to remediate weak spots and avoids over-compaction. It reduces the number of roller passes, the number of conventional proof tests, and provides a soil modulus at all locations where the roller has travelled. It gives a more uniformly compacted layer. Thus its system allows reducing conventional heterogeneities such as – a layer with varying material properties, varying water content and heterogeneous stiffness of the underlying layer, which naturally cannot be achieved with a roller applying a constant speed, frequency and amplitude.

Very detailed description of the principles of soil compaction using dynamic rollers and principles of continuous compaction control (CCC) is presented by

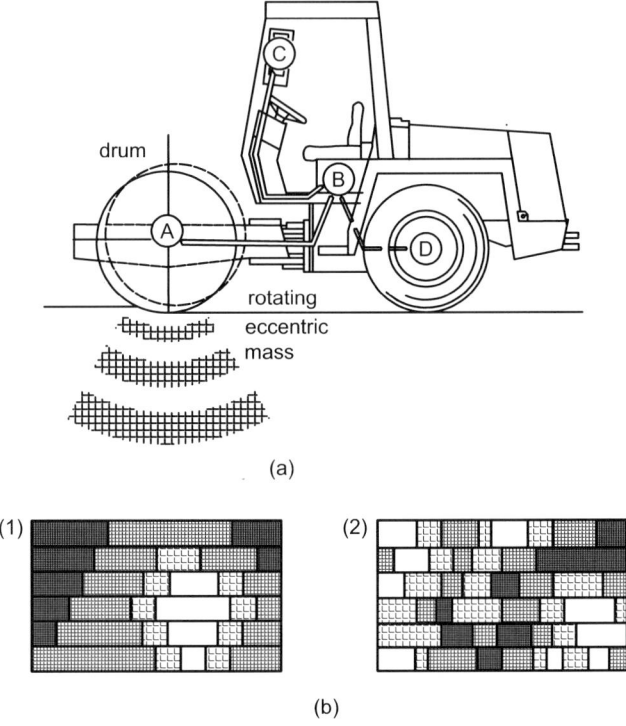

Fig. 2.33 Components of CCC systems (according to Brandl et al.). (**a**) Transducer A, processor B, memory C, GPS-based speed and location sensor D. (**b**) Driver's display – control plots of two compacted zones

Brandl et al. (2005). Next to theory they are using computer simulation (FEM) and field tests. Basically, CCC systems (Fig. 2.33a) comprise a transducer (A) mounted on the bearing of the drum; a processor unit (B) converting the signals received into measurement data; and the display and memory unit (C), which continuously displays and documents the measured data. A sensor (D) records the roller's velocity and the distance travelled. The measured values can thus be clearly identified and classified, and any deviations from the specified velocity can be documented. Most important output is shown on Fig. 2.33b – display in drivers cabin showing the area which is compacted and with results of degree of compaction. The driver is very well informed about places which need to be improved. On the Fig. 2.33b(1) the less compacted area is concentrated on one place, on Fig. 2.33b(2) week zones are randomly distributed and require comprehensive measures to achieve a sufficient and homogeneous quality, Brandl (2001b).

Brandl et al. (2005) states that for the vibratory rollers, four main types of CCC systems were launched on the market in recent years, as:

– the compactometer marketed by the Swedish firm GEODYNAMIK,
– the terrameter which was developed by the company BOMAG,

2.3 Soil Compaction

- KB value, which represents the spring stiffness of the ground at the interface with the drum, offered by manufacturer Ammann,
- The oscillometer CCC system developed also by the firm GEODYNAMIK, and represents a certain improvement of compactometer system.

The first two systems will be specified in more details, first is expressing the interaction of measured "cylinder deformation module" E_c with modulus obtained from dynamic plate test and the second one the interaction between vibration modulus E_{vib} with moduli obtained from static plate tests, $E_{def,1}$, $E_{def,2}$. For this type of interaction is the very important fact that the demanded values of compaction is very often prescribed by modulus of deformation, as for example 45 MPa for roads with lower traffic up to 120 MPa for motorways.

For the first case Thurner and Sandström (2000) specify that the drum of a vibrating roller exposes the soil to repeated blows – one per cycle of vibration. Analogous to a dynamic plate load test the blows from the cylindrical drum can be used as a load test of the soil. It can be shown that the force amplitude F of the blows is proportional to the first harmonic of the vertical acceleration. The displacement s during the blow can be approximated by the amplitude of the double integral of the fundamental acceleration components, Adam (1996).

Therefore it is relevant to express the "cylinder deformation module" E_c as the ratio of the force and the corresponding displacement as

$$E_c = \text{constant.} \ F/s = \text{constant.} \ \omega^2. \ A_1/A_0 \qquad (2.15)$$

where

ω – fundamental angular frequency of vibration
A_0 – acceleration amplitude of the fundamentals component of vibration
A_1 – acceleration amplitude of the first harmonic component of vibration.

These values can be registered by the measuring device thus arriving at the sought value E.

To express the relation

$$s = f(F, E) \qquad (2.16)$$

(Kröber et al., 2001) exploit the relations derived by Lundberg in 1939:

$$s = (1 - \nu^2)/E \times 2.F/l.\pi \times (1.8864 + \ln l/b) \qquad (2.17)$$

where l is the width of the drum and b the contact width, where for b the following equation holds true:

$$b = \sqrt{((16.R(1 - \nu^2)F)/(\pi \times E \times l))} \qquad (2.18)$$

where R is the radius of the drum.

Since the ground does not behave in a linear, elastic and isotropic way, the term E-modulus is, similar to the plate bearing test, not used for practical evaluation, but with respect to the oscillatory excitement and for the purpose of differentiation the modulus is designated as the vibration modulus E_{vib}. With the Vario control roller, (Kröber et al., 2001) performed compaction tests for the comparison of the obtained E_{vib} with the results of the deformation modulus $E_{def,1}$ and $E_{def,2}$ from plate bearing tests. Silty gravel was the material used. The results of comparison for different passes are shown in Fig. 2.34. Especially with the first pass the value $E_{def,2}$ is relatively high in relation to E_{vib}, because the load plate causes a local subsequent compaction during the initial load application. Since the vibrating machine always passes over new and uncompacted soil, the vibration modulus E_{vib} first converges towards the value of $E_{def,1}$. With an increasing number of passes this effect diminishes and the vibration modulus converges towards the value $E_{def,2}$.

To conclude this part, we may sum up that the proof of compaction quality by means of continuous compaction control was implemented into several national standards and wider spreading of this method is expected. Nevertheless, it is recommended even in this case to perform supplemental tests based on dry density control or on load plate tests. They serve not only for calibration but are also needed for checking the design assumptions – (Brandl, 2001b).

2.3.5 Soil Properties as a Function of Compaction Effort

The basic objective of compaction, as was already mentioned in the introduction, is to improve the properties of soil as a construction material. This mainly refers to the improvement of its shear strength, compressibility or permeability. These three fundamental mechanical-physical parameters will be in the centre of our attention, even though other characteristics may also be affected, e.g. capillary elevation or susceptibility to frost penetration.

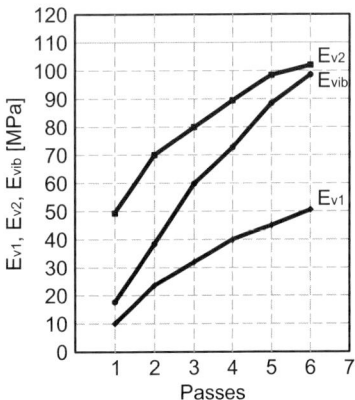

Fig. 2.34 Comparison of vibratory modulus E_{vib} with deformation moduli E_{v1} and E_{v2} during compaction of silty gravel with respect to the number of passes (according to Kröber et al.)

2.3 Soil Compaction

The most elementary idea of a change in the behaviour of soils with different porosity (and thus also dry density) can be obtained from a model of ideal spheres with an identical radius. In the case of the loosest arrangement of these spheres, the porosity $n_{max} = 47.6\%$, and in the case of the tightest arrangement $n_{min} = 26.0\%$. Although soil particles differ from ideal spheres, the porosity values determined in the laboratory, in particular for evenly graded sandy soils, are very close to ideal theoretic values.

Under static load and a loose structure of ideal spheres, the compressibility does not have to be big, but under even a slight vibration allowing the movement of the spheres practically in all directions, an arrangement of the spheres with significantly lower porosity arises. That is why mainly granular soils must be exposed to a prominent vibratory effect during compaction. On the contrary, under shear load the loose structure easily collapses, a vertical deformation occurs and, successively, shear resistance grows. This is called contractant behaviour. An ideal sphere with the tightest arrangement, on the contrary, exposed to shear stress, must overcome the resistance of mutual bonding and so the initial resistance results in lifting, which is called dilatant behaviour. The dependency between shear resistance and shear strain for both cases is displayed in Fig. 2.35.

The angle of shearing resistance of granular soils shows a roughly linear growth together with the relative density index I_d. The range of the angle of shearing resistance φ moves approximately in the band shown in Fig. 2.36. Here, so-called Mogami's formula may be mentioned – see Shimobe (1997):

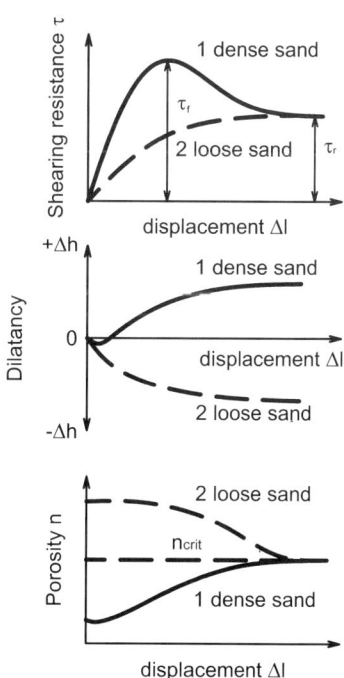

Fig. 2.35 Behaviour of loose and dense sand during shearing

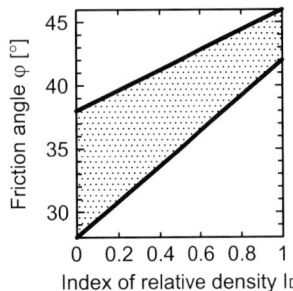

Fig. 2.36 Range of angle of shearing resistance ϕ related to the index of relative density and coarseness of soil particles (top limit – crashed coarse particles; bottom limit – fine rounded particles)

$$\sin\varphi = k/(1+e) \qquad (2.19)$$

where k – Mogami's strength constant, e – void ratio

Recommended characteristic values of the angle of shearing resistance for non-cohesive soils in ČSN 731001 are shown in Table 2.6. The real results will also depend on the shape of grains, angularity and moisture content. Recommended characteristic values of the modulus of deformation E_{def} for non-cohesive soils for basic ranges of the density index I_d according to ČSN 731001 are shown in Table 2.7.

To express a change in the behaviour of cohesive soils is a substantially more complex task. Under a constant value of porosity or dry density, the Proctor compaction curve may be used for obtaining two different moisture content values. Under lower moisture content, internal porosity of individual clods is lower, but there are relatively big macropores between them. Under higher moisture content, macropores hardly exist, individual clods being well interconnected, but higher porosity is given particularly by higher moisture content, the volume of water filling

Table 2.6 Recommended characteristic values of angle of shearing resistance according to ČSN 731001

Symbol	φ_{ef} [°]	
	$I_D = 0.33–0.67$	$I_D = 0.67–1.0$
SW	34–39	37–42
SP	32–35	34–37
S-F	28–31	30–33
GW	36–41	39–44
GP	33–38	36–41
G-F	30–35	33–38

Table 2.7 Recommended characteristic values of modules of deformation according to ČSN 731001

Symbol	E_{def} [MPa]	
	$I_D = 0.33–0.67$	$I_D = 0.67–1.0$
SW	30–60	50–100
SP	15–35	30–50
S-F	12–19	17–25
GW	250–390	360–500
GP	100–190	170–250
G-F	80–90	90–100

2.3 Soil Compaction

in the pores. In this respect e.g. (Barden, 1971) mentions a significant difference between the macro – a microstructure. He points out that from the engineering aspect macrostructure has a much greater effect on the behaviour of soil as a construction material.

For drier material, the deformation characteristics measured for an unsaturated sample are high, but after its saturation considerable settlement may occur. Therefore, compaction to even a higher compaction degree as compared to the maximum dry density value may not be appropriate if the moisture content is too low as compared to the optimum. On the contrary, the modulus of deformation reached for a sample compacted under higher moisture content is lower. Therefore it is stressed that the sample should be compacted under optimum moisture content as for the respective rate of compaction energy it afterwards shows not only the tightest particle arrangement (Fig. 2.37). Also from the viewpoint of moisture the sample is stable with respect to the deformation and does not matter if the samples are unsaturated or submerged.

A similar situation applies to shear strength. Undrained shear strength, determined for the wet side of the compaction curve, decreases with increasing moisture content, as is generally known. For the dry side (close to the optimum moisture content), undrained shear strength of unsaturated sample will increase with decreasing moisture content, but after watering the sample may be homogenized and its successive strength will be close to that of the sample from the wet side. Therefore, here again we can recommend to compact the sample as close to the peak of the compaction curve (optimum moisture content) as possible.

The significance of micro- and macrostructure (sometimes also referred to as a double porosity sample) is becoming most prominent in permeability measurement. Water preferentially flows through macropores and so permeability is generally

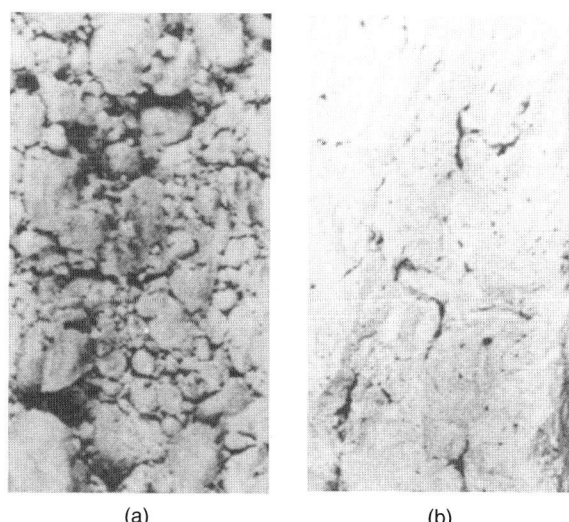

Fig. 2.37 Photos of samples compacted under dry (**a**) and wet (**b**) conditions

(a) (b)

Fig. 2.38 Permeability of compacted samples as a function of initial moisture content

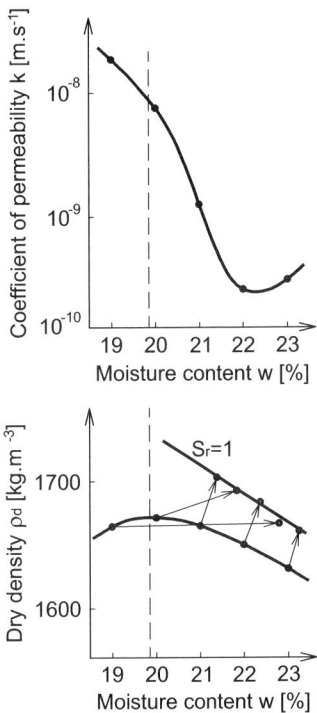

higher. The lowest permeability is reached under moisture contents slightly higher than the optimum – see Fig. 2.38, Vaníček and Pachta (1976). Before testing the samples were saturated to eliminate this fact. These problems will be more closely discussed in the chapter dealing with the sealing system of landfills.

Chapter 3
Geosynthetics in Earth Structures

During about the last 30 years, geosynthetics has become an integral part of earth structures. The term geosynthetics is generally used for geotextiles and other geosynthetic materials to be build into earth and similar structures. Therefore, it is necessary to acknowledge this and at least briefly specify their main uses. It is not the aim of this publication to fully cover the nowadays very wide subject of geosynthetics, just to show the advantages of using geosynthetics in earth structures. Generally, we can say that geosynthetics can:

- Replace certain functions that can be up to a certain limit managed by soil itself, but sometimes only under very strict conditions and detailed specification of its properties. Into this group we can include replacement of natural soil filters or drains by geosynthetic ones. The advantage that is often mentioned is very low weight of geosynthetic material fulfilling the required function compared with the natural soil filter or drain. The expenses for transport are smaller, quality is guarantied and in many cases it is possible to install geosynthetics even during poor weather conditions, e.g. during winter.
- Add certain properties that soil by its nature does not have. Into this group belongs mainly reinforcing function of geosynthetics. The geosynthetic reinforcing element adds significant tensile strength, while the soil itself has negligible tensile strength. Due to the significant importance of this function, the problem of soil reinforcement with geosynthetics will be covered in more detail also in the next chapter "Soil improvement".

The wide subject of geosynthetics is discussed at international conferences – Int. Conf. on Use of Fabrics in Geotechnics, Paris 1977, 2nd Int. Conf. on Geotextiles, Las Vegas 1982, 3rd ICG Vienna 1986, 4th ICG Haag 1990..., European Geosynthetics Conferences starting in Maastricht 1996, series of International Symposia on Earth Reinforcement in Fukuoka. It is also discussed in different journals, e.g. an official journal of the International Geosynthetics Society – Geosynthetics International; International Journal of Geotextiles and Geomembranes published by Elsevier, and special lectures such as the Mercer Lecture. From the first comprehensive references about the subject we can mention: Ingold and Miller (1988) Geotextiles Handbook – Thomas Telford London, Koerner (1990 – 2nd edn.) –

Designing with geosynthetics, Prentice-Hall or Fluet Jr (ed.) (1987) Geotextiles Testing and the Design Engineer – ASTM International Publication 04-952000-38, Van Santvoort (1994) Geotextiles and Geomembranes in Civil Engineering, Balkema (Taylor & Francis), Rotterdam.

3.1 Basic Types

Mainly the second half of the twentieth century is connected with a significant increase in the use of synthetic materials. Geosynthetics are produced from synthetic – thermoplastic materials produced from crude oil by distillation. General properties of thermoplastic materials are influenced by internal structure, chemical additives and supplements. From the most well known it is:

- Polyester   PET   (PES)
- Polypropylene   PP   (POP)
- Polyethylene   PE
- Polyvinyl chloride PVC
- Polyamide (nylon)   PA
- Polyvinyl alcohol   PVA
- Aramid   AR

One of the main advantages of geosynthetics is their resistance against typical soil conditions. Geosynthetics are resistant to the influence of moisture, rottenness and mycosis. Once built into the earth structure, their design life is high and corresponds to the design life of general earth structures. This assumption can be made even with respect to up to now relatively short practical experience with the geosynthetics. It is possible to confirm this statement by laboratory experiments that could simulate the influence of surrounding environment under extreme conditions that could develop in the distant future, or using prognosis of development in time from results of measurements within shorter time, nowadays of about 30 years maximum.

The exception is highly contaminated subsoil with high concentration of contaminants or direct contact of geosynthetics with chemical environment. These exceptional situations are asking for individual approaches and we can experience it mainly for use of geosynthetics in environmental geotechnics. Generally we can say that products made of polyolefins (polypropylene, polyethylene) and polyvinyl alcohol are highly resistant to organic acids and on the other hand materials from polyester, aramid and polyvinyl alcohol have high tensile strength and low creepability.

Though the design life of geosynthetics in earth structure is practically limitless, the raw polymers are during the design life subject to degradation due to long-term presence of light, weather conditions, water, chemistry of the environment and surcharge.

3.1 Basic Types

The highest degradation rate is from UV light before installation into the earth structure or when they are installed on the surface. The basic protection consists of reduction of exposure to the direct sun light, otherwise we have to select materials, which are after chemical treatment more resistant to UV light. This chemical treatment is becoming a standard for geosynthetic products on the market.

Products can be distinguished by their geometrical properties into:

- 1D – tapes, strips, individual yarns;
- 2D – planar products, with typical dimensions of about 3–5 m across and about 50–300 m along, while the thickness in the range of millimetres;
- 3D – volumetric products – in this group we can include mainly geocells, which have the third dimension (height) in the range between about 0.1 and 0.25 m; very often they are produced from strips that are locally connected and once stretched the 3D cells are created, e.g. in the form of honeycomb.

From the above division the most often used are the planar products and in between them it is possible to distinguish:

- Geotextile – permeable planar technical textile intended for installation into earth or similar structures in order to take advantage of their specific properties. We can divide them even further on:
  - Woven geotextile – technical textile produced by weaving on warp and weft;
  - Nonwoven geotextile – technical textile produced from endless or staple fibres connected mechanically (needle punching, weaving, stitching), thermally or chemically;
  - Knitted geotextile – technical textile that does not use classical weaving (for big elongations), but special techniques that give the final product high strength and small elongation in longitudinal direction;
- Geomembrane – planar impermeable membrane from polymers with thickness between about 0.5 and 3 mm;
- Geogrid – planar reinforcing product with mesh structure that can be divided based on the production technique as:
  - Extruded geogrid – produced from geomembrane that is regularly perforated and than stretched in one or both directions;
  - Woven geogrid – produced by laying reinforcing yarns across in predefined spacing that are finally connected using chemical, thermal or mechanical processes;
- Geocomposites – products created by combination from more than one geosynthetic product (geodrain, geomattress).

From the overview presented, it is obvious that the production of geosynthetics is preceded by a stage during which is produced:

- Extruded monofilament yarn of circular cross-section with the diameter of a fraction of a millimetre and infinite length;

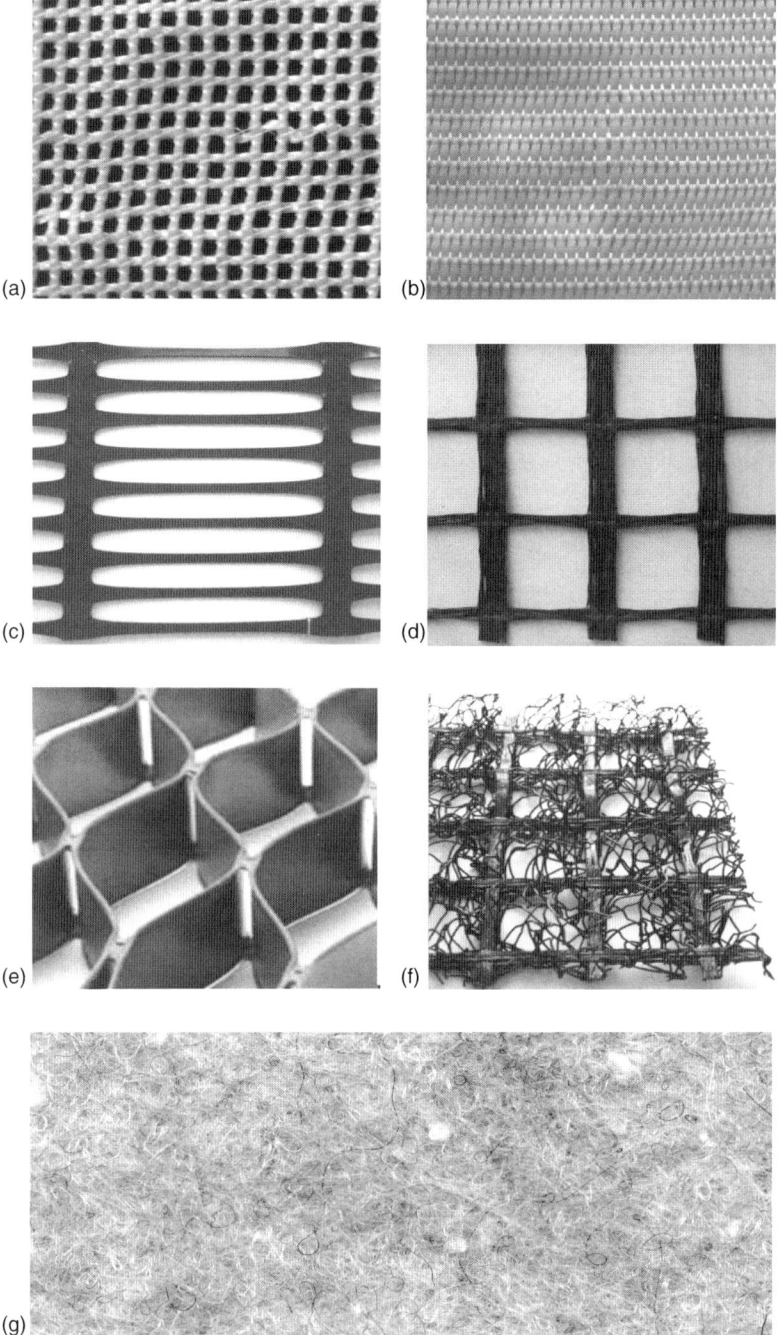

Fig. 3.1 Basic types of geosynthetics. (**a**) Knitted fabric. (**b**) Woven geotextile. (**c**) Extruded geogrids. (**d**) Woven geogrids. (**e**) Geocell. (**f**) Geocomposite – geomattress and woven geogrids. (**g**) Nonwoven geotextile

- Extruded tape yarn, strip of a few millimetres wide and a fraction of a millimetre thick of infinite length;
- Extruded sheet with a width of several metres and a thickness between a fraction of a millimetre and several millimetres and with infinite length.

The individual monofilaments can be used directly or connected together in multifilament yarns or cut into about 50 mm long fibres. Sheets are usually cut into short strips, which can be used directly or connected together.

The main types of geosynthetics are shown on Fig. 3.1.

From the definitions of each type of geosynthetics, as above, you can see that some similar products have been produced in textile industry for a long time now and are described as technical textiles. At the beginning of use of textiles in geotechnics, technical textiles with properties as close as possible to those for a given application in earth structure have been selected. Nowadays the production of geosynthetics is an independent branch and producers, specialists, are introducing an ever-increasing scale of products with specifically selected properties most appropriate for different purposes in earth structures. This change from technical textiles towards geosynthetics started in the mid 1970s and can be connected with the first congress on this subject in Paris in 1977.

A final remark is about the fact that under the general term of geosynthetics we can find not only products from synthetic yarns but also products – mainly geotextiles – from natural fibres (e.g. jute, cotton, straw, and coir). Their durability is much shorter as they are susceptible to degradation more quickly and therefore they are used on applications for which the short-term function is enough and degradation desirable. As an example we can mention function of temporary erosion protection of earth structure surface (e.g. slope of cut) until the natural vegetation (grass, rootlets) takes over erosion protection.

3.2 Properties

Properties of geosynthetics, as other building materials, are subject to a complicated process of defining certain properties and their validation. This can be done on one side by the manufacturer and on the other by an independent certified laboratory (institute). All parties in the construction process should therefore have a guarantee that the material used fulfils the properties assumed in the design. This situation, with respect to the problems of earth structures design mentioned in the introduction, is of benefit as the tolerances from the declared properties are generally smaller than those for the soil itself.

Verification has therefore three phases:

- The producer declares on the shipped product, using label in accordance with ISO 10320 "Geotextiles and geotextile-related products – Identification on site", its type and main information about the material and some of the properties. This label is not only on the packaging but also on the product itself, which is

generally delivered in rolls, and once unrolled it is therefore possible to do local checks not to interchange similar products. The user has the possibility to do simple (index) tests in order to do checks on site.
- The producer has the duty to collect samples from certain percentage of the production and test and archive them for future references. These tests are usually performed in manufacturer's laboratories and they should be used to check that the properties of given type are within the limits declared and certified by an independent testing laboratory.
- The contractor sometimes has the duty to test a certain percentage of installed materials in an independent laboratory accredited for the tests in order to prove that the installed material has the declared properties.

Monitored properties can be divided as for soil to the groups of commonly described as index properties and groups of specific properties for given type of use. In the groups of the specific properties, we can include not only hydraulic, mechanical or tensile properties but also properties that describe possibility of mechanical damage. Mechanical damage of geosynthetic structure is mainly caused by sharp edges of individual soil particles by both installation and subsequent compaction.

3.2.1 Index Properties

3.2.1.1 Unit Weight

Unit weight – m_g [g.m^{-2}] – defines weight of geosynthetics per unit area. This property is typical for nonwoven geotextiles, as they can differ mainly at this property despite the same material used and production process. For example, nonwoven geotextiles are typically produced in the range between 100 and 1 200 g.m^{-2}. By the increase of unit weight the capability of protection of other elements increases and resistance to puncture, increase of strength and reduction of characteristic opening size. As an example, we can mention protection of geomembrane against mechanical damage from surrounding soil or CBR test for puncture resistance. For precise determination of this property, a steel frame with a dimension of 0.3×0.3 m is used as a template for cutting and subsequent weighting. The user on site can easily cut 1 m long piece from the roll and weight it in order to check this property. Lighter geosynthetics, generally below 250 g.m^{-2} are usually woven geotextiles from tape yarns, thin nonwovens mechanically bonded by needlepunching or thermally bonded. Heavier ones are on the other hand mostly nonwovens mechanically bonded by stitching or wovens from multifilament yarns used as reinforcing geosynthetics.

3.2.1.2 Thickness

Thickness – t_g [mm] – perpendicular distance between top and bottom faces of geosynthetics measured for normal pressure of 2 kPa. For the determination of this property, we can use the above-mentioned template, which can be surcharged by

2 kPa and the thickness of geosynthetics measured. The definition of normal pressure therefore eliminates errors that could occur if the measurements are done using slide gauge with different compression of measuring tips. For nonwoven geotextiles the thickness is in millimetres and generally in the range of about 1 mm–10 mm.

Note that to this category we can include also control measurement of mesh sizes of geogrids and geonets, which can be done easily by slide gauge, as the mesh size is usually mentioned in the name of the product.

3.2.2 Hydraulic Properties

Into this type of tests, it is possible to include both measurement of hydraulic permeability and characteristic opening size of geosynthetics.

3.2.2.1 Characteristic Opening Size

Characteristic opening size – O_n [mm] – defined by the diameter of soil particle, of which $n\%$ can pass through the geosynthetic. For geogrids and geonets, as mentioned above, the determination of mesh size and hence characteristic opening size is not a problem as the mesh size is generally uniform. However only geogrids with very small mesh size can be used for filtration purposes. For woven geotextiles this procedure is unsatisfactory. Simplistically it is possible to use projection of the woven geotextile on the wall, by which way we magnify in a controlled manner the openings that can be than measured as a circle that can be inscribed into the opening between warp and weft yarns. The distribution of pore sizes is fairly narrow. Narrow distribution of pore sizes therefore allows quick determination of soil that can be well protected by this filter.

The procedure to determine characteristic opening size for nonwoven geotextiles is more complicated and their size is more spread. Most of the procedures suggest sieving precisely divided and declared fraction (in the narrowest band) through the tested geotextile. For dry sieving, the method is almost the same as for soils sieving through sieves. The vibration table is used and measurements are made for standard conditions after the defined time, e.g. after sieving for 5 minutes. At the end of sieving the proportion of particles passed through and retained above are determined. From these results, we can find out how many particles are retained and how many of them are trapped inside. For the sieving, it is possible to use sand and silt particles or spherical glass balls (ballotini). In the case when this geotextile would be used in the ash tailings, it is possible to use as selected fractions for sieving also this ash. The ash has more problems to pass through the geotextile as the particles have sharp and irregular edges and they are more easily attached to the fibres of nonwoven geotextile. Electrostatic forces that can obstruct the finest particles passing through can cause another problem, therefore the finest fraction used for sieving is in the range between 0.04 and 0.06 mm. For the elimination of electrostatic forces, wet sieving is preferred. For the elimination of different procedures, it is therefore used ISO 12956. The representation of the result can be in a form of classical

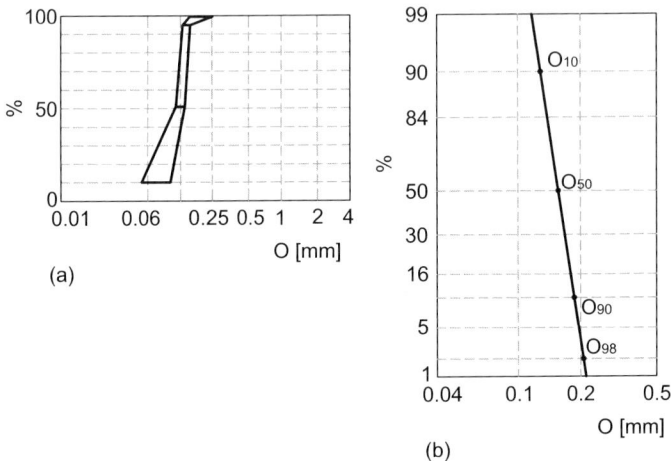

Fig. 3.2 Visualization of effective opening size distribution of geotextile. (**a**) Semi-logarithmic scale diagram. (**b**) Probabilistic scale diagram

semilogarithmic scale that is used for soil or in a form of probabilistic scale, see Fig. 3.2. The O_n defines the size of opening size of geosynthetic through which passed $n\%$ of sieved fraction. Therefore O_{95} defines that 95% of the particles from fraction O passed through geosynthetic and 5% have been retained (or trapped). The probabilistic scale representation of O_n, for which the relationship is linear, allows easier determination of finest opening sizes, if the product has them.

Note that characteristic opening size varies with the geotextile surcharge. For increasing surcharge the characteristic opening size becomes smaller, but it doesn't mean that in places it can't get bigger due to the direct contact with soil particles. Generally we can state that characteristic opening size is a function of fibre diameter, density of fibres, thickness of geotextile and type of bonding. There are some procedures showing how to determine the characteristic opening sizes from theoretical equations that involve the above mentioned influence factors. Nevertheless, the laboratory tests are preferred.

3.2.2.2 Filtration Coefficient

Filtration coefficient of geosynthetic – k_g [m.s^{-1}] – equivalent of filtration coefficient of soil as per Darcy law for water flow through soils $v = k \times i$. For the hydraulic gradient $i = 1$ the filtration coefficient is equal to filtration velocity v and has the same units. Permeability of geotextiles is high, filtration coefficient k_g is usually in the range between 1×10^{-3} and 1×10^{-4} m.s^{-1} and so matches the permeability of gravels and clean sands. For geodrains the permeability is about one order of magnitude higher. Values of filtration coefficient are used even when sometimes there are questions on validity of Darcy law. Darcy law is based on laminar flow and this condition doesn't have to be fully valid for some products and small hydraulic gradients.

With respect to direction of water flow in geosynthetics it is possible to distinguish filtration coefficient in the plane of material (k_{gp}) and filtration coefficient normal to the plane of material (k_{gn}). The differences are generally not so important, but sometimes these are used for determination of other hydraulic properties of geosynthetics as follows.

Permitivity ψ [s^{-1}] – which is a ratio between filtration velocity and hydraulic head difference on both sides of geosynthetic:

$$\psi = k_{gn}/t_g = v/dh \qquad (3.1)$$

and allows for comparison of flow capability between fabrics.

Transmissivity θ [m^2.s^{-1}] – defined by the equation:

$$\theta = k_{gp} \times t_g \qquad (3.2)$$

The main use of transmissivity is in the fact that filtration coefficient and thickness are functions of normal pressure. The design approach is to determine the required transmissivity θ_{reqd} and compare it to the laboratory determined value of the candidate geotextiles transmissivity θ_{act}, thereby forming a factor of safety:

$$F = \theta_{act}/\theta_{reqd} \qquad (3.3)$$

At the beginning the measurements of filtration coefficient k_g utilized different modifications of permeameters from soil mechanics. Three of those modifications are presented in Fig. 3.3. Figure 3.3a shows the simplest version, simulating permeameter with falling head (gradient) that measures filtration coefficient normal to the plane of geosynthetic. Its results are highly approximate and they can be used just for a quick comparison of results. In the other two cases the measurements are in the plane of geosynthetic under the falling head. In the apparatus on Fig. 3.3b the thickness of geosynthetic is given by the distance of side plates and this simulates different thicknesses determined for different normal stresses. The apparatus in Fig. 3.3c is a modification of triaxial apparatus, when normal pressure is defined by cell pressure via rubber membrane. The changes in permeability, filtration coefficient in the plane of geosynthetic as a function of normal pressure, are shown on Fig. 3.4. For the precise calculation of the results for the last case it was necessary to eliminate hydraulic losses, which are given by the resistance of the whole system, as these resistances are significant for high permeability of geosynthetics. Filtration coefficient of nonwoven geotextile decreases with increasing surcharge and is in the range between 1×10^{-4} and 1×10^{-5} m.s^{-1}. For more precise determination of filtration coefficient it is necessary to follow suggested procedures, e.g. ISO 11058 "Geotextiles and geotextile-related products – Determination of water permeability characteristics normal to the plane, without load" or ISO 12958 "Geotextiles and geotextile-related products – Determination of water flow capacity in their plane". The principle of ISO 11058 can be seen in Fig. 3.5.

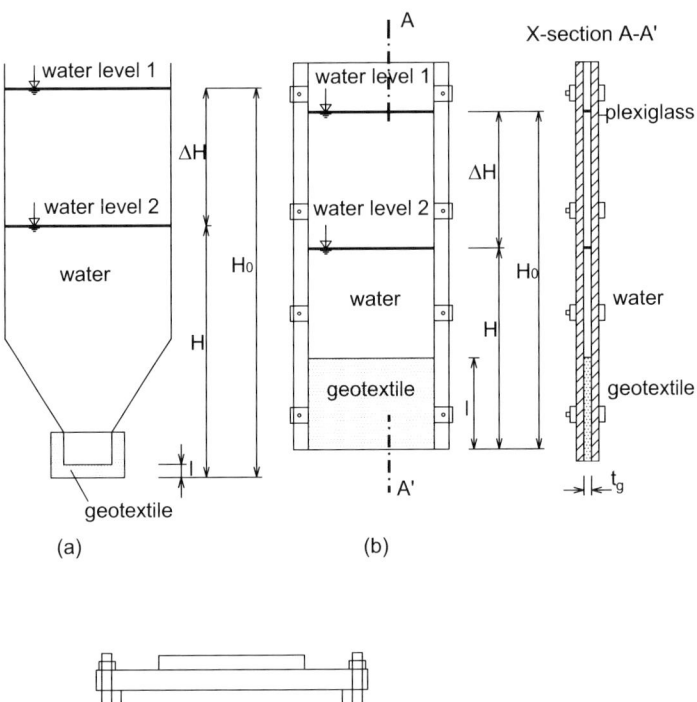
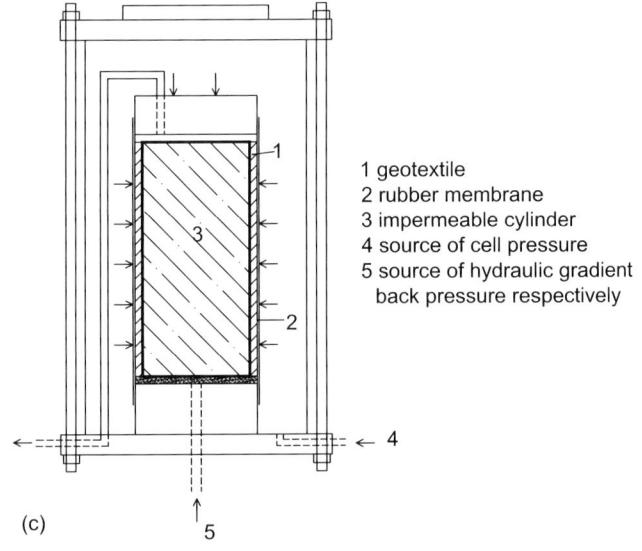

Fig. 3.3 Schemes of adapted permeameters for geosynthetics permeability measurement. (**a**) Normal to the plane of geosynthetic. (**b**) In the plane of geosynthetic. (**c**) In the plane of geosynthetic under surcharge using adopted triaxial apparatus

Fig. 3.4 Relationship between filtration coefficient and normal pressure for woven and nonwoven geotextiles

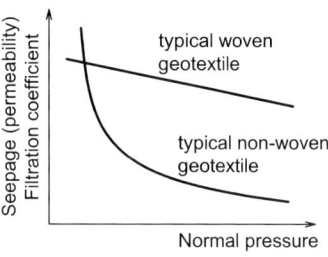

3.2.3 Strength Characteristics

For the use of geosynthetics as soil reinforcement, it is important to know their *behaviour under tensile loading*. As already mentioned soils do have very limited tensile strength and loose soils none. Cohesive soils compacted using the Proctor Standard energy have tensile strength under effective stresses in the range of few kPa and under total stresses in the range of tens kPa. On the other hand, elongation at failure under effective conditions can be up to few percent while under total stresses is lower by about one order of magnitude. In more detail, this issue is discussed in the chapter about cracks in sealing layers of rockfill dams.

Compared to soil, geosynthetics have high tensile strength though the differences are huge. The tensile strength is declared in kN per linear metre of the product. Some woven and knitted geotextiles have strength up to $1\,500\,\text{kN.m}^{-1}$ with elongation below 10%. Values of strength for nonwovens are significantly lower, most often in the range between 5 and $50\,\text{kN.m}^{-1}$ with elongation over 50%.

Tensile strength of particular geosynthetic depends on many factors, to name the most relevant:

- Type of loading – unidirectional or planar.
- Shape of tested sample – the classical tests in textile industry are utilising strip samples with a width of 50 mm. These samples are becoming narrower in the middle part under uniaxial load. This situation is significantly different from the situation of geosynthetics inside earth structures, where whole sheets are surcharged. A certain compromise is a testing sample of 200 mm width and 100 mm

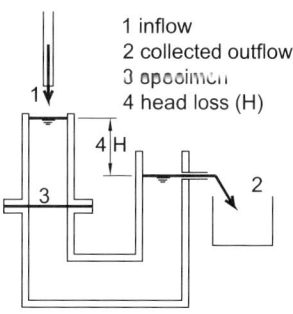

Fig. 3.5 Permeability apparatus according to ISO 11058

length. On this sample, the above explained narrowing is significantly reduced, and laboratory equipment does not have to have huge dimensions.
- Technique of testing sample gripping into the grips. In order to provide a transfer of very high loads the grip is important, however this can significantly damage the tested sample just next to the grips. The transfer of tensile load is arranged easier by wrapping the sample around the cylindrical grips and the measurement of elongation is realized by optical means on the sample middle part.
- Magnitude of transverse loading. Effect of which is increased strength and describes in principle the depth of geosynthetic reinforcement in the earth structure.
- Speed of test performance. This factor is very important for the design of earth structures, as it is very well known that synthetic materials are prone to creep, continuation of deformation even for constant load. Therefore, generally it can be stated that if load is applied slower, the strength is lower. This issue will be discussed in Chapter 4 in more detail.
- Temperature during the loading of geosynthetic reinforcement. It is necessary to take care of this factor during the application of reinforcement for example in landfills of domestic waste, where the temperature can reach up about 50–60°C.

Therefore, several tensile tests can exist, as none is capable of including all of the above-mentioned aspects. The test in ISO 10319 "Geotextiles – wide-width tensile test" describes a quick test in unidirectional tension without transversal load on strip of 200 mm wide and 100 mm long. This test is nowadays accepted as standard and can be used for basic comparison of individual products. Typical result of such a test is presented in Fig. 3.6, where T_f represents tensile strength at break and ε_f represents relative elongation (strain) at break. In the figure there is also indicated the design tensile strength T_d and corresponding design strain, that can be used during the design. In Chapter 4 it will be shown that only about 20–40% of maximum tensile strength can be used in the design depending on creep properties or due to the sensitivity of structure to relative deformation.

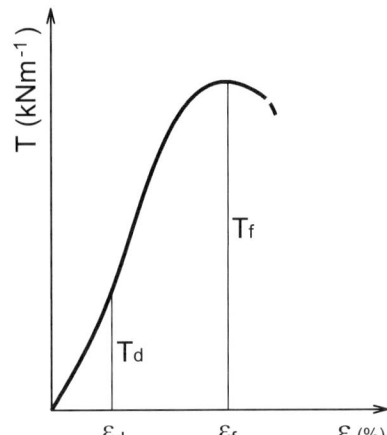

Fig. 3.6 Typical result of tensile test of geosynthetic.
T_f – ultimate tensile strength;
ε_f – elongation at break;
T_d – design tensile strength;
ε_d – design elongation

3.2.4 Friction Resistance of Soil – Geosynthetic Interface

Determination of friction resistance between soil and geosynthetic material is important for:

- Possibility to check the overall stability of earth structure – if the inclusion of geosynthetic hasn't created a weak point, along which slippage could occur;
- Determination, if the resistance would be able to transfer required tensile strength during fulfilling the tensile function, e.g. to determine required anchorage length.

In principle we need to answer the question, if the application of geosynthetics increases or decreases the shearing resistance of the whole structure compared with the case without geosynthetics. For this comparison we can use the same equation for the shear strength as for soil:

$$\tau = \sigma \times \text{tg}\, \varphi_{gs} + a \qquad (3.4)$$

where

τ [kN.m^{-2}] – shear resistance
σ [kN.m^{-2}] – normal stress
a [kN.m^{-2}] – adhesion between soil and geosynthetic
φ_{gs} [°] – angle of friction between soil and geosynthetic

Very often, and not only for loose soils, the adhesion is small and the most focus is on the angle of friction between soil and geosynthetic.

Examples of possible tests and different layouts of tests are shown on Fig. 3.7:

- Modified shear box test for soils is presented in Fig. 3.7a,b, while option (b) is generally preferred – nowadays suggestions prefer large box of samples with dimensions 0.3 m × 0.3 m.
- Pull out test – Fig. 3.7c can have several alternatives. For example when using big stand it is possible to monitor resistance of strip pull out, when normal pressure is given only by soil weight and also is possible to see the 3D effect – Vaníček and Hasaan (2000). Big attention was given by Schlosser et al. (1985) also to the dilatancy behaviour on the contact on the increase of resistance.
- Inclined plane test – Fig. 3.7d – this type of test is typical for testing of interface resistance of soils without adhesion. The angle of inclined plane at the moment of slippage start is the searched one for angle of friction on the interface φ_{gs}.

Again to avoid confusion it is necessary to perform the tests according to some norms, e.g. ISO 12957 "Geosynthetics – Determination of friction characteristics" or BS 6906-8 "Methods of tests for geotextiles. Determination of sand-geotextile frictional behaviour by direct shear".

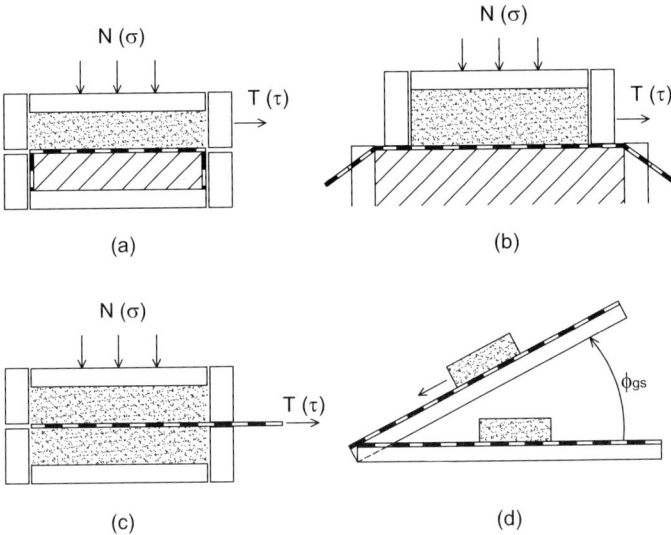

Fig. 3.7 Measurement of shear resistance on the contact between soil and geosynthetic. (**a** & **b**) Shear box test (apparatus). (**c**) Pull-out test. (**d**) Inclined plane test

Nonwoven *geotextiles* and most of woven geotextiles have on the contact with soil lower strength than soil itself. Typical values of ratio between tangent of angle of geosynthetic and soil friction tg φ_{gs} and tangent of angle of internal friction of soil tg φ are in the following ranges:

	tg φ_{gs}/tg φ	
– for dry sand	0.72–0.87	for peak strength
	0.68–0.71	for residual strength
– for wet sand	0.63–0.68	for peak strength
	0.59–0.63	for residual strength

Note, that lower values are for woven geotextiles from polypropylene tape yarns, whose surfaces are very smooth. For the most often used products nowadays, it is possible to assume the ratio in the range between 0.8 and 0.9.

In the case of cohesive soils, the shear test is performed under drained conditions (effective parameters) for the case of long-term stability, and under undrained conditions (total parameters) for the case of short-term stability. Approximately:

$$\varphi_{gs} \geq \beta \cdot \varphi \tag{3.5}$$

where β is a reduction coefficient in the range between 0.5 and 1.0 with typical value of around 0.75.

$$a \geq D \cdot c \tag{3.6}$$

where

> D is a reduction coefficient with typical values between 0.5 and 0.7,
> c [kN.m^{-2}] is cohesion of soil

Note 1: The equations (3.5) and (3.6) are generally valid, if the values on the right hand side will be effective soil properties than contact properties will be effective as well and the same is applicable also for total properties.

Note 2: The undrained conditions are appearing in reality very rarely, e.g. as already mentioned for polypropylene tape yarn geotextiles, otherwise geotextiles always allow for some drainage.

Shear resistance of the contact between soil and *geogrids* is much better than for geotextiles. The ratio of tangents of angles tg φ_{gs}/tg φ is generally above 1, which means that friction angle of the contact is higher than angle of internal friction of soil itself. This can be explained by possible interlocking of soil particles with mesh of the geogrids. The best is when the geogrids mesh size is about three times the typical soil grain size d_{50}, which is common for gravelly soils and coarse sands. For bigger grain sizes, but mainly for finer soils, the effect of interlocking is lowered and for clayey soils it is advised to use value of this ratio below 1.

Note: Similarly good properties from this point of view have also some knitted geotextiles. These geotextiles have the crossing of perpendicular yarns connected by means of strapping, which also has a positive effect on the interface shear resistance.

The lowest interface friction angle generally is between geomembranes and soils. *Geomembranes* typically have a very smooth surface and therefore the friction angle can be as low as 11°–13°. Therefore on the slopes, for example on the sealing of landfills of domestic waste, usually textured geomembranes are used. These geomembranes have implemented on their surface differently organised protrusions that increase the interface friction already during the production process. For the geomembrane applications, in some cases it is necessary to determine also the friction between geomembrane and protective nonwoven geotextile.

3.2.5 Mechanical Resistance

Above mentioned tests, mainly tensile strength test, are representing the ideal situation of undamaged geosynthetic. However during the practical application it is normal that certain damage is caused to the geosynthetic by coarse soil on their contact. The most sensitive is the application of compaction equipment, when the contact is loaded dynamically and as such that the movement is not only vertical but general. Nevertheless, mechanical damage is not limited to the period of construction, but also to the period of design life, if the contact is loaded by variable action.

The tests for this purpose can be divided into three groups:

- Penetration tests – which have a lots of alternatives and that are basically testing resistance of geosynthetic against penetration by loading element (cylinder, ball, spike) by constant speed or by dropping of the element on the geosynthetic material either undamaged or damaged;
- Tearing tests;
- Abrasion tests.

From the group of penetration tests the common one is the CBR test, during which a circular sample is attached to the 150 mm in diameter CBR cylinder and loaded by a cylindrical piston with a diameter of 50 mm by a speed of 0.83 mm.s^{-1}.

Tearing tests are monitoring the behaviour of geosynthetic after the initial damage. For example the Hoek test consists of a nail pushed through geosynthetic and then its tearing in the plane or trapezoidal test with EDANA specimen consists of a tensile test of trapezoidal shape with initial tear.

Abrasion tests do not have a common character; mostly they are tested in a laboratory similarly to the way geosynthetic will be loaded in the field. Prevail monitoring of abrasion by simulating impacts and/or cyclic loading in the plane of contact.

Possibility of mechanical damage and therefore reduction of geosynthetic strength, in time has to be accounted for in the design values. Proposed partial factors that take this aspect into account are based on the results from sites where geosynthetics have been excavated after serving a certain time in the earth structure. The excavated geosynthetics were then tested and the results compared with the results on the original material prior to installation. In most cases the reduction of initial strength by about 10%–20% is contributed to mechanical damage.

Examples of codified tests are ISO 90734 "Textiles – Test methods for nonwovens – Determination of tear resistance", ISO 13427 "Geotextiles and geotextile-related products – Abrasion damage simulation (sliding block test)" or ISO 10722-1 "Geotextiles and geotextile-related products – Procedure for simulating damage during installation – Installation in granular materials".

3.2.6 Chemical Resistance

Good resistance against typical soil environment is generally accredited to geosynthetic materials. They are practically completely resistant to moisture, decaying and mycelial microorganisms. A more complicated situation is when geosynthetics get into contact with chemicals of high concentration. However that is not a typical situation for soil reinforcement, but for example for use of geosynthetics during landfilling.

Individual polymers are degrading due to the long-term influence of daylight, weather, water, temperature, environment chemistry and loading, which affects their properties.

Gourc (1996) describes some results of hydrolysis of polymers. This is a case of chemical degradation caused by presence of OH^- ions and which happens more often in alkaline environment compared to acid environment. Alkaline environment can be present more often and represents environment with $Ca(OH)_2$. Except for a few naturally calcareous soils, these are soil modified and/or stabilized by lime or cement, around cement stabilized aggregates, concrete, etc. The first results suggest that polyesters are degrading in alkaline environments while polyolefins do not. The degradation affects tensile properties and resistance against puncture. The hydrolysis in polyester takes two forms. The first, alkaline or external hydrolysis, occurs in alkaline soils above pH of 10 and takes the form of surface attack or etching. The second, internal hydrolysis, occurs in aqueous solutions or humid soil at all values of pH. It takes place throughout the cross-section of the fibre. The rate of internal hydrolysis is very slow, such that the process has little effect at mean temperatures of 15°C or below in neutral soils, although it can be accelerated in acids. Therefore it can be stated that for polyesters of high quality, with surrounding temperature below 20°C and with pH in the range between 4 and 9, the design life of 50–100 years can be ensured.

3.3 Different Functions in Earth Structures

Geosynthetics can fulfil in earth structure different functions, from which the main ones are:

- *Filtration function* – guaranties water flow in the direction normal to the plane of geosynthetics while at the same time prevents movement of solids in the direction of flow;
- *Drainage function* – geosynthetic material in its plane drains water from the earth structure that has to be drained;
- *Reinforcing function* – the additional strength of earth structure is gained from geosynthetic;
- *Separation function* – separates two materials, in between which the geosynthetic is placed and therefore the mixing of those is avoided and the function of each is secured;
- *Erosion protection function* – provides protection of surface layers of earth structure against external effects like surface water, wind, etc.;
- *Sealing function* – significantly limits the leak of liquids (water, effluents) from place, where these liquids are stored, into the surrounding environment.

However geosynthetic very often does not fulfil just one of the mentioned functions, but combines some of them. The individual functions are hereafter specified in more detail.

3.3.1 Filtration Function

While geosynthetics fulfil this function they must provide stability of interface of two soil layers, avoid that from one side of the interface (so called protected, basic soil), from where water flows, so that soil particles would not be pulled and washed out into the second side of the interface.

Most often nonwoven or woven geotextiles or fine geogrids are selected for this function. With the exception of nonwoven geotextiles, the problem is rather simple. Only relatively uniform channel (pore) is in the way of the soil particle pulled by water current, which is given by the mesh size of geogrids or by the space in between weft and warp yarns of woven geotextile. Very often, the characteristic opening size is in the range of 0.1 mm–2.0 mm. Therefore, these are generally preferred in the case when the protected soil does not have many very fine particles. In the case of nonwoven geotextiles the channel is longer (even rather short as it is thickness of geotextile) and along this channel the diameter is changing. Some of the fine particles, that enter the geotextile, could therefore be trapped in the smaller pore and this way they partially block not only the passage of other particles, but also partially reduce the water flow.

Therefore, good function of geotextile as a filter depends on creation of soil filter on the side of protected soil. This situation for nonwoven geotextile is schematically shown in Fig. 3.8. Directly at contact with geotextile arched mesh of coarser particles is created thus allowing only small amount of finest particles to be washed-out from protected soil. Thin layer of soil filter is made of increasingly finer soil with lowering permeability. Once this layer is created, further washing out of soil is stopped and the whole system gets into equilibrium and can fulfil its function in the long-term.

During the design of geotextile filters, usually fulfilment of the following three conditions is required, e.g. Lawson (1982):

- Filters must allow permeation by water without significantly increasing water pressure in pores in front of the filter;
- Filters must stop movement of particles of protected soil (with the exception of small amount of fines next to the filter during the creation of soil filter);
- Filters should not become blocked by particles of protected soil.

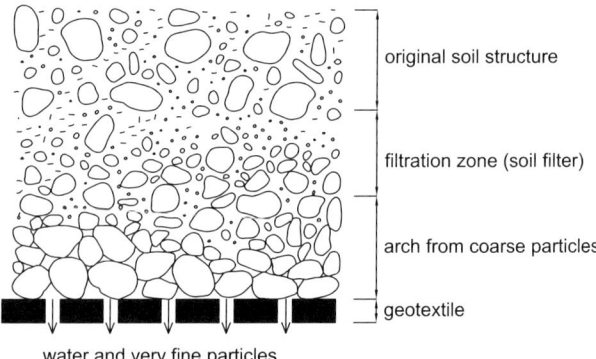

Fig. 3.8 Geotextile as a filter – creation of soil filter at the interface

The first condition requires geotextile with high permeability (i.e. with large pores), while the second one requires geotextile with small pores. In between these contradictory conditions there must exist a compromise design. Safe filtration function is fulfilled via permeability criteria, retention criteria and criteria for blockage of filtration pores (clogging criteria).

3.3.1.1 Filtration Criteria

There is not a big problem with the filtration criteria due to the high permeability of geotextiles. To fulfil the filtration criteria one can comply with direct condition:

$$k_g \geq 10.k \tag{3.7}$$

where

k_g [m.s^{-1}] – is filtration coefficient of geotextile
k [m.s^{-1}] – is filtration coefficient of protected soil

or with indirect condition for comparison between typical grain size and characteristic opening size of geotextile:

$$O_{90} \leq d_{15} \tag{3.8}$$

where

d_{15} [mm] is a diameter of soil grains of which 15% by weight is smaller than this diameter and it corresponds to 15% on particle size distribution curve,
O_{90} [mm] is size of geotextile pore, through which 90% of particles of this fraction size will pass.

3.3.1.2 Retention Criteria

Nowadays there are several filtration criteria that specify the conditions for maintaining the interface stability of soil layers. These are not based only on the type and production process of the geotextile but also depend on soil character (e.g. for noncohesive soil the coefficient of uniformity $C_u = d_{60} / d_{10}$) also on the type of flow – laminar or turbulent. Due to the complexity of this problem, only main principles will be presented here. The issue of geotextile filters including their limited applicability will be discussed in more detail in the chapter about limit state of internal erosion and more stringent filter specification for rockfill dams and tailings will be discussed in their respective chapters.

A fairly simple relationship is presented by Schober and Teindl (1979) for nonwoven geotextiles and laminary flow:

$$O_I \leq B \times d_{50} \tag{3.9}$$

where

O_I is geotextile characteristic opening size and authors suggest $O_I = O_{90}$;
d_{50} is average diameter of the protected soil particle for which 50% of particles by they weight are smaller than this diameter;
B is coefficient, value of which depends on the uniformity coefficient of protected soil and type of geotextile – see Fig. 3.9. The B values guaranteeing reasonably long-term performance (factor of safety 1.5) are for wovens and thin nonwovens (thickness smaller than 1 mm – usually thermally bonded) between 1.7 and 3; for thick nonwovens (thickness more than 2 mm – usually needle punched) between 3 and 5. For C_u values greater than 5, the coefficient B is decreasing and for $C_u = 20$ is equal to 1.

Čištín (1980) defines, for nonsuffosive soils and to ensure filtration stability, following condition:

$$O_{max} \leq a_1/(b \times m) \times d_{kl} \qquad (3.10)$$

where

O_{max} is maximum filtration pore ($\approx O_{98}$), which corresponds to the maximum grain size passed through geotextile,
a_1 – coefficient defining the type of geotextile (for nonwovens with thickness over 1.5 mm – $a_1 = 1.8$),
b – coefficient describing the change of filtration pores diameter due to installation and load tests ($b = 1-2$),
m – coefficient describing the influence of hydraulic loading of the installed geotextile ($m = 1$ for unidirectional flow),

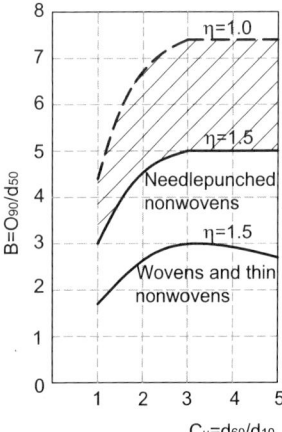

Fig. 3.9 Filtration criteria for thin (thermally bonded) nonwovens and wovens and for thick (needle punched) nonwovens with steady laminar flow (according to Čištín; Schober and Teindl)

η ... factor of safety

d_{kl} – diameter of protected soil particles that are creating arching interlocking (for uniformity coefficient C_u between 1 and 10 the d_{kl} approximates to about d_{50}–d_{60}).

Summary of other basic filtration criteria is presented by Fischer et al. (1990).

3.3.1.3 Clogging Criteria

Most often the suggestions for fulfilling this condition are arising from the following conditions:

$A \geq 5\%$ for woven geotextiles, respectively
$n \geq 30\%$ for nonwoven geotextiles

where

A – percentage area of filtration pores with respect to the whole area of woven geotextile,
n – porosity of nonwoven geotextile (porosity can be calculated from unit weight, geotextile thickness and density of yarns).

In this case it is the condition for mechanical clogging of the filter, which is not determined for granular filters, as the required condition on porosity of more than 30% is almost always fulfilled. Apparently the same seems to be also for geotextile filters, however Christopher et al. (1993) state that "designing for clogging resistance is the most problematic of all the criteria". By definition, clogging results when fine soil particles either penetrate into the geotextile blocking off pore channels or are deposited on the upstream side of the geotextile, thereby reducing its permeability. At the same time they emphasize that "clogging potential depends on the relation between the fines in the soil and their ability to clog or block a majority of the openings and pores in the geotextile. The key word is majority, because for the permeability and flow capacity to be significantly diminished, most of the pores must be filled with soil particles. That is because a geotextile even with a small porosity usually will still have a greater permeability than that of the soil, especially a fine-grained soil which tends to cause clogging problems". Also Schober and Teindl note that "the tests conducted with uniform and non-uniform non-cohesive soils and four different non-wovens geotextiles showed no tendency towards clogging. Only soils susceptible to suffusion showed disturbance in the transition zone. This would also be the same case with natural filters". Of course the sensitivity towards filter clogging by chemical and microbiological effects should be judged mainly for tailings and landfills of municipal waste.

Nevertheless during the application of geotextile filters, the laboratory verification of the filter could often be required not only for specific conditions, but also for relatively normal conditions. The Czech norms have this requirement for application of geotextile filter in tailing dams.

3.3.2 Drainage Function

The main aim of drainage function is to take away water from the drained earth structure in the required amount during the whole design life of the structure. The design of optimal drainage capacity of the drain is based on the defined maximal inflow into the drain, from its hydraulic capacity (permeability) for the given hydraulic gradient. Maximum inflow into the drain is determined from both the hydrological data during the design of collection drains of earth structures of transport infrastructure and from the calculations of water percolation through earth structure, for example from flow net of the worst case scenario.

The calculation of drain capacity is based on the Darcy Law, from the known coefficient of permeability of the drain, its thickness and hydraulic gradient:

$$Q = A \, k_g \, i \, t \qquad (3.11)$$

where

Q [m^3] is the flow rate for the given time interval t – the calculation is based on 1 m width for the planar drains and on the actual width of the band drains,

A [m^2] is cross-sectional area of the flow – thickness multiplied by width, thickness should be equivalent to the expected surcharge (depth of the drain within the earth structure),

k_g [m.s^{-1}] filtration coefficient of the drain in its plane,

i [–] hydraulic gradient,

t [s] time

For the most common thicknesses of the drains in the range between 5 and 20 mm, filtration coefficient of 1×10^{-2} m.s^{-1} and minimal hydraulic gradient of $i = 0.02$ (for the drained area with inclination of 2%) is the lowest drain capacity:

in the range between 1000 and 4000 mm^3 for the time interval of 1 second and in the range between 86.4 and 350 litres (0.0864 and 0.35 m^3) for the time interval of 1 day.

It is clear that the flow rates indicated are relatively low, however sufficient for the drainage of earth structures from soils with lower permeability or having a sealing core, when the drainage is only for water permeated through the sealing. Some geodrains with high porosity structure can have a flow rate up to 10–100 times higher. Vertical drains have flow rate capacity higher due to the higher hydraulic gradient and are acceptable in most cases, when the dissipation of excess pore pressures is required around the drains. During the application of the geodrains for drainage of large areas, these are necessary to combine with additional drain, e.g. with drainpipes.

The level of required safety of the geodrain design (e.g. via already mentioned ratio of transmissivities $F = \theta_{act} / \theta_{reqd}$) is for:

3.3 Different Functions in Earth Structures

- Long-term functionality 4–6 depending on structure importance
- Short-term functionality 2–3.

In very high level of safety for long-term functionality the following is considered:

- Potential danger in lowering of the filtration coefficient with time, e.g. due to already mentioned chemical and/or microbiological effects,
- Potential for lowering the thickness of the drain with time due to the compressive creep.

With respect to the last point it is necessary to comment that it is not enough to specify drain thickness for the expected surcharge, but it is recommended to monitor the deformation in time for constant surcharge (e.g. according to EN 1897 – "Geotextiles and geotextile-related products. Determination of compressive creep properties").

For the elimination of mechanical clogging of the drain it is necessary to pay special attention to their filtration protection. From this point of view we can divide the drains into, see Fig. 3.10:

- Composed of one main product (thick nonwoven geotextile, geomattress, geonet) when it is necessary to check that the pore distribution of the material will on its own fulfil the filtration stability;
- Composed from geocomposite consisting of permeable core (geomattress, geonet) and outer protective filtration layer (most often nonwoven geotextile);
- Vertical drains (geodrains, wick drains, band drains) composed of a composite of a strip of highly permeable core of a width of about 100 mm and thickness of about 5 mm, which is encased in filter geotextile.

From the planar drains the most common is the composite one, which is also more secure. The individual products are differentiated mainly by the specification of the drainage core.

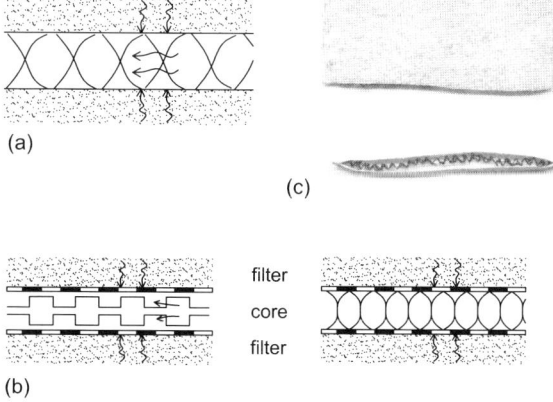

Fig. 3.10 Types of geodrains. (**a**) Simple geodrain (e.g. thick nonwoven). (**b**) Composite geodrains (3D core from geomattress or geonet with filter nonwovens on each side). (**c**) Vertical (wick) drain (3D core of about 100 mm wide surrounded by filter nonwoven)

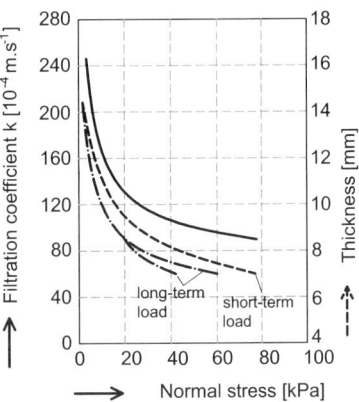

Fig. 3.11 Performance of geocomposite PETEXDRAIN $(1\,100\,\text{g.m}^{-2})$

Results of the permeability of the geodrain PETEX, whose core is from geomattress (3D structure of individual yarns) are presented in the Fig. 3.11. The permeability is very high, however a relatively free structure of the core is compressible and hence the thickness of the drainage core and coefficient of permeability are lowering with the increasing surcharge. Nevertheless, the permeability is still high. In the presented case, deformation under constant surcharge with time was also monitored and the results are compared on the already mentioned Fig. 3.11. From this point of view, the hard core of the drain has certain advantage.

3.3.3 Reinforcing Function

Designers of earth structures have generally two options for the design, use the soil as it is available and use it in classical way by compaction or improve its properties by other means. Soil has reasonably good properties when surcharged by compression as do other building materials, i.e. timber, steel or concrete. However compared with these materials soil has a huge difference between tensile and compressive strengths, the tensile strength is very low to nonexistent. One of the methods how to improve the soil is its reinforcement, as the reinforcement of soils by reinforcement of high tensile strength geosynthetic can add to the earth structure the missing tensile strength and that way significantly improve its properties.

Reinforcement of soil by geosynthetic materials is often divided into two groups – micro-reinforcement and macro-reinforcement. It is possible to see here a certain parallel with the development of concrete structures – mass concrete, reinforced concrete, prestressed concrete and fibre-reinforced concrete. For soil it is possible to distinguish compacted soil and reinforced soil. However, macro-reinforcement is equivalent to reinforced concrete, and micro-reinforcement is equivalent to fibre-reinforced concrete, and equivalent to prestressed concrete is preloaded and prestressed reinforced soil, see e.g. Tatsuoka et al. (1996).

3.3.3.1 Micro-Reinforcement

Micro-reinforcement is produced by mixing of soil with small, usually randomly oriented reinforcing elements as are for example continuous yarns (French patent Texsol – Leflaive 1985) or small cut-offs from geotextiles or geogrids – McGown et al. (1985). Individual micro-reinforcing elements are influencing relatively small volumes of soil in their surrounding and therefore a huge amount of this reinforcement is needed. Nevertheless, their mass content is usually below 1%. The principle of micro-reinforcement is based on the mechanical interaction and passive resistance. Application of such reinforcement has not yet been widely used. In summary, publications about soil reinforcement and mainly about practical examples the macro-reinforcement is dominating – e.g. McGown et al. (eds.) (1991) Performance of Reinforced Soil Structures, Thomas Telford, London. However, the research is still ongoing in laboratories on soils reinforced with short fibres of the length of about 40–50 mm, e.g. Michalowski and Čermák (2003) or Consoli et al. (2005) and hence we can expect development of this type of reinforcement.

3.3.3.2 Macro-Reinforcement

Placement of reinforcing geosynthetic materials that are large compared to soil particles in the earth structure is called macro-reinforcement. The reinforcing materials include not only 1D products like strips and 2D products like woven or knitted geotextiles or geogrids but also 3D products like geocells. Generally it is possible to apply also different geocomposites. Individual macro-reinforcing elements influence significant proportion of soil volume in the earth structure. Therefore, only a limited amount is needed. For example, the number of 2D macro-reinforcing elements starts with one or two members for embankment on soft subsoil (only for strengthening of this contact) and can go up to about 20 or even more for large reinforced soil slopes or retaining walls.

Figure 3.12 schematically shows the principles of micro- and macro-reinforcement. In further text we will concentrate only on macro-reinforcement and we will use instead of macro-reinforcement only term reinforcement.

The transfer of tensile force through reinforcement is possible only when the anchorage length is available. Surcharge is then transferred onto a larger volume of

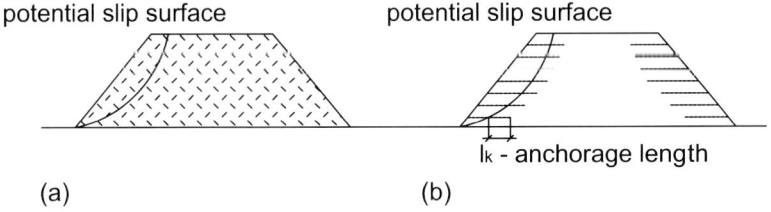

Fig. 3.12 Schematic representation of micro- and macro-reinforcement. (**a**) Micro- scattered reinforcing elements. (**b**) Macro- planar reinforcing elements

soil body and this is due to the interaction between reinforcement and surrounding soil. The main advantage of reinforced earth structure is the fact that it helps to:

a) Increase stability, bearing capacity of earth structure because potential shear surface must pass through the reinforcement or bypass the reinforcement area, which is kinematically more demanding and naturally, the stability is improved. However, during the design we have to prove which from these surfaces is the most critical. We speak about internal stability if the slip surface is passing through the area of reinforcement, respectively about external stability if it passes outside of the reinforcement area, or about combined stability if it passes partially through and partially outside of the reinforcement area.
b) Decrease both absolute and relative deformations due to the strength of the reinforcement and also by integrating a larger area into the load transfer.

The second factor is presented in Fig. 3.13 when the reinforcing effect is manifested in lowering the overall deformation under the surcharge, in decreasing of the stress acting in the level of the reinforcement due to incorporation of larger area in the load transfer.

We can conclude this introductory part on soil reinforcement that in principle the reinforced soil is a composite from two main materials – soil and synthetic material, whose properties are largely different and more over with the significant influence of creep. The development in the area of soil reinforcement is in the last three decades one of the stormiest not only in the area of earth structures, but also in the whole area of civil engineering. Very often, the praxis overcomes theory or research. This is due

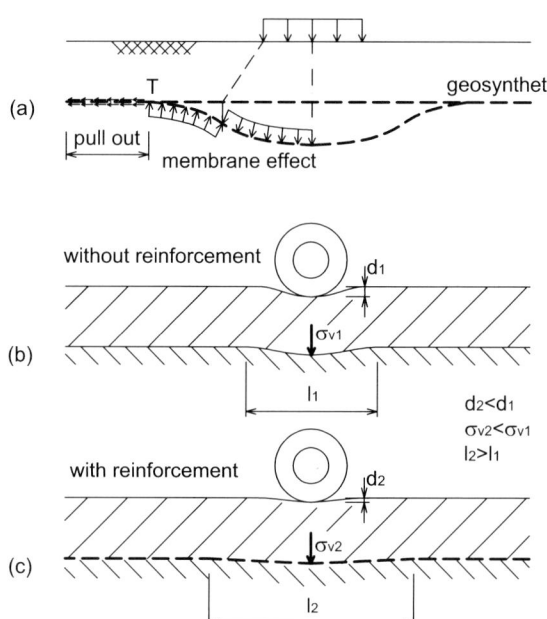

Fig. 3.13 Principle of reinforcing function in soil layer under loading. (**a**) Principle of membrane effect. (**b** & **c**) Comparison of soil layer behaviour without and with geosynthetic

to not only financial aspects and timesavings during reinforcement installation, but mainly because by soil reinforcement it is possible to solve problems that could not be solved using normal soils available for earth structures. Moreover, in Chapter 4 there will be additional information on this subject.

3.3.4 Separation Function

The requirements on this function are generally the lowest. The main task to fulfil the traditional separation function – to stop mixing of two different soils on their contact. The typical example is the storage of coarse granular material on the subsoil from fine-grained soil, when the inserted geotextile avoid degradation of the stored material. A similar case concerns earth structure, when the contact is made of two different soils, as subsoil soft clays overlaid by coarse-grained soil. The aim of the separation geotextile is to avoid pushing in of the fine soil into the pores of coarse soil by which to protect deteriorating of its properties, e.g. for capillary action and frost-ability. The examples of such a case are contact of embankment over soft soil (here the separation geotextile can supplement reinforcing geosynthetic) and separation of the subgrade of roads or railways from construction layers. In these cases, we have to take care of:

- Design life for separation function,
- Possibility of penetration of soil particles of soft fine-grained soil on the contact into the pores of separation geotextile, whereas d_{30} of protected soil is mainly checked.

The products that fulfil the requirements can have specification of only nonwoven geotextile with unit weight between 150 and 300 g.m^{-2} and short term tensile strength over 5 kN.m^{-1}. Anyhow for specific application the characteristics are specified and have to be attested. For example the EN 13250 "Geotextiles and geotextile-related products – Required characteristics for use in the construction of railways" recommends to specify for all conditions of use for separation purposes the following characteristics: tensile strength, elongation, damage during installation, static puncture (CBR), dynamic perforation and characteristic opening size. For specific conditions of use this information should be supplemented by tensile strength of seams and joints, friction, resistance to weathering and resistance to chemical degradation.

3.3.5 Protective Function

The most common application for protective function of geosynthetic is to protect a structure against damage by the backfilled soil, see Fig. 3.14. In this case we would propose to reflect:

- Tensile strength (usually ≥ 10 kN.m^{-1}),
- Strength against perforation,

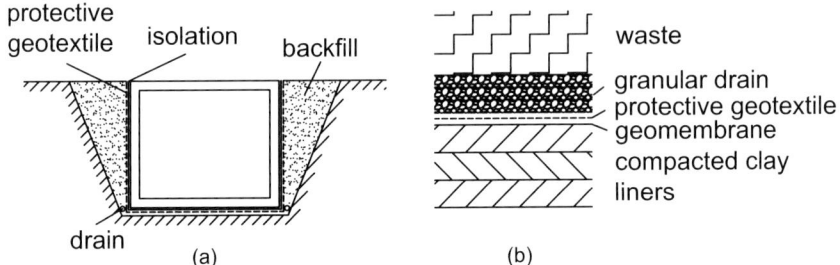

Fig. 3.14 Two examples of protective function. (**a**) Protection of structure isolation. (**b**) Protection of landfill geomembrane

- Thickness,
- Size and shape of soil particles in contact with protective geosynthetic,
- Expected normal pressure.

The final specification is usually unit weight for nonwoven geotextiles, mostly in the range between 500 and 1200 g.m^{-2}, respectively thickness of more than 5 mm for geomattresses and geodrains that should also drain water. If the specification is not given by concrete codes and specifications, it is useful to perform a verification test. For example in the case of the landfill sealing system it is proposed that drain from coarse-grained soil is composed of fraction 16–32 mm. In one case we had to review if the wrongly supplied fraction 32–64 mm should be replaced (which could to some extent damage the already placed sealing geomembrane) or if this coarser filter can be left in place. The contact has been simulated in the laboratory in large-size oedometer, the lower layer was from compacted soil sealing layer overlaid by geomembrane (1.5 mm thick) and protective geotextile (800 g.m^{-2}) under coarser filter and the surcharge was simulated in hydraulic press by a plate. Even so the selected surcharge was four times bigger (with some minor time variations) than expected, the coarse grains selected for the test were the sharpest than grains used for the landfill. There was no observed perforation within 8 hours of the test, only imprints of sharper edges were visible. Based on this test it was possible to suggest that in this very exceptional case it is possible to keep this coarser drain instead of replacing it. However, this statement was not at all reducing the full responsibility of the contractor for the long-term safe function of the structure.

The geocomposites are nowadays recommended in respect of elimination of the possibility to damage the sealing geomembrane. The geocomposites are usually composed of internal layer of woven geotextile and exterior layers of nonwoven geotextiles, see Fig. 3.15 for example.

3.3.6 Surface Erosion Protection Function

A typical example of this function is the protection against water erosion and only in the exceptional cases we can use geosynthetics against wind erosion, e.g. for tailings. In case of erosion protection of newly built embankments or cuttings, the

3.3 Different Functions in Earth Structures

Fig. 3.15 Comparison of geomembrane (SLM) protection by conventional protective nonwoven and by composite from reinforced nonwoven (courtesy of HUESKER)

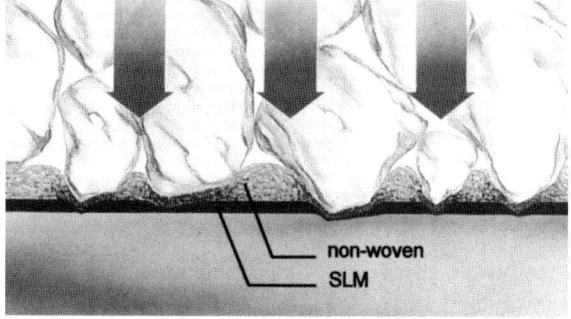

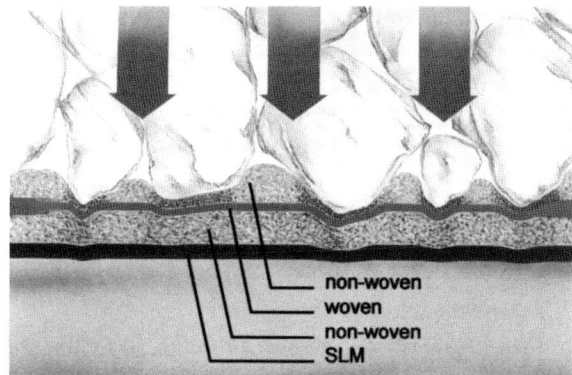

protection is mostly temporary until it is taken over by the natural option – grown grass or bushes. For this option, it is possible to use natural materials that degrade with time as already mentioned earlier. However, if the protection will remain functioning the situation is the most effective as the protection systems are working together, natural with root system combined with geosynthetic mattress.

At the beginning, there are rather conflicting requirements. Geosynthetic material with small opening size initially protects the surface better but it is more difficult for roots to grow through. For geosynthetics with larger opening size the situation is opposite. During the design, it is necessary to take into account:

- Angle and height of the slope, which determines the speed of surface water run-off,
- Characteristics of protected soil,
- Resistance of erosion protection against UV light.

From the products for erosion protection the best are special erosion protection mattresses, whose thickness is slightly bigger compared with traditional planar geosynthetics, i.e. thickness between 10 and 25 mm and have certain 3D character. The pores of mattress open structure from individual yarns can hold a layer of topsoil for better growth of vegetation. In order to eliminate the erosion, the erosion protection mattress must be properly attached to the underlying soil via pins or pegs. As one of the main requirements for good functionality is not only to avoid water flowing on the underside of the mattress, but also a good connection between mattress and subsoil via newly developing system of roots, see Fig. 3.16. As an alternative, it is possible to use geogrids with very small mesh size or nonwoven geotextiles with seeds and nutrients sewn in during production. When the need for high-level erosion protection is required from the beginning, it is possible to pre-grow in advance the erosion protection mattress together with vegetation and unroll it directly on site for immediate and highly natural protection. In the same way highly stressed areas are also protected, one of which is the place in front of a football goal; another is steep slopes of ski-jump ramps. In the case of ski-jump ramps, the erosion protection mattress should be combined with reinforcing element as the tensile forces in this application can reach high values.

A separate group is made of mattresses that are used for bank protection of the watercourses, canals and dams below water line, where vegetation has little chance to develop. In these cases the free volume of mattress is usually filled with stone chipping of about 2–6 mm fraction, which could be for highly stressed areas partially strengthened with a bitumen binder. The prefabricated permanent mat is flexible and permeable to water and allows vegetation to grow through. The main already mentioned types of erosion protection mattresses are presented in Fig. 3.17. Producers of erosion protection materials are suggesting standard available types for use based on the actual type and condition of the subsoil, and/or the speed of flowing water and required time of protection.

Special geosynthetics are generally behaving in the same way, e.g. geocells that are distributed on the slopes and canal beds with high-speed flowing water. The actual geocells are filled with gravely material that is unmoveable by flowing water. At the same time geocells as geomattresses allow permeation of water in both directions.

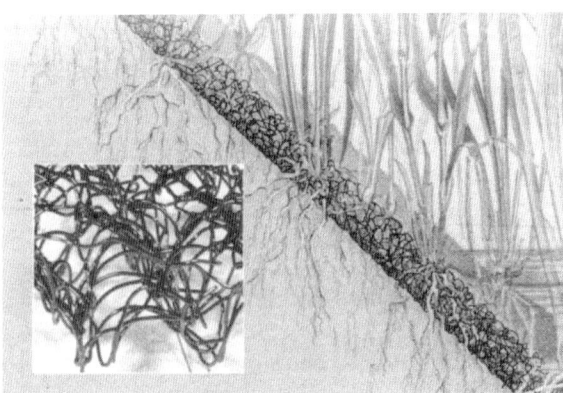

Fig. 3.16 Combined permanent geosynthetic and natural erosion protection system (courtesy of COLBOND)

3.3 Different Functions in Earth Structures

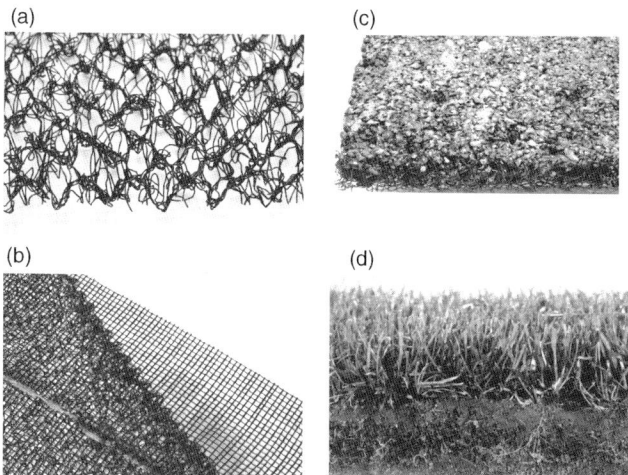

Fig. 3.17 Different types of erosion protection mattresses available on the market (courtesy of COLBOND). (**a**) Standard erosion protection mattress. (**b**) Reinforced erosion protection mattress. (**c**) Prefilled erosion protection mattress (standard type filled with gravel chippings and with a bitumen binder). (**d**) Pregrown erosion protection mattress

3.3.7 Sealing Function

Classical soil – clayey sealing is generally rather thick, for rockfill dams the thickness can reach several meters, for landfill sealing systems the minimal thickness is in between 0.6 and 1.0 m. The overall thickness of sealing layer is given not only by the coefficient of filtration of soil, but mainly by the requirement of minimal seepage.

The use of geosynthetics offer in principle two solutions, when the requirement can be fulfilled by a sealing with thickness in millimetres. Here we refer to the use of geomembranes, respectively use of geocomposite described as bentonite mattress or geosynthetic clay liner (GCL).

3.3.7.1 Geomembranes

In principle there are three categories of polymers that can be used for production of geomembranes – thermic elastomers, thermoplastic materials and bituminous materials. In the field of earth structures the thermoplastic products are used the most, though one example of bituminous application will be shown for a landfill of radioactive materials.

Under the term of thermoplastic materials belong polymers, that are softened by application of heat, but once the heat source is removed they return to the original (or almost original) structure. Nowadays the most used thermoplastic geomembranes are:

- high density polyethylene (HDPE) – with surfaces either smooth or structured,
- very low density polyethylene (VLDPE) – with surfaces again either smooth or structured,
- polyvinyl chloride (PVC).

The products usually contain small percentage of additives, carbon black, filler and/or plasticizer. Carbon black is usually made of crude oil or natural gas. Geomembranes from PVC are preferred for earth structures of hydraulic engineering and geomembranes from polyethylene are preferred for earth structures of environmental projects for their high resistivity to chemicals.

PVC membrane does not contain carbon black, is very flexible and is produced by calendering. Calender is made of a set of contra-rotating drums that create individual layers of geomembranes. Individual layers can be connected by invert calender and often a textile insert is added as reinforcement.

HDPE and VLDPE geomembranes are produced using hot rolling process. The width of produced sheet is often important in reducing the amount of joints during the installation on site. Current typical widths are about 5 m, but the extremes are about 10 m.

Products are differentiated by their flexibility – the HDPE membrane is much stiffer compared to the flexible VLDPE membrane. The differences in density are small despite the labelling. The density of HDPE is about 940 kg.m^{-2} and VLDPE has a density of about 920 kg.m^{-2}. Both types have density lower than water and hence are floating.

When describing individual geomembranes, their tested and declared properties are mentioned:

- *Physical (index) properties* – which mainly include thickness, density and unit weight.
- *Mechanical properties* – mainly tensile strength and elongation at break, but also other types of test on mechanical damage of geomembranes. In this group belongs also the already discussed friction between geomembrane and other geosynthetic (protective geotextile or geodrain) or with soil.
- *Chemical resistance* – is usually tested indirectly via swelling, ozone resistance, UV resistance, ... Direct resistance against chemical substances is also a subject of testing methods. Very often a resistance against landfill leakage is judged under the temperatures of $23°C$ or $50°C$ for the period of 30, 60, 90 and 120 days. After that comparative tests of mechanical durability are performed.
- *Biological resistance* – against bacteria and fungi. Resistance tests to rodents are also considered. Also here the resistance of HDPE is the highest.
- *Thermal properties* – influence of high temperature on mechanical properties are monitored, as creep has a big influence. On the other hand the materials are fragile under low temperatures, which lowers the value of elongation at break.

3.3.7.2 Geosynthetic Clay Liner (GCL)

The base of sealing geocomposite is a thin layer of dry bentonite protected from both sides by a geotextile or thin geomembrane. The amount of bentonite used

3.3 Different Functions in Earth Structures

in the mattress is relatively low, about $5\,kg.m^{-2}$ and the overall thickness of the bentonite mattress is in between 5 and 10 mm. Low permeability is guaranteed by the use of clay mineral from the montmorillonite group. It is one of the most active clay minerals with high swelling potential and extremely low permeability. After the contact with water thus starts swelling, but only in limited space by which any preferential flow path for water seepage is eliminated. The filtration coefficient is typically in the range between 5×10^{-11} and $1 \times 10^{-11}\,m.s^{-1}$. With respect to the expected seepage, this sealing is compatible with the sealing from compacted clay soil with thickness of about 0.6 m.

Typical cross-sections through bentonite mattress are presented in Fig. 3.18. While keeping the basic principle, the mattresses are differentiated by the system of connecting all the composite layers, which is usually a patented technology. This connection should in principle maintain uniform thickness of bentonite layer across the whole mattress area, so no weak areas exist which could act as preferential pathways for seepage. Another interesting solution is presented in Fig. 3.19. This solution is based on a highly porous layer called "aerofelt" into which a bentonite powder is placed during the production. This in effect partially reinforces the bentonite layer (as bentonite – active clayey mineral has very low shear strength) and at the same time maintains uniform distribution and durability of position of bentonite within the mattress even during handling and surcharge. Once the core is placed in between two polypropylene geotextiles the whole geocomposite is created by sewing. The use of aerofelt as reinforcement and sewing all three layers together very stable geocomposite is created, which can withstand even high shear forces. Therefore, these bentonite mattresses can be used even on relatively steep slopes and would not create instable shear plane.

Here we can place a small note relating to the clayey minerals. The properties of clayey minerals and mainly montmorillonite depend on the character of exchangeable cations. In soils, the most occurring cations are calcium Ca^{2+}, magnesium Mg^{2+}, sodium Na^+ and potassium K^+. Therefore we distinguish Na-montmorillonite, Ca-montmorillonite, etc. From the sealing and swelling properties point of view the best is Na-montmorillonite. This situation is sometimes expressed in the product name. However due to the type of soil environment there is possibility

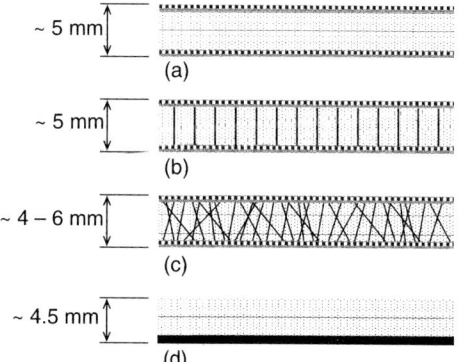

Fig. 3.18 Typical cross-sections through bentonite mattresses (GCLs). (**a**) Bentonite glued to both top and bottom geotextiles. (**b**) Reinforced by stitching of geotextiles. (**c**) Needlepunched between top and bottom geotextiles. (**d**) Bentonite glued to geomembrane

Fig. 3.19 Bentonite securely and evenly enclosed within geosynthetic clay liner

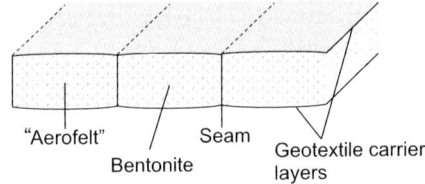

of cation exchange of one type by the other, which can negatively influence some properties. Hence where the possibility is real, some producers are offering bentonite mattresses with Ca-montmorillonite. They are produced with higher unit weight of bentonite in order to keep the same permeability (permittivity) properties. These are suggested in the areas where the presence of calcium is higher, e.g. in areas of soil stabilization, near ashes and/or concrete structures.

The concluding remark towards both types of sealing directs to the statement that the final quality highly depends on the technology of installation, in case of thin sealing systems mainly with respect to the connection of individual sheets. This issue will be covered in more detail while presenting applications of individual elements in future chapters. Nevertheless, here we would like to present brief comparison of both types of geosynthetic sealing systems with clayey sealing (compacted clay liner – CCL):

- Thickness significantly affects transport costs and also the available volume of the landfill.
- Speed and practicability of construction – for thin sealing systems the speed is more rapid including easier implementation.
- Vulnerability to mechanical damage is the highest for geomembranes, while GCL is capable of some self-healing due to swelling and comparing with CCL that is minimally vulnerable. Fault of thin element leads to immediate seepage through, while at improperly made CCL sealing the seepage is significantly retarded.
- Vulnerability to non-uniform settlement (e.g. surface of landfill) is smaller for thin more flexible sealing systems compared with clayey sealing sensitive to tensile cracks development.
- Price and availability – the price of thin sealing systems is relatively low, with limited variability for different projects and hence easily priced for, on the other hand the price of clayey sealing is highly dependable on the distance to the suitable borrow pit.

A relatively large number of advantages for thin sealing systems from geosynthetics is nowadays reflected in many regulatory approvals. Higher attention is paid to some details, see Daniel (1995) or Daniel and Bowders (1996).

3.3.8 Summary of Applications and Properties

In the Table 3.1 different functions of geosynthetics are summarised, as well as the summary of main properties of the product we should know for individual applications. The amount of crosses in the first half of the table indicate the importance of given product for given application, while in the second half it indicates the importance of the property determination for the given application.

3.4 Geosynthetic Handling on Construction Site

Geotextiles are dispatched to sites in rolls wound on to a robust cardboard, plastic or steel tube, wrapped in a protective layer of plastic and are in most cases transported by truck. As it was already mentioned, in the first instance it is necessary to check the delivered amount and confirm that it is the ordered material based on the roll stickers. If only the cover is damaged due to the handling or collection of sample for index testing, it is necessary to repair this area. The delivered rolls are this way protected against external moisture and UV light. Protection against moisture is relevant to nonwoven geotextiles and GCLs, as the additional moisture would increase the weight of the geosynthetic and worsen the workability and for GCL moisture can also negatively affect its function. For GCL functionality it would be better if the contact with water would happen after its installation. The protection against UV light is important for mainly reinforcing materials from polyester, as they are the most vulnerable to degradation by UV light. If it is not possible to store the rolls on site on slab under the roof, it is necessary to well prepare the storage area – allow proper drainage and remove surface irregularities such as tree roots or stones that could damage the stored material. The producer usually declares if the rolls can be stored in several layers and if so in how many layers.

The actual handling on site depends on the weight of individual rolls, which are in the range between about 25 kg for narrow rolls of geodrains and up to about 1 000 kg for very heavy rolls of geomembranes, GCLs or reinforcing geosynthetics. But most common are the weights between 75 and 150 kg. The weight determines if the material has to be handled by some lifting mechanisms or can be handled by hand. When lifting mechanisms are used, roll cores on which the material is placed, or another pipe or bar placed through the roll core, is used as a handling point.

The placement of the material requires some preparation and sometimes even since the time of order. Especially for the reinforcing geosynthetics it is necessary to avoid connecting of the product in the direction of acting force and the installed length would correspond as much as possible with the roll length, in order to minimise wastage. During the placement, care must be taken to the laying direction of the product. If the reinforcement is unidirectional, usually the strength in the main reinforcing direction is much higher than in the other direction and the product must by laid accordingly.

Unrolling from the roll is done either by hand or by using some mechanism, while the beginning of the layer is fixed in place (using pins, pegs, old tyres, small backfill). When installing reinforcing products the placement should be as such that there are no visible waves or folds, which would significantly delay the functionality of the reinforcement. Even some pretension is advised, which can be more easily done during mechanical placement when on the roll holder dynamometer can be

Table 3.1 Product importance and geosynthetic properties when fulfilling basic functions

Material type	Application					
	Filtration	Drainage	Reinforcement	Separation	Protection	Erosion protection
				Classical separation		
Woven geotextiles	X		XXX	XX		X
Nonwoven geotextiles	XXX	XX	X	XXX	XXX	
Knitted geotextiles	X		XXX	X	X	X
Extruded geogrids			XXX			
Woven & knitted geogrids			XX	X		XXX
Geonets	X	XXX			X	X
Geocells			XX			XX
Geomattresses		XXX			XX	XXX
Grass mattresses				XXX		XXX
Geosynthetic properties						
Unit weight	XX	XXX	XX	XXX	XXX	
Thickness	XXX	XXX		XXX	XXX	
Characteristic opening size	XXX	XXX		XX		XXX
Permeability	XXX	XXX		X		XXX
Tensile strength, elongation	X		XXX	X	X	X
Creep		XX	XXX			
Interface shear resistance	X	X	XXX	X	X	XX
Mechanical resistance	X	X	XX	XX	XXX	X
Chemical resistance	X	X	X	X	XX	XX
UV & weathering resistance	X	X	X	X		XXX

3.4 Geosynthetic Handling on Construction Site

Fig. 3.20 Photo of geogrid installation with certain pretensioning at Gröbers project

installed, which measures the tensile force activating in the just installing layer, see Fig. 3.20. In many cases the individual strips are laid next to each other with certain overlap, which is usually between 0.2 and 0.3 m only when required bigger. If fixing of the overlaps is required, this is done using pins or pegs penetrating about 0.3 m into the subsoil. In required cases, i.e. where the inaccuracy of overlap would have negative influence on the function, the designer can prescribe strengthening of the overlap, e.g. by sewing. The main advised type of sewing is placement of the next strip on the previous one and sewing of the edge and then overturning to its new position with seam on the underside. Connection of sheets of sealing elements will be discussed in more detail in the chapter about landfills.

If the length of the produced sheet is longer than required, the cut should be made precisely after placement using a special sharp knife or a knife with a heated blade when the cut is made also thermally. More problematic is shortening of the roll width (along the roll axis), and for this kind of cutting electrical saws can be used, but with extra care.

Generally, it is not allowed for mechanisms to drive directly over the laid material. When laying large areas this mean that the placed geosynthetic layers should be immediately covered by a certain layer of soil and mechanisms can run on top of this layer. Before the soil placement, it is the last moment to check the geosynthetic for potential local mechanical damages. If these are found, usually a patch is installed to fix the damaged location.

Chapter 4
Soil Improvement

The aim of this chapter is not to give an exhaustive survey of all methods of soil improvement, as e.g. Mitchell and Katti (1981), Van Impe (1989). The centre of our attention will be focused only on the basic ones which are in the closest relation to earth structures. Among them, there are methods leading to the speeding up of subsoil consolidation, soil stabilization with lime, cement or other bonding agents, and finally two basic methods of soil reinforcement, namely reinforcement of the structure under construction with geosynthetics or reinforcement of natural soil by means of soil nailing.

4.1 Speeding Up of Consolidation

The speeding up of subsoil consolidation results in faster dissipation of pore water pressures within subsoil, mainly due to an external load increase. To make the term of faster dissipation of pore water pressures clear, let us go back to fundamental chapters of soil mechanics. Figure 4.1 displays a 1D situation where a clayey, water-saturated layer with a thickness H is situated on uncompressible and impermeable subsoil. If this layer is overlaid with a permeable layer with a thickness z and a mass unit weight γ, then the total load increase on the surface of the clayey layer is:

$$\Delta \sigma = \gamma . z \qquad (4.1)$$

and in keeping with the principle of effective stresses and the 1D filtration consolidation theory by Terzaghi, the time $t = 0$ will produce a pore water pressure increase Δu in the clayey layer:

$$\Delta u = \Delta \sigma \qquad (4.2)$$

Immediately afterwards, however, dissipation of the pore water pressure starts, first at the boundary with the permeable layer where the water forced out of the pores is released due to the hydraulic gradient, so that it holds true for time t:

$$\Delta \sigma = \Delta \sigma' + \Delta u \qquad (4.3)$$

I. Vaníček, M. Vaníček, *Earth Structures*,
© Springer Science+Business Media B.V. 2008

Fig. 4.1 Distribution of total load increase on effective stress and pore water pressure along height H in time t

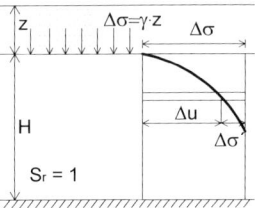

where $\Delta\sigma'$ is the effective stress increase. The shear strength of clayey soil grows with the increase in effective stress thus improving with time its resistance to shear. At the same time, however, the effective load increase increases the deformation of the clayey layer as under 1D conditions deformation is directly proportional to the effective stress increase. Also, the total volume of soil deformation corresponds to the volume of water forced out of its pores. In a theoretical time of infinity: $t = \infty$

$$\Delta\sigma = \Delta\sigma' \qquad (4.4)$$

when $\Delta u = 0$

i.e. the soil matrix takes over all load increase, complete dissipation of pore water pressures occurred, the shear strength reaches its maximum and deformation is completed if the solid phase grains are incompressible. The settlement development of a clayey layer in time t is presented in Fig. 4.2, where the total settlement of a clayey layer is

$$s_\infty = H\Delta\sigma'/E_{oed} \qquad (4.5)$$

where E_{oed} is the oedometric (constrained) modulus of deformation in the relevant stress range. The ruling equation of Terzaghi's uniaxial consolidation theory for a sudden load increase $\delta\sigma/\delta t = 0$ expresses the dependency of the pore water pressure (increase) as a function of position z and time t:

$$c_v \delta^2 u/\delta z^2 = \delta u/\delta t \qquad (4.6)$$

where c_v is the coefficient of soil consolidation which depends on soil permeability and compressibility:

$$c_v = kE_{oed}/\gamma_w \qquad (4.7)$$

Fig. 4.2 Consolidation curve – settlement as a function of time for 1D deformation (for incompressible grains)

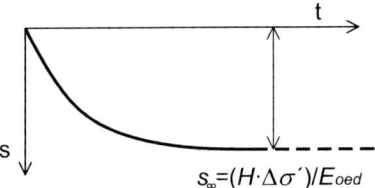

4.1 Speeding Up of Consolidation

There are several ways leading to solving this consolidation equation, e.g. numerical. Terzaghi himself proposed a solution using non-dimensional variables:

$$T = c_v t / H^2 \qquad (4.8)$$

resp.

$$Z = z/H \qquad (4.9)$$

where the depth z is measured from the surface of the consolidating layer and H is the layer height for a unilaterally drained layer and half the layer height for bilaterally drained soil.

T is a so-called non-dimensional time factor being the function of the degree of consolidation U:

$$U = f(T) \qquad (4.10)$$

The relationship $U = f(T)$ by Terzaghi for the respective case is depicted in Fig. 4.3.

Then, settlement in time t can be calculated as:

$$s_t = s_\infty U \qquad (4.11)$$

where U is the degree of consolidation reached in time t, which can be represented as the ratio of the area A' (where consolidation already occurred) and the total area A (see Fig. 4.4) for which it holds true that:

$$A = H \Delta \sigma \qquad (4.12)$$

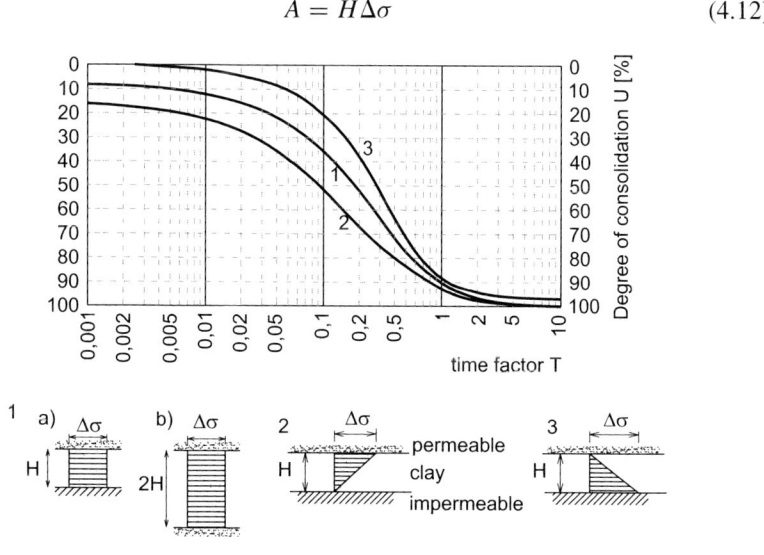

Fig. 4.3 Degree of consolidation U as a function of time factor T

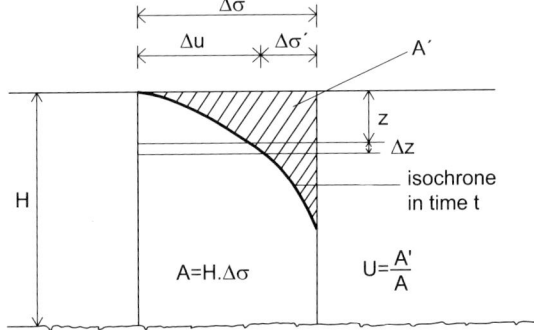

Fig. 4.4 Concept of degree of consolidation U in time t

Thus, the degree of consolidation U is governed by the relation:

$$U = \int_{z=0}^{H} \frac{\Delta\sigma' \Delta z}{\Delta\sigma H} dz \qquad (4.13)$$

For the three cases presented in Fig. 4.3, the equations for the degree of consolidation U can be written as:

$$U_1 = 1 - \frac{8}{\pi^2}\left(e^{-T} + \frac{1}{9}e^{-9T} + \frac{1}{25}e^{-25T} + \frac{1}{49}e^{-49T} + \ldots\right) \qquad (4.14)$$

$$U_2 = 1 - \frac{16}{\pi^2}\left(\left(1 - \frac{2}{\pi}\right)e^{-T} + \frac{1}{9}\left(1 + \frac{2}{3\pi}\right)e^{-9T} + \frac{1}{25}\left(1 - \frac{2}{5\pi}\right)e^{-25T}\right.$$
$$\left. + \frac{1}{49}\left(1 + \frac{2}{7\pi}\right)e^{-49T} + \ldots\right) \qquad (4.15)$$

$$U_3 = 1 - \frac{32}{\pi^3}\left(e^{-T} - \frac{1}{27}e^{-9T} + \frac{1}{125}e^{-25T} - \frac{1}{343}e^{-49T} + \ldots\right) \qquad (4.16)$$

The above-mentioned theoretical case 1 resembles construction of an embankment on soft, clayey subsoil, saturated with water. A special attention to consolidation equations gives Hansbo (2001).

The basic version of speeding up subsoil consolidation is *establishment of vertical drains in clayey subsoil* with a major aim of shortening the consolidation path H, as consolidation speed is proportional to the square of the consolidation path, Fig. 4.5, Hansbo et al. (1981). The result is another relationship $s = f(t) -$ see Fig. 4.6. Total settlement in infinity is identical for both the basic case and the case with vertical drains. For time t, when final touches of the embankment surface are supposed to be completed (e.g. road surface with a required grade) it

4.1 Speeding Up of Consolidation

Fig. 4.5 Shortening of consolidation path with the help of vertical drains

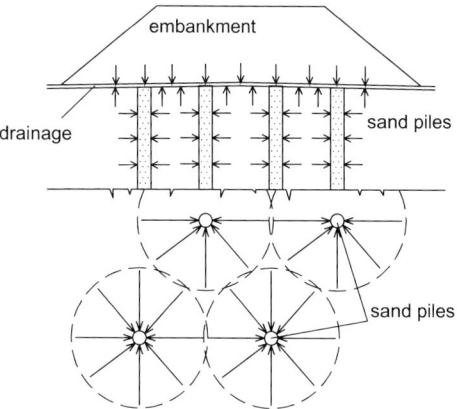

is always better to keep the post construction additional deformation as small as possible. This applies as well to the case with vertical drains, which is a positive with regards to the performance reliability of all structures situated on the surface of the embankment body. At the same time, however, shear strength will significantly grow in time t, which is relevant in terms of the subsoil stability and bearing capacity. An advantage of vertical drains is in fact that they instantly speed up the dissipation of excess pore pressures caused by the increase in surcharge from embankment construction. Therefore, the maximum value of the pore water pressure increase calculated from the surcharge could not practically occur.

The second basic alternative of how to speed up consolidation is a method referred to as "preloading technique", overfilling the embankment with a layer that will be subsequently removed. The preloading technique is more sensitive in terms of the maximum pore water pressures reached and thus also the stability and bearing capacity of subsoil.

Both basic methods will be more closely specified below, also in relation to this aspect. At this place, the above-mentioned, makes it clear that while applying both methods, monitoring – measurement of pore water pressures, will be a key factor. Their knowledge is important not only during the load increase, i.e. embankment construction, but also during their dissipation after the completion of construction. In this way, monitoring allows checking both the rate of embankment construction and the time when final touches on the embankment surface can be implemented.

An analogical situation to subsoil is that in the embankment body itself, e.g. Bishop (1957), Farrar (1978). For simplicity, it is presumed to be situated on solid

Fig. 4.6 Different consolidation curves for cases without (1) and with (2) vertical drains

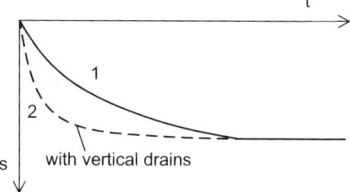

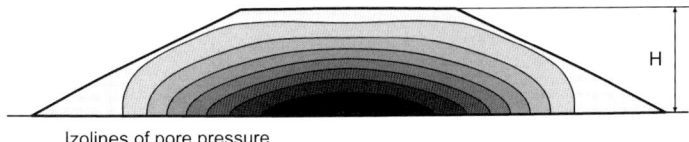

Izolines of pore pressure

Fig. 4.7 Distribution of pore pressures in the embankment at the end of construction

and impermeable subsoil. By gradual compaction of individual layers, the bottom layers are exposed to increased load, which also causes an increase in their pore water pressures. As soil after compaction is not 100% water saturated, but contains on average about 5% of air voids in its pores, which are more compressible after loading, and, besides, the situation is no longer a theoretical case of 1D consolidation, the increase in pore water pressures can be expressed from the equation by Skempton (1954):

$$\Delta u = B \Delta \sigma_3 + \bar{A}(\Delta \sigma_1 - \Delta \sigma_3) \tag{4.17}$$

where $\bar{A}$ and B are coefficients of pore water pressure. Thus the value of the increase in pore water pressures Δu may be predicted knowing the increase in the main pressures $\Delta \sigma_1$, $\Delta \sigma_3$ and also on $\bar{A}$ the basis of the coefficients of pore water pressure and B established in laboratory conditions. Then, the situation with pore water pressures within the respective embankment body can correspond to the situation in Fig. 4.7. Finding a solution to this situation for embankments of transport infrastructure is simpler, here all we have to do is to alternate more permeable layers within the embankment shortening considerably the consolidation path, at least at the contact with the subsoil. For hydraulic structures which impound water the problem is more complex, different types of drains can be used for homogeneous embankments, from toe via planar to chimney drains, see Fig. 4.8.

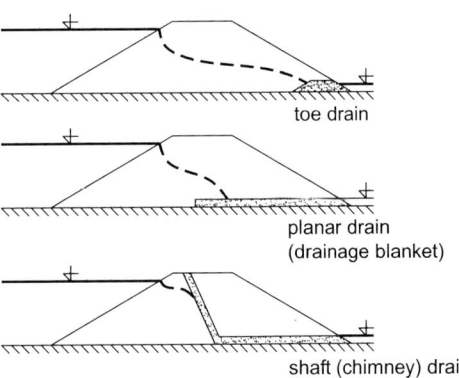

Fig. 4.8 Different types of drains used in dam construction

4.1.1 Preloading

This method, mostly known as preloading technique, but also as overconsolidation or precompression, has two principal subversions, Gatti and Gioda (1983).

The first (*precompression*) is preferably used for foundations on soft subsoil. This subsoil may be primarily exposed to external loading, e.g. with soil fill, left to consolidate, removing successively the fill and loading the subsoil with a structure sensitive to absolute and differential deformations. An example of this type was preconsolidation of subsoil (a pit) which was backfilled with spoil heap material to serve for subsequent "shifting of the gothic church in the North Bohemian town of Most"– Škopek (1979), Herle and Škopek (2003). The reason for the shifting was a huge brown coal deposit lying 6 m below the ground surface and reaching to almost 40 m in depth. The track of moving is shown in Fig. 4.9, where the most critical part was crossing the backfilled open pit mine over a distance of about 200 m, the depth of which was about 34 m. The levelling mechanism of transport carriages was able to compensate some differential settlement, but not the value of the calculated backfill settlement that was about 600 mm. Therefore, in order to minimize the risk of excessive and differential settlement during the transport of the church over this area, a pre-consolidation fill 5–6 m high was placed as a surcharge along the whole width and length of the track over the backfilled mine. The height of the embankment was supposed to cause a pressure increment on the backfill approximately equal to the pressure brought in by the moving church, i.e. 100 kPa. The backfill settlement was measured for 17 months (period of consolidation fill construction

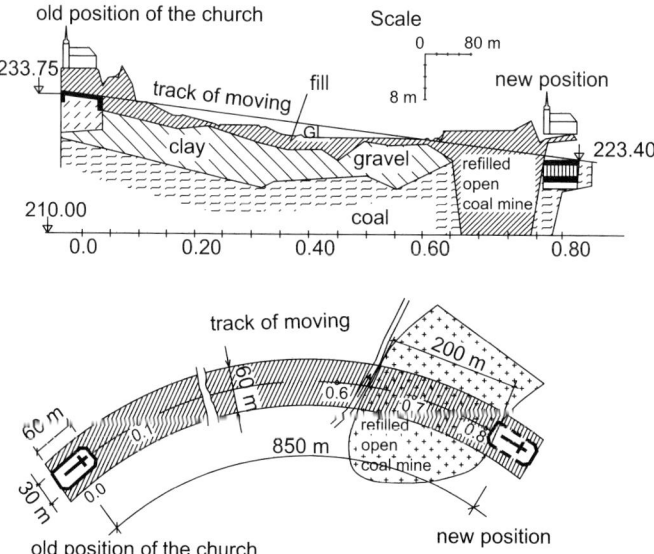

Fig. 4.9 Schematic plan and longitudinal section of the track for the shifting of the Most's church (according to Herle and Škopek)

and consolidation phase) and after the removal of consolidation fill for another 6 months. The maximum measured settlement of the backfill surface was 990 mm and the maximum measured lifting was about 100 mm. From the results the prognosis of settlement for the phase of church transfer was done and the calculated value of 100 mm was acceptable for the levelling mechanism.

The second subversion (*preloading*) is more typical of earth structures where the embankment body itself is built higher than presumed, left in place for a certain time, and only subsequently partly removed to reach the presumed value, e.g. Ache et al. (1983). This implies a certain problem concerning the subsoil bearing capacity. The process of speeding up of consolidation itself is aimed at limiting this hazard, while for higher embankments, the risk of losing the subsoil bearing capacity is even higher. Therefore, this method can be applied in situations where the loss of the subsoil bearing capacity is small (for example compressible subsoil is constrained at the sides), or the additional embankment must be built not only high up, but also to the sides applying more gentle slope gradients.

To explain the basic principles, we may again use the theoretical settlement curve of soft, water-saturated subsoil under 3D conditions, Fig. 4.10. Three separate components can be distinguished on this curve:

$$s = s_i + s_c + s_s \tag{4.18}$$

where

s – total settlement,
s_i – initial settlement under wet conditions where the vertical component acts due to the element's spread onto sides, the volume change being zero,
s_c – consolidation settlement occurring under drained conditions for the time when dissipation of the increment of pore water pressures occurs,
s_s – settlement due to secondary consolidation under constant effective stress as a consequence of grain strain at contacts.

Note: The value of secondary settlement is significant for clayey soils and soils with organic content. The magnitude of secondary settlement is characterized by the coefficient of secondary compression $C_{\alpha\varepsilon}$, defined as strain per log t.

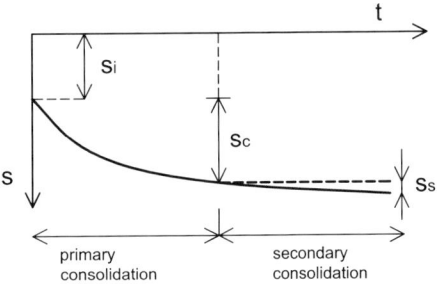

Fig. 4.10 Settlement curve for soft fully saturated soil under 3D conditions

4.1 Speeding Up of Consolidation

The first component can be, to a certain extent, separated from the others even in terms of time if we may theoretically presume that the external load increase is sudden. In practice, this is not feasible even though construction of an embankment up to several meters high can be a question of hours. The strain at grain contacts, as well, is in process already in the phase of consolidation settlement so that the above-mentioned classification is, to a certain extent, idealized, but still frequently used for practical considerations. In this respect, Jamiolkowski et al. (1983) state that "it is necessary to bear in mind the most relevant deviations from conventional assumptions", and, apart from those already mentioned, he adds "clay structural collapse and creep generated excess pore pressures".

Finding the solution to a practical problem consists in answering the fundamental question of how big additional load increase ($\sigma_2 - \sigma_1$) to choose and for how long it should be applied before unloading to the planned value σ_1. Generally speaking, after unloading to the planned value, the settlement curve character may change as shows Fig. 4.11. As a further recommendation, the most suitable curve profile is (b) where following a slight rise the settlement rate is stabilized at a practically zero constant value. In this way, elimination of not only consolidation settlement, but also the secondary consolidation component occurs, Veder and Prinzl (1983).

The condition of eliminating consolidation settlement can be, in a simplified way, expressed as follows. The increased stress σ_2 generally causes a greater increment of pore water pressures, which are gradually dissipated while increasing the settlement. If σ_2 is twice the value of σ_1 then it suffices to apply σ_2 for a time necessary for reducing the average degree of vertical consolidation U_v to one half. Generally, if one wants to avoid further consolidation settlements under the loading σ_1, it is necessary to remove σ_2 at time t_s at which the average degree of vertical consolidation U_v within the clay stratum is not less than:

$$U_v = \sigma_1/\sigma_2 \tag{4.19}$$

If this condition is satisfied upon the removal of ($\sigma_2 - \sigma_1$), the construction will experience only secondary settlement, Jamiolkowski et al. (1983).

The reduction of secondary settlement is principally given by the time during which increased stress at contacts of individual grains was acting. That is why a decision depends on the relevance of secondary settlement (using the value of $C_{\alpha\varepsilon}$ – coefficient of secondary compression), and the need for reducing this settlement.

In order to speed up consolidation in the form of preloading, precompression, not only additional earth fills, but also water pressure loading can be used where water

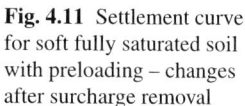

Fig. 4.11 Settlement curve for soft fully saturated soil with preloading – changes after surcharge removal

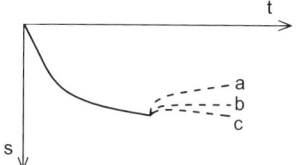

is used to fill reservoirs – Hegg et al. (1983), nowadays even in the form of pools of plastics.

Also, potential application of negative pressure in subsoil should be mentioned. In this case, the surface must be sealed (mostly with a geomembrane). This method is not very common, being connected with certain problems, but it can be recommended in cases where subsoil needs to be improved not only in terms of settlement reduction, but where there is also a need for reducing subsoil contamination – it is a combination of the method of speeding up consolidation with the treatment (remediation) method "Vacuum-Enhanced Recovery" – e.g. Suthersan (1997), Vaníček (2002d).

4.1.2 Vertical Drains

The main purpose of vertical drains is, as was already said, to speed up subsoil consolidation. This mostly happens not only by shortening the consolidation path, but also due to the fact that horizontal permeability of clayey soils tends to be greater so that horizontal flow into vertical drains is faster than that in the vertical direction only.

A traditional type of vertical drains are *sand drains*, sand piles, which can be implemented by two basic technologies:

– Boring a hole and filling it with sand – this method is preferred if the borehole walls retain stability for at least the time necessary for sand installation, otherwise casing must be used.
– Making a hole by pushing soil into sides and filling the free space with sand – the hole is produced by the penetration or vibratory driving of a steel casing pipe – filling it with sand can be performed simultaneously with pulling out the casing pile or after the borehole is free if its walls can retain stability.

The most common dimensions used are diameters between 0.2 and 0.6 m with lengths of up to 30 m. In the case of sandwicks the diameter is smaller. These drains are prepacked in a filter stocking and placed in a predrilled or pre penetrated hole.

In their plan view, vertical drains are usually arranged in a "triangular pattern or in a square pattern", see Fig. 4.12. Here, we can see that each vertical drain drains off a certain area defined by the effective radius r_e, where the drain radius is r_w and the axial distance between vertical drains is s. Jamiolkowski et al. (1983) give a detailed account of consolidation theory for various assumptions, in particular:

– Free vertical strain; the assumption is that the vertical surface stress remains constant during the consolidation process, and the resulting surface displacements are thus non-uniform,
– Equal vertical strain; the assumption is that the vertical surface displacement is constant throughout the drained area and the resulting vertical stress at the surface is thus non-uniform.

4.1 Speeding Up of Consolidation

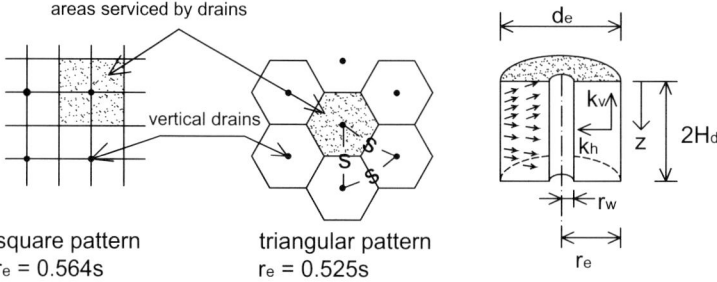

Fig. 4.12 Typical layout of vertical drains – square pattern and triangular pattern

Actual field situations are probably somewhere between free and equal strain conditions, depending on the ratio $r_e/r_w = n$, the thickness and the stiffness of the sand blanket and the ratio of the axial stiffness of the drain to that of the surrounding soil. In this respect, there exist more theoretically derived relationships expressing the relation $\Delta u = f(r,t)$, where t is time and r is the radial distance of the considered point from the centre of the drained cylinder.

For practical solutions in designing vertical drains we need to know the relation between the mean degree of consolidation in time t and the geometric properties of drains, i.e. their diameter d_e, or their ratio n. Here, the most fitting practice again is to use non-dimensional variables as was the case of the classic Terzaghi's consolidation theory. Then, for the non-dimensional time factor T_h, expressing the case of radial consolidation in the horizontal direction, it holds true that (e.g. Turček and Hulla, 2004):

$$T_h = c_h t / d_e^2 \tag{4.20}$$

where c_h is the coefficient of consolidation for horizontal flow:

$$c_h = k_h E_{oed} / \gamma_w \tag{4.21}$$

where k_h is the coefficient of permeability for horizontal flow.

For the sought relationship we may use either the relationship between T_h and U_h – Fig. 4.13 or the equation:

$$t = \frac{d_e^2}{8 c_h} \left[\ln \left(\frac{d_e}{d_w} \right) - \frac{3}{4} \right] \cdot \ln \frac{1}{1 - U_h} \tag{4.22}$$

which serves for calculating the time t necessary for reaching a certain degree of consolidation U_h and which is attributed to Kjellman and Baron, see Barron (1948) or Kjellman (1948).

Here, the need for establishing the value c_h or k_h respectively comes forward. In principle, an undisturbed core may be used cutting out a sample in the horizontal direction and testing it in a classic apparatus, preferably in the oedometer

Fig. 4.13 Determination of degree of consolidation for horizontal drainage

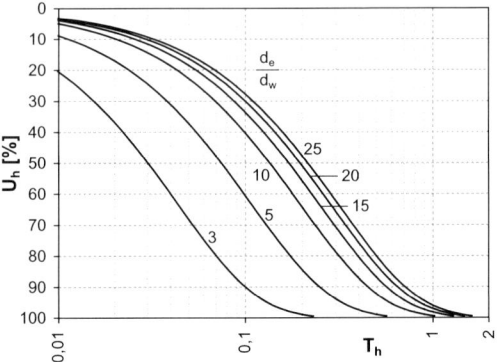

or the triaxial apparatus in keeping with the basic principles of soil mechanics or in modified apparatus using a sample with an original vertical orientation, e.g. Soranzo (1983). These are schematically shown in Fig. 4.14, where the first represents the oedometer with lateral drainage, the second the triaxial apparatus with lateral drainage, too, and the third case for in situ testing uses cone penetration testing.

Estimates are based on the coefficient of consolidation in the vertical direction c_v, which is multiplied by the constant in a range of 2–5 (2–10). Some authors recommend field determination of this value, e.g. Jansen and Den Hoedt (1983).

With regards to the links between the technology of vertical sand drain implementation and the theory applied for their design, the subject of discussion is mainly the effect of "driven displacement sand drains" on the contact stratum of the surrounding clayey soil. The authors stress the degree of the surrounding soil disturbance, which has the following consequences:

– Impact on permeability changes, by kneading the difference between permeability in the vertical and horizontal direction is impaired,
– There may be temporary reduction in shear strength, mainly for sensitive clays depending on clay sensitivity.

The most commonly applied drain type of today are *geodrains* produced by using geosynthetics, in principle geodrain strips described briefly in Chapter 3, Fig. 3.10c.

Fig. 4.14 Possibilities for determination of consolidation coefficient for horizontal flow (according to Soranzo). (**a**) Adapted oedometer with central horizontal drain. (**b**) Adopted triaxial apparatus with circumferential drainage. (**c**) Cone penetration test with measurement of pore pressure dissipation

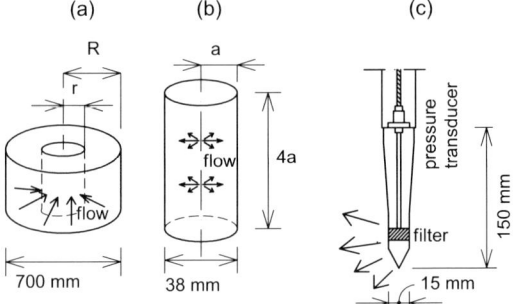

4.1 Speeding Up of Consolidation

The drain consists of a core of most diverse shapes and profiles, which is protected by geotextiles, e.g. Eriksson and Ekstrom (1983). The strip is mostly around 100 mm wide with thicknesses of up to 10 mm. They have practically replaced the original "cardboard drains". Small dimensions allow them to be installed by static penetration or vibratory driving of closed-end mandrels having limited cross-sections, see Fig. 4.15. Inside its profile and at its end, the geodrain is attached to a plate pushed into subsoil by a guide profile. Having reached the necessary depth, the plate, which has a slightly larger area than the guide profile's cross section, is fixed at the respective level together with the geodrain. The guide bar is successively pulled out; the geodrain is cut off and placed in the upper permeable layer serving as horizontal drainage. The installation depth depends not only on the required level, but also on the penetration resistance of the soil in which the geodrain is installed. The maximum depths reached are around 60 m.

In comparison to sand drains and their related design assumptions, the main factors discussed in relation to geodrains, e.g. Hansbo (1983) include:

- Equivalent diameter of drains – derived from the relation:

$$d_{eq} = 2(a+b)/\pi \qquad (4.23)$$

  where a = thickness and b = width of the drain, some authors propose to decrease the equivalent diameter by about 1.25, to a more realistic value, Fellenius and Wager (1977)
- Filtration properties of drains, mainly of filtration geotextiles around the drainage core; here not only the filtration coefficient, but mainly the filtration stability of the interface is the subject of assessment. In most cases, the filter tends to get clogged, but the permeability as compared to surrounding soil remains high.
- Discharge capacity of drains – is basically very important in the sense of the drain's ability to drain water due to on-going consolidation. The discharge

Fig. 4.15 Photo of vertical prefabricated drains installation

capacity of a prefabricated drain depends on the core structure and the form of the drain and the material and structure of the filter mantle. Another important factor is the lateral pressure of the soil against the drain filter.
- Similarly as for driven displacement sand drains also the pushing down and up of the mandrel can influence the smear zone around the drain. The net result is that effective horizontal permeability in the remoulded clay may become at least a factor of 2–4 lower than for the undisturbed clay.

In terms of mechanical and physical properties of the geodrain (with regards to its potential tensile failure during installation), the following are required:

- Minimum tensile strength of 0.5 kN,
- Strain under this force should be in the range of 2–10%,
- Tensile force of the filter mantle and the core should be mobilized at roughly identical strain.

Many of the above-mentioned factors affecting the function of geodrains are in more detail treated by Hansbo (1983), Vreeken et al. (1983), Jansen and Den Hoedt (1983), Tammirinne and Vepsalainen (1983). To conclude, a note can be made that drains usually function on a short-term basis, fulfilling the conditions for speeding up of consolidation, i.e. their performance must be guaranteed until the time considered as the end of consolidation settlement when the increment of pore water pressures has been eliminated to zero.

An interesting combination of vacuum preloading with vertical drains is described by Park et al. (1997), see Fig. 4.16. A note to this combination – while vertical drains can normally divert into the upper horizontal drainage stratum water whose hydraulic head is higher than the level of this horizontal drainage stratum, in the case of negative pressure applications this reduction can be substantially higher. In practical implementations, it is very useful to combine the application of vertical drains with preloading.

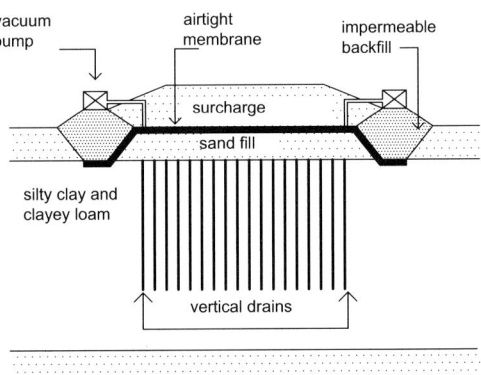

Fig. 4.16 Scheme of vertical geodrains with vacuum preloading (according to Park et al.)

4.2 Soil Stabilization

There are numerous methods that can be referred to as "soil stabilization". Nevertheless, for the needs of the construction of earth structures, the methods to be considered below are only those where the dominant field of application is an improvement of soils in layers or surface improvements. These include mainly earth structures of transport infrastructure including airports, but also the subbase of large halls, specifically if they are partly built in a cut and partly on an embankment. Here, the experience from the implementation of similar methods as e.g. "lime columns" is beneficial. The stabilization may be mixed in place or in plants. Depending on the soil characteristics and requirements, the following additives are used:

- Lime, cement, gypsum,
- Industrial by-products, mostly from the burning of coal, such as fly ash, energy-gypsum, slag (cinder),
- Additional soil – mechanically stabilized soil,
- Several chemical additives (resins, bitumen).

The use of chemical stabilizers in earth structures is limited, often not only by price but also by several potentially negative impacts. With some soil stabilization methods there is a risk that ground water may be polluted because the additives or the by-products may be harmful, concrete may deteriorate or the corrosion rate of steel members may significantly increase, Broms and Anttikoski (1983). A specific case are stabilized layers of road base courses where the gravel strata are stabilized with bitumen – in road construction this is referred to as asphalt-coated aggregates.

4.2.1 Mechanically Stabilized Soil

The purpose of using additional soil is to change the grain-size composition of the soil improving its properties. A common practice is adding coarse grains to existing fine-grained soil. The aim is to reach such composition that coarser grains will form the skeleton and the pores of this skeleton will by filled up with fine grains. The mutual contact of coarse grains ensures smaller compressibility, higher shear strength and lower susceptibility to frost penetration. In relation to the last mentioned aspect, however, we must keep in mind that frost susceptibility is not only given by the grain-size curve, but also by the character of fine particles, whether they are active clay minerals or non-clay minerals (neutral ones) as quartz, calcite etc.

An opposite case, on the contrary, leads to reducing the permeability of the soil by adding fine particles, such as clay minerals (montmorillonite), which allows reaching the required impermeability of the sealing layers of landfills or the sealing of rockfill dams.

A specific case of mechanically stabilized soil are mechanically strengthened aggregates. They, in fact, form the best-quality unbound base course of roads. They have more suitable properties compared with crushed or vibrated gravel. Mechanically strengthened aggregates is a layer composed of a mix of minimally two fractions of natural or man-made aggregates (e.g. slag) produced in a mixing plant, spread out and compacted under conditions ensuring reaching the maximum bearing capacity of the stratum. It is not made from a mono-fraction, but it must consist of individual fractions. The optimum composition of the particle size distribution curve is regulated in national codes (e.g. STN 736126 Road building – Unbound courses or equivalent EN 13285 Unbound mixtures – Specifications). The values recommended for the particle grading ranges of mechanically strengthened aggregates for fractions of 0–32 mm, or 0–45 mm are displayed in Table 4.1.

Apart from the required grading, however, the resulting mix must also comply with the minimum strength requirements, e.g. at least 80% CBR. If mechanically strengthened aggregates are used as the base course for roads carrying heavy traffic – motorways, I and II class roads – stricter grading and strength criteria are required. For the fraction 0–45 mm the recommended range of the particle size distribution curves under STN and the stricter criteria are shown in Fig. 4.17. The mix is in situ compacted to a minimum value of $I_D = 0.95$, and the value of the CBR should be at least 120%. It is recommended to verify the frost susceptibility of the mix by the frost susceptibility rate test. Under STN 721191 Testing of soil frost susceptibility, the frost susceptibility rate test is based on the relation:

$$\beta = \Delta h / \sqrt{I_m} \qquad (4.24)$$

where Δh is the maximum frost lift during the direct frost susceptibility test (mm), h = 117mm, I_m – frost index (°C) = sum of mean sample temperature gradient measured every hour (the temperature difference is about 6°C, +2°C at bottom and −4°C at top).

Table 4.1 Proposed ranges of particle size distributions for mechanically stabilized aggregates (STN 736126)

Particle size (mm)	Particle siftings (% weight)	
	0–32	0–42
63		100
45	100	92–100
32	95–100	79–100
16	71–93	54–96
8	50–72	37–77
4	35–55	24–59
2	25–42	16–44
1	17–33	10–33
0.5	10–25	6–26
0.25	6–18	4–20
0.125	3–12	3–15
0.063 *	2–8	2–12

* Sifting on sieve 0.063 mm should be below 60% of sifting on sieve 0.5 mm. For capillary water regime the amount of particles below 0.02 mm is limited below 3%.

4.2 Soil Stabilization

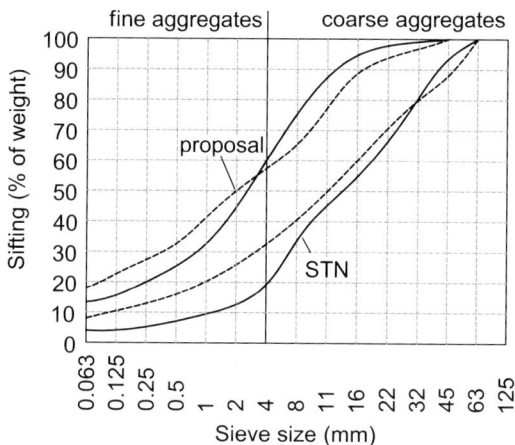

Fig. 4.17 Limiting range of particle size distribution curves for mechanically stabilized aggregates (fraction 0–45 mm) according to STN 721512 and with new proposal by Gschwendt et al.

Then, the frost susceptibility rate is assessed as follows, if:

- $\beta \leq 0.25$ is non-frost susceptible material,
- $0.25 \leq \beta \leq 0.5$ – is slightly frost susceptible material,
- $\beta > 0.5$ – is dangerously frost susceptible material.

4.2.2 Stabilization with the Help of Cement, Lime and Other Hydraulic Bonding Agents

The principal bonding agents are cement and lime. Stabilization with the help of cement is preferred for non-cohesive soils, while for cohesive soils lime is preferred – Moh (1962), Bryhn et al. (1983). Among other bonding agents we may include fly ash, dust from rotary cement kilns, granulated blast-furnace clinker, slag from steel mills or their combinations. Basic cases where stabilized soil is used include transport structures, the subbase of large halls, but also other applications, e.g. dykes; Serra et al. (1983).

The use of stabilization is most common *in highway construction* where it can fulfil several tasks:

- Stabilization of subgrade,
- Stabilization of base courses,
- Temporary stabilization of road surface or light road pavement surfacing.

The stabilization of natural subsoil principally involves its improvement, very often connected with the value of the modulus of deformation $E_{def,2} > 30$ MPa. The subsoil is predominantly formed by cohesive soils, which favour the use of lime. Highway construction codes usually subdivide stabilization measures into classes, e.g. STN 736125 "Stabilized subgrade" distinguishes three basic classes for which the following properties are required:

- Class S I – compressive strength 2.5–4 MPa, frost and water resistance minimally 3.5 MPa,
- Class S II – compressive strength 1.8–3.0 MPa, frost and water resistance 2.1 MPa,
- Class S III – compressive strength 1.0–1.8 MPa, frost and water resistance 1.2 MPa.

Note: Compressive strength is determined on cylinders (a diameter and height of 100 mm with a maximum grain size of 16 mm) placed in a humid environment with a minimum relative humidity of 95% for 28 days, while determination of frost and water resistance is based on the same test, but carried out only after the completion of a prescribed number of freezing cycles where the number of cycles and the maximum negative temperature are chosen with regard to the climatic zone of the site and the type of stabilization used (in protective layers to $-10°C$, in lower base courses to $-15°C$ and in upper base courses to $-20°C$, which is valid for the Czech and Slovak Republics).

Besides the basic classes, it is possible to distinguish the mix manufacturing procedure as the static plant mix method and the mix-in-place method using either a cutter or a heavy cutter. Rough composition of the particle size distribution curve of soils and their basic characteristics – plasticity are given in Table 4.2, where individual classes are also characterized by the processing methods. Recommended combinations of binders and their rough composition expressed in % of dry stabilized material (soil) are shown in Table 4.3.

With regards to the manufacturing procedure, the static plant mix method is the option for larger sites where the mixing plant is of the same type as in the production

Table 4.2 Reference characteristics of soils and materials suitable for soil stabilization (STN 736125)

Sieve size (mm)	Sifting (% of weight)							
	SI		SII		SIII			
	Mixing type							
	In plant	Cutter	In plant	Cutter	In plant	Cutter	Heavy cutter	
45 (63)	100	100	100	100	100	100	–	
32	83–100	83–100	83–100	83–100	83–100	100	–	
16	65–100	65–100	65–100	65–100	65–100	100	–	
8	48–85	48–100	48–85	48–100	48–100	83–100	–	
4	30–66	30–100	30–66	30–100	30–66	66–100	–	
2	24–50	24–100	24–50	24–100	24–50	50–100	–	
1	18–40	18–80	18–40	18–80	18–40	40–100	–	
0.5	10–30	10–60	10–30	10–60	10–30	30–100	100	
0.25	7–25	7–50	7–25	7–50	7–25	25–80	80–100	
0.13	6–20	6–40	6–20	6–40	4–20	20–70	70–100	
0.06	0–15	0–30	0–15	0–30	0–15	15–50	50–100	
0.01	–	–	–	–	0–6	6–35	35–70	
0.002	–	–	–	–	–	0–20	20–40	
Liquid limit w_L	–	–	–	–	–	–	<40	
Plasticity index I_P	0–10	0–17	0–10	0–17	0–10	10–22	17–27	
Sand equivalent E_P	28–36	28–36	–	–	>15	–	–	

4.2 Soil Stabilization

Table 4.3 Proposed amount of additives for different classes of soil stabilization (STN 736125)

Stabilization class	Soil type	Combination of additives	Reference amount [%]
SI	PSD suitable for mixing in plant	cement PC, SPC cement + slag	8 2C + 10T
	PSD suitable for mixing by cutter, slightly acid (pH 5 to 6)	cement PC, SPC lime + cement cement + slag lime + cement	12 2V + 10C 2C + 12T 2V + (8–10)C
	Sandy & silty ($I_P = 4$)	cement + ash cement + slag	3C + 15P 2C + 12T
SII	PSD suitable for mixing in plant	cement PC, SPC cement + ash cement + slag residue of cement production	6 2C + 10P 2C + 10T 12
	PSD suitable for mixing by cutter	cement PC, SPC cement + ash cement + slag residue of cement production	10 3C + 10P 3C + 10T 15
	Slightly acid (pH 5 to 6)	lime + cement	2V + 8C
	Sandy & silty ($I_P = 4$)	cement + ash cement + slag residue of cement production	2C + 10P 2C + 10T 12–15
SIII	PSD suitable for mixing in plant	cement PC, SPC cement + ash cement + slag residue of cement production	4–6 2C + 8P 2C + 8T 10
	PSD suitable for mixing by typical cutter	lime + cement lime + ash cement + ash	2V + 6C 2V + 10P 2C + 10P
	PSD suitable for mixing by heavy cutter (conditionally suitable)	lime + cement lime + ash lime residue of cement production	4V + 8C 4V + 10P 7 15

Legend: C – cement, V – lime, P – fly ash, ST – slag, PSD – particle size distribution, PC – portland cement, SPC – slag portland cement of concrete mixes, often fitted with continuous mixers and very precise batching of individual components. After transporting to the site, the stabilized mix is treated by compaction like classic soil. In the mix-in-place method, the layer of soil prepared for stabilization is first modified, then the bonding agent is spread all over with the batcher and successively mixed with soil using the cutter. Additional wetting is applied as necessary as mixing at moisture contents lower than optimum is preferred. A typical cutter for the production of soil stabilization can be seen in Fig. 4.18.

The recommendation specification states that cement stabilization is close to lean concrete.

For fine-grained soils, quick lime is used. Its adding basically serves two purposes:

– To improve workability,
– To improve strength.

Lime proportioning depends on the type of soil and its moisture content. It ranges from 2 to 6% of dry soil weight. It is roughly presumed that 1% of quick lime will modify the moisture content by 1–2% with a simultaneous increase in the optimum moisture content during the Proctor standard compaction test – see Gschwendt et al. (2001), Fig. 4.19. An example of the effect of adding 3% of quick lime on the change of the compaction curve for clayey soils (CI – clay with intermediate plasticity, $w_L = 36.9\%$, $w_p = 20.1\%$) for original and stabilized soils is shown in Fig. 4.20.

The above-mentioned practical outputs, however, are supported by some fundamental principles, mainly in the domain of lime stabilization as the reaction of lime with clay minerals is an issue known for a long time. Referring to "the Pyramids of Shersi in Tibet, which were built using a compacted mixture of clay and lime", Greaves (1996) mentions the age of 5000 years. In detail, this method

Fig. 4.18 Cutter for production of soil stabilization (Photo Bartozela)

4.2 Soil Stabilization

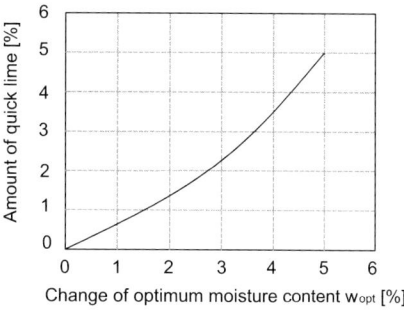

Fig. 4.19 Change of optimum moisture content of soils stabilized by quick lime for compaction (according to Gschwendt et al.)

has been developed mainly since the second half of the twentieth century. In general, in the branch of geotechnical engineering, it has applications both in highway construction, motorways, railways and airports, and in classic geotechnics in the form of lime columns, mostly to improve subsoil properties. Numerous publications in the area of lime columns are included in the European Conference SMFE in Helsinki 1983 or the proceedings of the seminar Lime stabilization at Loughborough University – Thomas Telford 1996 for the area of applications in highway construction.

Under the common term of lime we may imagine quicklime – calcium oxide, slake or hydrated lime – calcium hydroxide or agricultural lime – calcium carbonate. For lime stabilization, however, only the first two types are suited with a preference for quicklime as it produces less dust, faster reduces the soil moisture content, its hydration heat speeds up the growth in strength and, in general, its consumption is by about 25% less than in the case of slake lime. The stabilizing effect depends on the reaction of lime with clay minerals, resulting in:

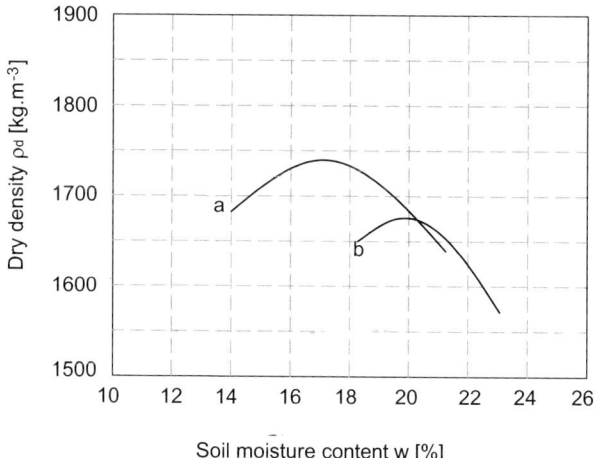

Fig. 4.20 Proctor standard compaction curves for soil without (**a**) and with (**b**) addition of quick lime (according to Gschwendt et al.)

- Growth in strength and the subsoil bearing capacity in general – due to the creation of cementing bonds resulting from soil-lime pozzolanic reactions, which, however, develop in time and therefore strength tests are performed with a time lag of mostly 28 days,
- Reduction in the susceptibility to swelling and shrinkage,
- Reduction in moisture content and thus workability and successive compaction – where the reduction in the moisture content is caused both by absorption, extraction of moisture from the surrounding soil after the lime type is changed into slake lime, and by hydration heat as was mentioned above. At the same time, however, the new reaction of lime with clay minerals leads to a change in plasticity thus affecting another factor relevant for workability.

With reference to the last point, it must be added that the growth affects the moisture content at the plasticity limit w_p while the moisture content at the liquid limit w_L remains practically the same. E.g. Greaves (1996) referring to Sherwood (1967) states that a 4% addition of lime resulted in an increase in the plasticity limit from 24 to 43%. With an additional growth of the lime percentage, however, no more plasticity reduction occurred. The relevance of this fact for workability is evident, for example originally plastic soil, which was practically uncompactable, becomes stiff with much better workability.

The development of cementitious bonds is attributed to the fact that "clay is a natural pozzolan containing silica and alumina. Free lime reacts with these elements causing gels to form very similar composites to those found in a cement paste". Therefore, in order to create cementitious bonds, soil must contain clay minerals. An indirect indicator of a certain content of clay minerals can be the plasticity index with the lower boundary of about 10%, e.g. Greaves (1996).

Based on the above-mentioned facts, we can state that lime application for cohesive soils is relatively well covered, and it is in our interest to apply lime stabilization as much as possible, as it brings about numerous advantages. Among them there are:

- Possibility of using soils previously hardly usable due to their high plasticity and moisture content – thus avoiding a necessity for transporting more suitable soils, sometimes over long distances, or avoiding deposition of unsuitable soils in landfills. Avoiding of both transporting and landfilling will have positive impacts on the environment and economy,
- Improved workability of soil used as a building material, not only in terms of transport and spreading, but mainly in terms of compaction even under relatively unfavourable climatic conditions,
- Improved strength due to newly formed bonds, which may limit the thickness of additional structural layers.

Nevertheless, it has been manifested that mutual bonds between lime and clay minerals in cohesive soils are more complex than might seem at first sight, and so an increased focus in the last decade has been on some other problem areas as follows.

4.2 Soil Stabilization

Lime modification × lime stabilization. As was already mentioned, in the phase of mixing lime with cohesive soil, soil is first modified before a successive chemical reaction, which in most soils leads to stabilization. Not all cohesive materials are stabilizable, but all are modifiable. Mixing quicklime with a wet soil immediately causes the lime to hydrate and an exothermic reaction occurs, the heat produced being sufficient to drive off some of the moisture within the soil as vapour and hence reduces the moisture content. The second effect of ion substitution results in a reduction in plasticity as the clay particles flocculate – e.g. Perry et al. (1996). Modified material after that has a coarser texture, is more friable and is less plastic. In most cases, there is a growth in the plasticity limit w_p and a less steep growth in the liquid limit w_L, nevertheless e.g. Chaddock (1996) also mentions a case where the growth in w_L was higher and thus there was a growth in the plasticity index I_p by 5%. The lime modification phase usually lasts for 24–72 h. After modification, a growth in strength can occur, although some authors add that these gains can be reversed. A highly important issue is what percentage of lime will set off the phase of modification but not yet stabilization. The most frequently mentioned range is 2–4% of lime.

The stabilization phase requires a larger amount of lime, which can be indirectly measured through a growth in pH. According to Holt and Freer-Hewish (1996) when an alkaline environment with a pH value of around 12.4 is maintained after the immediate reactions have taken place, the pozzolanic reaction and stabilization of the soil occurs. The high alkaline environment promotes the dissolution of silica and alumina from the clay particles which in turn reacts with the calcium ions from the lime to form calcium silicate hydrates and calcium aluminate hydrates. These reaction products bond the clay particles together and are similar in composition to those of a cement paste. The process is relatively slow because the available lime has to diffuse through the soil structure and the initial cementitious products to the reaction sites. Stabilization of the material also causes a considerable increase in strength and durability; although the gain in strength is both temperature and time dependent.

One of the exceptions where the strengthening process does not proceed in keeping with the above-mentioned procedure is a case where the soil or pore water contains sulphates or where sulphates arise through oxidation of "sulphides", or where the soil contains a large amount of organic matter. Sulphates such as gypsum react with lime and their reaction causes swelling and a successive reduction in strength. The detailed chemistry of the reaction between the lime, the sulphate and the soil minerals is very complex and depends greatly on the pH of the soil, the temperature and the water conditions. A closer look at this process was e.g. by Hunter (1988). Significant swelling of lime stabilized soils is attributed to the creation of a mineral (ettringite), one of the calcium-aluminium-sulphate hydrates. The hydration of ettringite is accompanied by large dimensional changes as the water of crystallization is incorporated into the mineral structure. Ettringite often occurs as long needle like crystals and, where the interparticle clay bonding is weak and the ambient stresses are low, these crystals can push the clay particles apart. In addition, the crystal growth will probably weaken the physico-chemical bonding between clay particles in the soil and permit them to swell under the low ambient stresses in the presence of water.

Note: Holm et al. (1983) mention a significant growth in the strength of clays stabilized with a mixture of lime and "gypsum" as compared to stabilization with pure lime in the application on lime columns. Tests were carried out with contents of the binding agent between 8 and 13%. The reliability performance of these columns, however, vastly differs from stabilized compacted soil as for the columns lateral expansion is beneficial. The expansion increases the lateral earth pressure in the soil and causes the soil to consolidate with a reduction of the water content as a result. The increase in the bearing capacity of the soil and the reduction of the settlement are caused partly by the reduction in the water content and partly by the hardening of an annular zone around the lime piles – Holeyman and Mitchell (1983). The authors also point out that the diffusion of lime through intact clay is very slow, about 10 mm/year.

Therefore, it is very important as early as the research stage to determine the amount of sulphates or their future creation. The most commonly stated boundary value is the sulphate content of 1%. The ability of ground water to penetrate lime stabilized soil is highly dependent on the permeability and hence the compaction is of fundamental importance. It is essential that the treated soil is compacted with no more than 5% of air voids and that close attention is focused on the moisture content at the time of compaction to ensure that this is achievable.

A great focus is generally placed on the technology of processing stabilized soil. Great emphasis is on the time between mixing lime with soil and final compaction referred to as – a mellowing or maturing period. Following each mixing pass the material is trimmed and lightly compacted. The requirement for light compaction is justified by Smith (1996) in the following way:

- It brings the lime into intimate contact with the clay,
- It minimizes evaporation loss,
- It reduces possible damage from rain,
- It reduces the risk of lime carbonation.

Carbonation occurs when the lime reacts with carbon dioxide in the air and reverts back to calcium carbonate on long time exposure, before it has had time to react with the clay. If this were to occur, it would obviously reduce the amount of lime available to improve the soil.

The maturing period follows, mostly required in a range of 24–72 h. During this maturing, the processes mentioned within the lime modification phase occur. Discussions concerning this problem often point out that there is no exact theoretical background for this requirement, which, moreover, is difficult to fulfil in some cases. An example is a repair of the railway track subgrade using heavy machinery performed only with partly lifted rails where individual reconstruction phases follow in a fast sequence.

After the maturing period, new mixing, addition of water to the required value and final compaction follow. For large-scale projects, the use of a compaction trial is worth while. Sheep-foot rollers can be used for thicker layers, but they must be followed by smooth rollers. The final surfacing is protected from drying e.g. by sprinkling with water. Traffic may be allowed on the treated layer, preferably after 2–3 days' curing, provided it is not rutted or distorted by the loading.

Brandl (1999) mentioned the possibility of not only using lime for fine soils but a combination of lime with cement. Lime prepares soil for cement and should therefore be added at least some hours before cement. The overall effect of this combined stabilisation is to produce more workable material by breaking down large clods of clay and reduce plasticity with lime and increase strength and freezing-thawing resistance with cement. Mostly, about 2–3% lime is sufficient; a smaller portion can hardly be distributed and mixed homogeneously. A combined lime-cement stabilisation is especially recommended if heavy clays should be stabilised with lime during late autumn and are afterwards exposed to frost. Fresh lime stabilisation may be even more susceptible than the untreated soil. But adding cement produces a material with more early life strength and frost resistance than treatment of soil by lime alone.

4.3 Soil Reinforcement

In accordance with the introduction to this chapter, in this part two main methods of soil reinforcement will be specified in more detail – geosynthetic soil reinforcement, which is in principle suitable for embankments and soil nailing reinforcement, which is used to reinforce natural soil as a block, mainly for stabilization of slopes in cuttings.

4.3.1 Geosynthetic Soil Reinforcement

Overview of the possibilities of geosynthetic soil reinforcement is very wide, e.g. Ingold (1995) or Jones (1996). As it is not the purpose of this publication to do the same, we will stress only the main types. At the same time it is necessary to mention that reinforced soil is a composite of soil and reinforcing element of totally different properties, respectively composite between soil, reinforcement and facing elements for the case of permanent geosynthetic-reinforced soil retaining walls. Therefore apart from the main examples, the attention will be focused also on these three components and for the reinforcement mainly from the strength and creep properties point of view.

4.3.1.1 Typical Application Examples

Main type of application is the *reinforcement of soil embankment*, e.g. Ingold (1983), Vaníček (1983b). In principle we are solving the alternative for use of available soil and or design of reinforced slope. Reinforced slope allows steeper slopes and hence reduction of the amount of soil required for the construction of

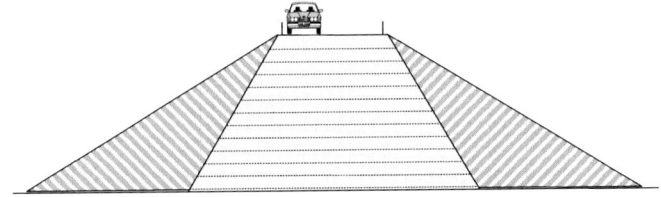

Fig. 4.21 The differences in cross sections for reinforced and unreinforced embankments

the embankment or utilization of soils that would otherwise be categorised as potentially suitable or unsuitable. The use of reinforcement is reducing the amount of fill material, which is sustainable for the case of reduction of the transport requirements. Nowadays, when the price of land is very high or its acquisition is difficult, it is not neglectable (sometimes even decisive) that the reinforced slope requires significantly lower appropriation of land, see Fig. 4.21. On the other hand we have to counterbalance these savings by the price of the reinforcement and its installation costs.

The form of reinforcement can have different character, see Fig. 4.22, the reinforcing elements are placed horizontally (a), or with wrap-around at the end (b), which helps with the protection against surface erosion. If the thickness of compacted layers is lower than the vertical distance between reinforcing elements, it is possible for half thick compacted layer use wrap-around only to the higher layer (c). While compacting the next layer, the reinforcement can be activated more effectively.

For the slopes, where the failure on shallow surface slips is critical, it is not necessary to install the reinforcement through out the whole cross-section of the embankment (so called internal reinforcement), but only close to the edges (so called edge reinforcement) or combine both alternatives, see Fig. 4.23.

The above mentioned solutions assume that the subsoil underneath the embankment is good enough that the critical slip surface that would pass through the subsoil has higher factor of safety than the one that passes only through the embankment. When constructing the embankment over soft subsoil, one of the alternatives how to realize the embankment is to *reinforce the contact between embankment and soft subsoil*. On Fig. 4.24 several alternatives of this application are presented. The effectiveness of the reinforcement is combined with gravely layer. In the last alternative the reinforcement in two layers is wrapped and overlapped and creates reinforcing distributing cushion. To this cushion on soft subsoil there are reservations based on field measurements, because lower reinforcement is much more heavily loaded

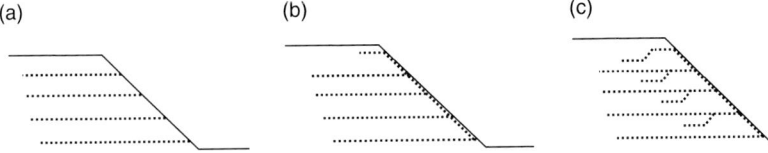

Fig. 4.22 Basic types of slope reinforcement

4.3 Soil Reinforcement

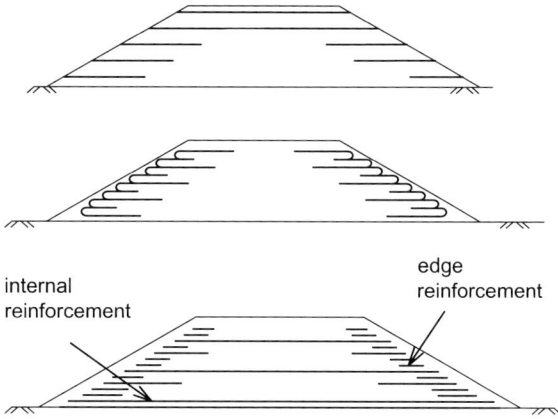

Fig. 4.23 Basic types of embankment reinforcement

compared with the higher layer and underestimation of this can easily lead to the rupture of the lower reinforcement.

The availability of constructing very steep slopes leads to the construction of *soil reinforced retaining walls*. Apart from significant savings on fill material and land, to their construction helps also so called dry method of construction, when during the construction only soil and reinforcements are used (Fig. 4.25a) and also already made concrete prefabricates. Concrete prefabricates are in this case used as facings and could be made of large elements requiring mechanization for their installation, or small elements that can be moved by a man (Fig. 4.25b). Both main alternatives have several sub-alternatives, some of which are described in more detail in Chapter 6. In some cases the emphasis is placed on the protection of wall face, especially if it is made of geosynthetics as it is more vulnerable to either natural or human activities. Certain combination of slope and retaining wall is the case

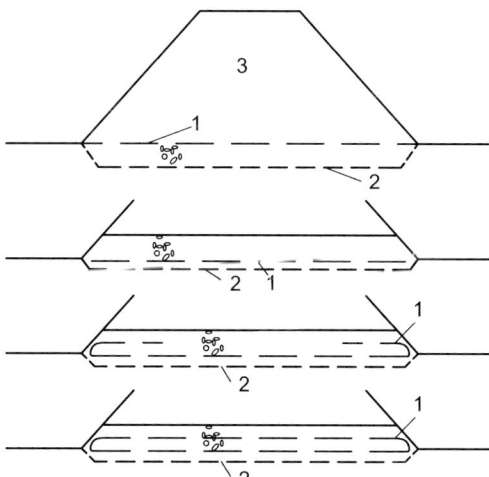

Fig. 4.24 Reinforcement of contacts between embankment and soft subsoil
1 – Geosynthetic reinforcement;
2 – Foundation level;
3 – Embankment fill

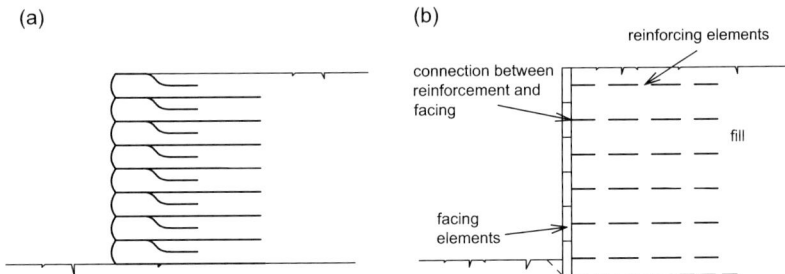

Fig. 4.25 Examples of vertical slope reinforcement

presented on Fig. 4.26a. Here the construction of the soil reinforced retaining wall allows installation of embankment on top of which road is situated. Another alternative shows the construction of the road in mountainous area, when one lane is made with the help of soil nailing and the other lane is constructed from soil reinforced embankment in which the soil from cutting for the first lane is used (Fig. 4.26b).

Specific example of vertical wall is *reinforced soil bridge abutment*, schematically shown on Fig. 4.27.

From several other cases of reinforcement it is possible to mention the use of reinforcement over openings, e.g. cavern, place where we can expect development of such opening above mined or karstic areas etc. Also in the area of environmental structures, mainly landfills, the reinforcements are used for the risk lowering of the slippage of the soil layers above smooth geomembranes. Even in the area of hydraulic structures there exist nowadays more examples of reinforcement use, however potential risk exists here that the contact between soil and reinforcing element would be the preferential pathway for water leakage, which can have a negative impact on the sealing role of the reinforced earth structure.

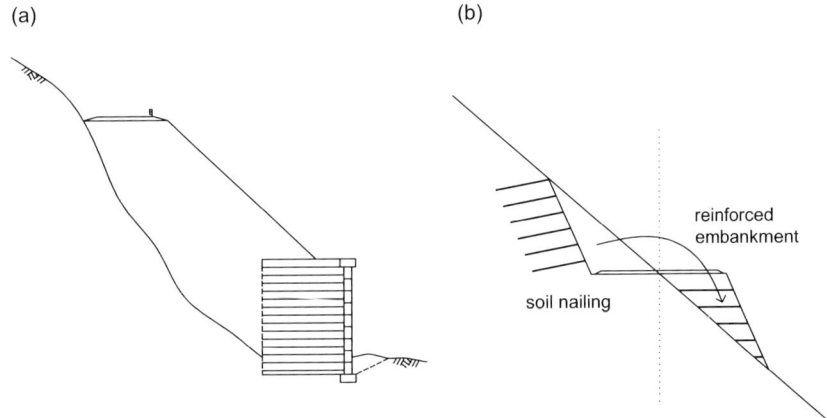

Fig. 4.26 Construction of the road in mountainous area. (**a**) Reinforced retaining wall supporting road embankment. (**b**) Combination of reinforced soil embankment and soil nailed cutting

4.3 Soil Reinforcement

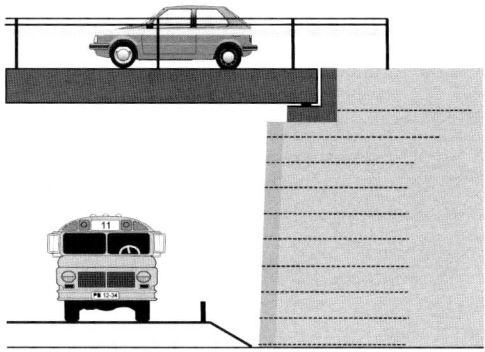

Fig. 4.27 Scheme of reinforced soil bridge abutment

It is possible to describe the example in more detail when reinforcement allows the construction of new structure on soft subsoil. Three main cases are presented in Fig. 4.28. The first two represent reinforcement of soil underneath shallow foundation and have been investigated in laboratory by the author using stereophotogrammetric method – Vaníček (1983a). The third example often occurs in cases when part of the large production shop is founded on embankment and structure columns are founded on piles. For reduction of floor deformation in between individual piles, reinforcing elements are installed there, which will increase deformation modulus of subsoil, obtained from plate load test, and hence help to reduce the differential deformations of the designed floor.

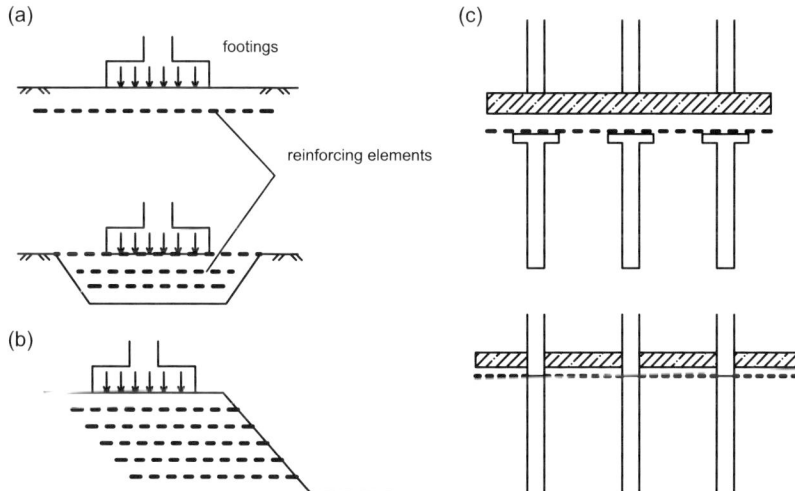

Fig. 4.28 Examples of shallow foundations on reinforced soil. (**a**) Footing on reinforced backfill of a trench. (**b**) Footing on reinforced slope. (**c**) Slab foundation over load transfer platform from piles and geogrid

4.3.1.2 Suitability of Soils

On the suitability of soils used for their reinforcement can be several views:

- With respect to the interaction with the reinforcement are more suitable soils that have properties similar to the reinforcement, mainly from strength mobilisation point of view. Generally noncohesive soils have higher deformation modulus as well as higher initial shear modulus which prefer them from this point of view and more over in most of the cases the shear strength of contact between soil and reinforcement is higher for coarser materials, which is especially true for geogrids.
- With respect to the economy it is better to use local soils or even waste materials although these soils are likely to have inferior properties, the main disadvantage of cohesionless fills is that it would usually be imported material and therefore, might, be costly.
- With respect to the importance they have on given composite. Here we have to realize that for the case of reinforced slopes the reinforcement has often complementary function, as without reinforcement the slope is close or just above indifferent stability and the reinforcement is increasing the factor of safety towards about 1.5 which is usually required. On the other hand for faced reinforced soil structures, where soil structure without reinforcement would be unstable, a better quality fill is likely to be specified.
- With respect to the drainage of the structure – coarse soils have higher permeability and do not require additional drainage system.
- With respect to the frost susceptibility again coarse soils are preferable.

From the above comparison coarse soils are preferred, while economical or environmental point of view can shift the soil type preferences. Practically these days there is no reason to exclude any soils, only we have to account for them in the design of earth structure. When selecting coarse soils it is possible to use apart from crushed gravel, river sands and gravels also soils with low proportion of fine grained soils, which should remain below 35%, which is the boundary when coarse particles are still in contact and affecting more the soil behaviour. In principle based on the classification from Chapter 2 these are soil types starting with symbols G or S. Of course it is better to use only sands and gravels that have proportion of fine particles lower than 15%, this is because of drainage and frost susceptibility point of view. Another advantage is in their higher unit weight, because reinforced earth structures are in principle gravitational structures.

The use of fine-grained soils in reinforced embankment, e.g. Battelino (1983), does not cause any problems, however reinforcing elements from geotextiles (woven or knitted) are preferred because of shear strength on contact over geogrids. Higher attention has to be paid when selecting fine-grained soils for geosynthetic-reinforced soil retaining walls. The reasons being, behaviour in short-term under undrained conditions, requirement for drainage system and need for protection against freezing. Post construction horizontal deformations can also have their influence. Tatsuoka et al. (1996) presents an extreme example. Here the backfill soil for the Nagano wall is a highly weathered tuff which was available on-site. This soil became nearly saturated soft clay after compaction with average water content of about 30% and a degree of saturation of 70%. The backfill soil was reinforced with a composite of non-woven and woven geotextiles, with short term strength at failure $35.3\,\text{kNm}^{-1}$ for which the tensile strain is 7%. The curiosity of this project was also highly compressible subsoil. Therefore for speeding-up of consolidation principle of preloading was used. Approximately double surcharge was applied over the reinforced embankment, which caused settlement of about 1 m. A full-height rigid facing was cast-in-place after 1 year after about 6 months of preloading.

From the high volume waste materials that can be used for reinforcement comes on force mainly mine waste and flying ash. In both cases it is necessary to chose an individual approach not only towards reinforcement but also on the effect of leachate from waste on reinforcing element. The environmental point of view will be explained in more detail in Section 6.8.1, while discussing the use of waste materials in embankments of transport infrastructure.

4.3.1.3 Suitability of Geosynthetic Reinforcements, Importance of Creep

When selecting the suitable geosynthetic reinforcement we have to account for:

- Type of soil that will be in contact, influence of contact friction properties, need for yarn protection from mechanical damage by soil (mainly from sharp edges), respectively assessment of chemical resistance if in soil or pore water are present chemical substances that can cause certain degradation of the reinforcing element. Due to the last two reasons, some products are impregnated by additional material, which is more resistible to mechanical and chemical influence.
- Stress-strain diagram of the reinforcement – relationship between tensile stress and strain – we do not compare only maximum strength at failure but we also acknowledge the deformation of the earth structure in relation to the maximum elongation at failure. The comparison in this matter is presented for different geosynthetic materials by Jones (1996), see Fig. 4.29. The highest short-term strength at lowest elongation have polyaramid (aramid) fibres, followed by polyester fibres, polypropylene strips and finishing with HDPE grids.
- Creep properties of geosynthetic reinforcements, because for long-term function of reinforced earth structure we have to proceed from reduced working diagram, as for slow, long-term load tests generally the tensile strength is lower for the same elongation.

Creep is defined as deformation of the material with time under constant load. Short-term tensile test is usually performed within minutes, while the reinforcement in the structure is usually fulfilling its function in long-term, about tens of years. Creep has different character not only based on selected material, but also based on the stress level, i.e. ratio between current level of load and the short-term strength from tensile test. It is possible to differentiate creep, Fig. 4.30a:

Fig. 4.29 Working diagrams of different materials used as soil reinforcement (according to Jones)

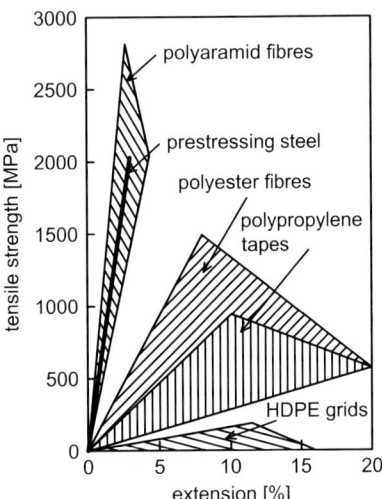

- Stable creep, when the speed of deformation is lowering with time, and
- Unstable creep, when the speed of deformation is increasing with time – up to failure.

Therefore, in order to fulfil the long-term function of the geosynthetic reinforcement, the creep must remain stable. Hence the tensile strength design value T_d must be in the range between about 25 and 50% of short-term tensile strength. The higher range is applicable for the reinforcement from polyester, aramid and polyvinylalcohol, while the lower range is applicable to the polypropylene and polyethylene. When using logarithmic scale for time (Fig. 4.30b), we can derive linear relation between time and elongation and the equation can be written as:

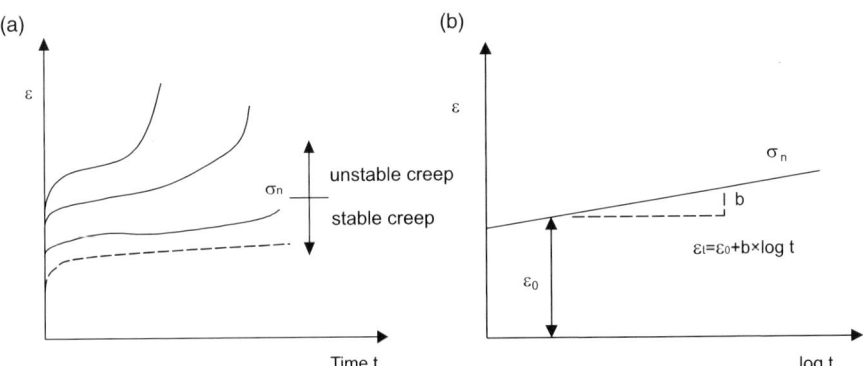

Fig. 4.30 Presentation of strain as a function of time. (**a**) Stable and unstable creep. (**b**) Scheme of creep coefficient b determination

4.3 Soil Reinforcement

$$\varepsilon_t = \varepsilon_0 + b \log t \qquad (4.25)$$

where ε_0 is initial elongation, t is time and b is creep coefficient. From this relationship it can be seen that we do not have to know only the initial elongation from the short-term tensile test, but we need to take care also about the long-term deformation based on creep coefficient b. The creep coefficient b depends not only on the raw material of the reinforcing element, but also on the stress level of the geosynthetic.

The comparison of the creep coefficient b for selected range of stress levels and four main raw materials that are used for production of the geosynthetic reinforcements – polypropylene, polyethylene, polyester and aramid is shown on Fig. 4.31 according to Jones (1996). From this figure it is obvious that creep coefficient increases with stress level increase and presented variation points out the importance of manufacturing for individual polymer. Polypropylene is the most sensitive to the manufacture. In principle, it is obvious that aramid will experience the smallest additional elongation, followed by polyester, polyethylene and polypropylene with worst creep properties. The similar conclusions are also presented by Watts et al. (1998), who realised huge amount of creep tests of geosynthetics. From this understanding we can conclude that if products from different materials have short-term strength of e.g. $100\,\mathrm{kNm^{-1}}$, then for polypropylene or polyethylene we can allow in the design for creep reduction only about 20%, i.e. $20\,\mathrm{kNm^{-1}}$ and for polyester about 40%, i.e. $40\,\mathrm{kNm^{-1}}$. Therefore we should not compare different products only via short-term strength, but via design strength already reduced for creep.

The danger of higher reinforcement elongations after their installation in the earth structure can be seen on the relationships between force and strain for different times, so called isochronous curves. The producer with longer tradition can already present results of such tests from the last up to 30 years, with very reliable extrapolation even for longer design times. On Fig. 4.32a are presented results for polyester geotextile, recalculated from information presented by Watts et al. (1998). The shape of these curves shows that to the normally considered stress level, i.e. about 40%, the

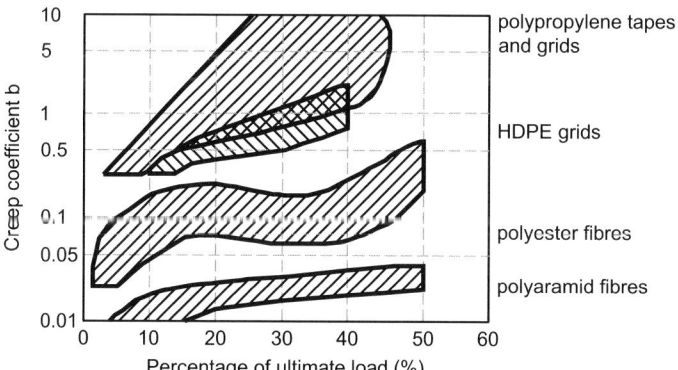

Fig. 4.31 Creep coefficient b versus percentage of ultimate load for various geosythetic reinforcing elements (according to Jones)

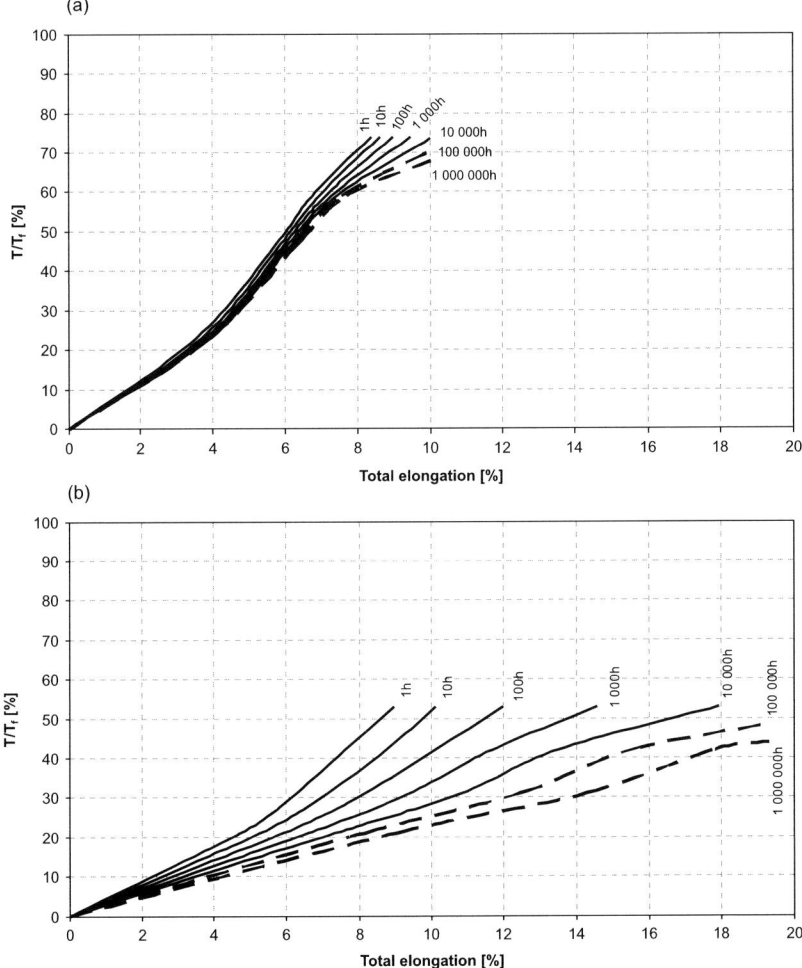

Fig. 4.32 Isochronous creep curves. (**a**) Polyester geotextile. (**b**) Polypropylene geotextile

difference in elongation is very small, practically below 1%. This means that once initial deformations are realised the additional ones will be small, within 1%. On the other hand for geotextile from polypropylene (Fig. 4.32b) the additional deformations are much higher and for the stress level of 20% they can be as high as 4%.

These values must be accounted for when selecting reinforcing material for certain type of structure, as some regulations have this requirement. For example BS 8006:1995 Code of practice for Strengthened reinforced soils and other fills requires maximum post-construction elongation of 0.5% for reinforced soil bridge abutments and 1.0% for reinforced soil retaining walls.

We would like to note here that in hotter climates than in moderate climate range it is necessary to include the influence of temperature on the creep behaviour of geosynthetics.

Another note we would like to make here is that in order to engage the reinforcement in action since the beginning of loading, it is necessary to lay down the planar reinforcement without visible waves and irregularities, otherwise the engagement would be delayed till certain soil body deformation, which is schematically shown on Fig. 4.33. Therefore it is preferable to take extra care in the installation and in important cases to even try to prestress the reinforcement. This prestressing can be realized systematically during the reinforcement installation (see also Fig. 3.20) on the compacted soil layer, when the reinforcement is tensioned and anchored or new soil is placed over the ends so the tension could be maintained. The result of such pretensioning on the working diagram change is also presented on Fig. 4.33. Certain level of tensioning can be gained by backfilling and compacting next soil layer over properly laid reinforcement over uneven surface of previous compacted soil layer.

Higher, focused form of prestressing is mentioned by Tatsuoka et al. (1996). He describes examples of reinforced soil bridge abutments that have been realized in Japan in significant numbers, however with the largest span of 13.2 m. In order to support longer and hence heavier beams the reinforced soil abutments would have to be stiffer. Whereas the reinforcement has mainly a horizontal effect on the surrounding soil, it is more difficult to increase the vertical stiffness even when using stiffer reinforcement in layers of small vertical distance. Therefore, new construction method that significantly increases vertical stiffness due to prestress of vertical reinforcement was proposed by Tatsuoka et al. (1996) and is shown in Fig. 4.34. Before the construction of the prestressed reinforced earth structure, in this case bridge abutment, the anchorage block together with four anchorage bars is built. Alternatively it is possible to use good quality subsoil for additional anchorage. Once the reinforced earth structure is finished, but still without full-height rigid facing, the top anchorage block is constructed as well. And then the abutment wall is preloaded and prestressed using the following approach:

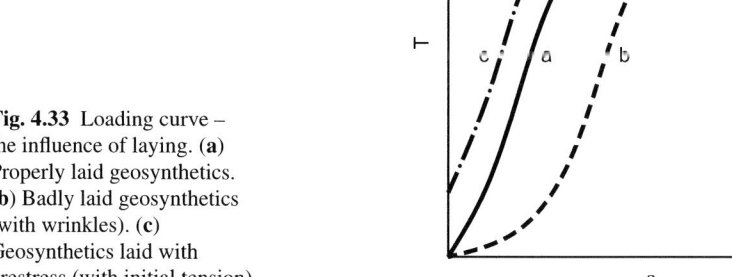

Fig. 4.33 Loading curve – the influence of laying. (**a**) Properly laid geosynthetics. (**b**) Badly laid geosynthetics (with wrinkles). (**c**) Geosynthetics laid with prestress (with initial tension)

Fig. 4.34 Typical preloaded and prestressed geosynthetics reinforced soil retaining wall (according to Tatsuoka et al.)

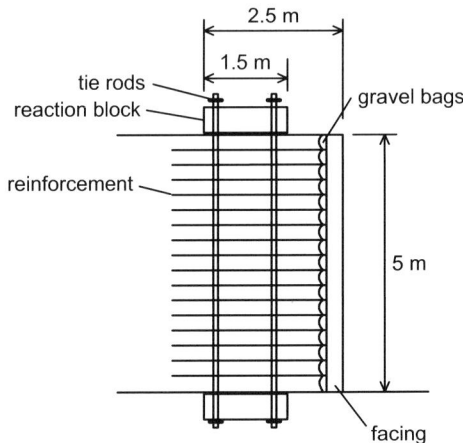

a) High enough preloading is generated by hydraulic press mounted on each of the anchorage bars. This relatively high preloading is possible because the earth abutment is reinforced. Preloading and subsequent partial unloading gives to the abutment almost elastic behaviour during the subsequent loading P_c on the top.

b) After the partial unloading to certain level, the hydraulic jacks are removed from the anchorage bars and they are fixed into the top anchorage block. The remaining tension in the anchorage bars acts as prestressing, which is equalized by tensile force on the top of the abutment. This prestressing gives to the abutment much higher vertical stiffness compared with the abutment without it. For long-term function of this advantage, it is necessary to maintain the prestress during the whole life-time of the abutment.

c) When applying external surcharge P_c on top of the anchorage block, the tension in the anchorage bar is reduced as the abutment is compressed. The anchorage bar acts as a compressive structural element against force P_c.

d) At the same time this vertical prestress activates additional force acting between horizontal reinforcements and surrounding soil, which enhances their reinforcing effect. This prestress can this way contribute to the additional integrity of the whole abutment.

After the model field tests, first application on railway line "Sasaguri Line" was planned. In Japan they expect also a beneficial effect from these structures during the earthquakes, as this type of structure can more easily transfer additional dynamic loading compared with other types of structures.

4.3.1.4 Facings

Certain need to protect the facing always exist for permanent geosynthetic-reinforced soil retaining walls. The main reason is to avoid surface erosion and also a certain requirement on architectural appearance. Among the main facing types we can include:

- Wrapped around from reinforcing geosynthetic,
- Face made of geosynthetic bag filled with soil, when the reinforcing element is placed in between the bags,
- Similar face but from gabions instead of bags,
- Face made of large concrete prefabricated panels,
- Face from small concrete units (sometimes using lightweight aggregates, e.g. expanded clay), weight of proposed units can be within the lifting capacity of one man,
- Concrete facing cast on full height,
- Prefabricated panel on full height that needs propping during the construction,
- Panel made of plastic materials or steel.

As an interesting example of the use of waste materials we can mention face made of old tyres.

When selecting the type of facing and its evaluation, we need to consider the requirements on mechanisms for construction, maintenance, sensitivity to additional deformations, vulnerability to failure (including vandalism), aesthetics, etc. Generally, large prefabricated panels that do not have continuous horizontal joint are mentioned as the most common type. For smaller heights the preferred type are small concrete blocks. As already mentioned, cast concrete facings on full height are very popular in Japan.

Concerning the horizontal deformations and stresses acting on the wall facing we can distinguish different types based on stiffness. Discussion on this subject is going in two directions. For the flexible facings, the deformation occurs mainly at the wall face and in the backfill immediately behind the wall face. This can be on one side unfavourable as the deformations on the face are bigger, while on the other side the deformation leads to lowering of the surcharge on the wall down to active earth pressure with very small force transferring to reinforcement at the connection. Conversely, for the rigid wall the earth pressure can be around earth pressure at rest. At the connection between facing and reinforcement it is necessary to achieve certain initial force. Tatsuoka et al. (1996) are convinced that the precast concrete on full height facing can accommodate both advantages. This is because the deformations of the wall face and surrounding soil are mostly finished before the facing is made and hence very good connection between concrete facing and reinforced soil can be achieved.

4.3.2 Soil Nailing

Soil nailing is generally classified as a method of soil improvement in cutting, during which quasi-homogeneous body is created. The method consists of the installation of nails into soil body. The nails are generally steel bars or pipes that are inserted into the drillhole and grouted or driven into the soil using different technologies.

Nails are usually installed subhorizontally. It is suggested to install the nails of the same length and if required for reduction of horizontal deformations to have the top ones longer. The installation the other way round is not suggested. Aside from these main cases, Schlosser (1997) or Brandl and Adam (2004) mention other alternatives for soil nailing, like slope stabilisation, stabilisation of tunnel excavations in soft soils or capacity increase for foundations. In Fig. 4.35 not only the main soil nailing applications are shown but also the type of soil nail loading for each application.

The first example of soil nailing was mentioned the 18 m high steep slope in Versailles constructed between 1972 and 1973. Advancement in understanding of the function and design is credited to the large-scale experiments in Germany – Stocker et al. (1979) and to the French research project "Clouterre" from 1991 and its English version from 1993, prepared by the team of Professor Schlosser. Most recently a document has been prepared in the UK by Johnson and Card (1998) and by Phear et al. (2005) – CIRIA C637 – Soil nailing: Best practice guide.

Stocker (1994) summarises 40 years of micropiles usage and 20 years of soil nailing, and describes their differences. He also summarises some general findings for soil nailing:

- In Germany alone they realised until 1992 about 140000 m^2 of soil nailed walls,
- Soil nails in retaining structure are transferring the load via tension,
- Even when the longest soil nails reached 26 m, the most common length is between 2 and 8 m with the spacing in the range between 1 and 3 m,
- Soil nails are often combined with anchors,
- Vibration caused by traffic loading on the top of the slope are not decreasing the slope stability,
- Eight soil nailed walls in California survived the earthquake in 1989 without problems even when the horizontal acceleration reached 0.47 g.

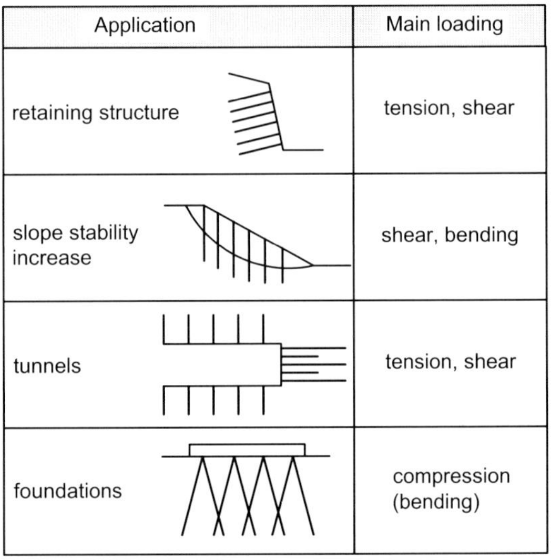

Application		Main loading
retaining structure		tension, shear
slope stability increase		shear, bending
tunnels		tension, shear
foundations		compression (bending)

Fig. 4.35 Main types of soil nailing applications (according to Schlosser, Brandl and Adam)

4.3 Soil Reinforcement

At the beginning soil nailing was thought to be mainly a tool for temporary stabilization of vertical face of cuttings. Driven steel nails have hence been fulfilling only a temporary function and there was no need to take care of the corrosion protection. Nowadays most used are soil nails of 20–30 mm in diameter inserted into the drillholes of 90–120 mm in diameter, which are grouted with anchor type head, the detail of which is in Fig. 4.36. Drillholes are routinely realised in soils that allow at least temporary stability of the hole walls. Grouting helps not only with the corrosion protection of nails but also improves the connection between steel and soil.

French soil nailing guidelines distinguish those two mentioned types.

The first one, sometimes called the Harpin method, is using driven nails with lower capacity (120–200 kN) and length of about 0.5–0.7 H and small (below 1 m) spacing in both vertical S_v and horizontal S_h directions.

The second one concerns grouted nails with higher capacity (200–600 kN) and length of about 0.8–1.2 H and larger spacing (max one nail in 6 m^2). If the spacing is even bigger, i.e. one nail per more than 6 m^2, the wall is not considered as nailed.

Corrosion protection of nails consists of grouting, nail bar protection, e.g. by galvanisation, insertion into corrugated plastic sheaths from PVC. Also location in the drillhole by centralisers can be seen as protection. The use of non-steel nails that do not corrode, e.g. glass fibre nails, is possible and the advantages are in higher tensile strength and lower weight, however they are still more expensive compared with steel ones.

Concerning grouting there are two procedures in practice, Gässler (1991):

- The borehole is fully grouted shortly before nail setting,
- The borehole is fully grouted soon after nail setting.

The first procedure is recommended when conditions in the ground are not stable. Soon after drilling, the hole should be filled with grout (ratio water/cement not more than 0.4). Immediately afterwards the nail has to be completely inserted into the grouted hole before the grout starts to set. If possible, the second procedure should be preferred. Beginning from the deepest point of the borehole the grout is filled in through a lance held by hand or through a thin pipe fixed along the nail as a whole. In

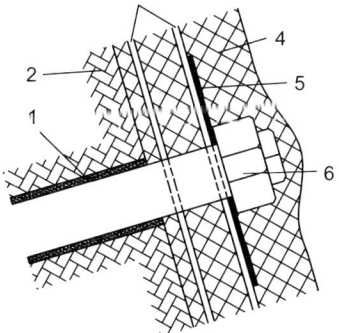

Fig. 4.36 Detail of grouted soil nail head
1 – cementitious grout;
2 – soil; 3 – steel mesh;
4 – shotcrete; 5 – head plate;
6 – nail head

special cases a second pipe can be fixed, through which a second cement injection is possible using higher pressure. According to Gassler this method of "post-injecting" used in ground anchoring, is sometimes used in Germany to increase the nail/soil bond in poor ground conditions.

Typical technological construction sequence (see e.g. French National Project Clouterre) is presented on Fig. 4.37 and has three main phases:

- Excavation, usually 1–2 m depth, typically limited also in width by the type of soil,
- Installation of nails usually sub-horizontally, but depending on the method used,
- Creation of wall facing to improve its stability in between nails using shotcrete or prefabricates.

In soils with small cohesion, when there is a danger of instability of the new section of excavation, it is possible to swap the phases, first stabilising the soil by shotcrete and afterwards installation of nails. However, usually the cohesion of soil around 4 kPa is already sufficient. Before application of shotcrete the soil is cleaned and reinforcing steel mesh with mesh size of 100 × 100 mm is fixed. One layer of reinforcing mesh is used for the concrete slabs of about 100 mm in thickness. For thicker slabs usually two layers of reinforcing mesh are used with minimum cover of 30–40 mm at both sides. The reinforcing mesh must be well laid (tensioned) in order to eliminate unwanted vibrations during shotcreting.

The drainage of the wall underside is very important, and this can be achieved by installation of different drainage geosynthetics before mesh installation or by subsequent drilling of the weepholes. Soil nailing technique is not suitable for permeable soils with ground water level above the wall toe.

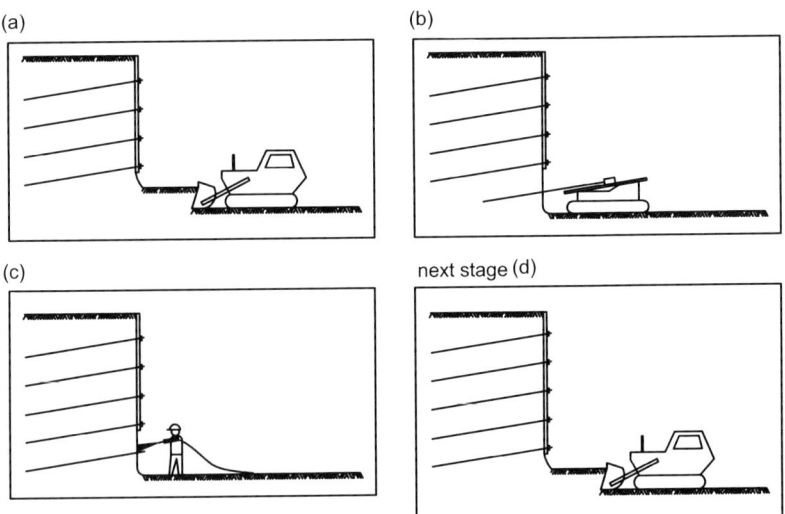

Fig. 4.37 Technological steps of the soil nail wall construction (according to Schloser et al. – FNP Clouterre)

4.3 Soil Reinforcement

Alternatively it is possible to secure the excavation face by steel fibre reinforced shotcrete. Special properties of such material as durability and flexibility, higher tensile strength, minimum tensile strength after fissuring and long durability against frost-thaw-effects, should be useful in soil nailing. According to Gassler the portion of fibres amounts from 90 to 110 kg per cubic meter, which corresponds to 4–5% of the total weight of one cubic meter of the composite.

Both explained cases how to secure the excavation facing assume that the vertical cut in soil will remain stable for only limited time and needs immediate support after excavation. If the situation is better we can also use the application of prefabricated facing elements. Much attention is attributed to this case by Pokharel and Ochiai (1997) and they call this method – PAN Wall Method (Panel and Nail). The construction procedure has the following steps:

- Soil is excavated only for single row of panels in each step to ensure easy attachment of panels and also to secure the overall stability of the uncovered soil until completion of the row.
- Panel positioning – panels are usually positioned in row by row steps. But, cutting of space even for a single panel and immediate panel positioning is possible for poor sites.
- Drilling, nailing and grouting – it is quite similar to the conventional soil nailing practices.
- Concrete filling – the contact between back of panel and the slope surface behind the panel is confirmed by filling the gap with concrete or mortar.
- Reinforcement tightening – the head of the reinforcing nails and panels are tightened by nut-bolt connections.
- Preparation for next row.

Schematic view and detail of nail connection is shown on Fig. 4.38. Prefabricates are allowing for better appearance of protected wall.

In the case where we are trying to stabilise existing slope or we are making it slightly steeper, it is possible to use soil nailing and the space in between the nails safeguard using erosion protection mattress, which can be strengthened by reinforcing element,

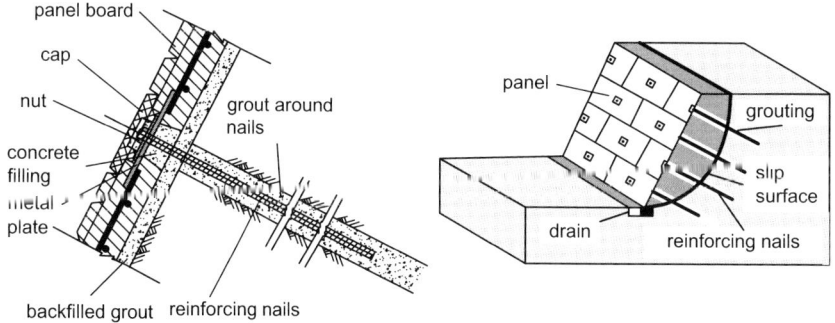

Fig. 4.38 Schematic view of the PAN Wall Technology and details of main components (according to Pokharel and Ochiai)

e.g. composite of geogrid with erosion protection mattress. 3D structure of the mattress is filled with topsoil to speed up the development of grass cover.

In principle the soil nailed wall behaves as monolith and basically it acts as a gravity wall. Therefore its design has two phases. In the first one, we have to design for external stability, i.e. stability of this monolithic block, checking the bearing capacity, overturning, sliding and overall stability along slip surface outside of soil nailed block. And in the second one we have to check the internal stability for tensile failure of nails and its pull-out from soil. Generally the soil nailed wall can fail also during the construction, for example when the lowest section prepared for shotcrete or facing panel is not stable enough and the soil fails – runs out. In theory we have to check also the case of shear failure between nail and grout and between grout and soil. Details of all these cases will be specified in future chapters. In principle the same checking scenario applies to the wall from reinforced soil. Even so both technologies, soil reinforcement and soil nailing, are very similar, significant differences can be seen in their deformation, mainly due to different technologies of construction, when the reinforced soil is constructed from bottom up and while soil nails the other way round. Figure 4.39 clearly shows the differences in monolith deformations. Soil nailed wall, constructed from top to bottom, has the biggest deformation both horizontal and vertical at the top of the wall, whereas geosynthetic reinforced soil has the biggest deformations at the bottom part due to the highest load from upper layers.

Most in situ measurements on walls are consistent in the opinion that horizontal and vertical deformations at the top of the wall are more or less equal and are within the range between 1 $H/1000$ and 3 $H/1000$, where H is the wall height. This fact is presented on Fig. 4.40 according to Schlosser and DeBuhan (1991).

There are no available routine calculation models for the calculation of deformations. Hence analogy from laboratory models is used or nowadays even more FEM (finite element method) is used. If we are afraid that the displacement would be bigger than allowed, it is possible to add anchor near the top of the soil nailed wall as is shown on Fig. 4.41.

Even thought the deformations of soil nailed and reinforced walls are different, it is possible to measure and find a point on the reinforcing element where the tension is largest. This point divides the nail on active part behind the facing, where the shear is directed outside, and passive part – anchorage length – where the shear is acting against nail pull-out. The highest tension on the nail is located about 0.3 H from the wall facing (Fig. 4.42).

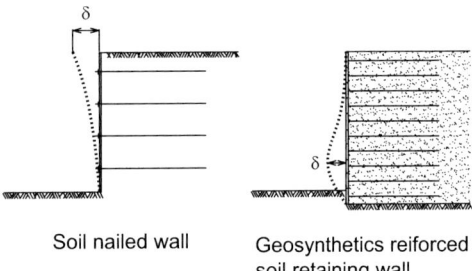

Fig. 4.39 Deformation profile comparison of soil nailed wall and geosynthetic reinforced soil retaining wall (according to Schlosser)

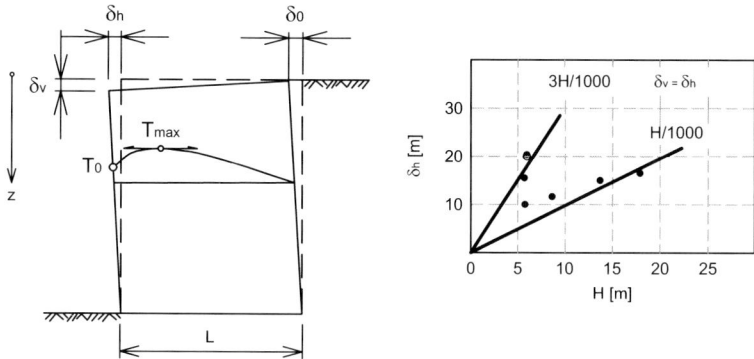

Fig. 4.40 Deformations and displacements occurring in a reinforced soil wall (according to Schlosser and DeBuhan)

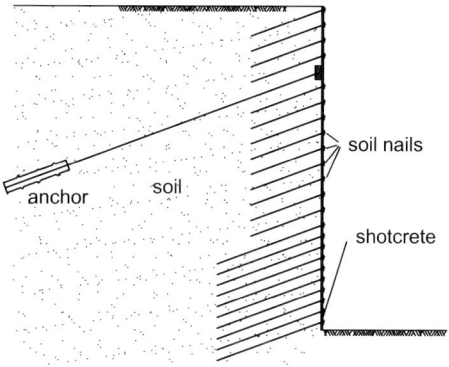

Fig. 4.41 Soil nailed wall supplemented by soil anchors for reduction of horizontal displacement

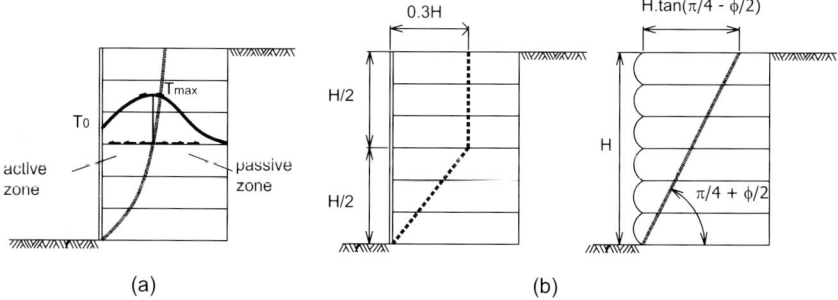

Fig. 4.42 Estimation of location of active and passive zones (according to Schloser). (**a**) For general case. (**b**) Idealization for soil reinforced wall design purposes

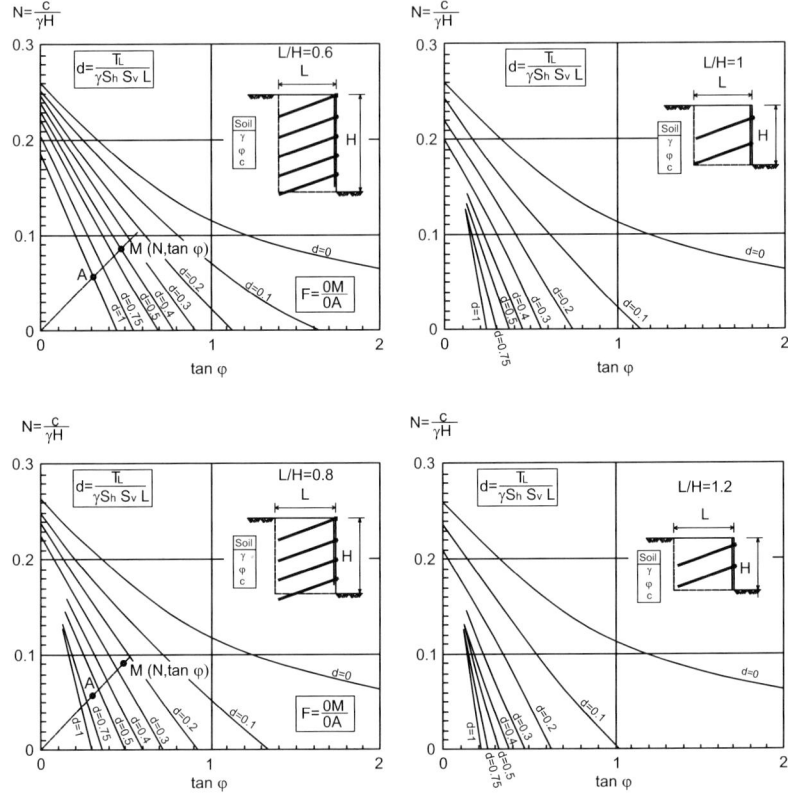

Fig. 4.43 Nomographs for preliminary design of soil nailed wall – French recommendation Clouterre

For preliminary design we can use graphs in Fig. 4.43, which are presented in French National Project Clouterre. Here for different soil properties (angle of internal friction φ and cohesion c) in relation to the wall height and recommended ratio L/H is given recommended "density" of soil nailing d, where:

$$d = T_L / \gamma \times S_v \times S_h \qquad (4.26)$$

where T_L is maximal force needed for nail pull-out that can be calculated as:

$$T_L = \pi . D . L_a . q_s \qquad (4.27)$$

where D is the diameter of drilhole for grouted nails or equivalent diameter of the driven nail (for arrow shaped nail the outer diameter of ribs is used), L_a is the length of the nail and q_s is unit shear strength on the perimeter, for grouted nails in sands and fine soils it is about 0.1 MPa, in gravels depending on injection pressure in between about 0.15 and about 0.45 MPa and for driven nails the value is lower, for sands about 0.08 MPa and gravels 0.04–0.06 MPa.

Chapter 5
Limit States for Earth Structures

5.1 Principles of Design Procedure for Earth Structures

The design of earth structures has gone through many changes in time. Roughly until the end of the nineteenth century, the designs that prevailed were based on practical experience gained to-date, i.e. on the experience of our predecessors, who had been building earth structures for centuries, millenniums. Their experience resulted in a set of different rules, recommendations, which gradually even found their way to some technical standards, technical codes. Very briefly said, we should exploit the "know how" of our predecessors.

These are, for example, recommendations concerning the allowable height of the vertical walls of earth excavations that can be built without protective measures, the recommended gradients of upstream and downstream slopes of low earthfill dams (e.g. up to 10 m) built as homogeneous of a certain type of soil, the measures to apply for erosion protection of slope both in embankments and cuts, or what measures to choose with a view to the protection of the earth structure from the negative effects of frost etc. This design procedure using the recommended measures is still respected, mainly in structures with low risks for their environment.

5.1.1 Risk Principle in the Design of Geotechnical Structures – Geotechnical Categories

Today, the basic criterion in the design procedure of earth and generally all geotechnical structures is that of risks. Thus, the risk principle comes to the foreground when considering the design of individual types of earth structures. This principle was applied in some Czech standards as early as 1967 and, in particular, in 1987 – see ČSN 731001 "Foundation soil under shallow foundations" where 3 basic approaches (geotechnical categories) were clearly defined, which combined both simple and complex foundation conditions, or less and more demanding structures. The first geotechnical category was the case of a combination of simple foundation conditions with an undemanding structure, the second geotechnical category referred to a combination of simple foundation conditions and a demanding structure, or an undemanding structure built in complex foundation conditions. Finally, the third geotechnical category was based on a combination of a demanding structure built in complex foundation conditions. These geotechnical categories were respected in the design principle as well, in the given case the design of spread footing; for 1st geotechnical category it was necessary to ensure that the contact pressure at the underside of footing would not exceed a value shown in a table for the respective soil and foundation depth. Second geotechnical category required verification through calculations of the basic limit states, the limit state of bearing capacity and the limit state of deformation, which, however, was done on the basis of characteristic soil parameters recommended for individual soil types and their consistency characteristics. It was only 3rd geotechnical category that required tests on undisturbed soil samples, and the sets of results then served for defining characteristic parameters of soil properties (mainly with a view to shear strength and deformation characteristics). These characteristic values were reduced using partial safety factors thus providing calculation values of soil properties to be used for the calculation of individual limit states.

Today, this principle is included even in Eurocode 7 Geotechnical Design – Part 1: General Rules of 2004. EC 7 e.g. states that: In order to establish minimum requirements for the extent and content of geotechnical investigation, calculations and construction control checks, the complexity of each geotechnical design shall be identified together with the associated risks. In particular, a distinction shall be made between:

– Light and simple structures for which it is possible to ensure that the minimum requirements will be satisfied by experience and qualitative geotechnical investigation, with negligible risk,
– Other geotechnical structures.

EC 7 does not specify individual geotechnical categories as explicitly as ČSN 731001, being based on a more general quantification. Geotechnical Category 1 should only include small and relatively simple structures, where there is negligible risk in terms of overall stability or ground movements and in ground conditions, which are known from comparable local experience to be sufficiently straightforward. In these cases the procedures may consist of routine methods for foundations design and construction. Geotechnical Category 2 should include conventional types of structures and foundations with no exceptional risk or difficult soil or loading conditions. It should also normally include quantitative geotechnical data and analysis to ensure that the fundamental requirements are satisfied. Therefore routine procedures for field and laboratory testing and for design and execution may be used. Further on, it gives examples of structures or their parts, including earth embankments. Also, it more closely specifies embankments for transport infrastructure and for small dams. Geotechnical category 3 then covers all other structures, which do not fall into Geotechnical category 1 and 2, these, in general, being "large or unusual structures, structures involving abnormal risks, or unusual or exceptionally difficult ground or loading conditions, structures in highly seismic areas or structures in areas of probable instability or persistent ground movements that require separate investigation or special measures". This also indirectly implies that high dams and earth structures of environmental engineering and some reinforced earth structures should be classified within Geotechnical Category 3. Discussion on the classification category should be led in the very first phases of the whole project as the extent and demands for geotechnical investigation are greatly adapted to this classification.

5.1.2 Ultimate Limit States and Serviceability Limit States

The main logic in designing earth structures and geotechnical structures in general consists of defining the following two conditions, which should guarantee long-term serviceability of the designed and implemented structure:

– The structure must be designed to avoid its total failure,
– The structure must fulfil its function on a long-term basis.

In the classic concept, this first condition means verification of the stability or load-bearing capacity of the earth structure, including its subsoil, and second verification whether its deformation, settlement, will be within admissible limits so as not to affect the serviceability of the structure or related structures. A more recent concept, which is based on design respecting limit states, specifies that in the first case a so-called first group of limit states is involved – ultimate limit states, while in the second case the second group of limit states – serviceability limit states. This specification procedure shows that there can be more cases leading to a failure or limited serviceability, and that these should not be limited solely to stability and settlement. The introduction to EC 7 sums this fact up into a simple sentence: For each geotechnical design situation it shall be verified that no relevant limit state is exceeded.

5.1.3 Actions, Design Situations

Loads. Load cases. In general, we may state that the stress state in the earth structure will not be constant, but variable in time. Therefore, it is necessary, in the first place, to split the load into two components, a permanent and variable load. The next step in specifying load components consists of distinguishing whether its effect is positive, having a stabilizing effect, or, on the contrary, whether it has a negative, destabilizing effect. For example, additional fill on a slope surface can have both a negative and positive impact on the slope stability – having generally a stabilizing effect in the lower part and a destabilizing effect in the upper part.

For earth structures in transport engineering, permanent load due to the weight of soil (filling) generally prevails, while variable load is, above all, conditioned on traffic loading. Nevertheless, short-term and long-term stability will be essential here, which is given by a different development of pore-water pressures during the construction of embankments and cuttings. During the construction of an embankment, pore-water pressures grow due to additional load, which fall (dissipate) in time, as fast as the favourable conditions for drainage exist. Therefore, short-term stability is of decisive importance, in terms of pore-water pressure development. In cuts, on the contrary, pore-water pressures decrease due to unloading but grow in time again, so the decisive factor will be the long-term situation. These problems will be explained in more detail in Chapter 6.

For earth structures in water engineering, water which is in direct contact with the earth structure will play a substantially more significant role. Therefore, for example, at least 3 basic loading states must be verified in earthfill dams, Fig. 5.1:

– Short-term stability immediately after the embankment completion – which is principally an analogy to short-term stability of a transport embankment,
– A state where the water level in the reservoir has been at the maximum level for a long time,
– A state where the reservoir is quickly being emptied.

In total, in defining loads and load cases, apart from the self-weight of soil, the following loads should be accounted for:

– Water loading, both in situations of steady and transient flow, or the force effect of seeping water affected by hydraulic gradient,
– Traffic loading,
– Snow, wind load,
– Additional loading due to vibrations and dynamic impacts,
– Ice, wave load,
– Additional loading or a change in loading due to creep, slide, settlement where mainly in earth structures built of materials with different moduli of deformation (e.g. dams with a central clay core) stress redistribution occurs in time due to so-called vault effect,

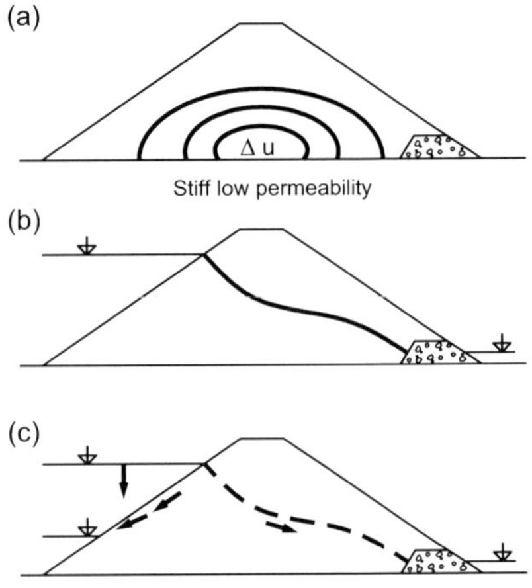

Fig. 5.1 Three main design loading situations for earth dam. (**a**) At the end of dam construction. (**b**) Normal situation with steady flow for maximum level. (**c**) Rapid drawdown

5.1 Principles of Design Procedure for Earth Structures

- Cyclic loading and loading with a variable intensity, they must be identified for special cases, e.g. for on-going movements, soil liquefaction, a change in deformation and strength characteristics,
- The effect of climatic conditions, in a dry period the earth structure dries up and negative pressure may develop in pores, which have a positive impact on stability, but after abundant rainfall the earth structure is re-saturated with water thus eliminating this effect,
- More emphasis has recently been placed on the assessment of the effect of extreme situations connected with natural disasters such as floods, extreme rainfall, earthquakes etc. The sensitivity of these effects in terms of the safety of earth structures, or a probability of their occurrence are therefore getting into the foreground of our interest.

In considering the worst load case, it must be assessed what combination of random loads to consider or for how long individual variable loads can act. This evaluation is very important as the assessment of the first group of limit states – ultimate limit states is based on short-term maximum loading, while the assessment of the second group of limit states – serviceability limit states must be based on long-term loading.

5.1.4 Ultimate Limit States

This first group of limit states in EC 7 includes the following cases:

- Loss of equilibrium of the structure or the ground, considered as a rigid body, in which the strengths of structural materials and the ground are insignificant in providing resistance (EQU);
- Internal failure or excessive deformation of the structure or structural elements, including e.g. footings, piles or basement walls, in which the strength of structural materials is significant in providing resistance (STR);
- Failure or excessive deformation of the ground, in which the strength of soil or rock is significant in providing resistance (GEO);
- Loss of equilibrium of the structure or the ground due to uplift by water pressure (buoyancy) or other vertical actions (UPL);
- Hydraulic heave, internal erosion and piping in the ground caused by hydraulic gradients (HYD).

For earth structures, the limit states with the greatest significance are those of GEO type, and for water structures also limit states of HYD type, even though the limit state of UPL type may also occur.

5.1.4.1 Limit States of UPL and HYD Type

Failures of UPL and HYD type fall under the category referred to as failures due to hydraulic effects – hydraulic failure. In situations where the pore-water pressure is hydrostatic (negligible hydraulic gradient), it is not required to check other than failure

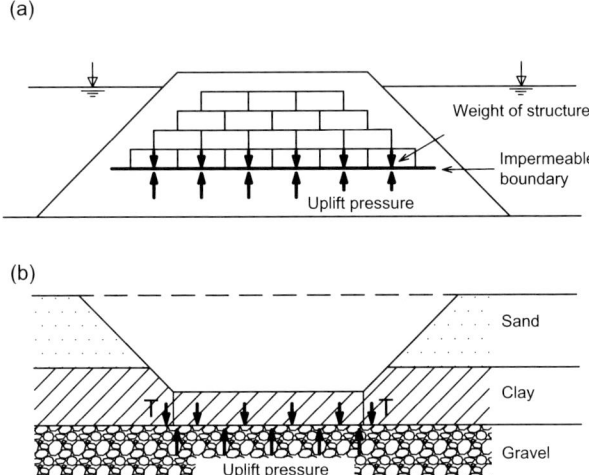

Fig. 5.2 Examples of situations where uplift might be critical (according to EC7-1). (**a**) Uplift of a lightweight embankment during flood. (**b**) Uplift of the bottom of an excavation

by uplift. In principle, stability at the impermeable level is assessed when overall soil gravity acts vertically downwards increased by potential friction resistance along the perimeter of the body susceptible to heave, and the magnitude of uplift by water pressure acts at this level from the bottom. The basic situations stated for this case in EC 7 which refer to earth structures are displayed in Fig. 5.2. The failures of HYD type include three cases – failure by heave, by internal erosion and by piping, when:

- Failure by heave occurs when upwards seepage forces act against the weight of the soil, reducing the vertical effective stress to zero. Soil particles are then lifted away by the vertical water flow and failure occurs (boiling).
- Failure by internal erosion is produced by the transport of soil particles within a soil stratum, at the interface of soil strata, or at the interface between the soil and a structure. This may finally result in regressive erosion, leading to collapse of the soil structure.
- Failure by piping is a particular form of failure, for example of a reservoir, by internal erosion, where erosion begins at the surface, then regresses until a pipe-shaped discharge tunnel is formed in the soil mass or between the soil and a foundation or at the interface between cohesive and non-cohesive soil strata. Failure occurs as soon as the upstream end of the eroded tunnel reaches the bottom of the reservoir.

Fig. 5.3 The condition for hydraulic failure expressed in terms of total or effective stresses

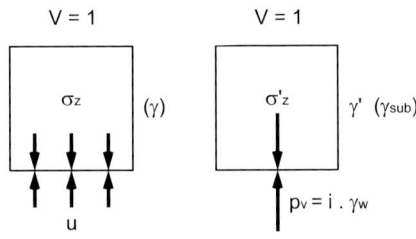

5.1 Principles of Design Procedure for Earth Structures

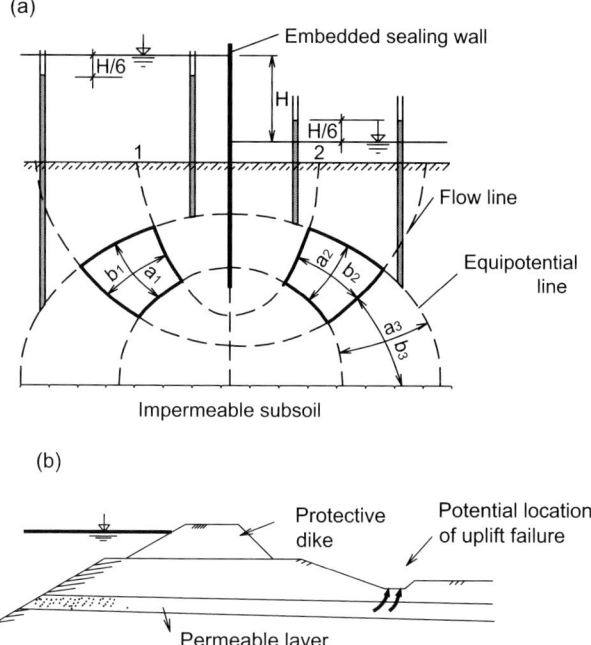

Fig. 5.4 Examples of situations where hydraulic heave might be critical. (**a**) Flow underneath embedded sealing wall. (**b**) Flow through more permeable strata underneath protective dike

For the first case – failure by heave – the solution of the soil element equilibrium may be carried out both in total and effective stresses. We either define the total gravity of the element against which pore-water pressure acts (analogically to the case of uplift) or solve the problem in effective stresses where the effective weight of the element is compared with the magnitude of flow pressure acting against it, Fig. 5.3. Here, in the first case, the total vertical stress σ_z and the water pressure in

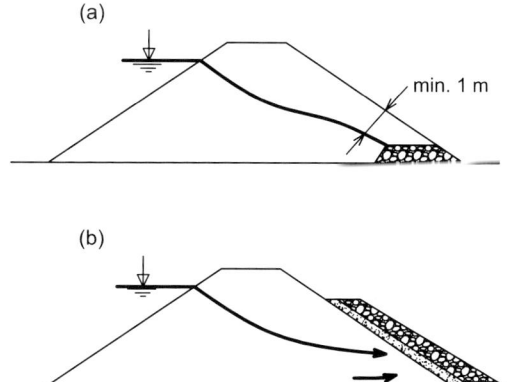

Fig. 5.5 Protective measures against progressive erosion.
(**a**) Toe drain.
(**b**) Downstream filter

Fig. 5.6 Failure of piping type for tailing dam. (**a**) Erosion started at the downstream face. (**b**) Collapse of the whole dam

pores u are compared, or, in the second case, the effective weight of a unit element with the magnitude of the flow pressure $p_v = i \cdot \gamma_w$, where i is hydraulic gradient. EC 7 mentions a classic case of failure by heave shown in Fig. 5.4a where water bypasses an impermeable vertical wall (e.g. a sheet pile), which is to secure the construction of foundations in the pit when the base of excavation is below the water level. Fig. 5.4b illustrates a case of a potential failure by heave type for flood protection barriers. At the time of an extreme water level, water seeps through the permeable horizontal layer running out onto the surface at a point where the overlying impermeable layer thins out or is damaged or does not resist uplift making the fine sand grains boil (boiling). Potential protection against this case of failure is building a vertical impermeable wall intersecting the horizontal permeable layer. As prevention, the space behind the dam can be monitored in floods, and at points where boiling starts to be shown remediation measures should be taken, e.g. establishment of a protective dam of sacks with sand around the seep.

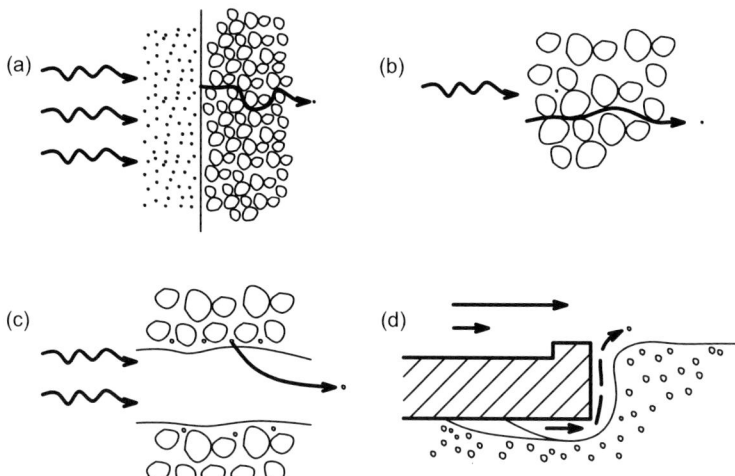

Fig. 5.7 Examples of situations where internal erosion might be critical. (**a**) Stability of interface layers. (**b**) Fine particles passing through pores of coarser ones. (**c**) Erosion around crack. (**d**) Erosion around structure

In the case of failure by internal erosion and by piping, the conditions are controlled by hydraulic gradient. At a certain magnitude of hydraulic gradient, the flow pressure acting on individual particles is so high that it may make them move in the water flow direction until the surrounding soil grains prevent this. Therefore, the most critical situation occurs if water flows out of the earth structure, then the grains at the boundary are easily washed out and the failure is propagated from this point in the direction inside the earth structure. Protection, in the first place, consists of ensuring that water does not flow freely out of the earth body, affecting principally the boundary flow lines by means of a suitable drainage system – Fig. 5.5a. Another possibility is to create a downstream filter Fig. 5.5b. In Fig. 5.6 is a photo depicting a typical case of failure of piping type where erosion started at the downstream face of a tailing dam (Fig. 5.6a) and where on-going erosion led to the collapse of the whole dam (Fig. 5.6b). The term piping reflects a rough shape of an eroded channel resembling a pipe in shape.

In the case of failure by internal erosion, several basic situations may exist:

– Failure of internal stability at the interface of two soil types with different grain-size curves when fine grains of base soil penetrate into the pores of coarse soil, Fig. 5.7a,
– Failure by internal erosion in soil whose grain-size curve is of Pg type – i.e. poorly grained soil with a gap where the missing middle fraction allows that fine grains are washed out from the basic skeleton through the coarse fraction pores, this phenomenon is also referred to as suffosion or internal stability, Fig. 5.7b,
– Failure by internal erosion by washing out grains along a crack arising in soil, Fig. 5.7c.
– Failure by internal erosion at the contact of soil and a solid element if the contact is not perfect, and some preferential pathway exists there, Fig. 5.7d.

Some other possibilities are mentioned e.g. by Schober and Teindl (1979). The basic measure against internal erosion is the establishment of a suitable filter, which will limit the danger "of material transport by internal erosion". These problems will be explained in more detail in Section 5.2.3 and in Chapter 7. Here, however, we may state that the soils most susceptible to internal erosion are fine non-cohesive soils. In coarse non-cohesive soils, the flow pressure is mostly insufficient to transport them, and in clayey soils there exist relatively strong bonds between individual grains preventing erosion of individual particles. An exception to this rule is dispersive soils whose specification will also be included in Chapter 7.

5.1.4.2 Limit State of GEO Type

Failure of GEO type. As has already been said, the failure of GEO type constitutes a basic case of failure for earth structures. These are the cases where the strength of the earth structure was exceeded. Shear strength of soil plays a decisive role here, and failures are of a type of shear failure along the slip surface. The nature of the failure may be called as "sliding" or "bearing" in the case that it is a failure of the bearing capacity of subsoil on which the earth structure was built, Fig. 5.8a,b. The case of sliding and bearing failure type has been subject to extensive publicity also historically, and so it will be used to manifest some elements of a classic solution applying the overall factor of safety F or F_s or FoS also referred to as the limit equilibrium method, or recommended more recent procedures based on limit states.

The classic factor of safety is generally defined by a ratio of the forces that can maximally prevent movement along the slipping surface to the forces that activate this movement. This procedure is documented by a fundamental example of stability of a homogeneous slope, Fig. 5.9. Here, the sliding failure of a spherical sector with a weight W is considered for a circular slip surface with a radius R with a centre O. The solution is conventionally carried out for a slope length of 1 m, i.e. a planar problem is solved. Based on the moment about the centre O we may determine:

- Active forces (activating slide):

$$W.x \qquad (5.1)$$

- Passive forces (preventing slide):

$$R.S \qquad (5.2)$$

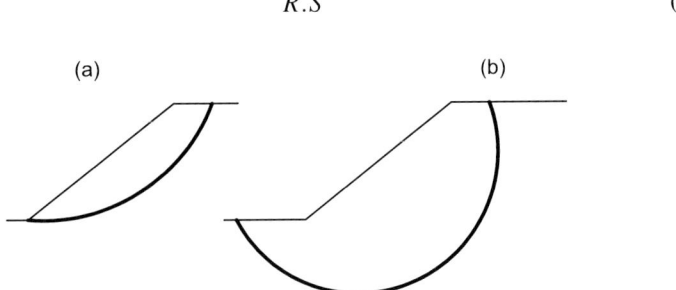

Fig. 5.8 Typical examples of failures along slip surfaces. (**a**) Sliding. (**b**) Bearing

5.1 Principles of Design Procedure for Earth Structures

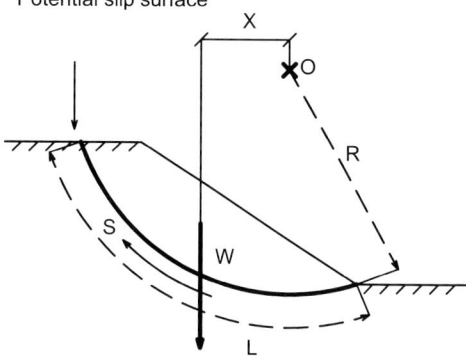

Fig. 5.9 Typical factor of safety analysis methods for homogeneous slope stability

where S is the overall shear resistance along the slip surface. Then, for undrained shear strength where $\varphi_u = 0$, it holds true that:

$$S = c_u.L \qquad (5.3)$$

Then F:

$$F = c_u.L.R/W.x \qquad (5.4)$$

Stability for the case of bearing type can be expressed analogically – see Fig. 5.10, where the moment about the centre O, lying at the boundary of a loaded area, valid for a circular slip surface with a radius B for active forces is:

$$c_u.\pi.B.B \qquad (5.5)$$

and the moment for passive forces equals:

$$\sigma.B.B/2 \qquad (5.6)$$

Then, the overall factor of safety F is:

$$F = 6.28.c_u/\sigma \qquad (5.7)$$

Note: In solving stability, another task is to search for a slip surface with the lowest factor of safety. For the above problem we may e.g. find another slip surface whose centre lies on a vertical line passing through the edge of a strip foundation, which provides still lower stability: $F = 5.5c_u/\sigma$. However, this does not have to be the most unfavourable case as the relation commonly considered for spread footings (based on Prandl theory) is $\sigma = (2 + \pi)c_u$, e.g. Terzaghi (1943), Vaníček (1982a), Atkinson (1993).

In using this method of expressing safety (stability) by one overall variable F, this is based on shear strength that is characterized, in the given case, by the undrained cohesion c_u, which can be denoted as a characteristic value for the specific

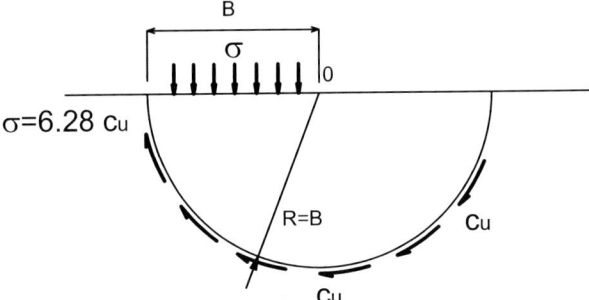

Fig. 5.10 Classical approach to the bearing capacity problem

soil medium. We may generally say that its magnitude is slightly lower than the average value determined from a set of samples; in EC 7 it is defined in more general terms: the characteristic value of a geotechnical parameter shall be selected as a cautious estimate of the value affecting the occurrence of the limit state. Its recommended value should be determined by an experienced geotechnical engineer for each respective project, for specific geological conditions, taking into consideration the quality and number of samples, the variability of measured parameters and potentially other factors. ČSN 731001:1987 allowed design of spread foundations in the case of Geotechnical category 2 based on characteristic values, specified for individual soil classes and their consistency characteristics in tables within the standard.

The overall factor of safety is also applied within a stability method referred to as a limit equilibrium method. Here, the equilibrium of active and passive forces is required, but the limit equilibrium on the slip surface is reached if we consider shear strength reduced by the factor of safety F:

$$s = \tau_{max}/F \tag{5.8}$$

then, when solving the problem in total stresses it holds true that:

$$s = c_u/F + \sigma_u \times \operatorname{tg} \varphi_u/F \tag{5.9}$$

resp. for $\varphi_u = 0$

$$s = c_u/F \tag{5.10}$$

resp. in solving the problem in effective stresses it holds true that:

$$s = c_{ef}/F + (\sigma_u - u) \operatorname{tg} \varphi_{ef}/F \tag{5.11}$$

The majority of classic slope stability methods were derived using the overall factor of safety, and they will be briefly characterized in the following chapter and supplied with a note how to adapt them to some of the solution procedures applying limit states. To make a brief summary of this discussion, it needs to be said that the overall factor of safety as a whole includes all risks which may be related to the design. It includes some safety concerning uncertainties connected with the determination of characteristic parameters of shear strength, the variability of load, in particular variable load and the uncertainty in the determination of pore-water pressures. Pore-water pressures are significantly affecting shear strength, for both steady flow and transient flow, during which pore-water pressures change.

5.1.5 Basic Logic of Design Approach Based on Calculation of Limit States

The principle of a design approach using limit states is trying to grasp the probability with which we are able to determine individual input parameters and to which procedure to adapt their selection so that the structures are safely designed, but not unnecessarily overdesigned. EC 7 in its principle considers mainly actions, geotechnical parameters and geometrical data. The basic approach will be shown on geotechnical parameters, namely on the undrained shear strength s_u. The distribution curve of measured values can be as shown in Fig. 5.11. Apart from the mean value, which is sometimes called the "most probable value", the characteristic value X_k is marked here. It serves for the calculation of the "design value" X_d from the relation:

$$X_d = X_k/\gamma_M \qquad (5.12)$$

where γ_M is a partial factor for a soil parameter (material property).

The selection of the partial factor should ensure that the mean value across the total slip surface considered will be lower than the design value only in exceptional cases. The most frequently mentioned possibility that this phenomenon will occur is in one case out of ten thousand. Saying this, we theoretically admit that a failure

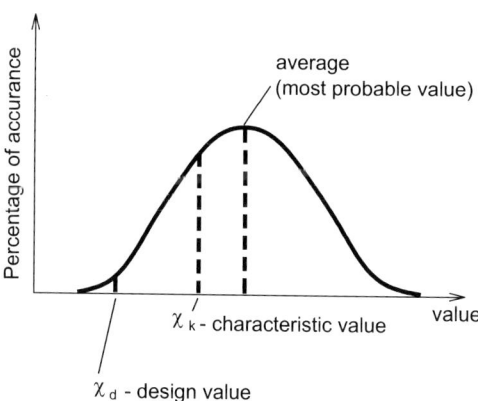

Fig. 5.11 Distribution curve of measured values of undrained shear strength s_u

may occur. The specification of this partial factor is, therefore, a highly complex process – under a still stricter requirement for a failure to occur e.g. in 1 case out of 100 000, it is evident that 99 999 cases will be overdesigned. Thus, the related economic damage may by many times exceed the damage due to one failure. Our considerations, however, cannot be governed only by economic aspects, but also by common sense, sociological or psychological aspects, mainly in the cases when one single failure may lead to a loss in human lives. At this point, let us mention a typical example not related to civil engineering – in response to certain problems with lift ropes at the beginning of their mass exploitation, the overall factor of safety was increased to nearly extreme values, to values of around 8–12. From a psychological point of view, it turned out that it is more sensible to use extremely high overdesigning. Unfortunately, even this measure failed to eliminate all accidents or losses in human lives in lifts, but these are connected with causes other than rope failures. And once again, we get to the concept of risk: if it is too high, our design principle should be more careful. This justifies the choice of a different approach to solving problems falling under geotechnical categories 1, 2 or 3, as mentioned above. The chosen approach, however, can also mean that the partial material factor need not to be so high, if another partial factor is applied for loading. In some areas of civil engineering, e.g. in designing of high dams, it is recommended to prioritize the risk factor also by adding another partial factor of the structure's significance. In this way, it is possible to ensure that for structures with high potential risks, mainly affecting human lives, by cumulating partial factors a higher overall safety can be achieved. A lower factor is then applied to more frequent, simpler cases where potential risks are low and mostly not related to human lives. It is, therefore, appropriate that design under EC 7 states recommended values, but each member state taking over this code can modify the partial factors. In order to allow their gradual, more precise specification and potential unification, individual failures on structures must be documented, assessed not only in terms of technical causes, but also in terms of the frequency of their occurrence.

5.1.5.1 Design Approaches According to EC 7

For failures of the type GEO, it shall be verified that:

$$E_d \leq R_d \tag{5.13}$$

where

E_d – design value of the effect of actions,
R_d – design value of the resistance to an action

The design value of the effect of actions is expressed in two ways; partial factors on actions are applied either to the actions themselves or to their effect. The design value of the resistance to an action is expressed in three ways, using partial factors either to ground properties or resistance or to both. As a result of different combinations, EC 7 proposed 3 "Design Approaches", whereas DA 1 has 2 combinations, from whose the worst has to be applied.

> DA 1 – combination 1: partial factors are applied only to actions or the effect of actions (recommended values – $\gamma_G = 1.35$ for permanent unfavourable action, $\gamma_Q = 1.5$ for a variable unfavourable action)
> DA 1 – combination 2: partial factors are applied for soil parameters and for variable unfavourable action (recommended values – $\gamma_M = 1.25 - 1.4$, $\gamma_Q = 1.3$)
> DA 2 – partial factors are applied to resistance and to actions (recommended values – $\gamma_R = 1.1 - 1.4$, $\gamma_G = 1.35$, $\gamma_Q = 1.5$)
> DA 3 – for slope and overall stability the approach is practically the same as for DA 1 combination 2, for other geotechnical structures not only material properties but also the structural actions are factored.

A very brief summary leads to a statement that DA 1 combination 1 and DA 2 do not apply a partial factor to soil parameters being closer in their concept to the classic concept using the overall stability factor.

DA 1 combination 2 and DA 3 are identical in the design of earth structures, being based on the reduction of soil parameters and increasing additionally random load. This approach is closer to the basic principle of the concept of limit states and to the concept which was already reflected in ČSN 731001:1987.

5.1.6 Other Methods of Limit States Verification

The above mentioned approach refers to the design of earth structures by calculation, and in particular to the limit state of failure of GEO type. EC 7 specifies design by calculation by means of analytical methods (see Sections 5.2.1 and 5.2.2 for more details), by means of semi-empirical methods and, finally, by means of numerical methods. As numerical methods can model the behaviour of earth structures, they will be more closely described in Section 5.3.

However, EC 7 admits that apart from the design and verification of limit states by calculation, it is possible to proceed using other methods as well or their combinations. In this context, we may proceed:

- By adoption of prescriptive measures,
- With the help of experimental models and load tests,
- By observational method.

5.1.6.1 Adoption of Prescription Measures

As was already said, the first possibility of "adoption of prescription measures" is generally acceptable for cases falling under Geotechnical category 1 with negligible risks. It is, therefore, mainly used "in design situations where calculation models are not available or not necessary". These involve conventional and generally conservative rules in the design, and attention to specification and control of materials, workmanship, protection and maintenance procedures. Design by prescriptive measures may also be used where comparable experience makes design calculations unnecessary. It may also be used to ensure durability against frost action and chemical or biological attack, for which direct calculations are not generally appropriate.

5.1.6.2 Experimental Models

Experimental models are very useful in situations where a new type of structure is applied with little practical implementation experience. Experimental models used in the first applications of reinforced soils, for example, very well helped to understand the mutual interaction of soils and reinforcement providing thus support to these applications. Experimental models are very often combined with numerical models, both for the verification of the structure's behaviour as a whole, or, in particular, for individual details. They also assist in parametric studies. In direct application of knowledge to an actual structure, the differences between in situ soil and the soil used in the model, between the time period during which the model can be tested and real conditions must be considered, and, last but not least, the model scale is also of great importance. Nevertheless, for earth structures the construction of a model can be well adapted to the future actual structure. A more detailed account of experimental models will be given in Section 5.3.

5.1.6.3 Observational Method

The observational method has presently reached the level of an official design approach to geotechnical structures. It may be efficiently used where prediction of the soil and rock mass behaviour presents great difficulties. The design itself is then prepared with a view to the most probable hypothesis with a simultaneous definition of boundary limits. In the case that the situation discovered on the site is better than the most probable hypothesis, the construction process can be simplified and improved. In the case that the situation turns out to be worse than the most probable version, but still within expected limits, the necessary measures are prepared in advance so that construction need not be interrupted in search for a solution, as it was previously prepared and approved by the participants of the construction project. At the same time, however, an alternative procedure must be prepared in case reality exceeds the expected limits. The exploitation of this method, therefore, requires good monitoring of the on-going construction and very good and fast feedback.

For earth structures, a suitable example of this type may be the construction of an embankment on soft subsoil where the development and dissipation of pore-water pressures were determined within certain limits. The verification of the actual state then first allows making decisions on the speed of construction (at slower construction pace the subsoil has a greater chance to consolidate), and second helps defining

5.2 Basic Limit States for Earth Structures

the scope of measures for accelerated consolidation. The method can also be applied in dam construction where continual monitoring checks whether the calculation presumptions are within the expected limits. The advantages and limitations of the observational method are more closely described by Peck (1969).

5.2 Basic Limit States for Earth Structures

Among the basic limit states of earth structures, the following may be included:

– Limit state of stability,
– Limit state of deformation,
– Limit state of internal erosion,
– Limit state of surface erosion.

EC 7 in Chapter 12, devoted to embankments of transportation structures and low earthfill dams, presents a more detailed specification of limit states, which, however, are always in some way related to the basic ones above. For the limit state of deformation e.g. the following are specified:

– Deformations in the embankment leading to a loss of serviceability, e.g. excessive settlements or cracks,
– Settlements and creep displacements leading to damage or a loss of serviceability in nearby structures or utilities,
– Excessive deformations in transition zones, e.g. the approach embankment of a bridge abutment,
– Deformations caused by hydraulic actions.

For this reason, the focus in the following part will be on the 4 basic above-mentioned limit states, and for the first two cases, the analytic calculation for their verification will be taken into account, while for the remaining two it will be the procedure based on adoption of prescriptive measures.

5.2.1 Limit State of Stability

The verification of stability in its simplest expression consists of finding equilibrium along the potential slip surface. Its shape varies depending on the type of soil of the earth structure; for non-cohesive soils planar slipping surfaces prevail (linear on section), for homogeneous cohesive soils the slip surface approximates by its character a cylindrical plane (circular on section), and in non-homogeneous soils we talk about general slip surfaces.

For a basic case of a circular slip surface, a calculation of the factor of safety was shown in the previous chapter from the moment condition for the whole annulus. In the case of solving this problem with effective shear parameters and knowing pore-water pressures from the steady flow mesh, this procedure presents problem in expressing the value of $(\sigma-u)$ along the slip surface, where shear strength $\tau = (\sigma-u)\operatorname{tg}\varphi_{ef} + c_{ef}$. Practical solutions, therefore, lead to the so-called method of slices where the spherical sector is divided into slices, which allow determination of the u value for each individual slice.

5.2.1.1 Conventional Method

This method of slices, frequently referred to as a Swedish method, solves stability solely from the moment condition, Fig. 5.12. The moment equilibrium about the centre of the shear circle O for the spherical segment as a whole is:

$$\Sigma W.x = \Sigma S.R = \Sigma s.l.R \qquad (5.14)$$

where s is mobilized shear strength. For a solution in effective stresses and knowing pore-water pressure u it holds true that:

$$s.l = \tau_{max}.l/F = [c_{ef}.l + (\sigma.l - u.l)\operatorname{tg}\varphi_{ef}]/F. \qquad (5.15)$$

as

$$\sigma.l = N = W.\cos\alpha \qquad (5.16)$$

and also

$$x = R.\sin\alpha, \qquad (5.17)$$

after substitution into the moment condition, the resulting expression is:

$$F = \frac{1}{\Sigma W \sin\alpha} \Sigma \left[c_{ef} \cdot l + (N - u \cdot l) tg\, \varphi_{ef} \right] \qquad (5.18)$$

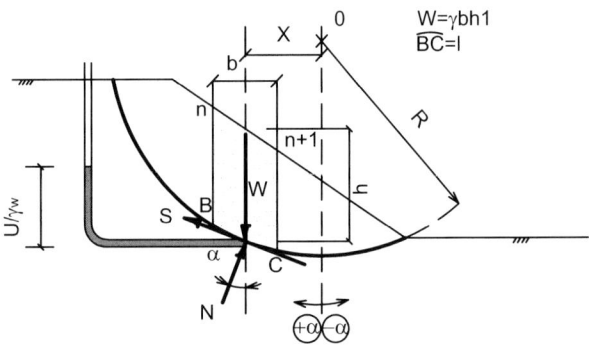

Fig. 5.12 Conventional (Swedish) method of slope stability

5.2 Basic Limit States for Earth Structures

For a solution in total stresses then it holds true that

$$F = \Sigma(c_u . l + N . \operatorname{tg} \varphi_u)/\Sigma(W. \sin \alpha) \tag{5.19}$$

and for the case where $\varphi_u = 0$ the resulting equation is simplified in the form:

$$F = \Sigma(c_u . l)/\Sigma(W. \sin \alpha) \tag{5.20}$$

The basic equation for effective stress may serve for the explanation of some more general calculation principles applying limits states.

5.2.1.2 More General Principles of Stability Calculation Applying Limit States

The initial situation to be used may be the principle expressed in ČSN 750290:1995 – "Designing earth structures of hydrotechnical constructions", which already applied the principle of limit states. It is based on taking into account design values of soil parameters, considering the uncertainties in loading, the design situation, the significance of the earth structure or the effect of the procedure (method) of stability calculation used. According to this code, the equilibrium of forces on a slip surface can be written in the form, I Vaníček and M Vaníček (2001):

$$\gamma_{sit} \cdot \gamma_n \cdot \psi_c \cdot \sum \gamma_{fai} \cdot S_{act,in} \leq \gamma_{stp} \cdot \sum \gamma_{fpj} \cdot S_{pas,jn} \tag{5.21}$$

where

$S_{act,in}$ = force influence of design loading values which act on the earth construction or its part above the slip surface to reach the limit state;
$S_{pas,jn}$ = force influence of design effective loading values (friction) and force influence of cohesion by which the earth structure or its part above the slipping surface resists to the limit state overrun (excess);
$\gamma_{fai}, \gamma_{fpj}$ = partial reliability factors of loading which are related to the loading which generates the force influences $S_{act,in}, S_{pas,jn}$;
γ_{stp} = partial factor of location stability which is usually taken as 0.9;
γ_n = partial factor of function (significance of earth structure), usually in the range of 1.1–1.2;
γ_{sit} = partial factor of design situation;
ψ_c = partial factor of loading combination.

For earth structures such as high dams, the determination of loading may be estimated with relatively high accuracy. Therefore, by considering only the location stability factor, which principally expresses the effect of the calculation method precision and the factor of purpose (significance of earth structure) taking into account mainly its potential risks for the surroundings, the equilibrium conditions may be put down in a simplified form:

$$\sum S_{pas,jn} / \sum S_{act,in} \geq \gamma_n / \gamma_{stp} \qquad (5.22)$$

and when substituting the values on the right hand side we get:

$$\sum S_{pas,jn} / \sum S_{act,in} \geq 1.22 - 1.33 \qquad (5.23)$$

Saying very simply when taking into account design values instead of characteristic ones to reach enough slope stability we can accept the safety ratio in the range between 1.22 and 1.33, according to the significance of the earth structure when passive forces are calculated on the basis of design parameters of the shear strength:

$$c_{d,ef} = c_{k,ef} / \gamma_{mc} \qquad (5.24)$$

$$\text{tg } \varphi_{d,ef} = \text{tg } \varphi_{k,ef} / \gamma_{m\varphi} \qquad (5.25)$$

Partial factor $\gamma_{m\varphi}$ – however applied directly on φ – (which for application on bearing capacity of spread foundations seems to be more appropriate) – see Vaníček (2005a) was recommended roughly in the range of 1.1–1.4 and γ_{mc} for effective cohesion is equal 1.6.

Note: EC 7 – for Design Approach 1 combination 2 and for Design Approach 3 – is recommending for solution in effective stresses to use for both partial factors the same value – 1.25. However this equity in recommended values of partial factors is not exactly with agreement of EC 7-1 itself stating that the greater variance of $c' (= c_{ef})$ compared to that of tg φ' shall be considered when their characteristic values are determined. The influence of additional partial factor on unfavorable variable actions $\gamma_Q = 1.3$ is relatively small when applied on overall stability. The consequence of this means that with recommended values for γ_{mc} and $\gamma_{m\varphi}$ of 1.25 the final result will be close to the classical solution with total factor of safety $F = 1.25$. This is a relatively great optimistic step forward because up to now generally demanded value was $F = 1.5$ based on Czech codes or 1.3 based on British Standard. This fact can lead to two basic approaches. First one will use rather conservative selection of recommended characteristic values of material properties, in the first place for cohesion. The second one was indicated before and will lead to the application of additional partial factors as for calculation method (e.g. 1.1) and as for significance of structure (e.g. in the range of 1.1–1.2). This second approach seems to the authors to be more adequate because it is taking into account the higher risk for more significant structures. Because EC 7-1 allows for certain flexibility in the National Annex when defining partial factors, this problem will be cleared with time.

5.2 Basic Limit States for Earth Structures

Now we can show how easy it is to implement the limit states design approach into the classical equation for overall stability based on limit equilibrium method. For the solution in effective parameters:

– The general approach is:

$$\Sigma[c_{d,ef}.l + (N-u.l) \text{ tg } \varphi_{d,ef}] \geq \gamma_n \Sigma(W. \sin \alpha)/\gamma_{stp} \quad (5.26)$$

– The approach from EC 7-1 DA 1, comb. 2 and DA 3:

$$\Sigma[c_{d,ef}.l + (N-u.l) \text{ tg } \varphi_{d,ef}] \geq \Sigma(W. \sin \alpha) \quad (5.27)$$

– But where $W = G + 1.3Q$ and G is permanent loading and Q is variable loading for individual slice.

5.2.1.3 Bishop Method

Bishop (1955) is taking into account the forces from the neighbouring slices and besides of moment equilibrium is counting also with balance of horizontal and vertical forces on individual slice, Fig. 5.13. From closed free-body diagram and from equilibrium in vertical direction the following equation (5.28) can be written.

$$N_{ef} = N - ul = \frac{W + (X_n - X_{n+1}) - l\left(u \cos \alpha + \frac{c_{ef}}{F} \sin \alpha\right)}{\cos \alpha + \frac{tg\varphi_{ef}}{F} \sin \alpha} \quad (5.28)$$

And after substituting into the equation for factor of safety from conventional method (5.18) and substituting $l = b \sec \alpha = b/\cos\alpha$, where b is width of slice, we are getting:

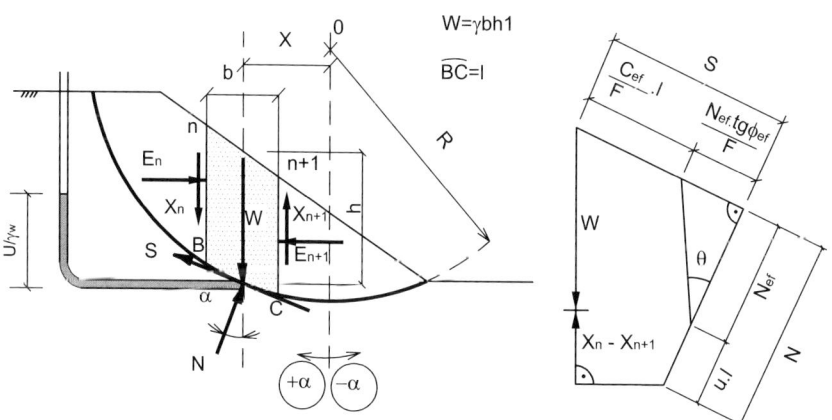

Fig. 5.13 Bishop method of slope stability analysis

$$F = \frac{1}{\sum W \sin\alpha} \sum \left\{ c_{ef}b + (W - ub + X_n - X_{n+1})tg\varphi_{ef} \cdot \left[\frac{\sec\alpha}{1 + \frac{tg\varphi_{ef} \cdot tg\alpha}{F}} \right] \right\} \quad (5.29)$$

This expression for factor of safety is called as the Bishop rigorous method. Its solution by iteration is rather complicated. Nevertheless Bishop itself found that neglecting the difference $X_n - X_{n+1}$ has little practical influence. After that we get the expression which is generally marked as the Bishop method (sometimes as the Bishop simplified method):

$$F = \frac{1}{\sum W \sin\alpha} \sum \left\{ c_{ef}b + (W - ub) tg\,\varphi_{ef} \cdot \left[\frac{\sec\alpha}{1 + \frac{tg\varphi_{ef} \cdot tg\alpha}{F}} \right] \right\} \quad (5.30)$$

Somewhere using different expression:

$$F = \frac{1}{\sum W \sin\alpha} \sum \frac{c_{ef}b + (W - ub)\,tg\varphi_{ef}}{\cos\alpha + \frac{tg\varphi_{ef} \cdot \sin\alpha}{F}} \quad (5.31)$$

Searched value of F is obtained with the help of iteration. When using limit state approach the design values of the shear strength as well as design loads are substituted in.

5.2.1.4 Planar Slip Surface

Planar slip surface parallel with the slope is roughly substituting the most of slip surfaces of translational movements. For $d/L < 0.1$ according to Fig. 5.14a it is possible to substitute this situation by the model shown in Fig. 5.14b which neglects the influence of lateral forces, e.g. Skempton and Hutchinson (1969). Here for equilibrium of forces in the direction parallel with the slip surface results in:

$$W \sin\beta = S \quad (5.32)$$

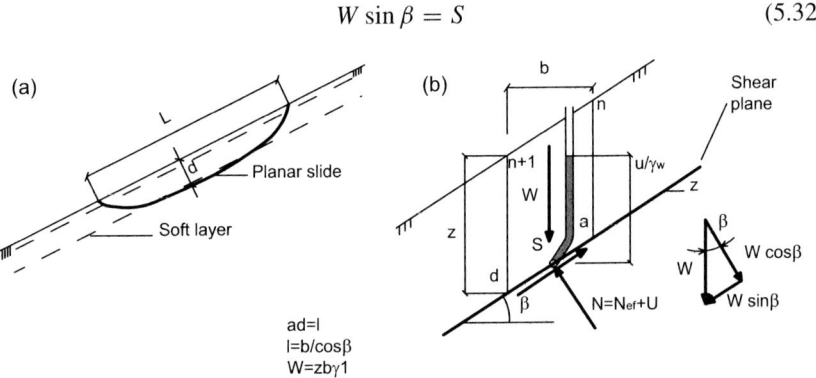

Fig. 5.14 Planar slides for slope stability analysis. (**a**) Real model. (**b**) Infinite slope

5.2 Basic Limit States for Earth Structures

Where $S = s.L$ and $s = \dfrac{c_{ef}}{F} + \dfrac{\sigma_{ef} tg\varphi_{ef}}{F}$, and therefore

$$W \sin \beta = \dfrac{1}{F} \left(c_{ef} \ell + N_{ef} tg\varphi_{ef} \right) \tag{5.33}$$

From this the factor of safety F is

$$F = \dfrac{1}{W \sin \beta} \left(c_{ef} \ell + N_{ef} tg\varphi_{ef} \right) \tag{5.34}$$

$$F = \dfrac{c_{ef} \ell + N_{ef} tg\varphi_{ef}}{bz\gamma \sin \beta}, \tag{5.35}$$

where $b = \ell \cos \beta$
and therefore

$$F = \dfrac{c_{ef} + \left(N_{ef}/\ell \right) tg\varphi_{ef}}{\gamma z \sin \beta \cos \beta} \tag{5.36}$$

For the equilibrium of forces in the direction perpendicular to the slip surface therefore for the factor of safety F applies.

$$N = N_{ef} + U = W \cos \beta \tag{5.37}$$

where $U = uL$

$$N_{ef} = W \cos \beta - uL \tag{5.38}$$

$$N_{ef} = \gamma bz \cos \beta - uL = \gamma zL \cos^2 \beta - uL \tag{5.39}$$

$$N_{ef} = L(\gamma z \cos^2 \beta - u) \tag{5.40}$$

By substituting the expression for N_{ef} into the equation (5.36) for F and after the adaptation of the general equation for slope stability in effective stresses for the slip surface parallel with slope is:

$$F = \dfrac{c_{ef} + \left(\gamma z \cos^2 \beta - u \right) tg\varphi_{ef}}{\gamma z \sin \beta \cos \beta} \tag{5.41}$$

Special Cases

a) $c_{ef} = 0$

$$F = \left(1 - \frac{u}{\gamma z \cos^2 \beta}\right) \frac{tg\varphi_{ef}}{tg\beta} \qquad (5.42)$$

b) $c_{ef} = 0$, $u = 0$ – stability of dry sand and gravel

$$F = \frac{tg\varphi_{ef}}{tg\beta} \qquad (5.43)$$

For cohesionless soils the standard values for the angle of internal friction $\varphi_{ef} = 30 - 42°$, so that the maximum slope inclination for the indifferent equilibrium is 1:1.25–1:1.75. For the demanded factor of safety $F = 1.3$ the recommended slope inclination is in the range of 1:1.5–1:2.25.

c) $c_{ef} = 0$ and water is seeping parallel with slope in height mz above shear plane.

$$u = \gamma_w mz \cos^2 \beta \qquad (5.44)$$

and then

$$F = \left(1 - m\frac{\gamma_w}{\gamma}\right) \frac{\tan \varphi_{ef}}{\tan \beta} \qquad (5.45)$$

d) for the case as c) when $m = 1$, it means that the ground water table is on the slope surface:

$$F = \left(1 - \frac{\gamma_w}{\gamma}\right) \frac{tg\varphi_{ef}}{tg\beta} \qquad (5.46)$$

For indifferent equilibrium when $F = 1$:

$$tg\beta = \left(1 - \frac{\gamma_w}{\gamma}\right) tg\varphi_{ef} \qquad (5.47)$$

and for $\gamma = 20$ kNm³

$$tg\beta \doteq \frac{tg\varphi_{ef}}{2} \qquad (5.48)$$

Last case is a certain simplification of slope cuttings in cohesionless soils with surface drainage. Inclination for such case is roughly 1:2.2 up to 1:3.6 for indifferent equilibrium and for $F = 1.3$ the inclination is rather gentle 1:3 up to 1:4.5. When this case occurs for clay soils after large shearing, when the strength drops to residual one ($c'_r = 0$), after that the slope can be unstable even for inclinations 1:7, 1:8 or even less.

e) $c_{ef} \neq 0$, water is running out of the slope in a horizontal direction and for pore pressure applies $u = \gamma_w z$ afterwards

$$F = \frac{c_{ef}/z + (\gamma \cos^2 \beta - \gamma_w) tg\varphi_{ef}}{\gamma \sin \beta \cos \beta} \quad (5.49)$$

For the solution in clays in total stresses when c_u = const., $\varphi_u = 0°$, the equation has the form of:

$$F = \frac{S}{W \sin \beta} = \frac{c_u l}{\gamma z b \sin \beta} = \frac{c_u}{\gamma z \sin \beta \cos \beta} \quad (5.50)$$

5.2.1.5 General Slip Surfaces

The above discussed circular slip surfaces are best fitted for homogeneous soils. However very often soil is heterogeneous, with significant bedding, with different zones as for fill dams and etc. For these cases a character of the slip surface is irregular, non-circular, general.

In principle the stability calculations for general slip surfaces can be divided into two groups:

- Method of slices creates a first group, where as for circular slip surface the slope above slip surface is divided into slices; however slip surface can be general one, composed from individual abscissas or from individual curvature parts,
- Wedge methods create a second group, which solve the stability of a specific block above the selected slip surface, which is loaded from both sides by active and passive blocks of soil.

Method of Slices

Into this group belongs for example the following methods: Janbu (1954, 1973), Morgenstern and Price (1965), Sarma (1973, 1979). Under certain conditions the conventional method of Skempton and Hutchinson (1969) can be used as well. Some problems of this method are given by redundancy (static uncertainty). Under the assumption that slope above slip surface is divided into n slices, total number of unknowns is $6n-2$ and number of possible equations is $4n$. Detailed analysis of this fact is given for example by Sarma (1979). In agreement with Fig. 5.15 the following unknowns can be distinguished:

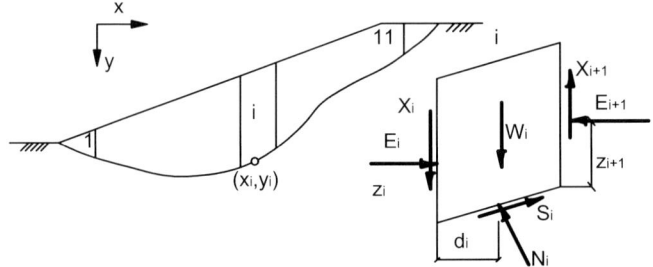

Fig. 5.15 Conditions of strip equilibrium for general character of slip surface

n	– number of $N_i(N'_i)$	normal forces
n	– number of S_i	tangential forces
$n-1$	– number of E_i	horizontal forces between slices
$n-1$	– number of X_i	vertical forces between slices
n	– number of d_i	horizontal distance of point of force N application from the slice edge
$n-1$	– number of z_i	vertical distance of point of force E application from slip surface
1	– F	factor of safety

$\sum = 6n - 2$

For each slice three static conditions of equilibrium are defined as well as relationship between normal and tangential forces:

n	conditions $H_i = 0$	– condition of equilibrium in horizontal direction
n	conditions $V_i = 0$	– condition of equilibrium in vertical direction
n	conditions $M_i = 0$	– moment equilibrium to the point of force N_i application
n	conditions $S = f(N_{ef})$	– Mohr-coulomb criteria for each slice

$\sum = 4n$

So we have $4n$ equations from which we have to count $6n - 2$ unknowns. Therefore we need $2n - 2$ independent assumptions and only $2n - 2$.

Rigorous solution is after that such that all conditions are satisfied under defined assumptions. Simplified solution does not fulfil all equilibrium conditions. Simplified conditions are used for simplified solution.

Acceptable solution is rigorous solution. It has supplementary criteria and assumptions in order to obtain acceptable solution with minimum effort.

The method of Janbu introduces the following assumptions: n – number of points of force N application and $n-1$ number of points of force E application. One assumption is extra and technically it is not a rigorous solution. However one assumption that is the point of application of force N is not used. If yes, moment equilibrium for the last slice ($M_n = 0$) will not be satisfied. This defect is rather a small one and is influencing only the position of the line of trust, and not the factor of safety.

The method of Morgenstern-Price introduces the following assumptions: n – number of points of force N application and $n-1$ – number of relative relationships between forces X and E. Because number of assumptions is by 1 higher, one extra unknown is introduced. It is a supplemental function $\lambda = f(x)$, where λ is a parameter which must be defined from the solution and $f(x)$ is a function, which must be specified.

The method of Sarma has the following assumptions: n – number of points of force N application and $n-1$ – number of relative values of forces X. It also introduces one extra unknown.

Skempton and Hutchinson (1969) indicate that the conventional method can be used as well. This method falls into simplified solutions:

$$F = \frac{\sum c_{ef} \cdot l + \sum (W \cos \alpha - ul) \, tg \varphi_{ef}}{\sum W \sin \alpha} \tag{5.51}$$

But this simplification can be used only for shallow slip surfaces without surface surcharge and for constant angle of friction.

Wedge Method

The principle of the wedge method is shown on the simple example of two wedges, see Fig. 5.16. Basic block *cbd* resists by mobilized shear forces along slip surface to the loading from the active block *abd*. If this loading is higher than the mobilized shear resistance, the basic block and subsequently also the active block move, which is generally accompanied by partial rupture of blocks on contacts and in some cases by detachment of wedges. Wedge method is based on field landslides observation. For solution in effective stresses with knowledge of pore pressures, free-body diagrams are constructed for both wedges and for selected initial factor of safety F_1.

For active wedge 1 we know the weight W_1, pore pressure acting on slip surface section ab–U_1 and pore pressure acting on separation plane bd–U_{12}. Also we know the mobilized shear resistance along the slip surface ab: $c_{1,ef} \cdot l_1/F_1$. From the other forces we know the direction of resultant of the shear forces along slip surface, which is diverted from vertical line by angle θ_1. The equation $tg \theta_1 = (1/F_1) tg \varphi_{ef}$ is valid. And finally we know the direction of resultant of forces $E_{1,ef}$ on separation plane bd, which is rotated from perpendicular line to this plane by angle δ. From the closed force diagram all forces are determined.

The similar procedure is realized for basic wedge 2. For the first step of approximation different values of forces $E_{1,ef}$ and $E_{2,ef}$ are obtained. For limit equilibrium both values should be equal and therefore the procedure is repeated with another

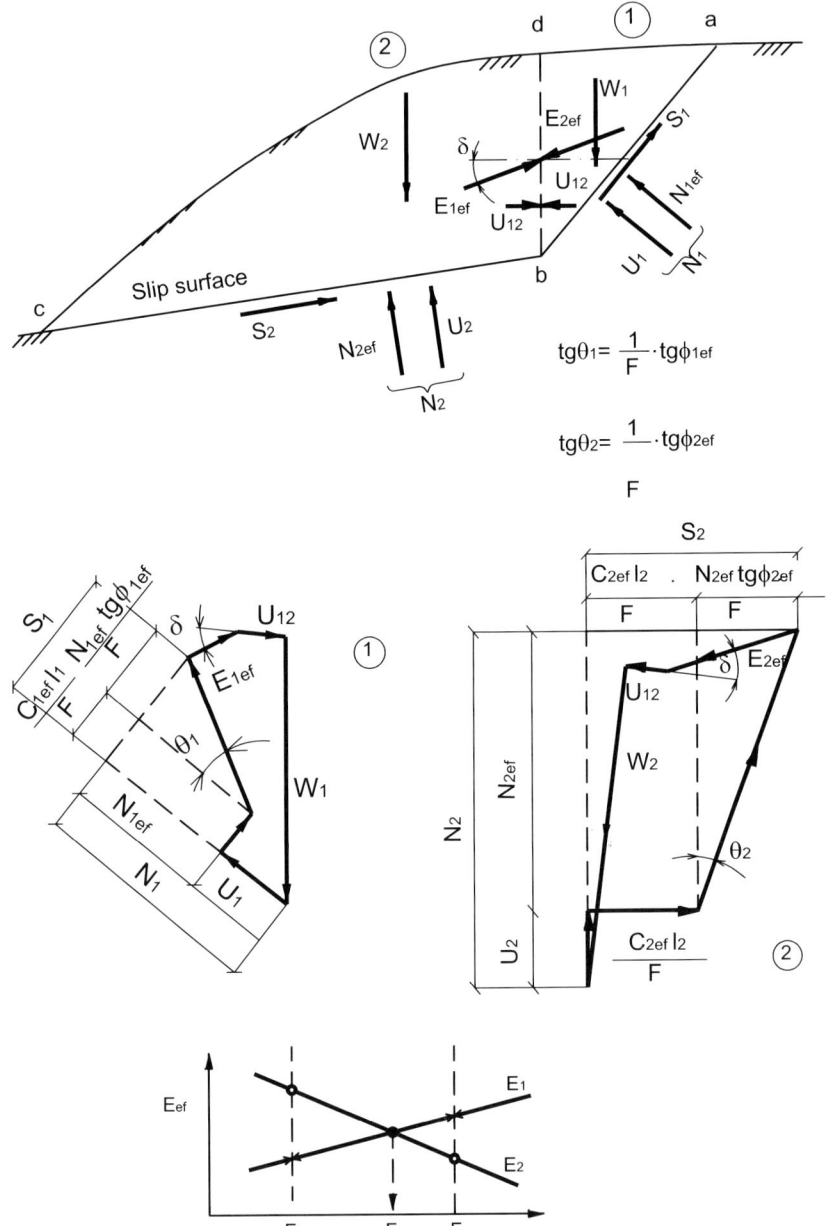

Fig. 5.16 Wedge method of slope stability analysis

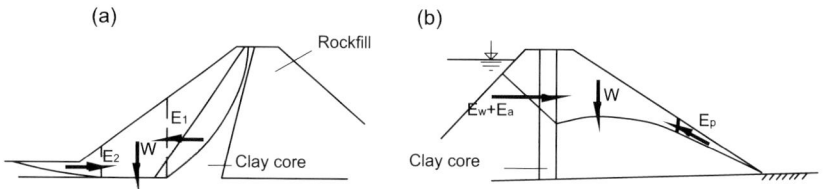

Fig. 5.17 Wedge methods applied for earth and rockfill dams. (**a**) Terzaghi approach. (**b**) Nonveiler approach

selected value of F_2. The determination of the final searched factor of safety value is visible from the Fig. 5.16.

Final value of F is rather sensitive to the selection of δ, generally factor of safety is increasing with higher value of angle δ. Presumption that $\delta = 0°$ undervalues factor of safety and therefore provides a pessimistic prognosis. On the other side a presumption that $\delta = \varphi_{ef}$ generally overestimates F and is therefore rather risky. Some authors recommend selecting angle δ corresponding to the average inclination of slope, in our case to the join of points *ac*. Nevertheless it is recommended to control the sensitivity of δ for individual case.

Wedge methods are more frequently used in dam engineering during control of limit state of overall stability for fill dams with clay core. For clay core closer to the upstream face the Terzaghi's method is recommended and for central clay core method of Nonveiler. Assumptions of these authors are shown in Fig. 5.17. For small dams on soft subsoil the methods with three wedges can be used, see Fig. 5.18. e.g. Vaníček (1973).

Very comprehensive overview of slope stability problems are presented by Fell et al. (2000).

5.2.1.6 Verification of GEO Limit State for Reinforced Soils

It is kind of paradox that the limit state approach for the design of limit state of stability is more often used for reinforced slopes than for unreinforced, natural, slopes. The main reason of this fact is the question which tensile force of the reinforcing element should be applied during the design. This question is very sensitive above all for slopes reinforced by geosynthetic reinforcing elements, which are sensitive to creep, e.g. Greenwood (1990). This problem led to the definition of the design value of the tensile strength of geosynthetic reinforcing element.

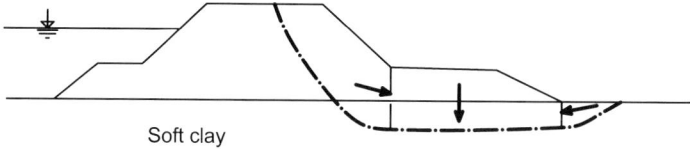

Fig. 5.18 Wedge method applied for small earth dams

Design Value of the Tensile Strength of Geosynthetic Reinforcing Element

There are different approaches, but very often they are similar – e.g. Viezee et al. (1990), BS 8006, Gourc (1996) and some problems are discussed by I Vaníček and M Vaníček (2001): In the simplest version the design tensile strength can be expressed by equation:

$$T_d = \frac{1}{F_{tc}} \cdot \frac{1}{F_{comp}} \cdot \frac{1}{F_{env}} \cdot T_f \qquad (5.52)$$

where

T_f is maximum tensile strength according to ISO 10319 – highspeed test;

F_{tc} – partial reduction factor for creep, depends on geosynthetic material and construction performance. Recommended values are for polyester $F_{tc} = 1.5$ (for short term structure – 7 years) or 2.25 (for long term structure – 70 years) and for polypropylene and polyethylene $F_{tc} = 3.0$ or 4.5;

F_{comp} – partial reduction factor for compacting effect (mechanical damage), ranging between 1.1 and 1.5 as a function of soil and geosynthetic types, upper value 1.5 is recommended for crushed gravel with high sharpness and for geosynthetics from polyester;

F_{env} – partial reduction factor for environment, recommended values are: $F_{env} = 1$ for temporary earth structures and $F_{env} = 1.1$ for permanent earth structures. This range is valid for pH between 4 and 9 and good protection against UV radiation, for more alkaline environment polyester is affected by hydrolysis.

It is useful to compare the design value T_d with relative deformation for tensile loading, see Fig. 5.19.

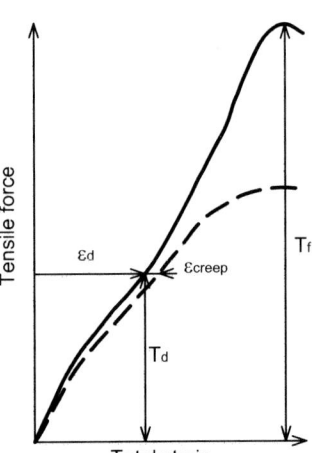

Fig. 5.19 Typical working diagram for tensile test of geosynthetics

5.2 Basic Limit States for Earth Structures

Relative deformation ε_d for this design value T_d is usually recommended between 2 and 6%, according to the sensitivity of the structure to the deformation. This condition can be used for the selection of the right reinforcing element material. Nevertheless the additional creep strain has to be considered as well.

The development of the strength in soil and the development of the tensile strength in reinforcing element depend on the relative deformation (tensile strain, shear strain) and by applying of partial reduction factors we are not sure that the design values are reached for the same deformation. Jewell (1990) uses for this optimum design the expression of equilibrium strain – see Fig. 5.20.

Instead of using partial reduction factor for creep from recommended values, the more precise way is to use cluster of tensile test curves obtained for different rate of loading (ISO 13431), see Fig. 4.32.

From this cluster the curve for expected service life can be easily selected.

Some authors however use even more extended expression for T_d definition:

$$T_d = \frac{T_f}{F_{tc} \cdot F_{comp} \cdot F_{env} \cdot F_{mat} \cdot F_{ost}} \tag{5.53}$$

where

F_{mat} – partial reduction factor expressing material factors,
F_{ost} – partial reduction factor expressing the significance of structure, material, numerical methods.

Comparison of individual partial reduction factors in different countries or according to different regulations (codes) is presented in Table 5.1, see Vaníček (2005b) in Turček et al. (2005). The approach to the calculation of design tensile value of the reinforcing element is a rather similar one. However Czech regulation for the application for motorways recommends that partial factor F_{ost} should be applied not only on geosynthetic reinforcing element but on the overall stability – see TP 97 – "Geotextile and other geosynthetic materials used in earth structure of roads and motorways" – Vaníček (1997). In this case the recommended procedure is probably

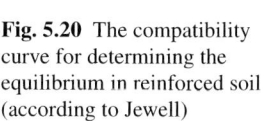

Fig. 5.20 The compatibility curve for determining the equilibrium in reinforced soil (according to Jewell)

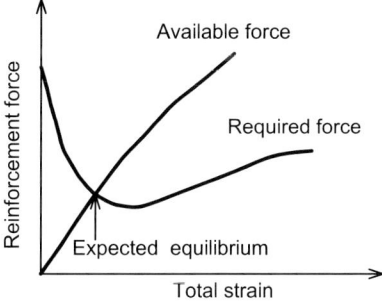

Table 5.1 Comparison of partial safety factors for geosynthetic reinforcement for different regulations (codes)

Regulation	F_{comp}	F_{env}	F_{tc}		F_{mat}	F_{ost}
			PET	PP ; PE		
FHWA (USA)	1.25–3	1.1–2	2.5	5	1.5	–
CFGG (F)	1.1–1.5	1.1	2.5	4.5	1.2	–
Gourc (F)(NFP – 38064)	1.1–1.5	1.1	2.25	4.5	–	–
DGGT (D)	1.1–1.5	1.1	2.5	4	1.75	–
TP 97 + ČD (CZ)	1.1–1.5	1.1	2.5	4.5	–	1.22–1.33

more precise because it includes not only the correction for reinforcing element, but also for soil as was recommended for unreinforced slopes in the previous chapter.

Implementation of Reinforcing Element Effect
into Slope Stability Calculation Method

Let us suppose a simple case of slope reinforced with one reinforcing element, see Fig. 5.21 and observe how this element contributes to the increase of slope stability for individual assumptions. According to the assumption ad (a) the reinforcing element, its design tensile strength, reacts in the horizontal direction. This additional effect from the reinforcing element is additional moment acting on cantilever y, which is the distance of the reinforcing element from the centre of the circular slip surface. This assumption is the recommended one in BS 8006:1995. It is obvious

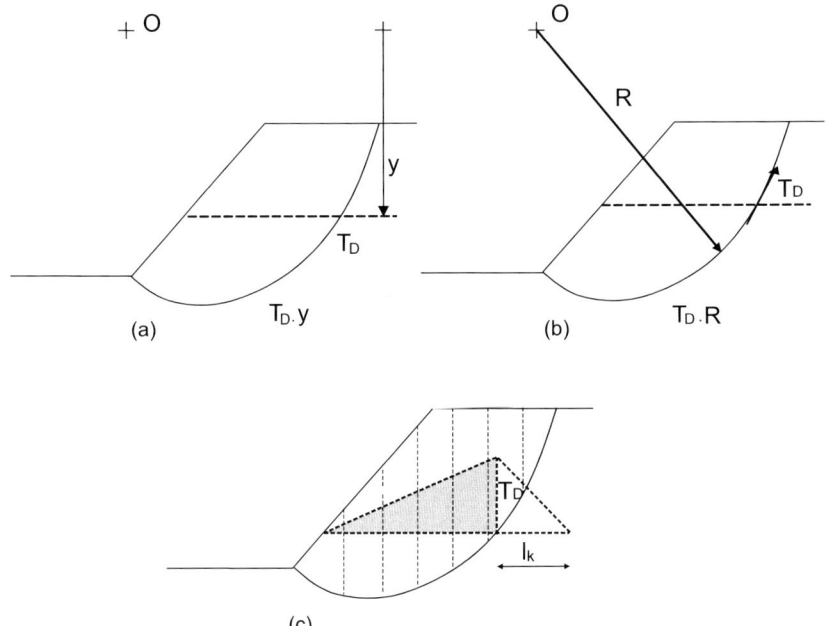

Fig. 5.21 Main options how to incorporate the reinforcing element into slope stability analysis

that in the upper part of slope the positive influence is lower than for the same element situated in the lower part of the slope. This approach assumes the maximum engagement of the reinforcing element without any deformation.

In the second case ad (b) the influence of reinforcing element is reflected as additional moment acting on cantilever R, which is the radius of circular slip surface. In principle this approach assumes that due to the development of shear plane accompanied by shear strain along the circumference of the circular slip surface the tensile force in the reinforcing element is mobilized also along this circumference. The influence of the reinforcing element is constant, independent of its position in the slope. This assumption represents another extreme; tensile force is activated after a significant shear strain in soil.

The third case ad (c) reinforcing element is acting as additional horizontal force, the maximum of which is in the point of intersection with slip surface; see e.g. Vaníček and Škopek (1989), Vaníček (2002c). This horizontal force is decreasing on both sides, in the direction of slope face side or in the direction of anchoring. The difference on interslice boundary is this additional force. In principle it is prestressing force between individual slices. The effect is variable, depending on the position in the slope, and is increasing with increasing area of the triangle which is bordered by design tensile force T_D and by length of reinforcement to the slope surface. Generally the highest effect is in the place where the tangent line parallel to the slope is touching slip surface; roughly in the lower third of the slope. The authors prefer this approach also for easier application with general shape of the slip surface.

Method of Janbu

Due to this assumption the method of Janbu (1973) was used for the calculation of the reinforced slope Vaníček and Škopek (1989), M Vaníček (2000). Janbu's method, which is adopted uses on each slice, to which the whole slope is divided, equilibrium equation in horizontal and vertical directions and momentum one. This approach is complex and accommodates all the forces, which act on each slice (see Fig. 5.22). This method solves the factor of safety F by the following equation:

$$F = \frac{\sum_{i=1}^{n} \tau_{fi} \cdot \Delta x_i \cdot \left(1 + \tan^2 \alpha_i\right)}{E_a - E_b + \sum_{i=1}^{n} \left[\Delta Q_i + (T_i - T_{i+1} + W_i) \cdot \tan \alpha_i\right]} \quad (5.54)$$

where the meaning of symbols is also explained in Fig. 5.22. The shear strength τ_f is for effective stress analysis defined as:

$$\tau_{fi} = \frac{c_{ef} + \left(\dfrac{T_i - T_{i+1} + W_i}{\Delta x_i} - u_i\right) \cdot \tan \varphi_{ef}}{1 + \dfrac{\tan \varphi_{ef} \cdot \tan \alpha_i}{F}} \quad (5.55)$$

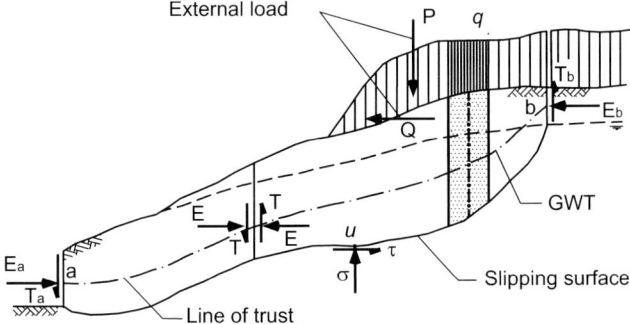

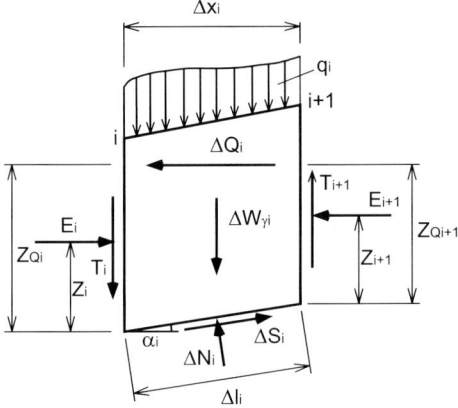

Fig. 5.22 General slipping surface and scheme of forces on slice for Janbu's method with marked basic terminology and quantities

where c_{ef} is design effective cohesion, φ_{ef} is design effective angle of internal friction. Looking at the equations it can easily be seen that the calculation is iterative.

From our own experience the biggest influence on the stability and calculation result is the position of horizontal inter-slice forces, which are in this calculation method assumed as known. These forces are acting on so called "line of trust", which should be in about 1/3 of the slice height.

Due to the fact that the calculation of this way modified Janbu's method is rather long for hand calculations and for the determination of the most dangerous slip surface with minimum factor of safety, a computer program SVARG (Slope Reinforced by Geosynthetics) was developed, see M Vaníček and J Vaníček (2000). The program can almost immediately solve factor of safety for selected slip surface or in a very short time find the worst slip surface. The slip surface can be general. This program also automatically checks the anchorage length of the reinforcement with the expression:

$$L_k = \frac{\gamma_L \cdot T_d}{2 \cdot (\gamma \cdot h \cdot tg\varphi_{gs} + a)} \qquad (5.56)$$

5.2 Basic Limit States for Earth Structures

where

φ_{gs} = angle of internal friction between soil and reinforcing element;
a = adhesion between soil and reinforcing element;
h = depth of the reinforcing element below the surface;
T_d = design tensile strength of the reinforcing element;
γ_L = partial factor for anchorage length, e.g. = 3 based on Czech requirements.

The assumption of the redistribution of the tensile force in the reinforcing element (see Fig. 5.21c) is adequate for the anchorage part and for the part to the slope face only if the face is without rigid elements. For the retaining structures, where the rigid elements on the face are used, the distribution of the force in the reinforcing element should have on face a value between 0.3 and 0.5 from the design tensile force.

To summarise this part dealing with stability of reinforced slopes it is useful to go back to Chapter 3, where internal and external stability of the reinforced block was mentioned. In many cases it is useful to combine both solutions and check also intermediate slip surfaces passing only through a smaller part of the reinforced block. From the reinforced block point of view, firstly for reinforced retaining wall, it is necessary to check also the cases of external stability, as potential risk to sliding, overturning and bearing capacity of subsoil. This problem will be discussed in more detail in the Chapter 6, especially for retaining walls with facing from small concrete elements.

5.2.2 Limit State of Deformation

The basic purpose of the calculation of deformation values of earth structures is their successive comparison with limit values. These, however, are mostly not defined in advance unlike e.g. the values for the settlement of spread foundations, which are relatively well defined where absolute settlement in a range of 0.06–0.12 m is usually admitted plus non-uniform deformations depending on the subsoil character and the type of structure ranging from 1/150 to 1/750 – Bjerrum (1963). Each individual case of earth structures usually calls for the need for defining special conditions, and analogically to the settlement of spread foundations, more distinct conditions are specified in relation to non-uniform deformations. For high, loosely packed earth structures without compaction, such as e.g. spoil heaps, the deformation can reach values of up to several metres. Non-uniform deformations are caused by a more abrupt change in the subsoil character, at contacts of different soil types with different compressibility and at the contact of the earth structure with another structure lying on less compressible subsoil or founded on deep foundations. Non-uniform deformation may show up as cracking, both shear, but mainly tensile cracks, which can successively become preferential paths for water flow with the resulting negative impacts – the failure of the earth structure due to internal erosion. Of great significance are also non-uniform deformations developed in time (as a consequence of on-going consolidation) as they may degrade the finish on the surface of the earth structure, e.g. by changing the gradient line of the final road surface or by crack development in this area.

Figure 5.23 displays typical cases of embankment deformation on uncompressible subsoil, deformation of compressible subsoil and heave of bottom of excavation pit. In embankments, we can see that not only vertical deformation applies here, but the embankment itself is also deformed in the horizontal direction. Thus, in general, we may find an element which is deformed only vertically (i.e. 1D deformation), even though the prevailing number of elements show general deformations (here we talk about 3D deformation, or 2D deformation in the plane), see Fig. 5.24. This distinction is of great importance in the assessment of immediate (initial) settlement – deformation, which can be most readily illustrated by the cases where the earth structure itself was built in a very short time and as total deformation (settlement), which occurs after the completion of consolidation, Fig. 5.25. In the case of uniaxial deformation of soils saturated with water, initial settlement (due to vertical load increase) is theoretically zero, while for triaxial deformation, even in the case of a fully saturated element, vertical settlement corresponds to the value s_i, even if the volume change is zero. The completion of settlement referred to as the consolidation settlement s_c corresponds to the time when the increment in pore pressures due to a change in the state of stress is zero. On-going settlement, however, may be recorded even subsequently, due to the deformation of individual grains at contact interfaces. For clayey soils or soils with greater contents of organic matter, this secondary settlement (consolidation) s_s can be of greater significance, while for gravely soils it is mostly negligible. In any time, when neglecting secondary consolidation, we have:

- for 1D: $s_t = s_c \cdot U$ where U is the degree of consolidation (for 1D – Terzaghi theory),
- for 3D: $s_t = s_i + s_c \cdot U_s$ where U_s is the degree of consolidation under triaxial conditions.

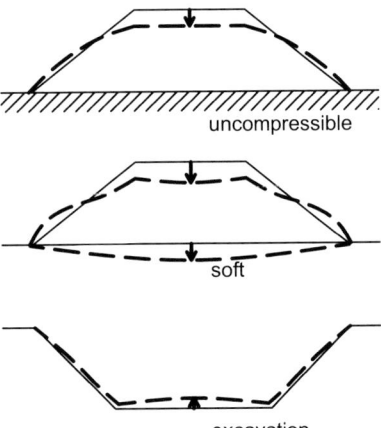

Fig. 5.23 Basic examples of earth structures deformations

5.2 Basic Limit States for Earth Structures

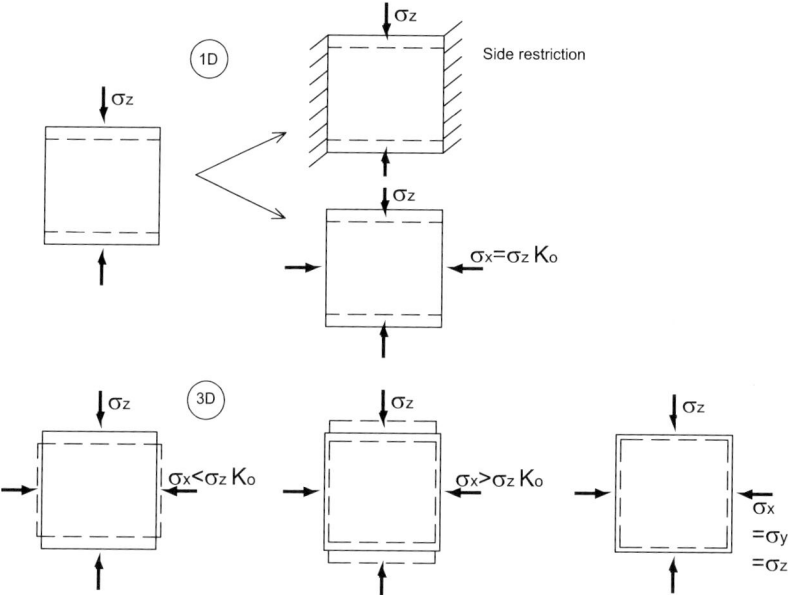

Fig. 5.24 Standard types of soil element deformations

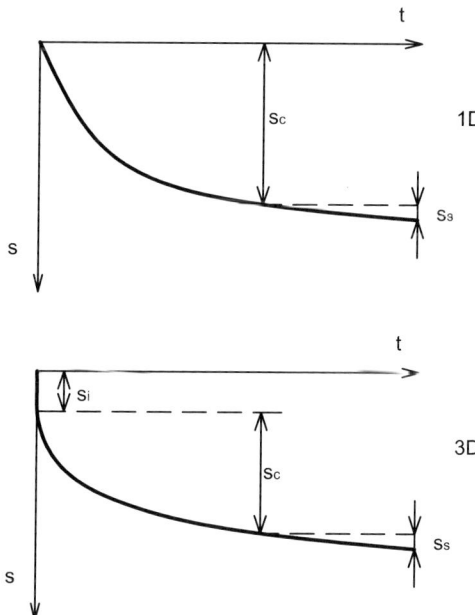

Fig. 5.25 Settlement in time for 1D and 3D conditions

This very fact clearly implies that the deformation characteristics applied in the theory of elasticity such as

- the modulus of elasticity (Young's) E
- Poisson's ratio v

cannot be generally applied in the domain of earth structures and soil mechanics.

5.2.2.1 Deformation Characteristics of Soils

Basic deformation characteristics for 1D and 3D deformation, or for undrained and drained conditions, are summarized in Table 5.2.

Another reason why deformation characteristics from the theory of elasticity cannot be applied is the fact that for soils the deformation is not linear to loading. Fig. 5.26 schematically shows the relative deformation ε as a function of the effective stress for the oedometer test, or deformation as a function of the difference in stresses for the test in the classical triaxial chamber. Principal differences between these types of tests are discussed e.g. by Janbu (1963). For the test in the oedometer, the procedure of determining the oedometric modulus can be seen, or the character of the deformation curve after the sample is unloaded and successively recompressed for triaxial test. For the test in the triaxial chamber, the cylindrically shaped sample is first subjected to the chamber stress σ_0 and successively to the vertical load increase $\Delta(\sigma_1-\sigma_3)$. Depending on whether bottom drainage is closed or opened for the second application phase of the deviatoric stress, undrained or drained conditions are distinguished.

Schematic pictures show the assessment procedure of the moduli where for 3D deformation the initial moduli can be assessed as both the tangent and the secant moduli, or the secant modulus for the selected stress range. At the same time, it is evident that for the loading phase and for the unloading phase (swelling), or even for the recompression phase, different moduli must be defined as the value of "springing" for the unloading phase is significantly lower – soils do not behave as elastic substances.

Table 5.2 Basic deformation characteristics for 1 and 3D deformation and for undrained and drained conditions

	Undrained conditions	Drained conditions
1D	–	E_{oed}
3D	$E_u; v_u$	$E_{def}; v_{ef}$

E_{oed} – Oedometric modulus of deformation,
E_u – Modulus of deformation under undrained conditions,
v_u – Poisson's ratio for undrained conditions (for $S_r = 1$; $v_u = 0.5$),
E_{def} – Modulus of deformation for drained conditions,
v_{ef} – Poisson's ratio for drained conditions.

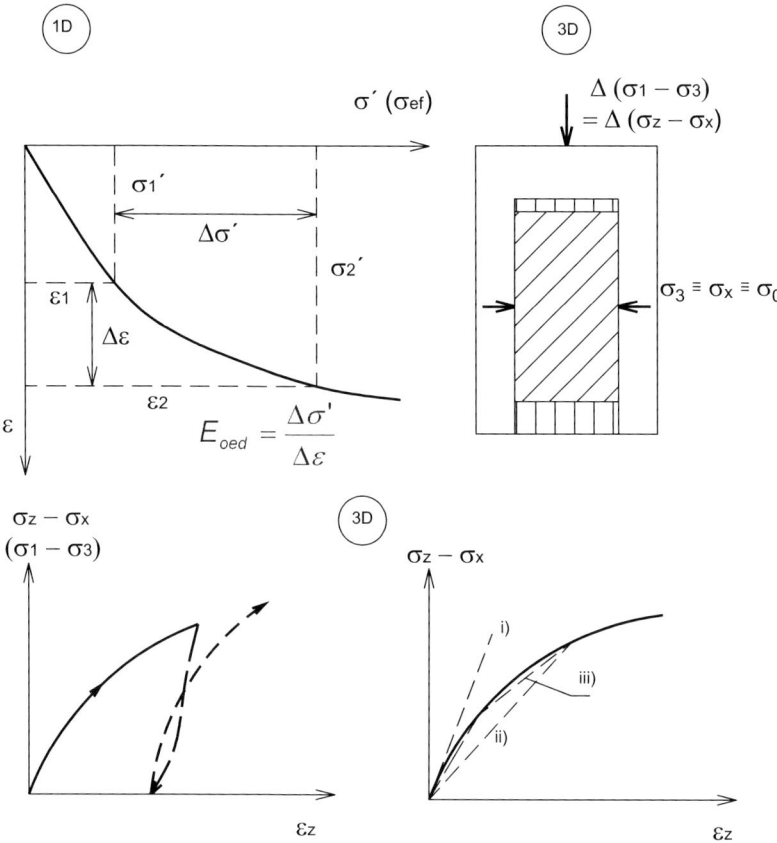

Fig. 5.26 Schematics for determination of deformation characteristics of soils for 1D and 3D conditions. (**i**) Initial tangent modulus. (**ii**) Initial secant modulus. (**iii**) Secant modulus for selected loading interval

Nevertheless, for a specific, strictly defined range of stress changes, the calculated modulus can be applied as a substitute modulus of elasticity, and used for subsequent related calculations.

The deformation curve determined in the oedometer can be obtained by means of step-by-step loading, i.e. during gradual loading with stresses such as e.g. 50 kPa, 100 kPa, 200 kPa, 400 kPa thus obtaining the deformation curve as a result of determined relative strains ε_{50}, ε_{100}, ε_{200}, ε_{400}, or by means of a test applying continuous loading where the oedometric box is placed in the press and gradually loaded. Provided that deformation in time t for individual loading degrees is well registered, the coefficient of consolidation U for 1D consolidation can be determined. The advantages and various methods of continuous loading are described e.g. by Janbu and Senneset (1979), or Vaníček and Šťastný (1981), Vaníček and Záleský (1984).

For 1D deformation, other deformation characteristics expressing the relation $\Delta\varepsilon = f(\Delta\sigma')$ and/or $\Delta e = f(\Delta\sigma')$ can also be used where e is the void ratio. Among the most commonly used ones there are:

– coefficient of volume compressibility: $m_v = \Delta\varepsilon/\Delta\sigma' = 1/\text{Eoed}$
– coefficient of compressibility: $C = 1/\Delta\varepsilon \times \Delta ln\sigma'$
– compressibility ratio: $a_v = -\Delta e/\Delta\sigma'$
– compressibility index: $C_c = -\Delta e/\Delta \log \sigma'$

An advantage of deformation characteristics based on the change in the void ratio e is some control over porosity during the whole loading process. It is natural that compressibility under the same stress range will be lower for more compacted soils. The deformation characteristics based on deformation relationships in a semi-logarithmic scale were preferred as their variability in relation to the loading range was small, this deformation characteristic is being counted as linear for large range of loading. However using the semi-logarithmic relation, the difference in the behaviour of normally consolidated and overconsolidated soils will be shown below. Figure 5.27 displays a typical deformation curve for clayey soils prepared from paste or sandy soils loosely poured into the oedometric box. In section a–b the relationship is of a linear type, referred to as a so-called primary, virgin line. After unloading, some springing occurs (section b–c). After a repeated increase in loading to the original maximum value, the deformation follows another curve (section c–d). The section of the deformation curve representing unloading and restoration of the original loading is called the "hysteresis curve" (section b–c–d). Following another load increase in section d–e, the deformation curve returns to its primary line. For overconsolidated soils, which have already gone through a history of higher loading with successive unloading, therefore, the typical deformation curve corresponds to the curve section c–d–e.

For 3D deformation, as was already implied, the test runs in two phases. In the first phase, the sample consolidates under omnidirectional loading $\sigma_0 = \sigma_1 = \sigma_2 = \sigma_3$, and the change in the sample's volume ΔV is measured (for a saturated sample

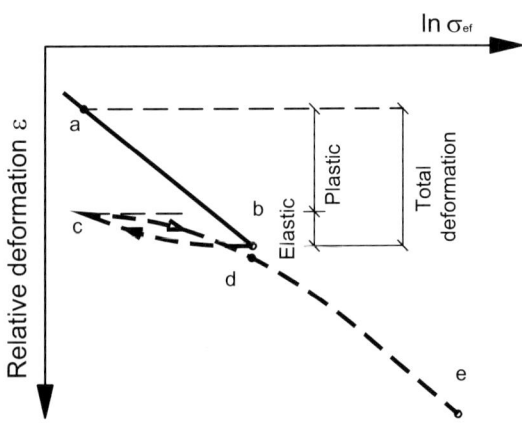

Fig. 5.27 Idealized stress-strain curve for oedometer test with the distinction of elastic and plastic deformations

5.2 Basic Limit States for Earth Structures

from the amount of water drained). The modulus of deformation for the given interval of load increase is determined from the relation:

$$E_{def} = V/\Delta V \times 3\,\sigma_0(1-2\nu) \tag{5.57}$$

At the same time, this phase may serve for the specification of the coefficient of consolidation for 3 D consolidation using the relation $\Delta V = f(t)$.

Poisson's ratio is determined from the second loading phase where the load increase acts only vertically, and the vertical strain $\Delta\varepsilon_z$ and the volume change ΔV are measured. In the theory of elasticity, it holds true that:

$$\Delta V/V = \varepsilon_z + \varepsilon_x + \varepsilon_y = \varepsilon \tag{5.58}$$

and then:

$$\nu = \frac{1}{2}(1 - \Delta\varepsilon/\Delta\varepsilon_z) \tag{5.59}$$

To conclude this part, we may sum up that in the case of 1D deformation the relative vertical deformation may be determined from the relation:

$$\Delta\varepsilon_z = \Delta\sigma_z'/E_{oed} \tag{5.60}$$

Then, for $\varepsilon_z = \Delta z/z$ the vertical deformation (settlement) Δz depends on the thickness of the compacted layer z and the mean effective stress increase in this layer $\Delta\sigma_z$':

$$\Delta z = \Delta\sigma_z' \times z/E_{oed} \tag{5.61}$$

For the 3D problem respectively the relative deformations may be determined from the relations:

$$\Delta\varepsilon_z = 1/E_{def}(\Delta\sigma_z - \nu(\Delta\sigma_x + \Delta\sigma_y)) \tag{5.62}$$

$$\Delta\varepsilon_x = 1/E_{def}(\Delta\sigma_x - \nu(\Delta\sigma_z + \Delta\sigma_y)) \tag{5.63}$$

$$\Delta\varepsilon_y = 1/E_{def}(\Delta\sigma_y - \nu(\Delta\sigma_z + \Delta\sigma_x)) \tag{5.64}$$

respecting all comments concerning the validity of the modulus of deformation.

5.2.2.2 Calculation of Deformation of an Element of Earth Structure

The calculation of settlement for 1D deformation or for vertical and horizontal deformations in 3D (or 2D deformation in planar problems) requires the knowledge of stress increases for individual elements (strata) and their corresponding deformation characteristics. Thus, the problem may be subdivided into two parts which, however, are very closely interrelated. The first part of the problem consists of determining the initial state of stress of the element and successively its stress increase, while the second part of the problem specifies corresponding deformation characteristics for this change in the state of stress. The calculation of the relative deformation of an element $\Delta\varepsilon_z$ or also $\Delta\varepsilon_x$, $\Delta\varepsilon_y$ is governed by the relations mentioned above. Prior to the specification of basic cases, however, it is desirable to point out that some ways of changing the state of stress of an element lead to its strengthening, while others lead to its failure. For example, among those mentioned above, the case of omnidirectional stress increase $\Delta\sigma_0 = \Delta\sigma_1 = \Delta\sigma_2 = \Delta\sigma_3$ or the case of 1D deformation and the stress increase implemented in the oedometer lead to strengthening, but the case of only a vertical stress increase leads to a failure. Therefore, the oedometric modulus of deformation grows with the growing stress and the element is strengthened, and, on the contrary, the modulus of deformation under 3D conditions determined from the vertical stress increase phase decreases, and the stress increase procedure leads to a failure. In terms of changes in the state of stress, 4 basic cases can be specified, which lead to a failure:

i) $\Delta\sigma_z = +$; $\Delta\sigma_x = 0$, (vertical load increase $\gg$ vertical compression)
ii) $\Delta\sigma_z = 0$; $\Delta\sigma_x = +$, (horizontal load increase $\gg$ vertical extension)
iii) $\Delta\sigma_z = -$; $\Delta\sigma_x = 0$, (vertical unloading $\gg$ vertical extension)
iv) $\Delta\sigma_z = 0$; $\Delta\sigma_x = -$, (horizontal unloading $\gg$ vertical compression)

In a simplified form, the case ad i) can be put down to the element of subsoil, which is additionally loaded by the foundation, embankment, the case ad iii) to the subsoil in which the cutting is built, e.g. the behaviour of an element on the bottom of a foundation pit, the case ad ii) to an element behind a retaining wall in case of passive earth pressure and the case iv) to an element behind a retaining wall in case of active earth pressure (see also Fig. 6.37). This also implies that the element's failure may also be due to tensile loading – and the calculation of such tensile failure and deformation in the tensile domain will require the definition of strength and strain characteristics of soils in tension. These problems will be separately treated in Chapter 7 within the monitoring of tensile cracking in earth fill dams.

Calculation of Initial State of Stress

An element of the subsoil of an earth structure or the subsoil in which an earth structure is built prior to its implementation is generally characterized by original, geostatic state of stress – σ_{or}, and it generally holds true that:

$\sigma_z = \gamma.z$ – for the total state of stress or $\sigma'_z = \sigma_z - u$ for the effective state of stress (for $S_r = 1$)
or $\sigma'_z = \sigma_z - (u_a - \chi(u_a - u_w))$ – for the case $S_r < 1$
and where χ is the function of S_r (e.g. Bishop and Henkel (1962) or Bishop (1959))
where z is the depth of the element below ground and γ – unit weight of soil is constant,
then:

5.2 Basic Limit States for Earth Structures

$\sigma'_x = K_0 \cdot \sigma'_z$ where K_0 is the coefficient of earth pressure at rest and
$K_0 = K_{0nc}$ – where K_{0nc} is the coefficient of earth pressure at rest – corresponding to normally consolidated soils and also the loading method in the oedometer ($K_{0nc} = \nu/(1-\nu)$) and in this case is lower then 1, or
$K_0 > K_{0nc}$ which corresponds to the situation of overconsolidated soils.

Therefore, the determination of the history of "deposition, erosion and groundwater changes as soil origin" is of utmost importance, and may be more closely specified by means of OCR – the over consolidation ratio, which is defined as the ratio of the maximum vertical effective stress by which an element was loaded in its history to the current vertical effective stress, see Fig. 5.28. The outline of direct and indirect methods of the determination of the coefficient K_0 is given e.g. by Feda (1977,1982). For soils in artificial embankments, behaviour similar to the behaviour of normally consolidated soils is presumed. An exception to this rule is the case where intensive compaction occurs behind a rigid bridge support with incomplete relaxation of horizontal stress in the compacted layer just after the completion of compaction by roller. Thus, the horizontal load acting on the rigid bridge support can be greater than what would correspond to the earth pressure at rest for normally consolidated soils.

The aim of setting the initial state of stress is to determine the initial point on the deformation curve from which a load increase due to external load will be derived and thus also the range of the state of stress for the determination of the modulus of deformation. The initial state of stress corresponding to normally consolidated soils is simulated in the oedometer, whereas in the classical triaxial apparatus it is first omnidirectional loading ($K_0 = 1$) and then a load increase due to vertical stress, Bishop and Henkel (1962). Triaxial apparatus with controlled stress paths is used for the general representation of both initial state of stress and for loading along arbitrary stress paths, Bishop and Wesley (1975). In the first phase, the term of sample reconsolidation is used, an effort to put a soil sample taken from the site to an effective state of stress in which the tested element existed before sampling. With a view to the equation for effective stress for partially saturated soils, it is necessary to know the degree of saturation S_r. A sample of cohesive soil is reconsolidated in the

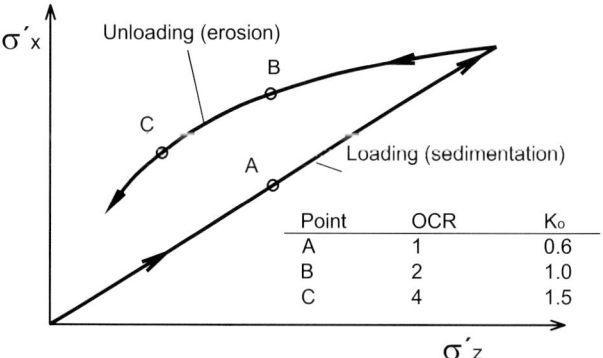

Fig. 5.28 Relationship of coefficient of earth pressure at rest K_0 on degree of overconsolidation

oedometer for the calculated effective vertical stress, and the sample is watered until $S_r \geq 0.9$, frequently even up to 0.8 including some inaccuracies in relation to pore pressures. At values lower than 0.8, the unsaturated sample is tested, considering, however, the fact that in the case of a potential higher degree of in situ sample saturation the sample may suffer from a settlement failure.

Calculation of Changes in State of Stress

The calculation of changes in the state of stress most commonly applies the theory of elastic half-space, even though the application of computational methods – the finite-element method, has lately prevailed. Other principal cases, however, cannot be omitted, either where effective stresses change due to the groundwater level changes. Thus, the lowering of the groundwater level leads to settlement, while its rise to heaving.

Figure 5.29 displays basic cases of external loading of elastic half-space, which are included in the majority of fundamental textbooks on soil mechanics, e.g. Myslivec et al. (1970), Kézdi (1964), Vaníček (1982a). Among special publications, this problem is described in detail by Poulos and Davis (1974).

For a concentrated force F representing point load – case (a), we have:

$$\sigma_z = 3F.z^3/2\pi.r^5 \tag{5.65}$$

$$\sigma_x = 3F.x^2.z/2\pi.r^5 \tag{5.66}$$

There are not many examples of point load in practice, but it may be applied as a form of simplified complex load, distributed into more concentrated forces, as the superposition principle may be applied for the elastic half-space.

For case (b) it holds true for the linear vertical load f:

$$\sigma_z = 2f.z^3/\pi r^4 \tag{5.67}$$

$$\sigma_x = 2f.x^2.z/\pi r^4 \tag{5.68}$$

Then, for stress at a general point M under vertical load on a strip – see Fig. 5.29c it holds true that:

$$\sigma_z = \frac{2 \cdot f}{\pi} \cdot \int_{\beta_1}^{\beta_2} \cos^2 \beta \cdot d\beta = \frac{f}{\pi} \left[\beta_1 + \frac{1}{2} \sin 2\beta - (\pm \beta_2) - \frac{1}{2} \sin(\pm \beta_2) \right] \tag{5.69}$$

$$\sigma_x = \frac{f}{\pi} \left[\beta_1 - \frac{1}{2} \sin 2\beta_1 - (\pm \beta_2) + \frac{1}{2} \sin(\pm \beta_2) \right] \tag{5.70}$$

Diagrams are often available to illustrate most commonly used cases. Figure 5.30 shows the stress distribution under the corner of a rectangular surcharge.

5.2 Basic Limit States for Earth Structures

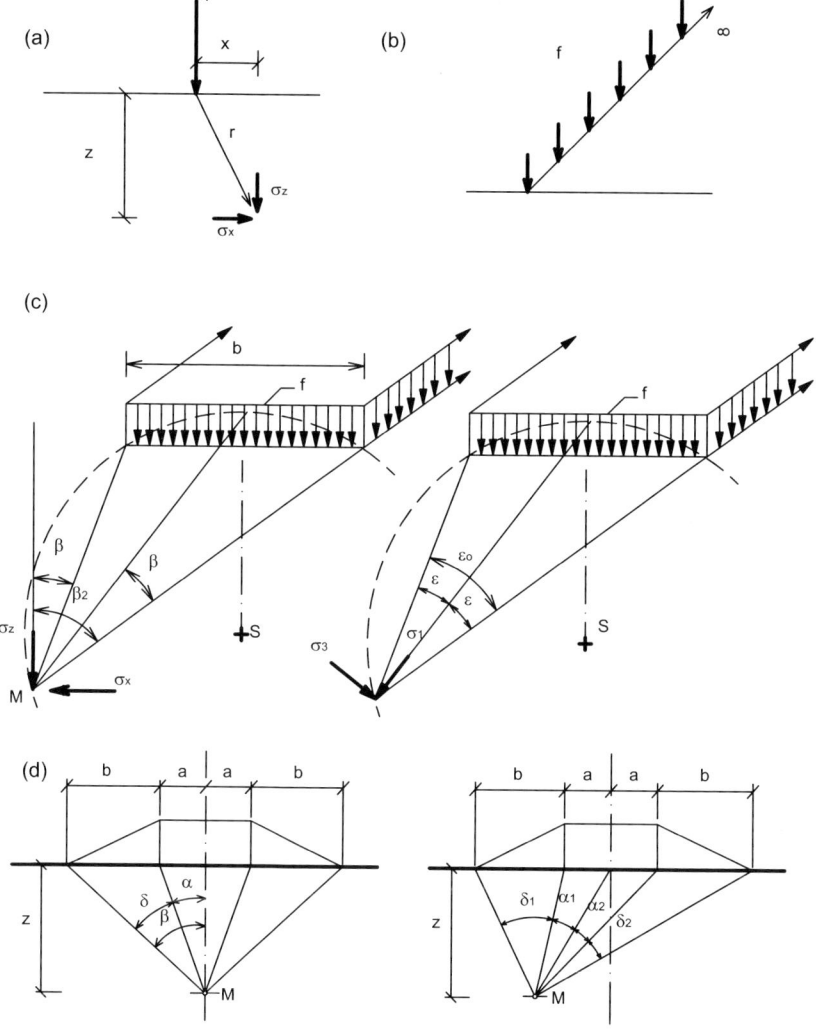

Fig. 5.29 Scheme for calculation of stresses in elastic half-space under basic load cases. (**a**) Point load. (**b**) Line load. (**c**) Strip load. (**d**) Trapezoidal load

The calculation of a case with a stress increase underneath an embankment (Fig. 5.29d), the load being trapezoidal in shape with maximum of q, may be solved with the use of the following relations:

$$\sigma_z = \frac{2q}{\pi}\left(\beta + \frac{a}{b}\delta\right) \quad (5.71)$$

$$\sigma_x = \frac{2q}{\pi}\left(\beta + \frac{a}{b}\delta - 2\frac{z}{b}\ln\frac{\cos\alpha}{\cos\beta}\right), \quad (5.72)$$

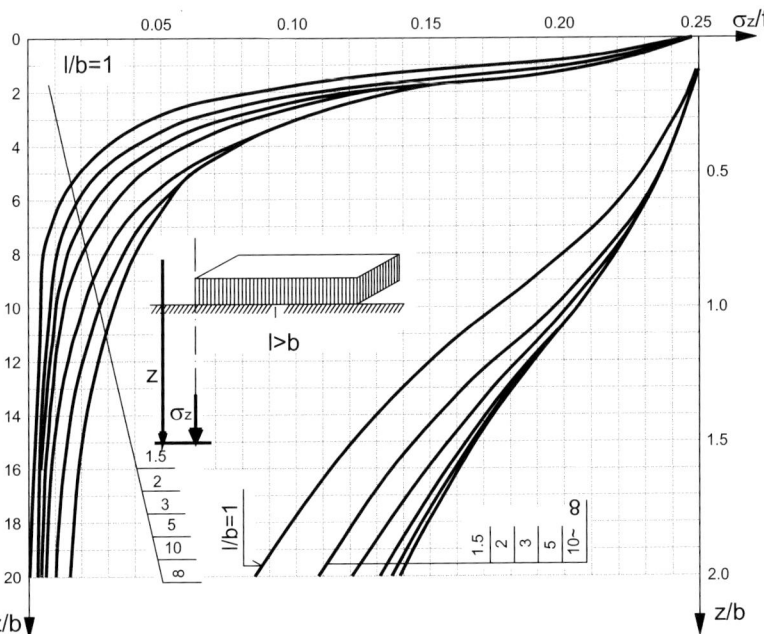

Fig. 5.30 Stresses in elastic half-space under corner of rectangular load

That are applicable for stresses on the axis of trapezoidal load and for the vertical stress at any point M, we have:

$$\sigma_z = \frac{q}{\pi b}[b(\alpha_1 + \alpha_2) + (a+b)(\delta_1 + \delta_2) + x(\delta_2 - \delta_1)], \tag{5.73}$$

The application of the theory of elastic half-space is confined to homogeneous subsoil. In the case of the occurrence of incompressible layer, e.g. bedrock, at real depths or the occurrence of multi-layered subsoil with different characteristics and inclined strata, significant limitations are encountered. Nevertheless, precise analytic solutions are very useful in the verification of the accuracy of approximate analytic and/or numerical methods. The exploitation of the finite-element method, therefore, represents a highly interesting possibility of solving this problem.

The finite-element method (FEM) is an efficient numerical method, suitable for the solution of planar and spatial problems in the field of mechanics of earth structures. The method is based on the discretization of the investigated domain into planar or spatial elements of finite dimensions which are thought to be continuous only in a finite number of points, so-called nodal points of a planar or spatial mesh thus formed. Three basic requirements of mechanics are then applied to individual finite elements and the system as a whole:

- fulfilment of conditions of equilibrium,
- mutual compatibility, i.e. condition of continuity,
- fulfilment of presumed constitutional relations (stress-strain relationships).

5.2 Basic Limit States for Earth Structures

The last-mentioned point in particular, i.e. the specificity and demands for setting constitutional relations for soils and rocks, is what distinguishes the solution from other problems of theoretical mechanics. Therefore, they will become the focus of at least brief attention in the next chapter. The tremendous significance of FEM for geomechanics consists of the fact that this method allows solving with high precision not only problems with complex boundary conditions and with a complex subsoil composition, but also domains with materials with arbitrary time-independent or time-dependent constitutional relations, each element being able to possess different characteristics. The solution results in components of stress, movement and angular rotation at all nodal points of the mesh. Thus, we can obtain an overview of stresses and deformations within the entire investigated domain.

Jesenák in Šimek et al. (1990) stresses the importance of FEM solutions even in a relatively simple case of a homogeneous, isotropic and linearly elastic subsoil model at a depth d subjected to uniform loading of width B. The result obtained from FEM is compared with analytical solutions by Kézdi (1964) and Škopek (1968), or with a solution recommended in ČSN 731001, and finally with an ideal case of loading on the surface of elastic half-space, Fig. 5.31. The comparison shows, on one hand, that the presumption of theoretical solutions is unreal (loading inside of undamaged half-space which is able to transfer tensile stresses above the point of load application), and a relatively good agreement with the approximate solution used in ČSN 731001 (stress at the underside of footing acts in its full value but falls faster with depth than if the load acts on the surface) on the other hand.

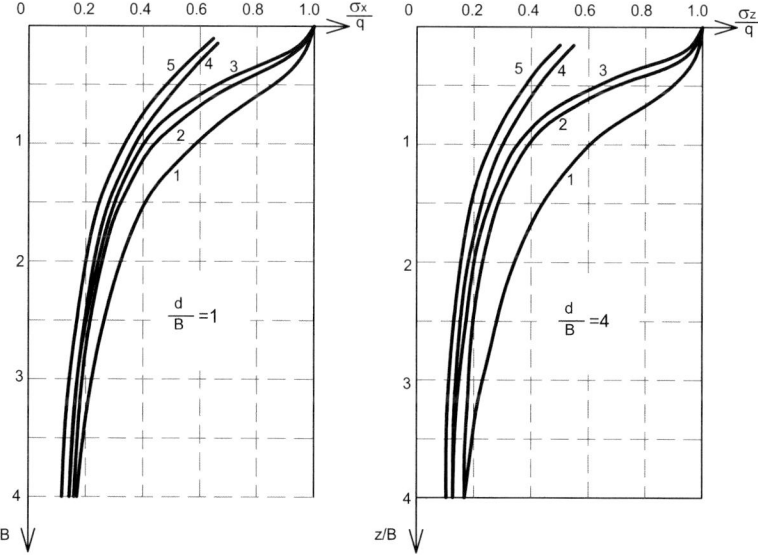

Fig. 5.31 Strip load – comparison of the influence of foundation depths using different methods. (**1**) Load on surface. (**2**) FEM. (**3**) ČSN 731001. (**4**) Škopek. (**5**) Kézdi

Calculation of Total Settlement for 1D Deformation

This case theoretically comes to mind in the case of large-area load. Then, the elementary thin layer in the subsoil with a thickness z is additionally loaded with a vertical stress corresponding to the surface load increase magnitude. Under a load increase with a limited area, this presumption is no longer valid as first the stress is spread into depth (the vertical stress increment getting smaller with depth) and, also, horizontal deformation cannot be shown any more. The original presumption is most closely valid in the area below the loading centre, and least evident under its edge.

Nevertheless, the 1D deformation principle is applied for the calculation of the settlement of spread foundations (which are loaded by spread load over a very limited area) as well. This is given not only by the simplicity of the solution, including the determination of deformation characteristics, but also by the fact that the loaded area (e.g. under the centre of a circular foundation) – see Fig. 5.32 can be subdivided into 3 parts where

- in the first part, the horizontal load increase is greater than would correspond to the 1D deformation state, approaching isotropic surcharge, hence the soil compressibility in the vertical direction will be smaller than would correspond to the 1D presumption,
- in the second part, the coefficient of earth pressure $K \triangleq K_0$ principally corresponds to the 1D presumption,
- in the third part, the coefficient of earth pressure falls rapidly, the loading procedure being nearly identical to the second phase of load increase in the classical triaxial apparatus applying only vertical load increase – the compressibility will be higher than would correspond the 1D presumption.

Thus, some inaccuracy and error compensation occurs. Therefore, the 1D deformation principle may be used even for the most frequent case of the calculation of the settlement of subsoil loaded with an embankment, and stress increments for individual points will be established by means of the methods mentioned above. Subsoil is principally subdivided into little layers, and the initial stress and load increase are calculated at typical subsoil points (e.g. under the embankment centre, under its edge). Undisturbed samples serve for the determination of the corresponding modulus of deformation for the specified stress range with the help of the deformation curve. The modulus of deformation, in turn, is applied for the calculation of the compression of any single thin layer at monitored points. Total settlement is then obtained as the sum of the settlement values of individual layers.

$$s = \Sigma \Delta z_i \qquad (5.74)$$

$$\text{where } \Delta z_i = z . \Delta \sigma_z' / E_{oed} \qquad (5.75)$$

5.2 Basic Limit States for Earth Structures

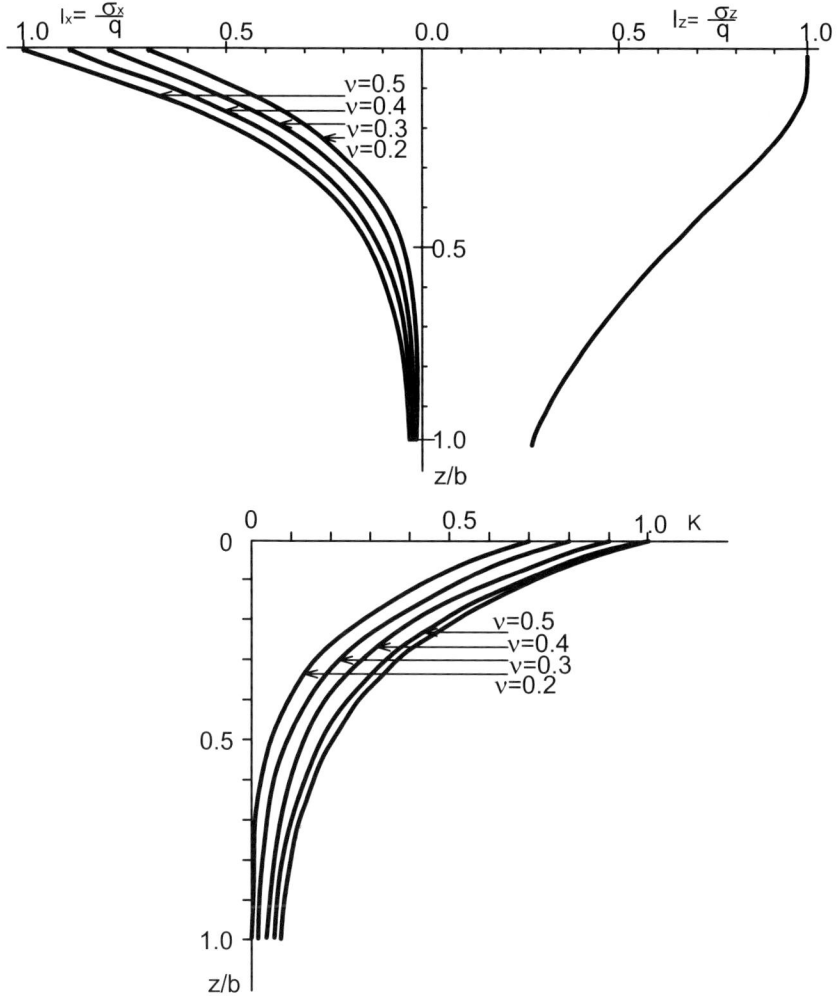

Fig. 5.32 Horizontal and vertical stresses under the centre of circular load and their ratio for different depth

As was already said, this is the total, consolidation settlement in accordance with Fig. 5.25. In order to calculate settlement in time t, the uniaxial consolidation theory by Terzaghi is applied and

$$s_t = s_c \cdot U \quad (5.76)$$

Calculation of Initial Settlement (Deformation) for 3D Conditions

The subdivision of subsoil into individual layers is analogical. Undisturbed samples are reconsolidated onto initial stress values, in this case onto omnidirectional load, and subsequently vertically loaded with presumed load magnitudes under undrained conditions. For clayey, water-saturated samples, the volume change in the second phase equals to zero, Poisson's ratio $v = 0.5$. The undrained modulus of deformation E_u is calculated from the known values of the vertical stress increase and vertical strain. The results are used for the calculation of deformations of individual elements with the application of Hooke's law, both for the vertical and horizontal direction.

Calculation of Total Settlement (Deformation) for 3D Conditions

The procedure is analogical to the previous case, with the only difference that drainage is opened for the second phase of the vertical stress increase, and the change in the sample volume is measured (for saturated soils from the change in drained water). The evaluation will produce the modulus of deformation E_{def} under drained conditions and Poisson's ratio v. Both values are used for the calculation of total deformation – the deformation of an element both in the vertical and horizontal direction.

Both procedures for 3D conditions show a certain degree of simplification, the initial stress state need not to be of isotropic type, and some simplification also occurs in the second loading phase as only a vertical load increase is applied. These problems can be eliminated by using the following procedure.

Calculation of Settlement (Deformation) of Earth Structures Under Stress Path Control

Laboratory determination of both the undrained and drained moduli will involve the triaxial apparatus with stress path control. Thus, the reconsolidation phase can be selected in general terms, without prioritizing isotropic loads or loads under K_0 conditions. The procedure can be based on the initial stress state values determined on the basis of more detailed investigation. The second loading phase, in particular, allows the application of both the calculated vertical stress and the horizontal stress increment. The loading process, as it is, can be of step-by-step type, simulating the construction process. The strain values obtained (or the evaluated modulus of deformation and Poisson's ratio) of the sample are then applied onto the strain of monitored elements. Thus, the aim of laboratory testing is as true simulation of the stress state change pattern for the selected element as possible applying the resulting strain values onto a real element.

The Effect of Structural Strength on Deformation Calculation

Numerous application problems show that in most cases the calculated settlements, deformations are greater than the real values measured. This difference is most often attributed to some type of sample failure during the process of sample extraction, sample handling before testing. In order to explain the above-mentioned difference, extensive laboratory and in situ tests were carried out. High-precision

5.2 Basic Limit States for Earth Structures

sensors registering vertical deformations were mounted under modelled real-life spread foundations. The results manifested that negligible vertical movements were measured in the subsoil despite that according to the theory of elastic half-space there should be stress increase and hence also deformation. The differences in results, however, were also determined for different types of soils, Havlíček (1978), Seyček (1995). The differences found were attributed to the structural strength of soil, i.e. the bond between individual grains, which develops in time for long-term acting initial stress. It is becoming evident that additional bonds arise in soils exposed to long-term action of geostatic stress, which are manifested by the appearance of overconsolidation stress even in normally consolidated soils or by its increase in overconsolidated soils. This fact is put down to secondary consolidation, or the appearance of additional diagenetic bonds (such as cementation). These bonds, sometimes also referred to as "cold welding", are of relatively fragile nature, and they may easily be broken even by mere sample unloading to zero external load. Some explanation can be drawn from the results of laboratory tests. In determining the deformation curve of soil prepared from paste, a linear relationship is obtained in a semi-logarithmic scale. If the sample is exposed, in the long-term, under a stress simulating the geostatic stress σ'_{or}, to the effect of secondary consolidation, with growing stress it is not deformed along the primary (virgin) line unless the stress increment exceeds a certain value, Fig. 5.33. This fact was already mentioned by Leonards and Altschaeffl (1964), who calculated the ratio of the difference between the overconsolidation stress σ'_c and the initial geostatic stress σ'_{or} as of about 0.4. This value was also obtained from other results for various clays prepared under different conditions. At the same time, however, Leonards and Altschaeffl pointed out a very important fact, i.e. that this strengthening is lost with the sample removal (i.e. after its unloading and reloading). Only in the case of so-called ideal removal where the sample was removed and reloaded to the initial stress within 15 minutes, this strengthening was also registered. This stresses the importance of sample extraction and its laboratory testing. In other words, under small stress increments above

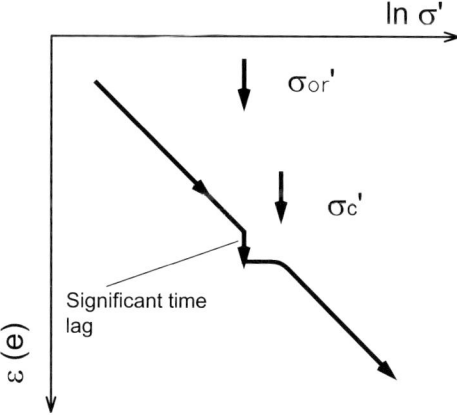

Fig. 5.33 Deformation of normally consolidated clay with significant time delay for loading σ'_{or}

geostatic stress, the deformation is very small, in practical terms even insignificant as the modulus of deformation in this stress increment range is many times higher than after the failure of these bonds. This phenomenon is sometimes also called as "small-strain stiffness".

Havlíček, therefore, recommended including the above-mentioned behaviour, called structural strength, in the calculation of the settlement of spread foundations. The solution accounts for the fact that a layer starts to be deformed if the load increase exceeds the initial geostatic stress to which the soil has already consolidated to by about 10–40%. The coefficient of structural strength $m = 0.1–0.4$ is allocated to individual soil types as follows:

- $m = 0.1$ – for fine-grained soils of type F normally consolidated, of soft to stiff consistency with a modulus of deformation up to 4 MPa, (all three attributes must be fulfilled),
- $m = 0.2$ – for fine-grained soils failing to fulfil the above-mentioned conditions, clean sands and gravels under groundwater level,
- $m = 0.3$ – clean sands and gravels above groundwater level, loamy or clayey sands and gravels,
- $m = 0.4$ – eluvia.

Extreme values of the coefficient m (0.6) were determined for loess and loess loams above groundwater level, which, however, may be used only if their future water-saturation can be eliminated. In 1987, this procedure became part of ČSN 731001, which uses for the calculation of 1D settlement following equation:

$$\Delta z_i = (\Delta \sigma_{z,i} - m.\sigma_{or,i}) \times z_i / E_{oed,i} \tag{5.77}$$

resulting in a substantially greater agreement between calculated and measured settlements.

This fact can be taken into account for subsoil settlement under the embankment construction. For the embankment soils this structural strength can develop after some delay, but maximum value of $m = 0.1$ can be recommended for.

5.2.3 Limit State of Internal Erosion

The limit state of internal erosion is one of the most monitored factors during the construction of embankment dams. Therefore, this case will be the centre of special attention mainly in Chapter 7 where potential non-standard cases will also be solved. Now, the focus is on basic stability cases of interfaces between layers, which can also occur in transportation and environmental engineering during the protection of drains.

5.2.3.1 Theoretical Stability Model of Interfaces Between Layers

Considering the fact whether small particles of covered base soil can pass through the pores of a filter due to hydraulic gradient, the scheme displayed in Fig. 5.34, can be used, which is applied for the explanation of basic forces acting on a small grain e.g. by Vaughan (1976) or Vaníček (1988). The stability of a base soil particle at the interface between layers, which decides if the particle is shifted into the adjoining filter, will depend on the size and direction of 3 forces which act on it:

- W – particle weight acting vertically,
- P – motion force acting, on average, on a particle in the flow direction, the motion force being proportional to hydraulic gradient,
- A – the resultant of attractive forces between individual particles of base soil acting on the mean normal of contact surfaces.

If a particle exposed to the effect of these forces is stable, then in principle there is no need for a filter. If, however, the particle is unstable, then a filter is necessary, and it must avoid the outwash of this particle. This is possible only if the continuous channel between the pores with a diameter δ is smaller than the particle in question. In practice, however, base soil contains some particles so small that they can pass through this continuous channel. These particles, therefore, are eliminated from the interfaces between layers, and their place is taken up by other shifted particles. Some of these particles can be so big to block the entrance into smaller seepage channels, which progressively blocks up other entrances with larger diameters with bigger particles. The effective size of the entry pore is thus getting gradually smaller until no more particles are able to enter into the filter. In this way, the particle movement further away from the interface of both layers is limited. The described process of equilibrium establishment goes on until even the smallest particle of base soil stops moving. The efficiency of a designed filter will depend on the size of filter grains, on the acceptability of particle movement and on material loss which occurs before the establishment of stable conditions.

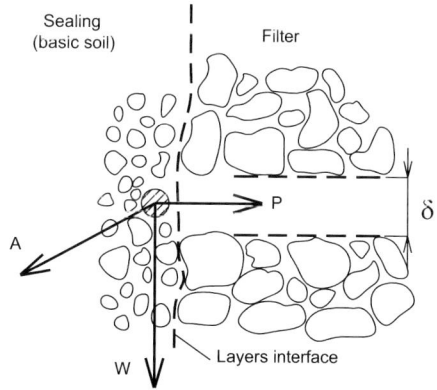

Fig. 5.34 Scheme of stability for individual particle (δ – dia of continuous channel in filter)

There is a fundamental difference between cohesive and non-cohesive soils as in non-cohesive soils the magnitude of the attractive force A between individual particles is practically negligible. In cohesive soils, a crack may arise, which can remain open for a certain time as the magnitude of attractive forces will ensure the stability of the crack's vault. In non-cohesive soils, under undrained conditions a crack may arise but after saturation with water it collapses. Therefore, there is no significant possibility for water to make this crack gradually larger and the situation worse.

Non-cohesive Soils

Because of a negligible effect of the force A, the forces decisive for further considerations are W and P. The effect of the flow direction is described by Zwek and Davidenkoff (1957). They discovered a structural vault effect between grains with a horizontal interface and vertical flow. In their tests, they used both even-(uniformly) graded filters and base soil. Under vertical degressive flow, the D_{50}/d_{50} ratio (D_{50} being the filter grain diameter, d_{50} the diameter of the base soil grain) necessary for effective filtration was independent of the hydraulic gradient and the base soil grain size.

Under ascending flow and under horizontal flow (vertical interface of layers) they found out that the D_{50}/d_{50} ratio is inversely proportional to the hydraulic gradient and proportional to the d_{50} average. Ascending flow was safer than horizontal flow. This effect of flow and the direction of the interface of layers is usually neglected by the filtration criteria. Still, the authors proved that the USBR criteria (US Bureau of Reclamation), which will be presented in the next section, are safe for all flow directions, with just a small exception of coarse-grained base soils under high hydraulic gradients. The USBR criteria are generally recognized as reliable, allowing only a minimum movement of particles for setting up stable conditions. These criteria are suited for critical situations in which the interface is exposed to long-term load with water seepage. The US ACE (US Army Corps of Engineers) criteria are less strict, and they may be used at places where some migration is allowable or where the flow is transient.

Undisturbed Cohesive Soils

In these soils, the force W as compared to forces A and P is relatively insignificant. The force A, however, also depends on the chemical composition of clays and pore water, as will be subsequently manifested for a specific behaviour of dispersive soils. The force A as the resultant of attraction forces of clay particles in clays is equivalent to the effective tensile strength σ'_t. Under the effective tensile strength, a stable vault in base soil is formed at its contact with two filter grains. This stable vault effect was experimentally verified (e.g. by Kassif et al. (1965), Wolski (1965), Istomina et al. (1975)); there was even an effort to express this fact by mathematical means. The attempts, however, show that the hazard of disturbing this vault in cohesive soils and the resultant risk of internal erosion are very low even when using relatively coarse filters.

5.2.3.2 Filter Design – Filtration Criteria

The first effective filter was designed by Terzaghi (see e.g. Terzaghi and Peck – 1948) more than 60 years ago. Other results of laboratory tests were published by Bertram (1940). Their designs were intended for non-cohesive materials so that the base soil and the filter consisted of sand or gravel (of different particle size distribution). In multi-layer filters and/or base soil they suit the character of sandy-silty soil with low plasticity. Still, their individual designs were gradually applied even to soil compositions consisting of cohesive soil. The fact that this type of application is ambiguous (with diversions in both directions) is supported by previous considerations. Therefore, individual applications below will be dealt with separately.

Base Soil is Non-cohesive

Terzaghi distinguished two basic conditions – filtration criteria - in the design of a filtration layer. The first condition ensures the interface stability, while the second ensures at least 5-times to 10-times greater permeability of the filtration layer.
First condition:

$$D_{15}/d_{85} < 4 \qquad (5.78)$$

where
  D is the grain size for the filter,
  d – the grain size for base soil.
Second condition:

$$D_{15}/d_{15} > 4 \qquad (5.79)$$

Bertram verified these conditions for various soils and found them adequate. They are safer for a combination of even-grained filters and base soil than for a filter of crushed aggregates and for poorly graded base soil.
  Sherard et al. (1963) state the first condition for non-cohesive soils in the form

$$D_{15}/d_{85} < 5 \qquad (5.80)$$

And in relation to permeability there is a condition

$$D_{15}/d_{15} > 4; D_5 > 0.074\,\text{mm} \qquad (5.81)$$

Moreover, they add the following conditions:

– The maximum grain of base soil allowed is 25 mm,
– The grain-size curves of the filter and base soil should run roughly parallel to each other,

- The maximum grain size of the filter allowed is 75 mm to prevent the segregation of the filtration layer during its deposition.

US Army Corps of Engineers – US ACE from 1941 states two conditions in the form:

$$D_{15}/d_{85} < 5 \text{ (for rounded grains)} \tag{5.82}$$

$$D_{50}/d_{50} < 25 \tag{5.83}$$

There is another requirement for base soil and the filter to possess grain-size curves similar in shape. Regarding their use see the note above.

USBR – Earth manual (Karpoff, 1955) states the following conditions for non-cohesive soils:

- for even-grained soils ($D_{95} < 8 \times D_5$):

$$5 < D_{50}/d_{50} < 10 \tag{5.84}$$

- for well graded soils with a filter of rounded grains:

$$12 < D_{50}/d_{50} < 58 \tag{5.85}$$
$$12 < D_{15}/d_{15} < 40 \tag{5.86}$$

- for well graded soils with a filter of crushed grains:

$$9 < D_{50}/d_{50} < 30 \tag{5.87}$$
$$6 < D_{15}/d_{15} < 18 \tag{5.88}$$

Plus the following additional requirements:

- $D_5 > 0.075$ mm,
- The maximum grain of base soil is 25 mm,
- The maximum grain of the filter is 75 mm,
- Similarity of grain-size curves in the bottom part for fine particles.

Some specifications of these criteria, mainly with regards to the wide range of the D_{50}/d_{50} ratio are introduced by Čištín (1973). He recommends selecting this ratio on the basis of the coefficient of uniformity C_u, both for base soil and the filter – Fig. 5.35. Čištín's proposal is commented by Schober and Teindl (1979), who claim that it includes 1.5-times higher safety against reaching critical state, and the criteria apply to the majority of cases. For this reason, they found these specifications

5.2 Basic Limit States for Earth Structures

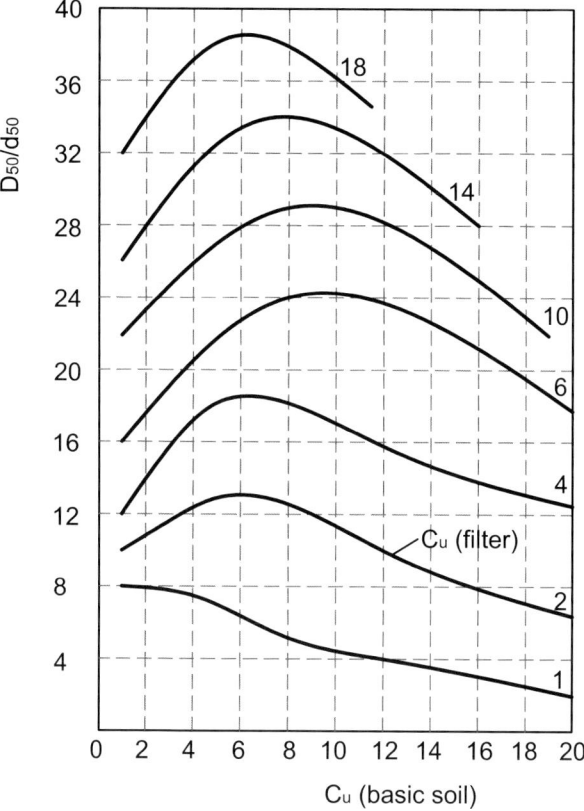

Fig. 5.35 Specification of filter criteria (ratio of D_{50}/d_{50}) based on grain size uniformity C_u for both filter and base (protected) soil (according to Čištín)

essential even in the design of filtration criteria for filters based on geotextiles – see Section 3.3.1.

The filtration criteria stated to-date compare a typical grain size of both covered base soil and the filter. The selection of a typical grain size itself can cause problems with regards to the dispersion of grain-size curves. This, in turn, can be of importance not only for covered base soil, but for the filter as well, even though the filter must sometimes be prepared "artificially" in grading plants to fit specific requirements. In tipping the delivered material of the filter onto the placement site, significant segregation of the material can occur with the known tendency where the largest grains are found along the perimeter of the tipped pile, while the finest grains are at the top of this pile. Consequently, it is easy to imagine that the filter is non-homogeneous with places with a coarser or, on the contrary, finer grain-size composition. Thus, this factor adds another uncertainty into the whole principle, and the above-mentioned condition is trying to at least partially eliminate it. In order to limit segregation of the filter material, therefore, the frequently mentioned condition of a maximum filter grain size of 75 mm is recommended. An interesting point of view concerning the variability of base soil and the filter is presented by Burenkova and Afinogenova (1974). On a statistical basis, they investigate a specific grain diameter, e.g. D_{10} or d_{50}, searching for the mean value and values providing the required safety, e.g. 99% safety according to Student distribution. They go on to make mutual comparisons of extreme cases found in a set, e.g. the smallest grain d_{50} and the largest grain D_{10}.

The issue of the thickness of the filtration layer has not been mentioned yet. There exist some recommendations, related mainly to safe discharge of water flowing into the filtration layer – e.g. Cedergren (1960). Wilson and Squier (1969) mention older recommendations by Terzaghi for a minimum width of 8 feet, i.e. 2.4 m for filters used in dams. At the same time, however, they point out that if the filtration criteria cannot be observed, some engineers opt for a greater thickness of the filtration layer. They presume that the probability of entrapping fine particles in the filtration layer would be greater. This way, however, need not lead to success as for a particular filter there exists a certain minimum pore value, which is limiting in assessing the passage capacity. Then, a further increase in thickness may not be effective at all. The widths are sometimes deliberately chosen so as to facilitate pouring and compaction.

Another view of these problems is presented by Silveira (1965). He stresses the significance of the filtration layer width, which is subdivided into two parts. The first part includes the area which can be choked and clogged up with fine particles of base soil, while the second part is necessary for water removal. Applying the calculus of probabilities, Silveira assesses through how many pores of the filtration layer a base soil particle with a certain diameter must pass before it encounters a pore with a smaller diameter which entraps it. In this respect, he emphasizes the importance of the probability curve of the filtration layer pores. This probability curve is established by him with the help of the particle size distribution curve of filtration soil. In the process of derivation, Silveira presumes that:

– the grains are rounded in shape,
– the filter is in a state of maximum compactness,
– the relative grain position is random.

Pores are formed at contacts of only three particles – Fig. 5.36. For each potential group, a pore is formed with a diameter of d^0 with the probability of occurrence $\bar{p}$.

Fig. 5.36 Assumed filter soil particles layout and determination of pore size d^0 (according to Silveira)

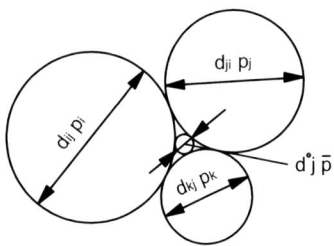

5.2 Basic Limit States for Earth Structures

The pore diameter d^0, as it is, will be a function of only three diameters of individual 3 grains d_i, d_j, d_k : $d^0 = \mathrm{f}(d_i, d_j, d_k)$
and its probability of occurrence

$$\bar{p} = \frac{3!}{r_i! r_j! r_k!} p_i^{r_i} p_j^{r_j} p_k^{r_k} \tag{5.89}$$

where r_i, r_j, r_k give the number of cases when d_i, d_j, d_k occur in a group.

Silveira converts a smooth particle size distribution curve of the filter into a stepwise curve, into m parts each of which is represented by the mean diameter d_i. By means of these substitute grain sizes the diameter of pores and the mean probability of its occurrence are calculated.

The solution results in obtaining the probability curve of the size of pore diameters – Fig. 5.37. This probability curve is then compared with the particle size distribution curve of base soil. Here, we may presume that all parts of base soil with diameters smaller than that of the smallest pore of the filter d_{min}^0, will pass through this filter independently of its thickness S. All particles greater than d_{max}^0, on the contrary, will not be able to enter the filter at all. For a particle with a diameter d_i the percentage of pores of a greater diameter equals p_i^0. If m is the number of encounters of particles d_i with pores in which $d_i^0 > d_i$, the probability that they will be entrapped is given by the formula

$$P' = 1 - (p_i^0)^m \tag{5.90}$$

where $m = \log(1 - P')/\log p_i^0$

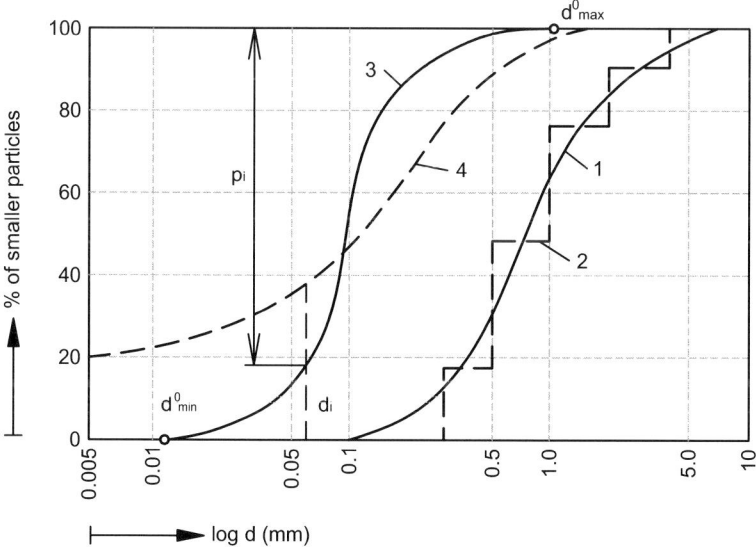

Fig. 5.37 Particle size distribution curve of filter (1) and its voids curve (3), (2) discontinuous particle size distribution curve, (4) particle size distribution curve of base soil (according to Silveira)

The length of the path to the point of encounter of each particle with any pores adopted by the author is approximately equal to the mean diameter of the filter particle D_{50}, so that the path in general – the depth of particle penetration into the filter is governed by the relation

$$S = \log(1 - P')/\log p_i^0 \times D_{50} \qquad (5.91)$$

The solution results in defining the path S as a function of d and P', as shown schematically in Fig. 5.38. It shows, for example, that particles with a diameter of 0.02 mm will be in 60% entrapped at a depth of 0.04 m, in 90% at a depth of 0.155 m and in 99.5% at a depth of 0.4 m. This means that only 0.5% of particles of this diameter can get further away from the interface of layers than 0.4 m. In general, we may conclude that e.g. in designing a filtration layer 0.7 m in thickness the probability of its pores being choked in the last 0.3 m part is so small that this part may be regarded as uncloggable and its function considered fully in terms of fulfilling the requirement for infiltration water discharge.

Silveira's theory is regarded as a general theory, but because of input conditions (e.g. rounded grain shape) it more corresponds to non-cohesive soils, and that is why it has been included in this chapter. At the same time, a case has been manifested of how to handle filter design in another way than only with the help of the filtration criteria based on geometric conditions. In 1993, Silveira presented a new method for determining the void (pore) size distribution curve for filter materials. It is based on previous theoretical developments, and the curves are obtained by points through laboratory tests conducted accordingly. The method is envisaged for all kinds of filter materials, including geotextiles.

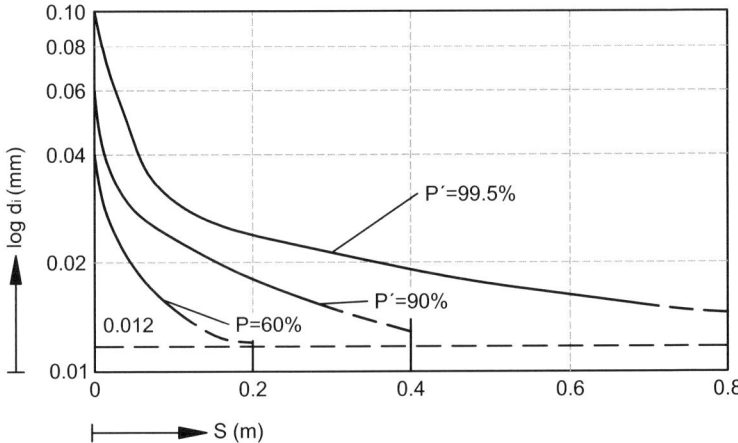

Fig. 5.38 Probability of retention of particle d_i in filter thickness S (according to Silveira)

Base Soil is Cohesive, Undisturbed

Figure 5.37 shows that a filter composed of particles greater than 0.06 mm (as is usually required) with its minimum pore diameter $d_{min}^0 \cong 0.01$ mm is unable to entrap finer particles. Practice, however, shows even in using a coarser filter that base soil is not washed out into the filtration layer. This was also verified in the laboratory. It is becoming evident that due to tensile strength a type of vault of base soil is formed between two greater filter grains, which is relatively stable.

Wolski (1965) measured the base soil deformation due to water seepage at the contact with a membrane with an aperture of 50 mm, see Fig. 5.39. He came to a final conclusion that erosion has two phases. The first erosion phase does not cause any destructive effect, with just a small vault being formed, and the particles directly adjacent to the pores, unsupported by the vault, are separated by flowing water. The second phase is characterised by the loosening of soil behind these vaults, a change in the structure and, finally, by the formation of concentrated seepage paths. The start of this stage is set off at the hydraulic gradient values of $i = 10$–20. At the same time, however, the author points out that this process is slow and under a critical gradient value failure occurs only after a certain, relatively long period.

In designing the interface of clayey soil and a coarse-grained filter, Kassif et al. (1965) define the safety factor F by means of the following theoretical formula:

$$F = b.\sigma_t'/\gamma_w.J.d \tag{5.92}$$

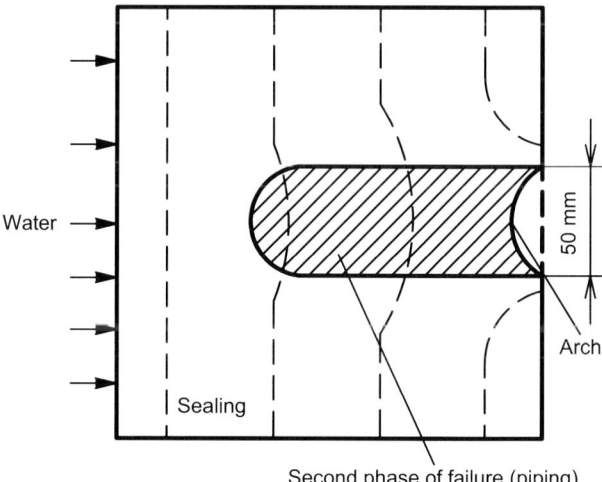

Fig. 5.39 Scheme of the seepage erosion in the silty clay sample at the moment of failure (according to Wolski)

where

σ'_t is the tensile strength of the clay, estimated from a drained triaxial test;
J is the actual exit gradient (different from the measured average gradient i);
d is the mean size of aggregates separated from the clay surface;
b is a dimensionless soil constant of the order of unity, depending on the mean size and the geometry of the aggregates washed out.

Both of these tests proved that the conventional filter design criteria based on the grain size, when applied to cohesive soils, may result in an uneconomical design.

Nevertheless, Vaughan (1976) comments on Kassiff's formula saying that it is of a relatively little practical value as the input parameters are not simple to obtain. He recommends observing the US ACE requirement setting the relation

$$D_{15}/d_{85} < 5 \tag{5.93}$$

plus another recommendation by this organisation applied to clayey soils of medium and high plasticity:

$$D_{15} < 0.4\,\text{mm}; \; D_{60}/D_{10} > 20 \tag{5.94}$$

A very detailed design of a filter for the considered case of undisturbed cohesive base soils was developed by VNII VODGEO in Moscow. It is included in Russian recommendations for the calculation of reversed filters of 1977, see e.g. Vaníček (1988). According to these recommendations, the stability of the filtration interface will be secured if the allowable value of the hydraulic gradient J_d causing the infiltration of clayey soil into the filter pores is greater than the design value of the flow pressure gradient at the interface of the sealing and the filter J_r, or if they are equal:

$$J_d \geq J_r \tag{5.95}$$

The value of J_r recommended is:

- for dams with a central sealing core $J_r = \text{tg}\,\theta$, where θ is the angle of the downstream sealing face with the horizontal,
- for dams with an upstream sealing – J_r is determined on the basis of a flow net and for J_d respectively:

$$J_d = 2.\sigma_t/\gamma_w.D_s.F \tag{5.96}$$

where

D_s – the design vault diameter formed in sealing soil at the contact with the filter,

F – a safety factor selected in accordance with the dam importance – $F = 5$ for class I, $F = 3$ for class II and $F = 2$ for class III and IV,
σ_t – tensile strength roughly related to the compressive strength σ_d, w_p and w_L.

The design vault diameter D_s is defined from the relation

$$D_s = D_r(1 + \alpha_n)\lambda_i \tag{5.97}$$

where

D_r – the design filter diameter, for dam class I and II $D_r = D_{90}$, for class III and IV $D_r = D_{60}$.

α_n – a coefficient expressing the relation between the size of grains of the filter d_i and for the same percentage i the corresponding size of the pore diameter d_i^0. Here, unlike Silveira, a simplified solution is applied, and the value of α_n can be seen in Fig. 5.40. The lowest value $\alpha_n = 0.155$ corresponds to a case of 3 spherical grains with the tightest placement possible with the porosity $n = 0.259$, here the inscribed circle of this pore has a radius of $0.155\,d$.

λ_i – the grading coefficient equal to the ratio.

$$\lambda_i = D_i^r/D_i \tag{5.98}$$

where D_i^r is the diameter of a graded filter and D_i the diameter of a non-graded filter; it depends on the coefficient of uniformity of the filter and the installation procedure – see Fig. 5.41. The recommendations contain principles how to limit the separation. By separation, accumulation of coarser and finer particles occurs so that the actual local filter composition can considerably differ from the mean composition.

5.2.3.3 Recommendation for the Limit State of Internal Erosion

The overview presented above makes it clear that the problems of a safe filter design involve many basic general principles. These principles are valid both for granular filters and for geotextile filters. Therefore, a separate definition of these principles for geotextile filters in Chapter 3 and for granular filters in this chapter is given not

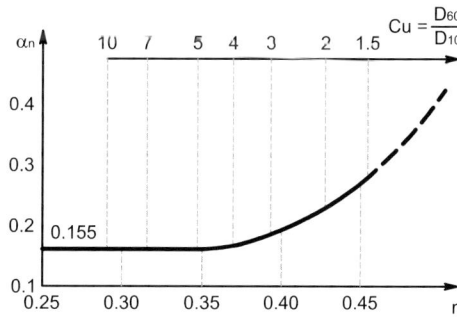

Fig. 5.40 Relationship of coefficient α_n representing ratio between void size and particle size based on porosity n and grain size uniformity C_u (according to VNII VODGEO)

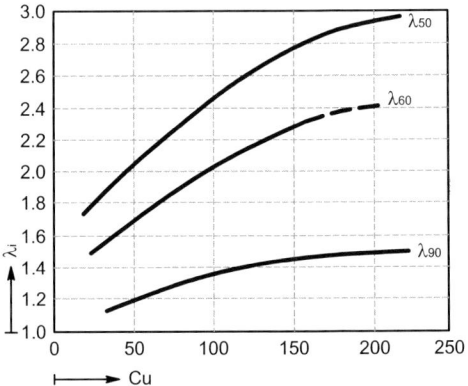

Fig. 5.41 Relationship of ranking coefficient λ_i on grain size uniformity C_u (according to VNII VODGEO)

only by the intended structure of the publication, but, to some extent, also by their independent historical development. Silveira (1993) rightly observed that here we may proceed with general validity. On the other hand, there are many simplifications applied for different recommendations, and the need for further development in the field of filter problems in geotechnics and hydraulic engineering needs further investigation. International conferences Geo-filters starting in 1993 and some papers presented there are confirmations of this fact, e.g. Schuler and Brauns (1993), Heerten (1993), Giroud (1996), Indraratna and Locke (2000), Mlynarek (2000) or in ASCE Special publication No. 78 – e.g. Bhatia et al. (1998).

Moreover, it was demonstrated that there are different ways to treat these problems, both empirically as well as theoretically, and also including probabilistic approaches. Nevertheless, the case discussed above has been the basic one, solving stability at the interface of layers where the base soil is undisturbed. A more complicated situation may occur if base soil shows internal instability or a crack may arise inside it. By internal instability, we understand potential motion of (the smallest) particles in the structure of base soil due to the action of flow pressure. For explanation of these cases in more detail see Chapter 7 as they apply mainly in the design of earth fill dams.

At this point, we may briefly state the following:

- for simple cases with low risks (e.g. in case of first geotechnical category) the filtration criteria applying geometric conditions, recommended and well verified in practice, may be used,
- for more complex cases we recommend considering scattering within the grain-size composition of base soil and the filter itself, or evaluation of a possibility of more radical separation of the filter during its installation, for uncertainties the recommendation is to verify the design in laboratory conditions,
- in the most demanding cases with high risks, it is necessary to consider if the design itself incorporates protection against internal instability and cracking.

5.2.4 Limit State of Surface Erosion

Designs aimed at limiting the limit state of surface erosion are a typical example based on "adoption of prescriptive measures or from experimental models". This limit state may result from extreme rainfall, concentrated surface water run-off, heavy seas or an overflow of the entire earth structure under extreme water levels during floods. In these cases, the term water erosion is used. In specific cases, e.g. in mud-settling ponds, wind erosion may also occur, and this case will be specified below. The principle of protection against erosion basically takes two forms – to concentrate on limiting the effect which may cause surface erosion, or to take measures which will increase the resistance of the earth structure's surface to the respective impact.

In limiting the effect, the greatest possibilities are in limiting concentrated surface run-off. At the embankment crest of an earth body of a transportation structure, for example, it must be ensured that water e.g. storm water is discharged via a surface system into collecting drains and safely removed. This prevents its running onto the earth slope where an especially dangerous case is water running down locally at points where surface erosion may create "deep erosion gullies on the slope". Water run-off from the transportation route surface onto an earth slope can be allowed, but only to specific locations which are both reinforced to resist erosion and limit water velocity – in the basic case by means of steps. Fig. 5.42

Protective measures limiting the effect of sea waves on coastal erosion fall under a special category, but this problem is outside the basic framework of earth structures.

It generally holds true that the higher the velocity of flowing water at the contact with an earth structure, the greater the danger of surface erosion. Therefore, at places where possible, one of the basic measures is limiting the water velocity. Further more, we can distinguish protection against erosion under wet conditions and under dry conditions. Whereas wet conditions are defined for the case where an earth

Fig. 5.42 Photo of surface water channel on embankment slope

structure is in contact with flowing water in streams, watercourses, ditches or water storm drainage channels, under dry conditions there is no direct contact with flowing water, and rainfall is the main reason for surface erosion. Under wet conditions, the key issue is mainly the velocity of flowing water at the contact or the velocity and magnitude of waves beating against the earth structure. Under dry conditions, it is mainly the intensity of rainfall, its amounts and the height or grade of slope.

The second decisive factor of surface erosion, as well as of internal erosion, is given by the size of soil particles and their inter-particle forces. In this relation, the most susceptible material to erosion are fine sands to silt size particles. In finer particles, internal forces preventing the extraction of the finest particles will prevail, while, on the contrary, in coarser particles the flow pressure will not suffice to extract and transport them, Fig. 5.43. The exception for dispersive soils will be described in Chapter 7.

5.2.4.1 Wet Conditions

A special case is that of earth structures which fulfill a sealing function, such as feed ditches, reservoirs for storm and drinking water, reservoirs collecting water that can be contaminated, e.g. water from parking spaces from motorway surfaces before its successive treatment. Here, the fundamental sealing element can be a concrete slab, "asphaltic concrete", but also sealing based on geosynthetics, such as sealing using geomembranes or bentonite mats – "geosynthetic clay liner". The basic condition in the respective case is a good drainage system under the sealing layer to prevent the limit state of "hydraulic uplift". It prevents the rise of water pressure on the sealing reverse and thus the action of the buoyant force on the reverse side. Geomembranes and GCL sealing mostly need surface protection of the sealing face. This is not only protection against mechanical damage, but also protection against climatic effects. GCL require protection against drying and protection against frost penetration.

In other cases, protection principally fulfills a double function – protection against surface erosion and protection against grain outwash from the earth structure. This may happen when the groundwater level in the earth structure is higher than the water level in the riverbed or reservoir. This may occur due to a rapid drawdown in the water level in the riverbed, reservoir or due to heavy sea waves. The

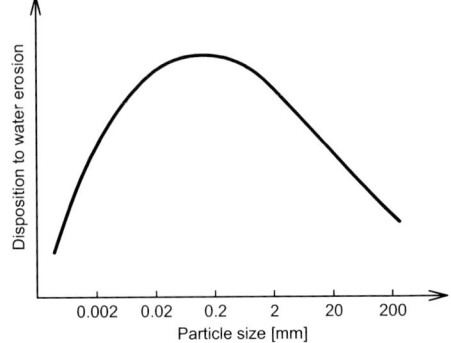

Fig. 5.43 Approximate relationship between size of soil particles and susceptibility to surface water erosion

composition of such protective layer is of an inverted filter type. At the contact with finer soil of the earth structure a filter is found, which is protected from outwash by coarse material whose grain size is able to resist surface erosion. Riprap is placed depending on the height of waves. The average grain size is 250 to 500 mm, and the thickness of 300–900 mm. Analogically, the filter thickness ranges from 150 to 250 mm, Fig. 5.44. Here, hand-placed riprap gives a pleasing finished appearance. The interesting commonly stated fact is that the effect of the slope gradient for standard values of 1:2–1:4 is not of much significance. Nevertheless, with a view to the riprap grain size, the filter as well must be relatively coarse to avoid its outwash. Stability of both interfaces, therefore, must be assessed. If need be, a two-level filter must be used whose second layer in contact with base soil can be a geotextile-based filter. Hand-placed riprap is now often replaced with concrete blocks e.g. hexagonal in shape with a hole inside. This hole is filled with coarse material allowing water passage in both ways. As an alternative, compacted no-fines concrete of sufficient permeability can be used. Another possibility is to use lattice-type mat composed of geosynthetic which is filled with concrete and encompasses open spaces. The latter can be filled with top soil and seeded or plant boxes can be placed in it immediately after installing the mat – see Fig. 5.45.

The application of geosynthetics has brought about a relatively interesting, cheap and technologically simple method of protection. It applies, above all, anti-erosion matting and geocells – see Chapter 3. Here, materials heavier than water (e.g. polyamide) are more suitable while materials lighter than water (e.g. polypropylene or polyethylene) will float in water and are difficult to install. In simpler cases, polyamide matting with a flat structure on the underside can be used to enable 2–6 mm stone chippings (with which the mat is filled after installation). The filament core structure securely holds the stone chippings and provides protection below the water line. In order to increase resistance to surface erosion, some mats are prefilled with a mineral filter of 2–6 mm chippings with a bitumen binder. This mat is permeable, flexible and is usually used at and below the water level on the banks of watercourses. The manufacturers of mats provide recommendations concerning the water flow velocity requiring protection by matting and how thick it should be – the standard is between 10 and 25 mm. To ensure an evenly distributed contact with subsoil, mats must be secured with pegs or pins.

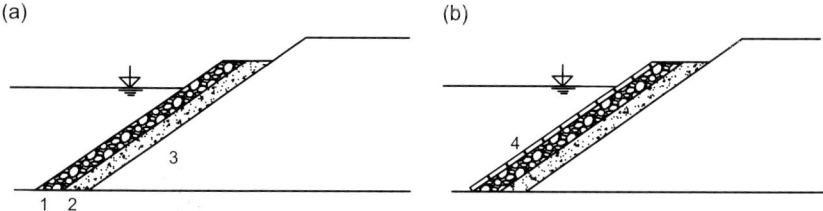

Fig. 5.44 Upstream slope protection against surface erosion under wet conditions. 1-Riprap; 2-Filter; 3-Protected soil; 4-Concrete blocks

Fig. 5.45 Photo of geosyntetic encased concrete crib erosion protection structure – Incomat crib (courtesy of HUESKER)

Geocells anchored to the slope surface and filled with coarse material have a similar function. Individual cells allow the application of smaller grains than those commonly used for riprap.

Everywhere that allowed by the earth structure's character protection applying plants is preferred. Green banks are an ecologically acceptable barrier which blends harmoniously into the landscape. Our predecessors would abundantly use protection by interconnected canes of woody plants which even grew in water such as a willow. Today, water plants may be applied in combination with geosynthetic anti-erosion matting whose seeds are placed inside the mat filling or directly onto the subsoil prior to the installation of matting.

5.2.4.2 Dry Conditions

In embankments of earth structures, as well as in slopes of cuts, natural slope protection, such as planting vegetation, is preferred. This, of course, requires specific climatic and geometric conditions. These are mainly the slope height and inclination but also soil types more resistant to surface erosion. The most sensitive period is the phase immediately after the final shaping of the earth structure. On the one hand, we need sufficient rainfall for the vegetation ecesis, mainly during the first month, while, on the other hand, there are justified worries that heavy rainfall causing erosion may come before the seeds germinate.

Therefore, additional measures to increase resistance to surface erosion may be of the following type:

- Temporary effect until the time of adequate growth of natural vegetation which will then suffice for surface protection,
- Long-term effect where an additional measure ensures slope resistance for the whole period as natural vegetation is not able to ensure this function on its own – these are mostly cases of steep slopes (e.g. due to their reinforcement).

In both cases, anti-erosion matting and geocells are of major importance, and in simpler cases also grids or unwoven geotextiles with sewn in grass seeds. After anti-erosion mats germinate, their structure is filled with fertile soil and grass seeds. The tough filament core structure can keep the fertile soil fill and the germinating seeds in place, preventing seeds from being washed out by heavy rain and encouraging active plant growth. If only temporary protection is needed, geotextiles of natural materials will do, which will allow mainly good ecesis of vegetation and then degrade. The material suitable for their production is e.g. waste cotton, flax. In some countries, coconut palm mats or jute-based products are used.

5.2.4.3 Final Remarks

In selecting and assessing the suitability of additional protection, the tensile strength of anti-erosion elements must also be taken into account. This effect is essential mainly in steep slopes, e.g. slopes of cuts in relatively strong materials. As a basis, the equilibrium of forces on the slipping surface at the contact of the anti-erosion element and subsoil is calculated. The shear strength of the contact may fall down to zero. This case may occur e.g. in using geocells, which are filled with loosely poured soil. Over winter, the soil freezes easily, increases its volume and during the spring thaw and rainfall it changes up to pasty consistency with practically zero strength. Therefore, it is necessary to ensure not only a good contact with subsoil, including fixing with clips, but also at least gentle compaction of the soil filling in the cells. In order to prevent the breakage of the additional element and its slipping, sometimes the space structure of the matting must be reinforced with additional reinforcing geosynthetic of a geogrid or woven geotextile. Placed under wet conditions, the lower reinforcing layer may also fulfil a filtration function. The combination of a matting structure with a high strength grid helps dense vegetation develop under the often unfavourable circumstances on rocky, eroded slopes. A mat forms a flexible grid layer in which soil is retained and seeds are encouraged to germinate. On very steep slopes the mat can be filled and seeded hydraulically. To prevent its pulling out, the whole mat, including the reinforcing element, must be well anchored in the upper part of slope, see e.g. recommendation of Colbond on Fig. 5.46.

For serviceability reasons, there must be good interconnection of all layers through the root system of vegetation. Here, the space structure of mats is a key issue, which is grown through by roots and thus interconnected with the subsoil (see e.g. Fig 3.16a). The importance of this interconnection is essential mainly where the earth structure may be overflowed, e.g. a pond dam, flood control barriers along water streams, dams of polders. If flowing water penetrates underneath the additional protective element, it gradually starts separating it from the subsoil with starting strong surface erosion. The significance of this hazard and the need for elaborating this detail comes into the foreground in situations where we deliberately want to transport large volumes of water. To increase the resistance of structures in terms of flood protection, it must be considered that these are often 2D structures which fail at the weakest point. This point may be even the lowest level of a flood control dam where water first starts overflowing. Therefore, it is sometimes more efficient to choose this point, lower the spillway crest against the remaining part there and ensure adequate protection of this crest and also the adjoining section of the downstream slope.

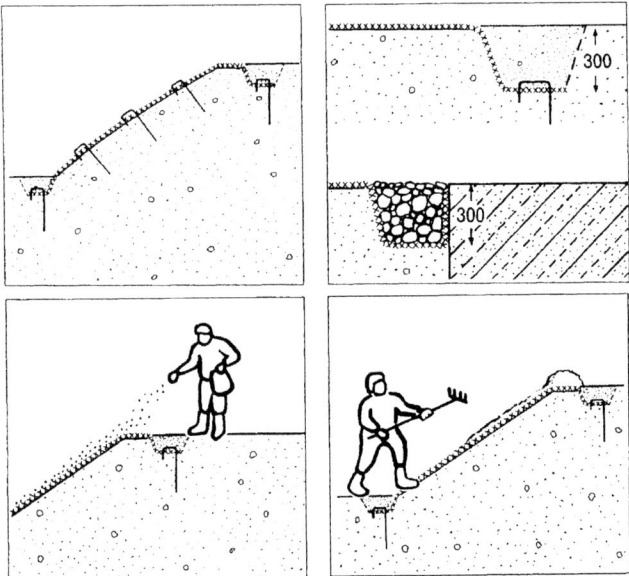

Fig. 5.46 Scheme of securing the erosion protection mat on the slope surface

The role of vegetation on slopes is prominent not only in the environmental context. It was shown above how little roots may strengthen the contact of additional anti-erosion protection with subsoil. Interception and transpiration as modes of water loss each contribute to the reduction of soil moisture. Moisture depletion enhances cohesion due to the surface tension forces in partially saturated soils. The reinforcing effect of a root system may also help during extreme rainfall limiting large-scale shallow slides, mainly on subsoil with a growing strength with depth. To assess the effect of a root system on shear strength, tests were implemented on undisturbed soil samples with a deep root system using 4 perennial graminaceous plants imported to Italy – Cazzuffi and Crippa (2005). The tests proved a growth in initial shear strength – apparent cohesion. Samples without roots showed cohesion of ca 10 kPa, while samples with a root system of up to 25 kPa, and the importance of roots was evident even at a depth of 0.6 m. The roots submitted to a tension test proved high strength values – for an average of 1 mm the tensile strength ranged at 20–40 MPa. This manifests that targeted exploitation of vegetation growth can add a new, significant element to the design of anti-erosion protection, but also protection against shallow surface slides.

5.3 Modelling of Limit States

Modelling the behaviour of earth structures is essential for closer understanding of the behaviour of earth structures, for their design and for the verification of a potential hazard of reaching limit states. EC 7, for example, respects this approach, and apart from the verification of limit states by calculation where we may also include verification using numerical methods, which principally model the behaviour of earth structures, it also allows verification with the help of experimental models. Therefore, verification by calculation, by numerical methods is presently becoming a common tool in design, mainly for more demanding structures in combination with more complex foundation situations. Nevertheless, for new types of structures, for the verification of some new presumptions, it is always beneficial to combine various modelling methods. The most common combination is a numerical model with a laboratory model or a real-scale model.

Modelling itself is aimed at the following:

- Modelling the stress-deformation behaviour of the structure – at individual points not only deformations are monitored, but also stress states, which can be successively compared with various failure criteria thus monitoring a gradual development of a failure in the earth structure – the results of monitoring serve for the verification of mainly the limit state of stability, the load-bearing capacity of the earth structure and the limit state of deformation (serviceability),
- Modelling the forces in individual structural elements which are part of the earth structure – most often in reinforcement – geosynthetic reinforcement or in soil nails – the results serve for the assessment of the limit state of internal stability,

- Modelling water flow in the earth structure, in the basic case for steady flow where the most common form of display is a flow net enabling verification of limit states activated by water in general and hydraulic gradient in particular.
- In modelling environmental problems, modelling of contamination spread within an earth body comes into the foreground. Its solution e.g. allows evaluation of how much the earth structure, its sealing part protects the surroundings from contamination – the results are compared with limits defined by the "authority", e.g. the Ministry of Environment.

Modelling may involve not only modelling of the entire earth structure, but also some specific detail. This detail whose modelling is easier allows elaboration of a parametric study monitoring the effect of a certain variable on the resulting behaviour.

5.3.1 Numerical Models

The main numerical method of today is FEM – finite element method, but of course there are others available like BEM – boundary element method. As already said above, it is not the aim of this publication to go through individual solutions, as there are special publications from Zienkiewicz (1967), Zienkiewicz and Taylor (2000) or Potts and Zdravkovic (1999). Nowadays FEM is the basis of many computational programs from different software companies and their use is becoming more common even for not so difficult problems. Therefore hereafter we will show only the basic principles and problems, which we can solve with them. Some effort will be made also to show by what questions the software users should be guided during the selection and results evaluation when they use certain software, which is always some kind of simplification.

As already mentioned, the basis of the method is discretization of area in question into planar or spatial elements of finite dimensions. These elements are allowed to be continuous only in finite amount of nodal connections, so-called nodal points of the created planar or spatial mesh. The first question is the replacement of the soil body by a mesh of those nodal points and thus elements that can be arranged in plane into triangular or rectangular mesh and what density this mesh should have. We have to acknowledge that in places with rapid changes of properties, the mesh should be much denser. However, with the increase of nodal points also increases the amount of equations that have to be solved. Next question is on the definition of relationships that govern the transition of properties from one nodal point to the other. We are talking here about shape functions. Generally using the shape functions we can determine the value of the property in any particular point inside one element based on values of that property in nodal points. In FEM the shape functions are generally selected in a form of a polynomial. The order of the polynomial depends on the amount of nodes per element, or more precisely the amount of mid-side nodes.

5.3 Modelling of Limit States

The selection of shape functions has to satisfy compatibility conditions, i.e. continuity of the displacement field, especially across elements, represent rigid body movements and constant strain rates.

Isoparametric finite elements are most often used in the programs for geotechnical calculations. Their advantage is that the real (global) elements are transferred in so called parent elements in natural coordinate system, which means that x and y real coordinates are represented by natural coordinates s and t that are in the range between -1 and $+1$, see Fig. 5.47. Another advantage is that interpolation functions N representing the transformation from global to parental elements are the same as for the shape functions that define geometry (x, y) and displacements (u, v) across elements:

$$x = \sum N_i x_i; \quad y = \sum N_i y_i \qquad (5.99)$$

$$u = \sum N_i u_i; \quad v = \sum N_i v_i \qquad (5.100)$$

Once the elements are defined and the modelled area is divided, it is necessary to define physical-mechanical properties to individual elements or more often to the blocks of elements. When this is done the program can prepare a mechanical model of the problem to be solved. The next step is the definition of boundary conditions that are essential for any analysis.

The term "boundary conditions" is used to cover all possible additional conditions required to fully describe a particular problem to be solved. Boundary conditions are mainly defining displacements at the area boundary and external loads applied on the modelled area. The displacements boundary conditions are essentially fixing the problem in space.

5.3.1.1 Solution of Stress-strain State

This type of problem is based on this initial stress state of the soil body, which we know, and subsequent solution of stress and strain changes due to additional surcharge. In principle, in this way, we are solving the relationship of strain as a function of surcharge based on the valid deformation characteristics for this area. This relationship we can write as:

$$\Delta \varepsilon = E(\varepsilon, \sigma) \Delta \sigma \qquad (5.101)$$

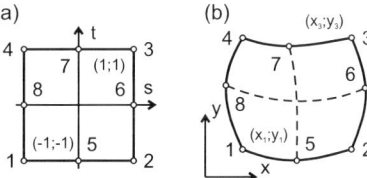

Fig. 5.47 FEM – eight noded isoparametric finite element for 2D applications. (**a**) Parent element in natural coordinate system. (**b**) Real element in global coordinate system

Nevertheless the solution to a certain extent can simulate the process of construction, surcharge can be applied either instantly at once or step by step and for each step specify the change of stress state and deformation. It is in fact a kind of simulation of stress path that has been described in previous parts of this chapter. The main difference in solution is made in the assumption of stress-strain relationships, so called constitutive relationships. In Fig. 5.48 are presented 4 main alternatives. First basic model (a) assumes linear relationship between stress and strain – it represents linear elastic model. Second alternative (b) shows plastic deformations and third (c) elasto-plastic deformations. This alternative is close to the behaviour of cohesive, normally consolidated soil or loose sand – graph (d). Model of the real cohesive overconsolidated and dense noncohesive soil is shown as (f) and can be approximated by elastic model with softening that is shown as (e).

The behaviour of normally consolidated soil (d) is therefore possible to model via bilinear relationship c) or by hyperbolic relationship proposed by Kondner (1963):

$$(\sigma_z - \sigma_x) = \varepsilon_z/a + b.\varepsilon_z \tag{5.102}$$

The other alternative uses the division of initial and final state into several stages, for which (given stress level) defines pseudoelastic parameters derived from real relationship between stress and strain, either by tangent moduli (Clough and Woodward

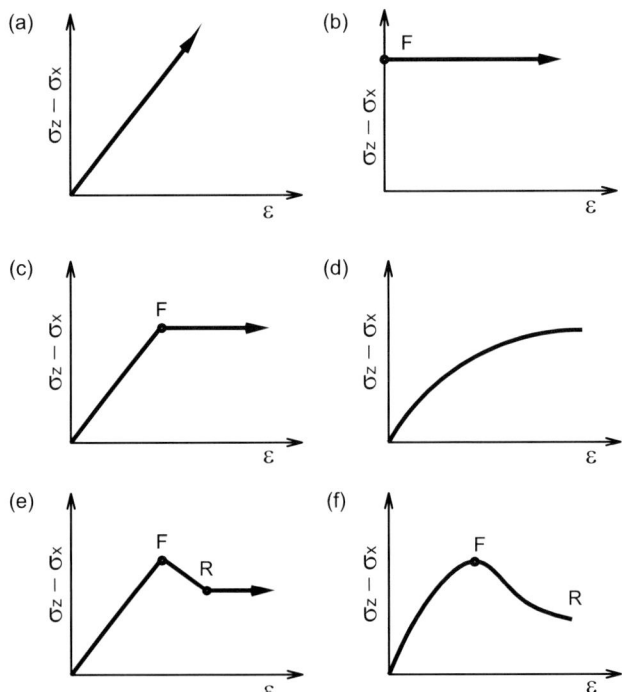

Fig. 5.48 Constitutive relationships for soils used in numerical modelling

5.3 Modelling of Limit States

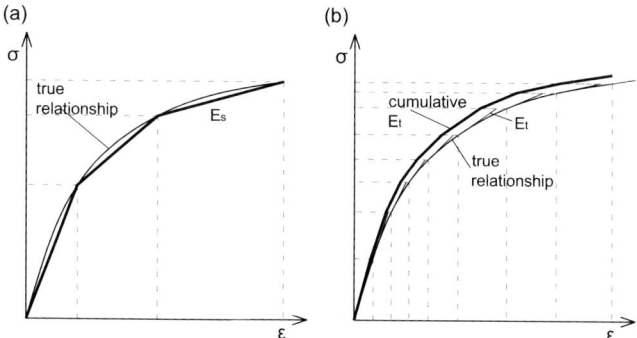

Fig. 5.49 Substitution of real soil behaviour by set of tangential and secant moduli. (**a**) Secant moduli. (**b**) Tangent moduli

1967) or secant moduli (Girijivallabhan and Reese 1968), Fig. 5.49. From triaxial test the deformation characteristics are based on the following relationships:

$$\Delta(\sigma_z - \sigma_x)/\Delta\varepsilon_z = E \tag{5.103}$$

$$\Delta(\sigma_z - \sigma_x)/\Delta\varepsilon_x = -E/\nu \tag{5.104}$$

It is always useful to check the results already during the analysis, if the calculated change of stresses relates to the loading or unloading, as the moduli for each case are different. The results of stress state in individual nodal points are checked against the stress state at failure, while the main criterion is the one of Mohr-Coulomb:

$$\sin\varphi = \frac{\sigma_1 - \sigma_3}{2 \cdot c \cdot \cot\varphi + \sigma_1 + \sigma_3} \tag{5.105}$$

no doubt that it is possible to use also other criteria for failure, e.g. Tresca, Drucker-Prager, Von Mieses, Lade, Matsuoka-Nakai.

The finite element method this way allows localisation of the areas of failure within the soil body, eventually propagation of this failure further towards the overall failure. Another advantage of this method is the possibility to represent more realistically the influence of initial geostatic stress state, e.g. influence of overconsolidation via variable coefficient of earth pressure at rest K_0. Advantages of possible construction simulation in stages are not irrelevant either, as the final results can be significantly different.

One of the possible results of step by step pit excavation modelling, see Fig. 5.50. Dunlop and Duncan (1970) used bilinear relationship between stress and strain and once the peak strength is reached, it becomes constant. The areas of plastic failure are enlarging with the continuous enlargement of the pit. At the top part of the slope are shown areas with tensile stresses. Even when the plastic zones exist, the overall

Fig. 5.50 Development of failure (plastic) zone during the process of excavation and comparison with classical slope stability analysis (according to Dunlop and Duncan)

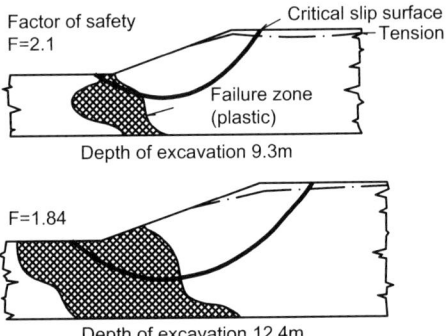

factor of safety F, determined using limit equilibrium method on circular slip, is still higher than 1, as in the elastic zone there is still a large enough reserve of strength.

Overall the advantages of the method of stress-strain state modelling except those mentioned above could be summarised in the following points:

- Step by step overview of the stress state in individual elements allows to monitor development of areas of failure and also the directions of main stresses and stress paths and this way influence the execution of laboratory tests in the future. Those should simulate as close as possible the in situ process. At the same time it is possible from the stress increment to determine more precisely pore pressures.
- In non-homogeneous materials we can follow the redistribution of the stresses. For example for rockfill dams with clay sealing to determine the risk of development of the areas of reduced state of stresses in which the risk of tensile crack development exist.
- During the deformation analysis the most important is the information about overall character of earth structure deformation, because sometimes the deformations are a better indicator for stability evaluation and structure execution than the factor of safety, e.g. for already mentioned rockfill dams.
- Numerical simulation and localisation of critical areas of earth structure from the view point of stresses and deformations, positively influences the location selection of different monitoring devices that should confirm the real behaviour of earth structure.

The advantages of finite element method are relatively obvious, but always we have to be aware of the fact that its precision depends on:

- Input data provided by the soil mechanics laboratory; we should mention that for planar analyses the best would be to use apparatus with planar loading and for spatial analyses to use real triaxial apparatus, while we are happy with just axi-symmetric one.
- The type of substitution of real behaviour by its model.
- The way of approximation of the whole soil body by finite elements and their type.

5.3.1.2 Modelling of Forces in Structural Elements

Soil reinforcement brings a new element to the modelling of earth structures. It includes both geosynthetic elements and reinforcement with soil nails. The principal problem here is a relatively small size of the reinforcing element in comparison with the surrounding earth medium. In the classic subdivision of the whole domain into individual elements, their numbers disproportionately grow, as subdivision in the reinforcement domain must be very fine. In terms of numerical modelling, this fact requires a slightly different approach to the logic of the distribution of the solved area into nodal points. The second important problem is modelling the interaction between reinforcement and surrounding soil. One of the possible solutions, the use of steel-spring mounts, is mentioned by Andrawes et al. (1982), Fig. 5.51. In reinforced supporting structures, the model may additionally include an interaction with the facing element.

For this reason, further efforts led to various other alternatives. Their overview is given e.g. by Rowe and Ho (1988). Two different models were used: a composite and a discrete model, Vaníček (2001a).

Composite models were created so as to preserve the simple principle of planar finite elements still applied in geotechnical engineering. Reinforced soil was approximated by means of homogeneous anisotropic material, and the effect of reinforcement used was added by means of superposition to the stiffness matrix of standard finite elements. In using this type of numerical modelling of reinforced soil, mutual movements between reinforcement and soil could not be directly respected. This technique was relatively simple, but did not provide the necessary information on the distribution of stress and deformation at the reinforcement-soil contact. Therefore, the composite model of a structure of reinforced soil could be applied only in situations where there was no movement between reinforcement and surrounding soil.

Discrete models differ from composite models in that for each part of a structure of reinforced soil a special finite element is used or a combination of these

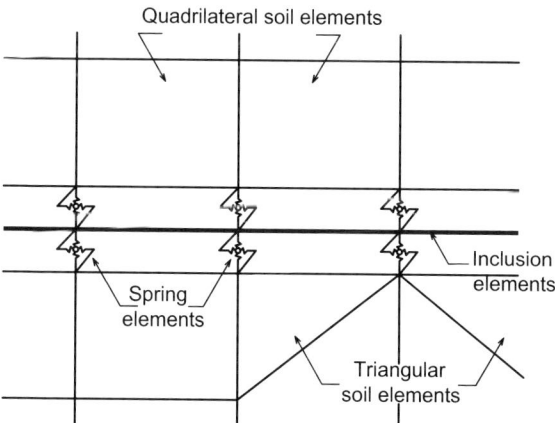

Fig. 5.51 Modelling the interaction between soil and reinforcing element with the help of spring elements (according to Andrawes et al.)

elements. Special elements are created so as to enable expressing real behaviour of all parts of the soil body exposed to load. For this reason, apart from standard finite elements modelling soil, discrete models of reinforced soil use special elements for reinforcement, complemented with contact elements modelling the interaction of both reinforcement surfaces with surrounding soil. The thickness t of the contact element is mostly small and its deformation characteristics are defined to specify the modelled contact in the most precise way. The lining of the structure's face of reinforced soil, which can be composed of reinforced concrete panels or geotextiles, is modelled in a similar way.

The advantage of a discrete model of reinforced soil is mainly the fact that it enables getting a quantitative idea of the stress and deformation of reinforcement and its interaction with surrounding soil. As compared to the composite model, however, the discrete model requires a greater number of finite elements for the modelling of a specific construction of reinforced soil, and as a discrete model applies special elements, the computer programme used for the finite-element method must also be more sophisticated. Setting a numerical model is more difficult and the preparation of input data for a discrete model requires substantially more time.

The development of FEM application in solving structures of reinforced soil went on in several directions:

- More general constitutive laws better expressing real soil behaviour were used for the modelling of stress and deformation of soil,
- Special finite elements were used for the modelling of reinforcing grids from geosynthetics and panels on the vertical face of the structure,
- Special elements were also used for the modelling of the mutual interaction of reinforcement with soil.

The use of the finite-element method for the analysis of structures of reinforced soil requires the application of a program allowing soil modelling as non-linear elastic material and simulating gradual loading during the embankment placement. Moreover, special contact elements must be available allowing the interaction modelling between the reinforcement and surrounding soil, or between soil and concrete prefabricates on the vertical face of the soil body. When the earth structure is founded on highly compressible subsoil, the programme must respect high deformation values.

It is a characteristic feature of present-day programmes aimed for solving reinforced soil using the finite-element method that they use both triangular and rectangular elements, with higher approximation polynomial from the family of isoparametric elements. For reinforced earth bodies we may, as a rule, predict that thanks to their dimensions they are in a state of planar deformation. Then, it suffices to assess a typical cross section of the structure, so that instead of 3D elements planar ones can be used. In this way, the application of the finite-element method is substantially simplified, contact elements being only non-linear and their definition on a plane is simple.

5.3 Modelling of Limit States

Improved numerical models were verified on experimental structures in a 1:1 scale, usually equipped with a large number of measuring devices; see e.g. Karpurapu and Bathurst (1994). The results of these tests showed that a numerical model requires not only a careful choice of boundary conditions, but the construction process of the structure must also be modelled. The deformation of a structure of reinforced soil also considerably depends on the rheological properties of soil and reinforcement used. Certain views of the FEM development for reinforced soils are presented by Yashima (1997).

5.3.1.3 Modelling of Water Flow in an Earth Structure

Water flow in an earth structure can have several basic versions. The most commonly assessed state is steady flow where a change in the amount of water W_w flowing into a unit element and out of it is zero in time (a case when $S_r = 1$ and $e =$ const.):

$$\Delta W_w / \delta t = 0 \qquad (5.106)$$

The focus of attention below will be on this case suitable for the assessment of a hazard of limit states caused by water in general and hydraulic gradient in particular.

In some problems, however, we may come across a case when this equation is not valid as:

- The soil is not fully water saturated – S_r changes in time – an example is the problem of saturation of a protective flood control barrier along the river under a high water level during a flood where the subject of monitoring is the velocity of river water seepage through the dam, which has been only partially saturated with water so far,
- Soil skeleton is compressible – the void ratio e changes – a typical case of soil consolidation.

The case of steady flow may be expressed by a partial differential equation for a 2D problem in the form:

$$\partial/\partial_z(-k_z.\partial h/\partial_z) + \partial/\partial_x(-k_x.\partial h/\partial_x) = 0 \qquad (5.107)$$

which, for constant filtration coefficients, which are, moreover, equal, i.e. for an isotropic domain, changes into the basic form of Laplace equation:

$$\partial^2 h/\partial z^2 + \partial^2 h/\partial x^2 = 0 \qquad (5.108)$$

By solving this equation for the given boundary conditions of a soil domain through which water flows, we may obtain the total height h as a function of position – coordinates z, x. From Darcy's law we may then calculate the flow velocity at individual points using the relation:

$$v = -k.\text{grad } h \qquad (5.109)$$

and the flow direction is given by the velocity vector at this point. The graphic form of the solution output is a flow net consisting of equipotentials – lines for which it holds true that $h = $ const. and of flow lines following the trajectory of water particles. In terms of boundary conditions, a case without a free surface may be distinguished – e.g. for a case displayed in Fig 5.4a, or a case with a free surface, which is more demanding to solve. An example of a flow mesh in a mud-settling pond is shown in Fig. 5.52.

If a flow net is composed of curvilinear squares, the following may be calculated from it:

- Total seepage through the earth body: $Q = k.H.m/n$, where m is the number of flow line channels and n the number of squares in one channel,
- Total height at any point, i.e. water pressure in pores, or water pressure (uplift) acting on an impermeable boundary,
- Hydraulic gradient for individual elements, curvilinear squares and so e.g. a hazard of the bottom liquefaction – the limit state of hydraulic heave type.

The flow net in Fig. 5.52 may be used for the stability calculation of the downstream face of a tailings dams, and analogically for dams: for one of the basic loading states – a case with a maximum water level in the reservoir. In general, this is a case of the flow net application for the assessment of the limit state of failure – stability loss. The flow net, however, also serves for the assessment of limit states caused by water flow – hydraulic failure, or for the assessment of serviceability of the structure, if the seepage through its body fulfils the respective requirements related to seepage. Note: The flow net can also be sketched by hand, but only for the basic case of an isotropic domain. FEM modelling allows solving flow problems in non-homogeneous and anisotropic domains. More aspects of numerical modelling for ground water flow studies are described e.g. by Kazda (1990).

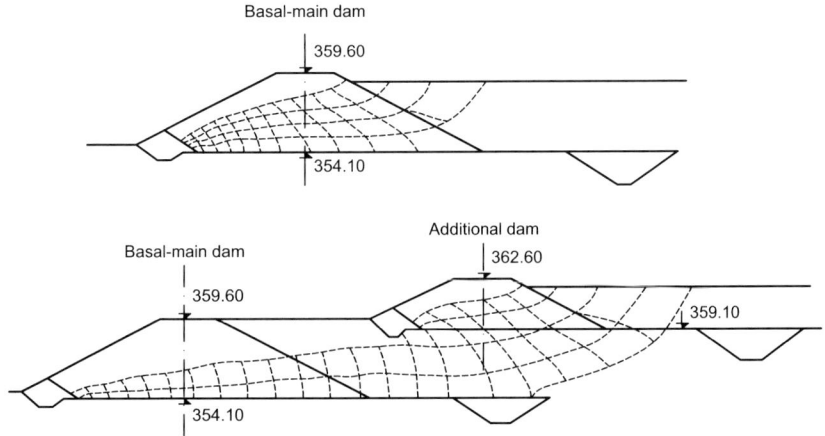

Fig. 5.52 Flow net demonstration in tailing dam

5.3.1.4 Modelling of Contamination Spreading in an Earth Body

Environmental conservation is presently a subject of great concern. Therefore, Chapter 8 also includes earth structures in environmental engineering such as landfills, tailing dams and spoil heaps. A specificity of these earth structures is that their design must comply with all limit states applied to earth structures, and specified e.g. in EC 7. Moreover, it must comply with an environmental condition that protection of the environment against contamination is sufficiently ensured. Here, one of the basic tasks is the assessment of the sealing system, and how it is able to protect the surrounding environment. A similar problem, however, may be encountered in transport structures: embankments may also be constructed with the use of waste materials, such as ash, and our intention is to assess whether the leakage from this ash may represent a hazard for the surroundings. As this question will be discussed both in Chapter 8 and 6, the following part provides only brief information.

The basic equation for a 2D problem of contamination spreading in an earth domain may be written in the form:

$$n \frac{\partial c}{\partial t} = n \cdot D_x \frac{\partial^2 c}{\partial x^2} + n \cdot D_z \frac{\partial^2 c}{\partial z^2} - n \cdot v_x \frac{\partial c}{\partial x} - n \cdot v_z \frac{\partial c}{\partial z} - \rho_d \cdot K \frac{\partial c}{\partial t} \quad (5.110)$$

which expresses a change in the concentration c of a certain substance in time t in relation to the porosity n and the groundwater flow velocity v in the direction z and x. Further on:

D_z, D_x – are coefficients of hydrodynamic dispersion in directions z, x,
K – distribution coefficient,
ρ_d – dry density of soil

This equation incorporates contaminant spreading through the following processes:

- Advection – contaminant movement with flowing water, the decisive factor being the water velocity v
- Diffusion – expressing contaminant movement from a point of a higher chemical concentration to points with a lower concentration, the decisive factor being the effective diffusion coefficient D_e
- Dispersion – which includes "mixing" and dispersion due to the non-homogeneity of the aquifer, the decisive factor being the coefficient of mechanical dispersion D_m
- Sorption – i.e. mechanisms which remove contaminants from the solution, the decisive factor being the distribution coefficient K.

Although the dispersion mechanism differs from the process of diffusion, for practical reasons it may be mathematically modelled in the same way, and so both processes are often connected and characterized by the transport parameter D, referred to as the coefficient of hydrodynamic dispersion:

$$D = D_e + D_m \tag{5.111}$$

When process of dispersion is usually modelled as linear function of velocity:

$$D_m = \alpha.v \tag{5.112}$$

where α – is dispersivity

Example of contaminant spreading modelling from a landfill in time t is presented on Fig. 5.53 – Vaníček and Kazda (1995). There stored municipal waste will be surrounded by clayey backfill fulfilling pit after brown coal excavation. Low coefficient of permeability, roughly $5.10^{-10}\,\mathrm{ms}^{-1}$ can be expected due to high overburden pressure. After the end of pumping in the pit water table will rise and spreading of contaminants from the stored waste material can start. Two dimensional transport was solved by finite element method and the following processes were taken into account:

- advective transport – $v = 0.02$ m/year,
- diffusion – coefficient of diffusion $D_e = 0.02\,\mathrm{m}^2/\mathrm{year}$,
- mechanical dispersion – longitudinal dispersion $D_L = 10.0$ m, cross dispersion $D_T = 1.0$ m.

Process of sorption capacity of clay material was not taken into account in this case so the solution is on the safe side. Solved area was covered by the mesh from isoparametric elements with 8 nodes – total number of elements is 1350 and nodes 4195. Different possibilities of the result presentation are shown in Fig. 5.53b and Fig. 5.53c. They show that from transport processes the diffusion has the greatest influence. The influence of advection is slightly visible only after 100 years – izolines are little bit shifted from the centre of the area with waste material in the direction of water seepage. The pollutant concentration drops in the centre of waste material to 89% after 100 years and affected area out of the stored waste material is still relatively small one, concentration of contaminant drops down very quickly. More aspects of the numerical modelling for contaminant spreading are described by Kazda (1997).

5.3.2 Laboratory Models

Laboratory models are used not only because in some cases they can be applied as experimental methods for direct design of earth structures but very often predominates the possibility just to more closely understand the processes which take place in earth structure, or it is possible to observe a certain detail and for this detail to perform a parametric study. Pedagogic effect is also very important. A certain specificity of the laboratory model for the design of filter protection was shown previously.

5.3 Modelling of Limit States

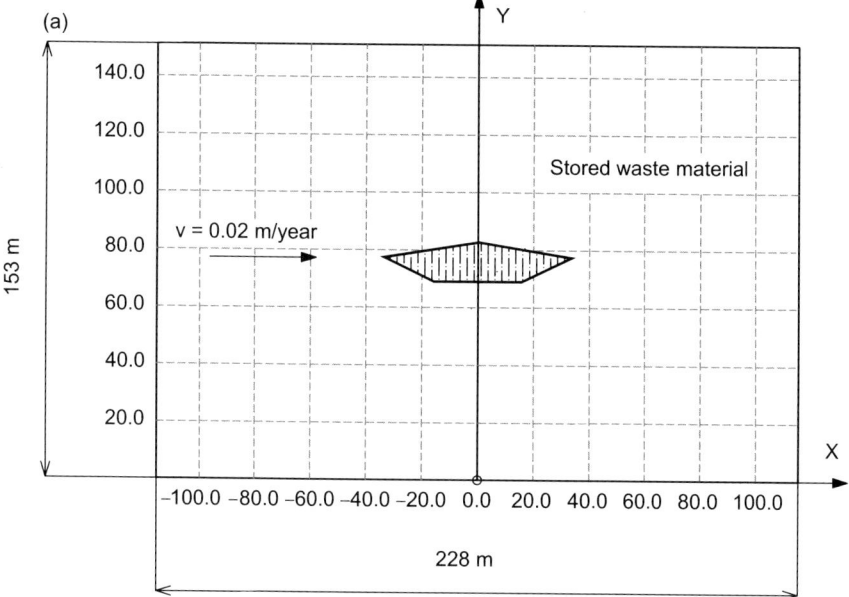

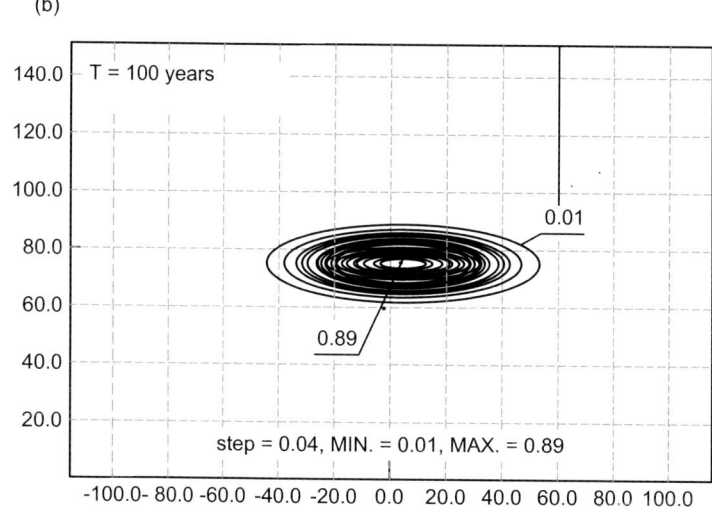

Fig. 5.53 Contaminant transport modelling. (**a**) Boundary of solved area. (**b**) Izolines of relative concentration after 100 years. (**c**) 3D presentation of relative concentration after 100 years

(c)

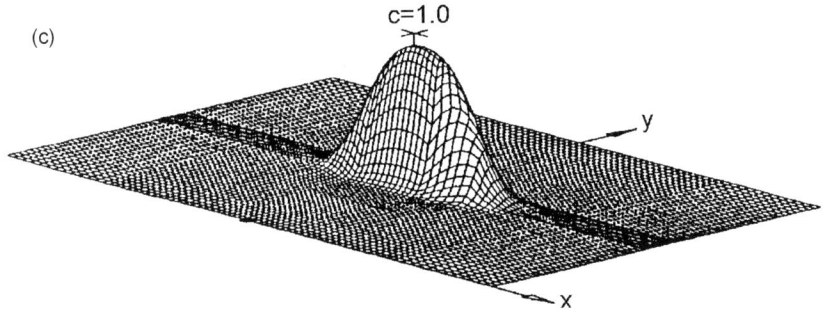

Fig. 5.53 (continued)

There are more reasons why it is not possible to directly use the laboratory modelling for design of earth structure, but mostly it is due to model scale. From the view of the model scale it is necessary to evaluate other possibilities of laboratory modelling.

5.3.2.1 Models Observing Stress-strain Behaviour

Fundamental model falling into this section is observing the bearing capacity of earth structure – in principle is observing how much this earth structure can be loaded by external loading. Different models can simulate this situation as model observing bearing capacity of the surface of the reinforced soil, model observing possibility of loading very close to the upper edge of slope, observation of the optimal situation of reinforcing element and many others, e.g. Scharle et al. (1983).

In the simplest case it is possible to speak about observation of behaviour of subsoil during loading by spread foundations. Here, with continuous loading, deformation (settlement) is measured. The obtained relationship $s = f(\sigma)$ is a rather simple output but very useful one. Not only bearing capacity can be determined but also settlement for which the deformation is not linear to the applied loading. If there are doubts about direct application on in situ example, the obtained results can be used only for the comparison of results. In Fig. 5.54 the tested example observing the influence of the depth of reinforcing element under spread foundation, Vaníček (1983a) is illustrated. The results show that for the given case the most positive influence of the reinforcing element is in the depth 0.375 to 0.75 B, where B is width of foundation. However if more outputs are needed, e.g. how exactly the subsoil behaves, there are still some other possibilities.

For the presented case the laboratory model is situated in the laboratory stand with front glass. 2D problem is modelled. For observation of movement of the individual grains of subsoil behind the glass, two different versions of photographical monitoring can be used. During the first of them loading of the model is rather quick, leading to loss of bearing capacity. During this entire phase the camera is opened. On the final photograph the grains which are not moving are easily visible. On the

5.3 Modelling of Limit States

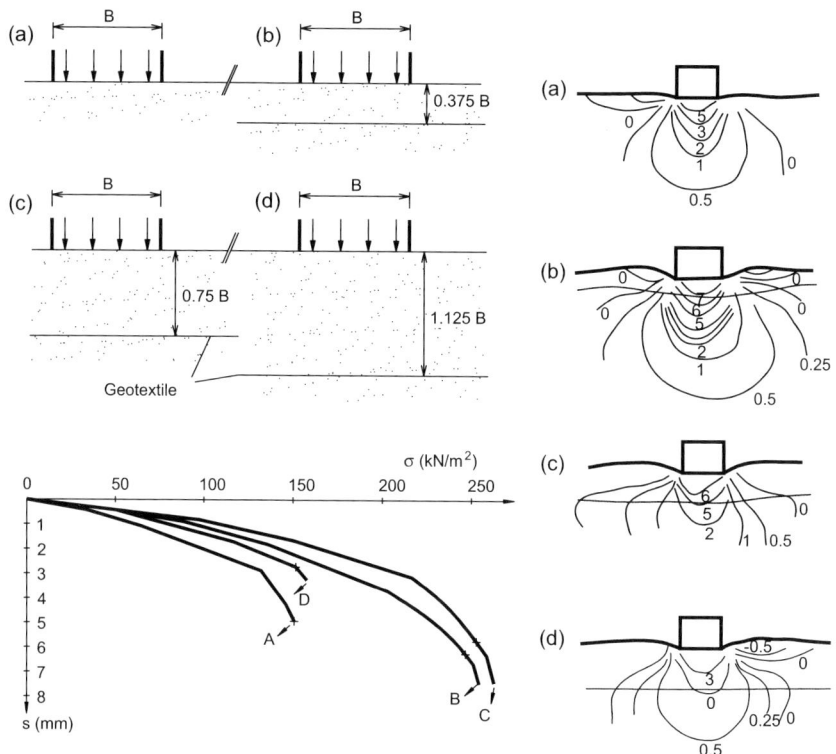

Fig. 5.54 Laboratory model studying the influence of subsoil reinforcement; presentation of vertical displacement contours for indicated load phase (+) close to failure

contrary the grains which moved during the phase of loading are not clear but their trajectory is easily recognized, see Fig. 5.55.

During the second of them more photos are taken in different steps of the model loading. Subsequently two pictures are compared in stereo comparator. We are getting 3D picture, where the third dimension corresponds with magnitude of grains movements. To the individual "contour line" it is possible to attribute individual movement and this movement is subsequently graphically interpreted. This type of interpretation is used even in Fig. 5.54 where for selected steps 0 → + "vertical displacement contours" are depicted. From this type of interpretation we obtain some additional information, such as significance of reinforcing element situation, where especially for case (c) it is visible how reinforcing element brings into action a bigger part of the earth structure.

In the second case the influence of the geotextile on the behaviour of a railway embankment was observed from the point of view of deformation and limit failure, Vaníček (1983a). A model of railway embankment was constructed from sand particles with dia 0.1–2.0 mm. This embankment was covered by rail-ballast from coarse sand 2–4 mm. Loading was applied through small loading bridge and stiff

Fig. 5.55 Trajectory of sand grains during lost of bearing capacity for spread foundation

rubber. The approximate model scale chosen was 1:50. On Fig. 5.56 the results are presented for the case when geotextile was embedded in 0.5 m (10 mm) below rail-track. For the phase of loading from zero to the point just before the failure not only izolines of vertical and horizontal deformation but also trajectory of selected grains is shown. Similar results were obtained for the model situated close to the upper edge of slope. Stereophotogrammetric method proved to be a useful tool for observing deformation of planar model from sand. The main advantage of this method is the opportunity to observe the displacement of any grain at any time for selected interval of loading. A certain disadvantage is the fact that we can observe only sand grains which are in contact with frontal glass wall and can be affected by this wall. In more detail the stereophotogrammetric method is described in soil mechanics literature by Butterfield et al. (1970) or Vaníček (1980). The disadvantage connected with wall friction can be eliminated by x-ray radiography. Small lead balls (shots) are placed in the model. They are used as displacement markers, movements of which are monitored by x-ray pictures. From the tensile test point of view – see Chapter 7 this method was used by Ajaz and Parry (1975a), for study of soil reinforcement by Bourdeau et al. (1991).

We can include the models mentioned up to now in the group of conventional model tests. Discussion on their importance took part mainly during the European conference in Brighton 1979 – Basset (1979), Ovesen (1979). On one side the models have a number of distinct technical advantages over full-scale work such as:

- Because of their small size they should be relatively cheap and quick to carry out,
- They can be taken to failure and beyond with no catastrophic or expensive results,
- The soil properties can be chosen and to an extent controlled and overall parameters can be assumed to be known,
- The tests can be repeated to provide statistical scatter or can be modified under the experimenter's deliberate control to high-light one variable at a time.

5.3 Modelling of Limit States

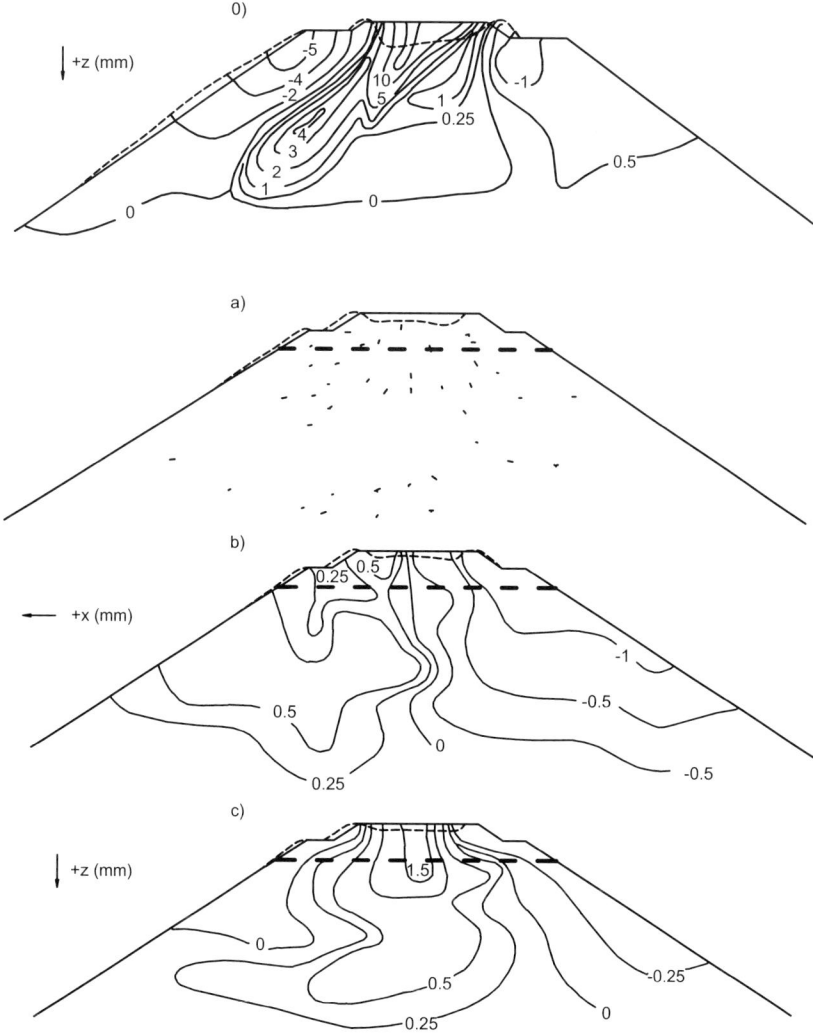

Fig. 5.56 Laboratory model studying deformation of reinforced railway embankment under loading with the help of stereophotogrammetric method. (**0–a**) Vertical displacement for load case without reinforcement and grains trajectory. (**b–c**) Load case with reinforcement–horizontal and vertical displacement

On the other side they are connected with certain problems, as with structure of a cohesionless soils sample during preparation procedure, with soil layering, local inhomogeneity, consolidation stresses, stress history etc.

Ovesen (1979) compares conventional models with centrifugal model tests and from that comparison the centrifugal tests are more credible if they are done in a scale 1:n. Whereas for conventional model tests for cohesionless materials we can get only good agreement for porosity and the angle of interparticle friction of the sand grains, for centrifugal tests the disagreement is only for the average grain size of individual particles. The specification of centrifugal models are described e.g. by Schofield (1980) or by Taylor (1995). Model of earth structure is located on a lever arm of centrifuge and loaded by the acceleration higher then gravitational acceleration g. If for prototype is valid, that the stress at depth h_p is:

$$\sigma_{vp} = \rho.g.h_p \tag{5.113}$$

than the same stress σ_{vm} is reached in centrifuge with the acceleration N times bigger than is the earth gravity in depth h_m:

$$\sigma_{vm} = \rho.N.g.h_m \quad \text{where} \quad h_m = h_p/N \tag{5.114}$$

This way the model can be N times smaller, Fig. 5.57. Scale factor (model: prototype) is 1:N for linear distances. As the model is in liner scale to the prototype, also the deformations will be at the same model scale of 1:N. Therefore we have that the strain has a model scale of 1:1 and hence part of the stress-strain curve mobilised in the model will be identical to prototype.

Ovesen (1983) presents the results of model study of reinforced embankment on soft subsoil using centrifuge in the scale 1:50 – see Fig. 5.58. In here the undrained strength of soil increases linearly with depth. The author describes problems with preparation of subsoil clay sample, reinforcement (a gauze which is normally used for medical purposes has a stress-strain curve which fulfils with a reasonable accuracy the model requirement) and with fill embankment, which consists of a dry sand with a rather uniform grain size distribution and an average grain diameter of 0.4 mm. Based on the test series, in which the input parameters have been changed, i.e. subsoil strength, reinforcement, and height of model, the author presents some general conclusions. The biggest difference was shown for the case of continuous contact reinforcement compared with divided reinforcement. For the case with continuous reinforcement the author summarizes:

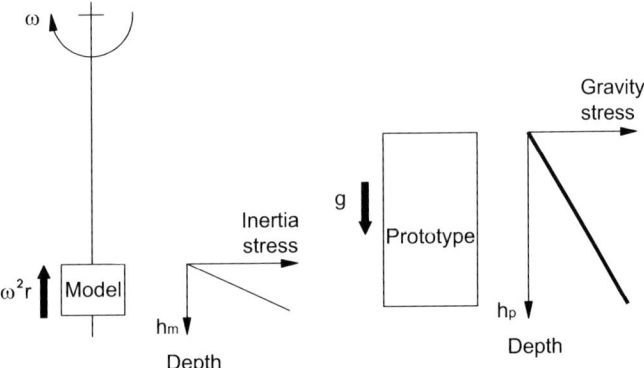

Fig. 5.57 Comparison between prototype and centrifugal model test (according to Taylor)

5.3 Modelling of Limit States

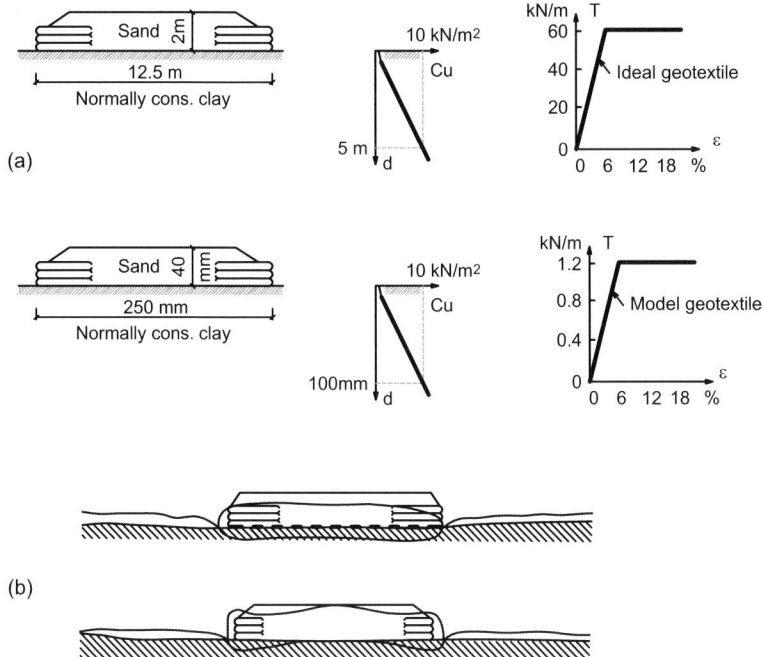

Fig. 5.58 Prototype and centrifugal model of reinforced embankment on soft soil (according to Ovesen). (**a**) Comparison between real and model properties. (**b**) Different settlement with and without reinforcement on contact with subsoil

- The embankment settles as a whole; the vertical movements of the surface of the embankment fill are uniform,
- The stability of the embankment is improved considerably by the reinforcement.

In none of the tests failure occurred in the geotextile-reinforcement, even though some of the tests were performed with geotextile-reinforcement having a tensile strength lower than that of most of the geotextiles available on the market.

5.3.2.2 Modelling of Interaction of Reinforcement with Surrounding Soil

In modelling structural details, the experimental scale may be up to 1:1. Pull-out tests of reinforcement, for example, may provide information not only on the magnitude of reinforcement friction with soil, but also on the distribution of friction along the reinforcement. This is, in principle, a test where part of reinforcement is embedded in the earth structure and exposed to tensile load – pulling out of reinforcement from the earth structure. Thus, a detail of reinforcement anchorage behind a potential slip surface is simulated in a certain way where this anchorage length is loaded for pulling out. The authors used an apparatus schematically shown in Fig. 5.59 – Vaníček and Hassan (2000). For the case of horizontal reinforcement embedding,

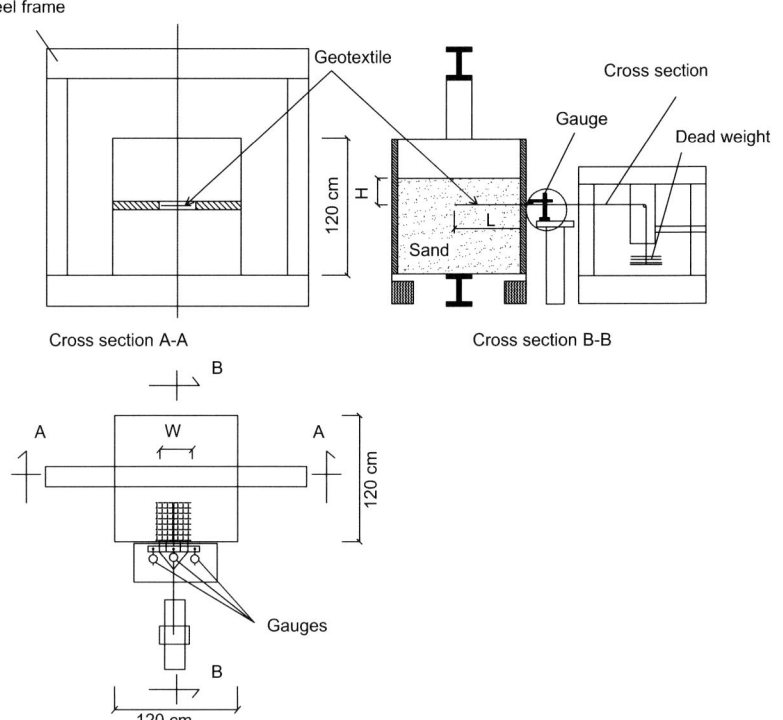

Fig. 5.59 Layout of the pull-out test

the effect of the reinforcement length L, width W and embedding depth H was monitored. The geogrid used had a tensile strength for a longitudinal and transverse direction $35 kNm^{-1}$ and a mesh size of 35 mm (Fortrac type 35/35-35). The geogrid was situated into medium to coarse sand, with a dry density $16.8\,kNm^{-3}$, a relative density 0.67 and an angle of internal friction $35°$.

Pull-out load was applied by manual load through a disc, wire and clamping plate fixed with the geogrid. The front displacement was measured by three-dial gauges and the average was taken. The results obtained are as follows:

- The face movement grows with a growing tensile force, for lower load levels nearly independently of the length of embedded reinforcement, at a higher load level (approaching the failure of the shortest element) the movement for shorter element grows signaling the coming failure – Fig. 5.60a. For longer elements, there is still a significant reserve and so the dependencies are practically linear,
- The mean shear stress at reinforcement-soil contact for the state of failure slightly grows with the falling length of embedded reinforcement L (for identical normal stress). These results indirectly support the basic precondition of non-linear distribution of tensile forces along reinforcement, – Fig. 5.60b,

5.3 Modelling of Limit States

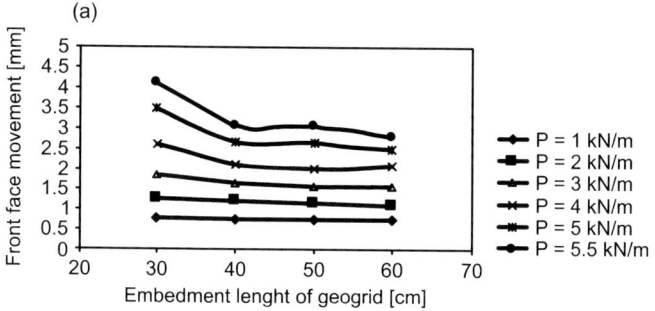

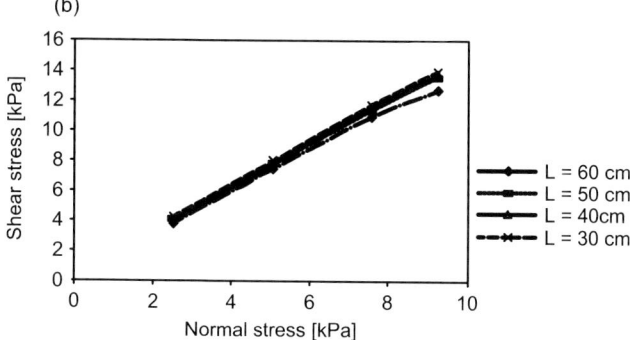

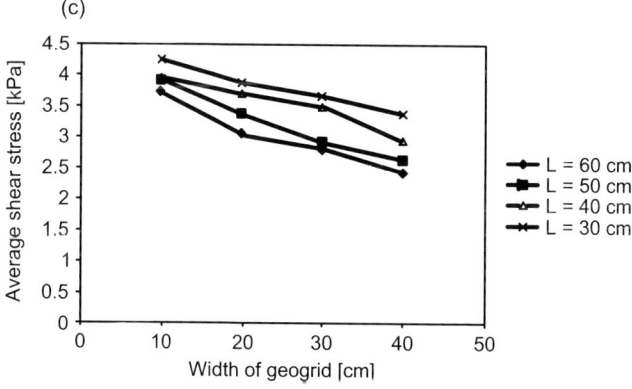

Fig. 5.60 Some results of the pull-out test. (**a**) Elongation as a function of different mobilisation of tensile force in reinforcing element. (**b**) Average shear stresses along the contact soil – reinforcing element for different normal stresses. (**c**) Average shear stresses along the contact for different width of reinforcing element

– Presumptions on a growing resistance of the contact for narrower strips were supported due to dilatation behaviour and also due to a 3D effect – see e.g. Schlosser, Magnan and Holtz, (1985). The narrower the strip, the more significantly the 3D effect is manifested, – Fig. 5.60c.

5.3.2.3 Laboratory Models of Groundwater Flow

Experimental methods model the filtration area of water flow through an earth structure in a certain scale.

Hydraulic models built in a certain scale are highly instructive for students. A soil medium model made in a model stand may preserve non-homogeneity and anisotropy. The model allows following flow lines e.g. by dying water at the input into the earth structure with potassium permanganate grains. Then, the seepage paths – flow lines – are shown in colour on the model. Total heights can be read with small built in piezometers. A free surface is read visually. The problem of a model scale is connected with capillarity, which is of much more importance in a model than in reality. Sometimes this problem may be solved by using a liquid with higher viscosity – glycerine, oil – to lower the capillary height.

The methods of analogy apply the fact that Laplace equation describes not only steady flow through soils, but also the flow of heat, electricity, magnetic field around a magnet etc. For practical reasons, the most suitable analogy is that with electricity flow in a conductive medium. Comparing Darcy's and Ohm's laws we get:

$$v_z = -k_z . \delta h/\delta z \tag{5.115}$$

$$I_z = -1/R_z . \delta V/\delta z \tag{5.116}$$

It can be seen that the gradient of total hydraulic heights h corresponds to the gradient of electric voltage V, and the electric resistance R_z of the medium through which electric current flows is analogous to $1/k_z$ – i.e. the permeability of a soil medium.

The models using this analogy, which is referred to as an electro-hydro-dynamic analogy, are basically of two types – Vaníček (1982a). The first type is a so-called continual model, while in the second case, a substitute model consists of individual discrete points (nodes) which are mutually interconnected by resistance conductors. The first case is more illustrative, while the second more general, more precise and prior to the development of FEM it was manufactured.

In a continual model, the conductive medium selected is conductive paper whose electric resistance is analogous to soil permeability. It applies direct current where electrodes are placed at the boundary of the investigated medium represented by an equipotential line of a high or low level, and the difference in levels in meters is replaced with analogous electric voltage in volts. The third electrode – a tip – is used to determine the voltage at different points of the conductive medium. By connecting the points of selected identical electric voltage, it is possible to draw directly equipotential lines in white pencil on black conductive paper. Flow lines may be obtained

5.3 Modelling of Limit States

by reversing the problem, as flow lines and equipotential lines are perpendicular to each other. Electrodes, in this case, are placed along impermeable boundaries or between the impermeable boundary and a free surface (for a case of water flow with a free surface). The method also allows modelling a medium consisting of two or more homogeneous soil types, but with different filtration coefficients. This may be achieved by interconnecting conductive sheets of paper with different electric resistances corresponding to the required permeability. Havlíček and Fiedler (1969) also used this method for the determination of pore-water pressures in an earth dam considering a state of capillary saturation and the effect of long-lasting rainfall.

5.3.2.4 Laboratory Modelling of Surface Erosion

In hydraulic structures, the limit state of surface erosion cannot be eliminated. Therefore, some laboratory models monitor the importance of erosion protection and its various combinations, mainly natural protection by grass plantation combined with erosion protection geosynthetic matting. An Ninh (2000) e.g. modelled various combinations in an experimental trough with a 45° grade where he simulated the effect of different compositions of erosion protection and surface water flow velocity. At the trough bottom, two holes were made and filled with soil to be covered with erosion protection matting, and the effect of surface water flow on soil outwash was monitored for various combinations:

- Protection solely by erosion protection matting of Enkamat or Raltex type,
- Protection by erosion protection matting where grass seeds had been placed in the subsoil in advance, and the test was carried out only after grass plantation had been formed,
- Erosion protection matting was filled with fertile soil together with grass seeds, and the test was carried out only after grass plantation had been formed,
- Analogous test to the preceding one plus erosion protection matting was fixed to the subsoil with small anchors.

The tests explicitly proved that the best results are reached in the last case. Mutual interconnection of erosion protection mating with subsoil is essential, both by means of small anchors, and mainly by means of a root system which grows through the matting as well as subsoil. A very important element is eliminating water penetration under the matting as, together with vibration, it accelerates erosion. Sandy soils eroded faster. The boundary velocity of surface water at which erosion was set off ranged from 1.0 to 1.5 ms^{-1}.

5.3.2.5 Laboratory Tests for Solving Contaminant Transport

Laboratory models in this case provide a certain image of contamination spread in a soil medium, but in practice they are rather applied for the determination of input parameters for a monitored interaction of soil with a specific contaminant so that these may be used in solving the contamination equation – see Section 5.3.1.4. These are the column leaching test and batch test.

5.3.3 Modelling of In Situ – Monitoring

Verification of the behaviour of an earth structure in a 1:1 scale is principally the most suitable method. Behaviour verification can be done on a model built in a 1:1 scale for the needs of investigation, or for the verification of some new technologies. This method is relatively exceptional as it requires a testing locality and represents the most expensive modelling type. On the other hand, it may provide the best information. Due care must be given to the construction technology to respect reality, and not only in terms of potential differences in the structure of laid soils. Sometimes careful configuration on a model is considerably better than that in a real construction. A more common procedure, therefore, is verification of the behaviour of an implemented earth structure which is part of an existing construction. There are two reasons applied. First, the verification is a proof that the respective earth structure was built in accordance with the requirements and through monitoring this fact is confirmed. In the second case, monitoring becomes part of the design, providing feed-back and is an inseparable part of designing geotechnical structures using the observation method.

The most common application case of the observation method is connected with the design and implementation of an embankment on soft subsoil where subsoil stability to a great extent depends on the magnitude of pore-water pressures developed. Through feed-back, by measuring pore-water pressures, the presumption of pore-water pressure development is confronted with reality, which allows adapting the speed of embankment construction or its total height. Analogous measurement of pore-water pressures in an earthfill dam under various loading states, on the contrary, verifies whether the original presumptions concerning the filtration coefficient, anisotropy, permeability or serviceability of the drainage system are in keeping with the initial presumption and that the dam operation is within the presumed limits.

5.3.3.1 Brief Overview of Monitoring Methods

This chapter will include only a brief overview of monitoring methods as their detailed description would exceed the framework of this publication. Some monitoring methods and related problems will be mentioned in Section 5.3.3.2 while describing exemplary cases of modelling in a 1:1 scale, or while describing some specific constructions in Chapters 6–8.

Deformation measurement is the basis for the monitoring of the behaviour of earth structures. In the first place, there is monitoring of the settlement (vertical deformation) of the surface of earth structures, mainly as a function of time. Thus, it involves not only absolute values, but also time-related changes in deformation necessary for the prediction of further settlement. As was already said, the majority

5.3 Modelling of Limit States

of earth structures will be deformed three-dimensionally, and so not only vertical, but also horizontal deformations must be monitored. In this case, observation is focused mainly on the bottom part of the slope where the greatest horizontal deformations may be expected. The question is whether the magnitude of horizontal deformation really represents deformation of a soil element under 3D conditions, or whether it is already the case of slipping surface propagation. In this way, timely assessment of deformation can avoid situations which would lead to a failure by slope slipping. Today's surveying methods are highly accurate and the monitoring of the deformation of the earth structure's contours does not present problems.

A more complex problem is the monitoring of deformations inside an earth body. Here, various level marks inside the earth structure are used, e.g. penetrated height finding probes. Inclinometers are the most common devices for the monitoring of a potential development of a slipping surface. A flexible casing pipe embedded in a vertical bore hole deforms at the point of a slipping surface development and this change is registered as a function of time. An inclinometers based on a similar principle can also be mounted horizontally to get information e.g. on the deformation of the subsoil-embankment contact.

The measurements of relative deformations inside bodies are done with various types of extensometers, mounted at fixed points whose distance is monitored etc.

The requirement for monitoring total stresses in soil is less common. Vertical stresses may be relatively well assessed by means of geostatic stress, the total weight of overburden above the investigated point. A more complex problem in assessing vertical stresses is connected with points where stress transfer from one zone to another can occur, as in the case of rockfill dams with a central clay core. Horizontal stress depends, to a greater extent, on the soil type and placement method. Various pressure cells are used where the soil pressure is measured through the deformation of a membrane in the pressure cell and the pressure of a liquid or air which is inside.

The measurement of water pressure in pores is very useful for the calculation of effective stresses, which decisively affect deformation and strength characteristics of soils. It consists in precise determination of water pressure in pores at a point where total stress may be well assessed. Therefore, the choice of open boreholes or wells, protected with a perforated casing pipe, is not the right solution in the majority of cases – as there is a danger of interconnection of horizons with different water pressures resulting in an inaccurate reaction to changes. Proper measurement is realized by piezometers, which measure water pressure at a concrete, closely specified point. The condition is good isolation of the sensor from the surrounding medium. The simplest type is open piezometer, which operates well in permeable soils. The basis is a piezometric tube of a small diameter, fitted with a short perforated cap with filtration packing. The piezometer cap is installed in a well at a required depth, and the probe stem is sealed above the measured horizon. After some time, acceptable for permeable soils, water from the cap's vicinity gets into the piezometric tube and water pressure is directly given by the water column height. Direct reading of a water level in the piezometric tube is carried out by measuring bars, a hydrogeological pipe or an electric measuring tape with light or sound indication.

One of the basic disadvantages is the fact that the measurement requires a relatively large amount of water which must flow from the earth structure inside the piezometric tube. This means that the results can be obtained only with a time lag and so alerting to sudden changes in pore-water pressures due to external load increase in poorly permeable soils, is problematic. This problem is, to some extent, solved by the enclosed piezometers where water pressure is read by means of a membrane of the piezometer cap. The effect of pore-water pressure is read in different ways. The effect of water pressure on an elastic membrane, for example, is eliminated by equivalent back pressure whose magnitude is read with the needed precision directly on the pressure manometer. The advantage of mechanical measuring systems is their relatively high reliability and precision. Their disadvantage is a reduced possibility of remote readings, problems with automation. Electrical measuring systems convert the deformation of the membrane of a piezometric cap into electric variables with easy transfer into central measuring units. Like open piezometers, enclosed piezometers can be installed in bore holes, or the measuring cap is forced into the earth medium – e.g. in CPT – cone penetration test. The choice of a suitable piezometer depends on the soil type, activation speed, device sensitivity, a possibility of registering only positive pressures or also negative ones (suction) and in electric measuring systems on the method used (string, induction, resistance...). A change in the magnitude of pore-water pressure giving a good response can also be a significant indicator of the development of a slipping surface in the earth structure.

The amount of water flowing through the earth structure is important mainly for hydraulic structures. Here, the seepage caught by various collecting drains is monitored as a function of time. The character of entrapped water is also monitored if it contains larger amounts of fine washed out particles.

In earth structures of environmental structures, numerous other variables are also monitored, from the chemical composition of groundwater, leachate quality from landfills or quality of air contained in the pores of unsaturated soil, to the temperature or composition of landfill gas, pH, conductivity and many more variables, which have a certain informative value in relation to subsoil contamination, Pecková (2005). Various additional sensors may currently be applied in CPT, as is stated e.g. by Lunne et al. (1997).

5.3.3.2 Practical Examples of In Situ Modelling – Monitoring

Modelling of Stability Problems of Temporary Slopes

A very interesting problem was described by De Beer (1969). For the construction of the ventilation buildings of the road and railroad tunnel under the Scheldt River at Antwerp, the fundamental question was to design the slopes of the excavation in such a way that over their whole height they should be stable over a period of about 1 year. Total depth of excavation was nearly 30 m and last 15.5 m in the Boom clay. The Boom clay belongs to the oligocene series (Rupelian stage). According to the geologists, at the beginning of the continental Pleistocene erosion it was covered by

5.3 Modelling of Limit States

about 40 m of neogene sand (Antwerpian). The Boom clay should never have been submitted to larger loads than those corresponding to these 40 m.

Liquid limit amounts to about 80%, and the natural water content of 25–32% is very near the plastic limit. The Boom clay in the upper 35 meters exhibits horizontal layering and has a medium to high degree of fissuring. Many of the fissures have a slickenside appearance. Therefore the Boom clay in its upper part has to be described as "stiff fissured and layered clay".

The following shear parameters were obtained:

- Peak shear strength on $1\frac{1}{2}$" samples, adopted for the sake of security:
  $-c' = 15$ kPa, $\varphi' = 22°$,
- Residual strength – samples not previously cut in torsion apparatus:
  $c' = 0, \varphi' = 22°$,
- Residual strength – samples previously cut in torsion apparatus:
  $c' = 0, \varphi' = 15°$,
- Residual strength – samples previously cut in direct shear apparatus:
  $c' = 0, \varphi' = 11°$.

De Beer stated at this stage that cohesion vanished for sufficiently large distortion and that the lowest value of the residual angle is about 11°. Nevertheless with these data it was very difficult to answer the question about economical design of temporary slopes which are to last for 1 year. Therefore, a test pit, bordered by slopes with different inclinations was realized. Behind the slopes piezometers, and inclinometers were installed. In the upper sand layer the water was lowered by filterwells.

Altogether three different slopes were tested – slope I with vertical slope in the Boom clay, slope II with an inclination of 2:1 and finally slope III with inclination of 1:1.1. The first slope failed after 50 days. First tension cracks were observed after 14 days, when the horizontal deformation on the top of slope was 31 mm. Total failure occurred when this reading was about 88 mm. The distortions measured with the help of inclinometer were concentrated at the bottom of slope. The reading of piezometer indicated that after completion of the slope the pore-water pressure gradually decreased, which means that the zone of capillarity tensions gradually increases. However when the big movements started to occur, the pore-water pressure started to noticeably increase in the region of the potential rupture surface. Slope II was still stable after a period of 6 months and will be described in more details. Finally slope III was under observation for a period of 9 months and inclinometers showed a steady increase of the dial changes with time but very small one compared to the previous cases.

The cross section for slope II is shown on Fig. 5.61. The slope II runs with an inclination of 2:1 over height of 13.5 m in the Boom clay. The data of the inclinometer observations are gathered in Fig. 5.61a. The inclinometer IW5 shows at level – 18.3 (roughly at the level of the bottom of the test pit) a steady increase of the dial changes with time. However at the end of observation period (after 6 months) the dial changes were still much smaller than those which were observed in inclinometer for slope I, just before the vertical slope collapsed. At the end of

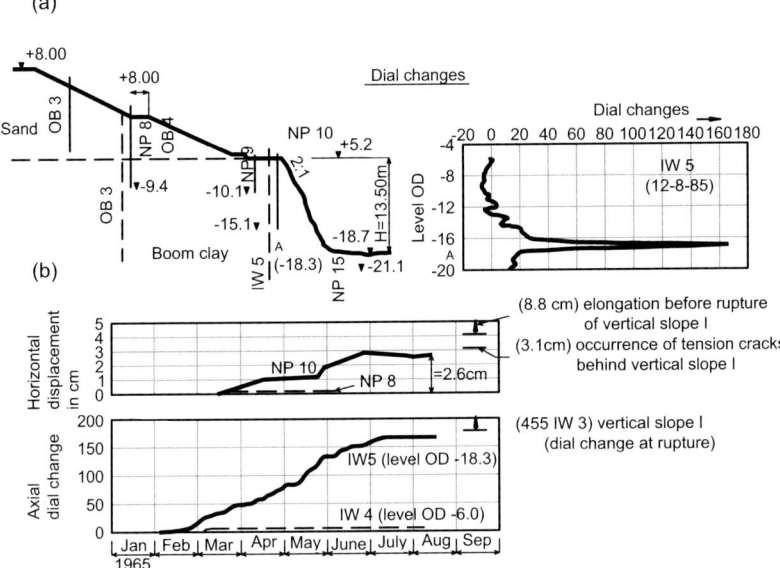

Fig. 5.61 Cross section of the test pit II and measured data in the Boom clay in Antwerp (according to De Beer)

the observation period the tops of the piezometers NP10 and NP8 show respectively horizontal displacements of 26 mm and 3 mm – no tension cracks were observed even if the elongation between these points was 23 mm.

Back calculation was performed for adopted value of friction angle $\varphi' = 22°$ and for water-pressure from measured values for both the end of excavation and for the moment of rupture, where it occurred. The data are gathered in Table 5.3, which gives the needed cohesion in order to have a safety factor $F = 1$. Therefore, if at the considered moment the slope has been just stable, the real mean cohesion of the clay was larger than the adopted value 15 kPa. Recalculated value of cohesion is about 35 kPa which is about the highest value which was obtained from the triaxial tests. It is also worthwhile to observe that although the slope III is much flatter than the slope II, the needed cohesion is larger due to the much higher pore-water pressures measured for this case. For the vertical slope I, assuming that the tension crack is filled with water, the cohesion c' to be introduced to have a safety factor $F = 1$ is 15 kPa. Thus, because of the large distortions the mean cohesion along the slip surface has already dropped from more than 35 kPa at the start, to 15 kPa at the moment of rupture.

After relatively sophisticated discussions which is out of the scope of this chapter De Beer recommended to count with cohesion 25 kPa for calculation of final stability after one year. So it was possible to design effective slopes for this case with the help of in situ test modelling. Nevertheless for permanent slopes de Beer recommended to count with following parameters and factors of safety:

5.3 Modelling of Limit States

Table 5.3 Test pits in Boom clay in Antwerp – the needed cohesion in order to have a safety factor $F = 1$ with assumed angle $\varphi_{ef} = 22°$ (according to De Beer)

Slope	Mean inclination	Height in the Boom clay [m]	Needed cohesion to have $F = 1$		Mean distance of the sliding surface to slope surface [m]
			Immediately after excavation [kPa]	At the end of the lifetime of the slope [kPa]	
I	Vertical	11	35	15 (rupture after 50 days)	5
II	2:1	12.5	17	17	
III	1:1.1	12.5	29	29	

$$\varphi' = 22°, \quad c' = 4\,\text{kPa}, \quad F = 1.2$$
$$\varphi' = 22°, \quad c' = 0\,\text{kPa}, \quad F = 1.0$$

and to count with conservative values for pore-water pressure, assuming that the flow-lines are horizontal and equipotential lines vertical. Finally the results are discussed and compared with experiences gained from slopes of the different canal cuts in Belgium.

Full-scale Trial Model of the Reinforced Earth Walls

In the early nineteen-seventies the Transport and Road Research Laboratory in the UK started to investigate the methods of design and construction of reinforced earth walls – Boden et al. (1979). The broad objective of the TRRL programme was to provide guidance for the design, construction, maintenance and repair of reinforced earth structures under the conditions likely to be encountered in the UK. The experimental work has followed the usual progression of increasing scale and soil complexity. In scale, the research has ranged from triaxial and shear box experiments investigating the soil-reinforcement interaction under carefully controlled laboratory conditions, going through laboratory model experiments to pilot scale experiment and finally to the full-scale trial model. Laboratory models used either centrifugal modelling or classical laboratory model which was 1 m in height. Before proceeding with the full-scale trials, it was considered prudent to investigate the short term behaviour of a reinforced clay wall at pilot scale, which was after that used as a basis for the design of the full scale trial. The pilot scale reinforced clay was built at the experimental facility having approximately 3 m height and 5 m width. The reinforced zone varied from a length of 2.3 m at the top to 1.7 m at the bottom of the wall. The soil used was a silty clay and facing units consisted of 600 mm high hexagons, manufactured from glass reinforced cement – see Fig. 5.62. The final step was connected with full-scale trial model constructed at Laboratory grounds at Crowthorne. The dimensions of the model were really large, length 45 m, width 14 m and height 6 m. To attempt to separate effects due to variations in the type of reinforcement and facing from those arising from the nature of the fill, it

Fig. 5.62 Pilot-scale wall towards the end of construction (according to Boden et al.)

was decided to build the structure in three separate layers. This arrangement resulted in vertical profiles which consisted of the same reinforcements and facing units from top to bottom, but which were constructed in three different layers of fill material. The first layer consisted of sandy clay, the second was constructed with a free-draining granular material and the final one was composed from silty clay with higher clay content than that selected for the first layer. But cohesion soils were not compacted directly to the facings, vertical drainage curtain 300 mm thick was applied along all facing surface. During those days more attention was devoted to the steel reinforcing elements (stainless steel, galvanised mild steel, aluminium coated mild steel, plastic coated mild steel), but also two geosynthetic reinforcing elements (fibre reinforced plastic, consisting of glass filaments embedded in polyester resin, plastic strips formed from polyester filaments embedded in polyethylene – trade name Paraweb) were applied.

But most attention was devoted to the reinforced earth structure monitoring to monitor and measure tension in the reinforcing elements, vertical and horizontal earth pressure, pore-water pressure, soil temperature and settlement within the fill mass. Additionally locating studs were inserted in selected columns of facing units to enable measurements to be made of their horizontal and vertical displacement using a theodolite and optical level respectively. The total instrumentation schedule is summarised in Table 5.4. The use of bonded gauges proved to be a suitable means of determining the distribution of tension in metal and fibre plastic strips. However, in case of the "Paraweb" strips, the difficulty in obtaining a bond on the material precluded the possibility of direct strain measurements, and it was therefore necessary to incorporate short metal plates on which the strain gauges were mounted at appropriate positions along the instrumented strips. Location of instruments installed in the section where "Paraweb" reinforcing strips were applied is shown on Fig. 5.63.

5.3 Modelling of Limit States

Table 5.4 Full scale trial of reinforced soil – Instrumentation schedule (according to Boden et al.)

Number used	Instrument	Purpose
39	Piezometers: twin tube hydraulic type	Porewater pressure measurement
114	Pressure cells: pneumatic type	Total earth pressure measurement
3	Pressure cells: hydraulic Bourdon tube type	Total earth pressure measurement
476	Strain gauges – electrical resistance type	Measurement of tension in reinforcing elements or anchor plates NOTE: Mounted in pairs on either side of the strips to take account of longitudinal bending
33	Settlement plates: reed switch and magnet type	Measurement of settlement within earth fill NOTE Arranged in three vertical profiles of 11 plates each
364	Locating studs mounted on facing units	Measurement of horizontal and vertical displacement of wall facing
9	Thermo-couples	Measurement of soil temperature

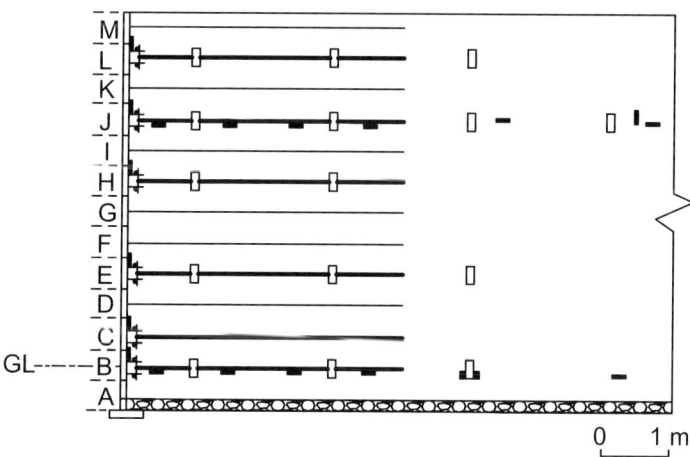

Fig. 5.63 Location of instruments installed in the full-scale reinforced earth structure to evaluate the performance of Paraweb reinforcing strips (according to Boden et al.)

Fig. 5.64 Full scale structure at completion of the second stage (according to Boden et al.)

From the dimensions of the full-scale model (Fig. 5.64) and the extent of instrumentation it is clear that the obtained results formed a significant step forward in our knowledge of the new type of earth structure.

In Situ Monitoring of Distribution of Tension in Geosyntetic Reinforcing Element

In the previous case the problem of bonding of strain gauges to the geosyntetic reinforcing element was mentioned. In the frame of the research project, different types of modelling were compared – Vaníček (2001b). One part of this project was connected with modelling of stresses along the reinforcing element made of geosynthetic geogrid from polyester fibres impregnated by other polymers, number of individual fibres in one rib is 7, total width of rib is 6 mm, mesh 23×23 mm or 38×38 mm. Maximum tensile strength at failure $55\,\text{kNm}^{-1}$ or $30\,\text{kNm}^{-1}$. Two basic methods of measuring were discussed – one with the help of – strain gauges – bonded on ribs or with the help of mechanical extensometers fixed in selected distance to the ribs in the direction of main ribs with maximum tensile strength. The first method was selected not only due to the financial aspect but also due to problems connected with the protection of extensometers from the parasitic effects. Defined demands on this system are as follows, Vaníček (2002c):

- Strain gauge should be situated directly on high polyester yarns covered by an additional protective layer of PVC,
- Width of this strain gauge should be smaller than 6 mm,
- Long term stability for minimum 1 year,

5.3 Modelling of Limit States

Fig. 5.65 Testing the applicability of strain gauges on individual geosyntetic yearn, in background the compensating strain gauge

– Range of measurement should be in the range of expected elongation, it is roughly 4–5%
– Measured value of tensile force is independent of bending of rib, of temperature and surrounding humidity.

Details of the strain gauge selection are described in the above mentioned publications. Calibration was done in the laboratory, where measured elongation -tensile force – was compared with the value measured by hydraulic jack. Compensating strain gauge was used as well; see Fig. 5.65. After that the strain gauges were installed on the reinforcing strip applied for improvement of the slope stability of a constructed sanitary landfill. This landfill is situated in old quarry where the owner is trying to use as much space as possible. The reinforcement was proposed to eliminate the possibility of landslide and also tensile cracks development. In Fig. 5.66

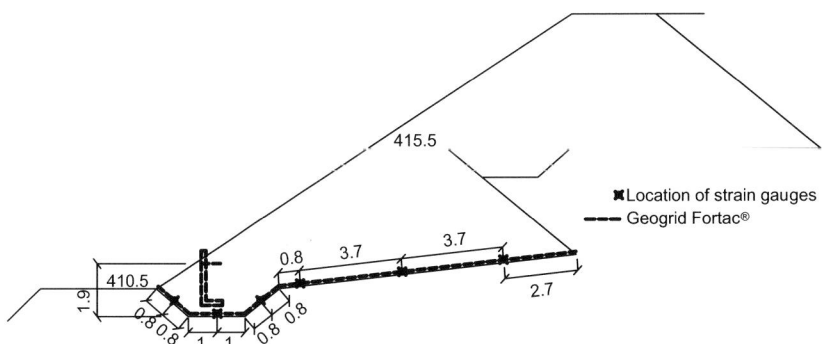

Fig. 5.66 Cross section and the detail of the strain gauges arrangement

is a cross section and the detail of the strain gauges arrangement. In Fig. 5.67 the reinforcing strip layout and measuring unit is shown. The results of the field measurement were compared with numerical modelling, using either limit state approach or FEM, Plaxis software with different assumptions for Mohr Coulomb model and hyperbolic model. The comparison was a rather good one, and a slightly better agreement was obtained for FEM method with hyperbolic model.

(a)

(b)

Fig. 5.67 The reinforcing strip layout and measuring unit

As a conclusion we can state that the method of measurement was successful and the measured values are in relatively good agreement with numerical model. The investor in this case obtained good approved results that the installation fulfilled the demanded aims.

More examples will be presented during practical case description in Chapters 6–8.

Chapter 6
Earth Structures in Transport Engineering

Transport engineering is currently undergoing relatively rapid development. This is foremost indicated by the development of an international highway network and by the construction of many new airports and these moreover in very difficult circumstances. Also at the same time there is a reconstruction of the railway network for higher speeds and entirely new construction of routes for high speed trains. Also the amount and area of dedicated car park space grows. For example Michaud (2000) states, that in Switzerland with the construction of 1600 km of highway network between the years 1960 and 2000 there has been a significant increase of transport on the roads network – fivefold for private cars and sevenfold for freight vehicles. Therefore attention must be directed to a philosophy of sustainable mobility. One of the proposed steps which will combine this philosophy with such increased traffic is the approval and realization of two long tunnels through the Alps: the first one, the Lotschberg, 37 km long and the second, the Gotthard, 57 km long. These two new rail tunnels will allow the transit of goods and especially of trucks on special trains linking Northern Italy with Southern Germany without using the Swiss road network. *The philosophy of "Sustainable Construction" has been getting primary attention in recent years. This new approach shows that an excellent technical solution is a necessary precondition, but not a sufficient one. Other aspects to which the modern project must apply itself to, are the environmental, sociological and architectural ones, and of course also an economic perspective because the final solution should be economically competitive, Vaníček* (2006).

From the point of view of general principles of transport engineering we can identify the following specific points:

- The total area for construction of transport infrastructure has a pronounced tendency to grow and impacts significantly on land appropriation, especially land already marked as greenfield sites, and this can be considered a negative factor,
- Earth structures in construction of transport infrastructure today present a significant potential for using various secondary materials, waste, by products in the process of construction – and in this they differ from earth structures in hydraulic engineering or environmental engineering works where their use is substantially more limited.

- In response to a growing concern for environmental protection the requirements placed on earth structures in constructions of transport infrastructure are increasing with special regard for possible crashes by different transport methods and the escape of any transported dangerous substances into a surrounding area. This also relates however to common products involved in any operation, be it oil drops, fuels, vehicle engine products etc.

These specific points go on to influence conceptual approaches to the design of new, and the reconstruction of existing transport infrastructure. Some views will be discussed in the following chapter.

6.1 Basic Assumptions – Situation of Traffic Network, Longitudinal Section

The above mentioned specific points influence the total approach to construction of transport infrastructure, and for linear projects it manifests itself in situating their route, or respectively this fact can come up in longitudinal profile.

Land protection changes a perspective on situating the route even when it may significantly complicate the actual design of earth structure. It involves

- Greater utilization of brownfields for situating new routes – an example could be a section of the highway D8 Prague – Dresden, which in the surroundings of Ústí nad Labem goes through the section of old dumps and sludge beds with very complicated foundation conditions. Situating the road, railway on the surface of uncompacted dumps about 130 m high near Ervěnice will be dealt with in more detail in Section 8.1.
- Situating of a route of transport infrastructure to an area with difficult foundation conditions – it involves especially territory with very soft subsoil, with a subsoil structurally unstable or territory that is sliding. To an even greater extent this applies to routes situated in mountain areas – the last example could be the railway route Peking – Lhasa.
- Situating the route away from the areas with significant supplies of drinking water or with a detailed specification of its protection.

With new approaches that also change perspectives on the longitudinal profile of the route. In the classical interpretation a balance between embankments and cuttings was in favour. From the present point of view it is obvious, that generally there exists a surplus of materials, which could be used for embankments and that is why the condition of balanced capacities loses its significance. At present with urban development the surplus of mined soils increases, for example as a result of underground construction, tunnels, enlarged capacities of excavations of construction pits for the use of building land also under the terrain level etc. The next area is production of a significant volume of waste rock in the mining and processing of raw materials, recycled materials from construction demolitions, products created from processing and utilization of raw materials – for example power station and heating plant fly ash from coal burning. And so we could continue in this elaboration, because it is possible to anticipate continually new suggestions.

Besides the balance of capacities there is of further interest two aspects regarding transport infrastructure embankments. With an improved possibility for the design of even steep embankments with both the help of new technologies of soil processing as a building material and with the help of its reinforcement the border edge moves when it is more appropriate to create transport infrastructure embankments or yet when to direct the route across a bridge structure. For example Brandl (2001c) describes cases of embankments 100–120 m high, which proved in the high mountain terrain of the Austrian Alps more economical than bridge structures. The second aspect is a certain contrast between directing the route in an embankment or in a cutting near a built-up area. Protection against noise and light can favour cuttings all the way to cut and cover tunnels. The solution however is to provide protective bunds along transport infrastructure routes or in the surrounds of airports, which again consume significant amount of materials.

6.1.1 Application of Waste and Recycled Materials

The detailed description of possibilities with the use of existing large-volume waste is given by Head et al. (2006) with reference to conditions in the UK. Besides the dumps from coal mining (the estimate for Wales circa 500 Mt), circa 100 Mt of spent oil shale in West Lothian is mentioned, which can be a useful construction material provided that due account is taken of the potentially elevated sulphate contents and a potential vulnerability to frost heave in near-surface applications. Other possibilities come from

- Quarry industry – quarry by-products; it has been estimated that there are 730 Mt of slate waste, around 2 Mt of sandstone quarry surplus.
- China clay waste – in Cornwall – around 45–100 Mt,
- Construction, demolition and excavation waste (CDEW) – 90 Mt in England in 2003,
- Industrial by-products such as steel slag and pulverised-fuel ash (PFA),
- Recycled and secondary aggregates (RSA)

Nevertheless certain restrictions are also mentioned for using these materials such as:

- Lack of specifications for their use, which in some cases led to inappropriate use of materials such as unweathered slag which is prone to heave in roads or to the use of CDEW with large quantities of wood,

- Price was sometimes higher – especially in areas where there is enough high quality primary aggregates (as in Scotland and Wales), or where cost of transport exceeds the price differences.

In spite of that, in the last 20 years there has been a significant improvement, with the additional help of various taxes and levies. As a mechanism for encouraging increased recycling of materials, the UK government established the Waste and Resource Action Programme (WRAP) in 2001. This programme is supported by quality protocols for the production of aggregates from inert waste (from the processing of inert CDEW), for pulverized-fuel ash (PFA) and for slag and incinerator bottom ash (IBA). The result is positive – 65 Mt of RSA out of total of 275 Mt aggregates is used each year in the UK, representing around 25% of aggregate use.

How seriously these problems are taken is further attested to in a published summary – cited as Geotechnical Special Publication No. 127 "Recycled materials in Geotechnics" by ASCE 2005 or WASCON Conferences – Waste materials in Construction – WASCON 2000, Pergamon, Amsterdam. For example Saathof and Ketelaars (1999) describe the research project "Reuse of soil from bored tunnels". The reason for this research is the fact that they suppose that in the coming years around 1 million m^3 per year will be excavated in TBM (tunneling boring machine) drives in the Netherlands.

From all the above mentioned it is obvious then that the situation is being changed in a positive direction from the view of "Sustainability" although all these changes bring with them increased demands on earth constructions. At the same time it involves both mechanical – physical qualities influencing the behaviour of an earth construction and environmental qualities – one of these demands being also a judgement on whether leakage from spare materials cannot endanger the living environment – Kamon et al. (2000). These problems will be therefore specified in Chapters to come – see Section 6.8.1.

6.1.2 Landslide Prone Areas

The interaction of a new route of a transport infrastructure body with a landslide prone area brings a potential danger of deterioration of the newly built structure. So it does not involve the basic case, when a slide – limited state of stability – is caused by the construction of an earth structure in an undamaged area. This case will be solved both for embankments and for cuttings independently in the following chapters.

As a landslide area we can characterize such an area, where there have been landslide movements:

- In the distant past – the area has present long-term stability, however this balance can be disturbed by the new earth structure,
- In the recent past there has been a slide but now the area is stable – the movement can be reactivated by a new route, especially in the case that the old slide had not

6.1 Basic Assumptions – Situation of Traffic Network, Longitudinal Section

been recorded. On the old slide area there has been in the past a certain arranged tendency of clay particles as a consequence of big slide movements. In such a slide area it is necessary to consider residual parameters of shear strength. If the old slide has not been identified and "peak" shear strength parameters are introduced into the calculation, as is common for primordial slides, this fact can have a negative impact on total stability, see Fig. 6.1.

– It involves active slides, but very slow ones of a creeping character, where even this slow movement can disturb the new earth structure.

In this regard it is obvious that the identification of slide prone areas is very useful and why more detailed information about the character of an old slide movement is so important.

In the first phase of planning of the road and/or railway line the geotechnical investigation uses all up to date available information supplemented by site observation. Here a study of archive materials and different maps comprises the most important part of this phase.

Very important sources of information can be obtained from a set of Geo – environmental maps in the scale of 1:50 000. In the Czech Republic this set is composed from up to 16 maps, starting with geological, hydrogeological, engineering-geology maps, going on to maps of mineral deposits, rock geochemical activity, soil or geophysical maps, maps of surface water geochemistry and ending with maps of geofactors of the environment, protected areas etc. So this means that not only initial information about the land slide areas can be obtained but also how this area might be sensitive with regard to the protection of underground water and mineral resources as well as environmental protection.

In some countries special maps of areas of slope instability exist. For example Baliak et al. (1996) present the map of Slovakia with the marking of several hundred of places where there is a coincidence of slides with transport infrastrucure routes. They specify 4 basic areas from the view point of geological development. Among the most significant there is the area of the Carpathian flysh, where landslides appear

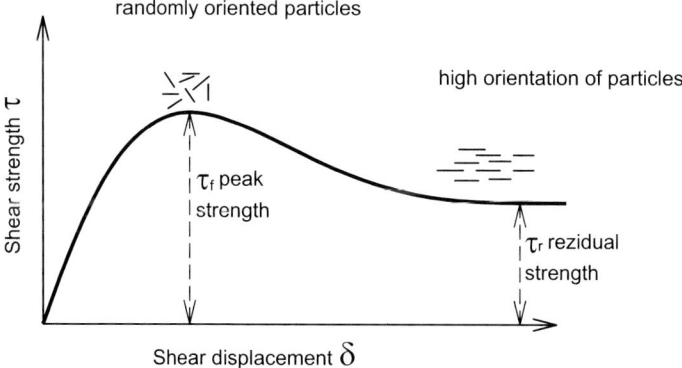

Fig. 6.1 Typical working diagram – shear resistance as a function of shear displacement

especially in the areas with clay development, in areas with an occurrence of microrythmical flysh and on slopes where there has emerged an accumulation of geests by deluvial creeping processes. The second significant area is a region of neogenic tectonic depressions. Here the frequent occurrence of landslides is conditioned by a geological structure of slopes on which in the upper parts there lay lake or terraced spatulas and in lower parts relatively soft neogenic claystones. All this information is a very important sign as to the extent to which the given area is prone to slide movements.

For a greater specification it is necessary to move to a detailed geotechnical investigation and then to recommend the most appropriate route. Pašek and Kudrna (1996) describe also a relatively complicated case of a route solution for the highway D8 Prague – Dresden across the České Středohoří mountains. The neovolcanic České Středohoří mountains suffer very much from mass movements. Huge blocks of basalts laying on plastic cretaceous marlstones moved down in Pleistocene into the valley of the river Labe. These fossil deformations are covered by masses of fossil and recent landslides, which are nowadays dormant (inactive). While slope movements of a block type cannot be practically influenced by embankments or drainage cuts (volumes are as regards the size of a slope phenomenon almost insignificant), fossil and recent landslides can be reactivated by embankments and drainage cuts. While recent landslides are in principle stabilized, there is still some fading movement, and these movements are measured in terms of mm up to tens of mm per year. It is possible to make a judgement here also from the sloping of older trees. Figure 6.2 gives an example of a typical profile through the observed area where alternatives of a route under consideration are also marked. With the cooperation of a geotechnical engineer and a highway designer it was possible to successfully link together various techno-economic, geotechnical and environmental demands. At the same time it was possible to use the experience with lining the

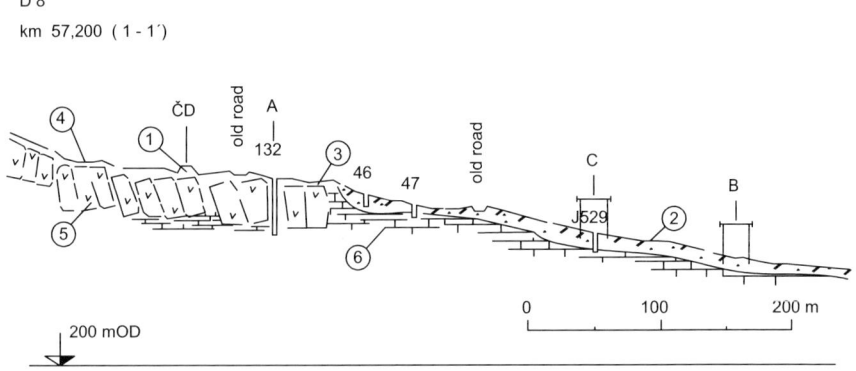

Fig. 6.2 Schematic cross section through land slide prone area with recommendation of new motorway line D8 (according to Pašek and Kudrna). 1 – fill; 2 – soils displaced by slope movement; 3 – slope debris; 4 – boulder debris; 5 – touchstone; 6 – marlite; A,B,C – proposed options for new motorway line; ČD – old railway

6.1 Basic Assumptions – Situation of Traffic Network, Longitudinal Section

railway completed in this area more than 100 years ago. The resulting route goes safely along recent landslides. By the choice of a situated and high lining route the highway body will thus contribute to increased stability of slopes in the given area.

From the viewpoint of a direct interaction of an earth body with an old landslide it is possible to use the principle of a neutral line (point). According to Hutchinson (1977) this line (point) is defined as a place, according to which a change of stress (caused either by an embankment or by drainage cut) will not cause a change of stability as a whole (stability F_1 after an application of a change of load is the same as the original stability F_0). The author derives the neutral point on the basis of a conventional method for a common slide area, namely on the basis of Bishop's method. The position of a neutral point depends on the angle α_n of the slide area from the horizontal one, Fig. 6.3:

$$\operatorname{tg} \alpha_n = (1 - \bar{B}.\sec^2\alpha_n).\operatorname{tg}\varphi/F_0 \quad (6.1)$$

Where the parameter of a pore pressure $\bar{B}$ directs the change of pore pressure:

$$\Delta u = \bar{B}.\Delta\sigma_1 \quad (6.2)$$

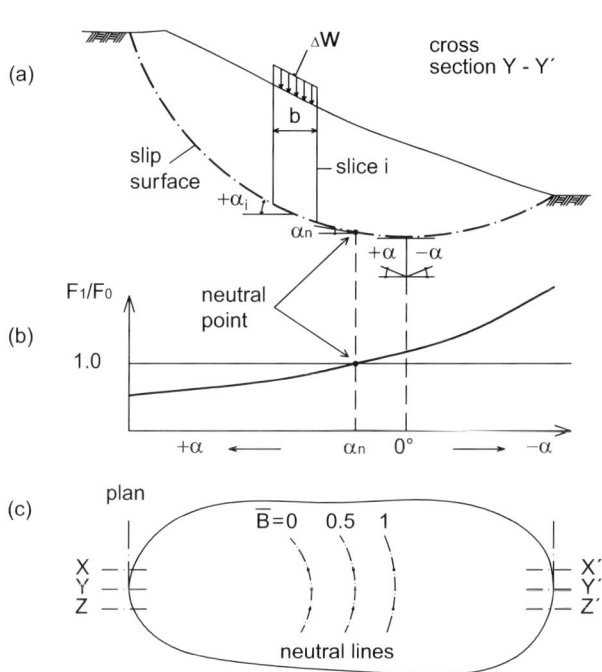

Fig. 6.3 The influence of additional loading on slope stability – explanation of neutral lines and neutral point (according to Hutchinson). (a) Cross-section of a landslide with an influence of load acting; (b) Influence line (diagrammatic) for the effect of load on overall factor of safety of the landslide; (c) Plan of the landslide, showing diagrammatically the position of the netural lines for different $\bar{B}$ values

From the above mentioned it is obvious, that α_n descends with a growing parameter $\bar{B}$. For $\bar{B} = 0$ it is valid that tg α_n = tg φ'/F_0 (for an old landslide φ'_r) namely for $\bar{B} = 1 \rightarrow \alpha_n = 0$. From this view Hutchinson puts a big stress on the size of a parameter of a pore pressure, because it significantly influences the position of a neutral line. That is why the embankment has a positive effect if it is realized under the neutral point, and in contradistinction the drainage cut is positive in effect above this point. The condition in the second case is the prevention of surface infiltration – saturation by water of a landslide area.

6.2 Transport Embankment – Fill

The construction of the embankment for a transport infrastructure is one of the basic tasks of geotechnical engineering. There comes to the fore then a definitive state of stability and deformation the solution of which is connected with the basic tasks of soil mechanics and consequently augmented for a definitive state of surface erosion.

Further two basic cases based on the quality of subsoil will be distinguished. For the first basic case it is assumed that the subsoil is of relatively good quality, potential landslide areas go through the body of the embankment and also the extent of deformation is regulated by the embankment deformation. In the second case soft subsoil is assumed and that is why both basic limit states depend on the quality of this subsoil.

6.2.1 Classical Case – Embankment on Good Quality Subsoil

The actual design of an embankment on relatively good subsoil depends above all on the quality of soil which is at disposal – see also the classification of soil given in Chapter 2. Nevertheless the whole process has several common steps:

- Foundation of the embankment must secure a good connection of the subsoil and material of the embankment – the basic measure is the removal of vegetation and organic topsoil in the whole width of the structure;
- The contact of subsoil and the embankment secures a drainage layer – for its dewatering there must be a gradient of at least 3% – on a horizontal area the roof-like gradient is applied – on a sloping terrain it is possible to make also a benching of the contact;
- For the prevention of a drainage layer clogging it is necessary to apply filtration criteria – the standard solution is the application of a sand layer (sandy gravel), in necessary cases the application of a filtration geotextile or directly of a geodrain;
- In simple cases (tentatively 1st Geotechnical category) it is possible to design the gradient of a slope according to the recommendation of existing standards, for example Czech standards recommend up to the height of the slope of the

embankment below 3 m slope 1:2.5, at the total height of the slope up to 6 m in the level 3–6 m slope 1:1.5, if the slope is higher than 6 m then in the level 3–6 m slope is 1:1.75 and slopes higher than 6 m are made in 1:1.5;
- In more complicated cases (2nd and 3rd Geotechnical category) it is necessary to make a design in accordance with the limit states calculations;
- In using fine-grained soil for the embankment (often determined as appropriate under certain conditions) the development of pore pressure in the construction of the embankment must be taken into consideration;
- It is appropriate to combine fine-grained soil with more coarse grained material – the scheme of the construction of a stratified embankment – sandwich shows Fig. 6.4;
- Slope surface is covered by a sub-arable layer overlaid with 0.2 m of topsoil, which allows the grows of vegetation, fulfilling consequently the erosion protection function;
- In any need for applying less appropriate materials into the embankment (that is also some secondary materials) or with any need to reduce land take for the embankment, it is possible to design reinforced embankment with the help of geosynthetic reinforcements;
- When solving short – term stability it is possible to use both the solution under undrained conditions (using total parameters of shear strength) and the solution with effective parameters (on the ability to evaluate pore pressures at the end of the construction).

The majority of cases are of non – reinforced embankments of transport infrastructure and with these there is associated much practical experience from the second half of the nineteenth century. At that time the great development of railway transport and the construction of big navigation canals began. Next there will a more focused description of only one extreme case showing how progress in this area developed.

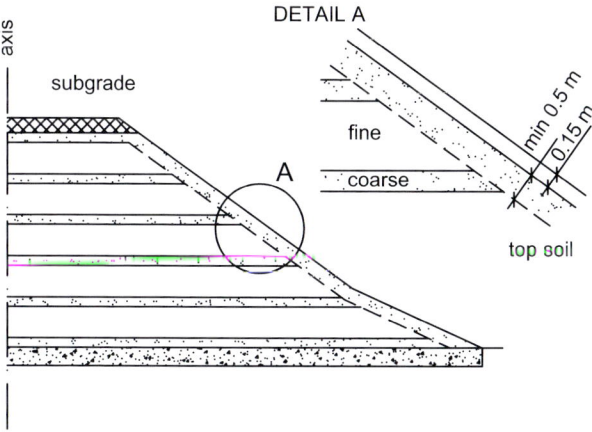

Fig. 6.4 Scheme of the arrangement of stratified fill

New interesting problems arise with cases of either steep reinforced embankments or embankments at which there are applied secondary materials. Both these cases will be consequently specified in this chapter and also in the chapter on embankment widening, where priority is given to steep embankments, or in the chapter on the interaction of earth structures of transport infrastructure with the environment. What impact on the environment can the use of waste materials have?

Brandl (2001c) describes extreme examples of high fill embankment with total height of 100 resp. 120 m. These embankments in principle substituted the originally projected bridge structures. Brandl states, that modern compaction equipment, optimization and control have opened the possibility of constructing high embankments instead of bridges for roads, highways and railways alike. In most cases, such alternatives reduce the construction costs and facilitate the utilization of local fill material. Furthermore, embankments can be vegetated therefore making them environment-friendly and keeping their maintenance costs much lower than for bridges. The long term behaviour of high embankments improves with time whereas that of bridges (especially of pre stressed reinforced concrete structures) worsens. Differential settlements between bridge abutments and adjacent embankment fills are minimized. Finally, high embankments may serve as a counterweight thus to increase stability of unstable slopes – whereas bridges would require very costly foundations and protective measures against local slope failure. Hence, slope embankments instead of slope bridges have proved especially advantageous.

Figure 6.5 shows a 100 m high embankment in a narrow valley which substituted for an originally designed bridge of 300 m in length. The embankment now supports an unstable slope as counterweight in its toe zone. The construction of the high embankment required the removal of the original river bed and a federal road, hence a 75 m high slope cut on the opposite side of the valley.

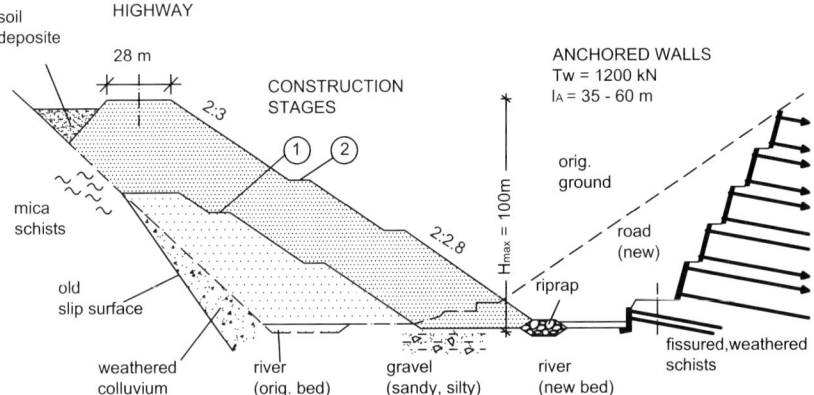

Fig. 6.5 A 100 m high embankment instead of a bridge for a highway in a narrow valley. Also indicated are the necessary slope cut and stabilizing measures at the opposite side of the valley (according to Brandl)

6.2 Transport Embankment – Fill

The valley floor consisted of river gravel with sand, cobbles and boulders overlaying mica schists which contained mylonitic zones with residual shear angles down to $\varphi'_r = 9°$. The fill material came from several slope cuts along the highway section under construction. Therefore its properties varied relatively widely; rock (mica schists, sometimes heavily weathered and decomposed), sandy gravel, fine sands, wide-grained colluvium and, locally, even (clayey-) sandy silts. Consequently, the fill material was placed sandwich-like, whereby weaker material was located in such zones which had no relevant influence on the stability of the embankment.

A heavy roller was used and the thickness of the fill layers ranged between 0.2 and 1.1 m, depending mostly on the quality of materials and climatic conditions. The maximum soil mass filled and compacted was about 130 000 m^3/month. Construction in two stages was made necessary by the relocation of a river and an old federal road.

Due to shortage of time it was necessary to continue with filling also in the winter time. Because of the sometimes low temperatures down to $-25°C$ the earth fill material had to be excavated in small parts so that the soil did not freeze during transport and placement. Frozen clods of head size (maximum) were accepted, but the energy of necessary compaction had to be increased. Soil layers were not allowed to be left loosely overnight but had to be compacted completely to the required degree until daily finish of earth work; otherwise the frozen layer had to be removed next morning. Therefore, in a cold period earthwork continued in day- and night shifts, whereby isolating mats were sometimes placed locally on fresh fill layers to avoid freezing. Frequently, before filling restarted after an unavoidable stop, thin covers of snow and ice had to be roughened. Thick layers of ice which might have caused long-term deformations or potential slip surfaces were thawed by salt. The thawed surface of a layer permitted sufficient bond with the next layer.

Settlements of the subsoil below the embankment bottom and of the embankment itself were measured in the bottom and in several levels of the fill. The settlement of the embankment itself was only about 0.1–0.2% of its height; the settlements of the subsoil were several decimeters. Both were running out relatively quickly, so that after a consolidation period over several months (during the winter), the sub-base, base and bitumen pavement of the highway could be placed without any special measures. Long-term observation of the embankment indicated excellent behaviour. The differential settlements along the bottom of the embankment (caused by heterogeneous ground) were widely levelled out over the fill height; hence the crest of the embankment remained very even due to the stiffness of the 100 m high fill. The positive impact on the environment can be seen comparing photos shortly after the end of construction (1979) and in 2004 – see Fig. 6.6 – Brandl 2004.

The pavement surfacing of the highway was constructed at a level of max. 100 mm higher than the theoretical design, in order to compensate future settlements. This reserve was used up within 5 years, and the additional long-term settlements have been negligible since then. Consequently, since the opening of the highway in 1980 no re-levelling of the pavement has been necessary.

Fig. 6.6 Photo of 100 m high embankment – photo Brandl. (**a**) Shortly after the end of construction (1979). Slope erosion protection with rhombus-shaped bio-reinforcement (willow branches etc.) that led to stable vegetation. (**b**) Detail to (**a**) – the embankment has become part of the natural environment

6.2.1.1 Fill Reinforcement

In this chapter we will refer back to Section 4.3, where typical types of reinforced earth structures have been discussed with their respective cross-sections, and also back to Section 5.2.1.6, where available design methods have been briefly described.

6.2 Transport Embankment – Fill

Reinforcement of the fill is also used in many other applications that will be explained later in this chapter, like embankments on soft subsoil, retaining walls and bridge abutments, but the design procedure is generally the same.

The design, of course, starts with definition of the geometry and first estimate of the reinforcements needed, based on either experience, tables or graphs. Then calculation of slope stability with the influence of reinforcements is performed and the results are compared with standards requirements. Here the designer is going to finish the design, if all the requirements are fulfilled or will amend the design in order to meet the minimum level of safety or optimize the design.

The geometry of the problem can involve the typical cross-sections as shown on Figs. 4.22–4.27, but also their combinations (reinforced steep embankment on soft subsoil) and other more complicated ones, like with benches and other features.

Now we will discuss a bit more about the design calculations in order to explain what the designer during the calculations has to take into account. And this will be shown on a recent example from design of motorway embankment on inclined layered soft subsoil, but not special treatment of it was needed. This design was done in accordance with Czech regulations TP97 using the author's own computer program SVARG, which calculation engine was included in a suite of geotechnical programs GEO 4 from company FINE. These Czech regulations require factoring of soil properties as was proposed in ENV version of Eurocode 7, define the design properties for the geosynthetic reinforcement using reduction factors (see Eqs. 5.52 and 5.53) and more over the design is considered as safe if factor of safety (utilization ratio) is at least 1.22 (see Sections 5.2.1.2 and 5.2.1.6).

The advantage of the authors software is the use of already mentioned Janbu's method of slope stability with the force from reinforcement implemented as additional horizontal interslice force (Fig. 5.23) and the possibility to analyse general slip surface. And of course the program automatically checks for the possibility of reinforcement pull-out, which is controlled by anchorage length and does not allow overstressing of the reinforcement in this case. The importance of the possibility to analyse general slip surfaces compared with circular ones will be shown as well, as this can lead to dangerous designs.

This design was rather unusual as there were several issues with this embankment next to the main issue of reinforcement of the embankment body, also the embankment was intended to be built on inclined clayey subsoil. The first set of calculations showed that the embankment on its own is not stable without any treatment to the subsoil. The agreed scenario required replacement of about 30% of the subsoil to treat it by good quality gravel in trenches extending underneath the whole embankment along the slope to provide additional drainage. The overall (global) slope stability analysis for the final scenario is presented in Fig. 6.7.

Once the external stability is guaranteed it was necessary to define the extent of soil reinforcement required in order to maintain also internal stability of the embankment. The first analysis was to determine the number of layers and the needed strength of reinforcement, see Fig. 6.8a. In order to optimize the design each layer of reinforcement was checked for its required length and this is presented on Figs. 6.8b–6.8f.

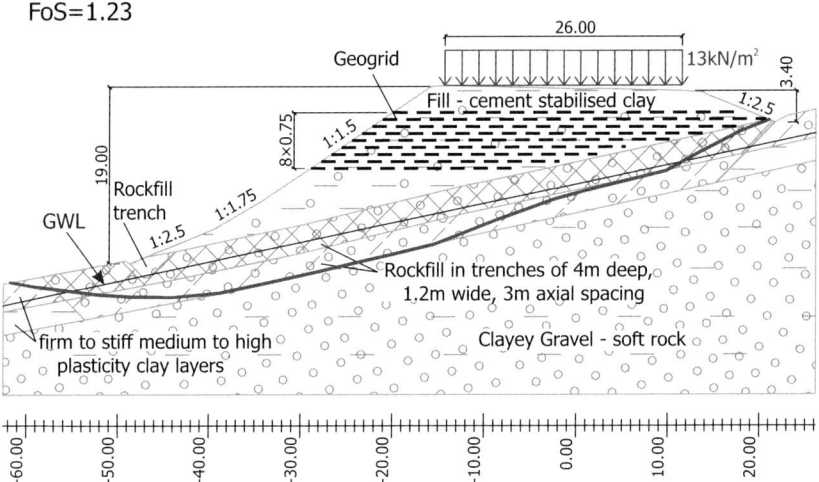

Fig. 6.7 Overall stability of motorway embankment on inclined subsoil

The promised comparison between polygonal and circular slip surfaces for this example is shown in Fig. 6.9, here you can see that circular slip gives over 8% more optimistic design.

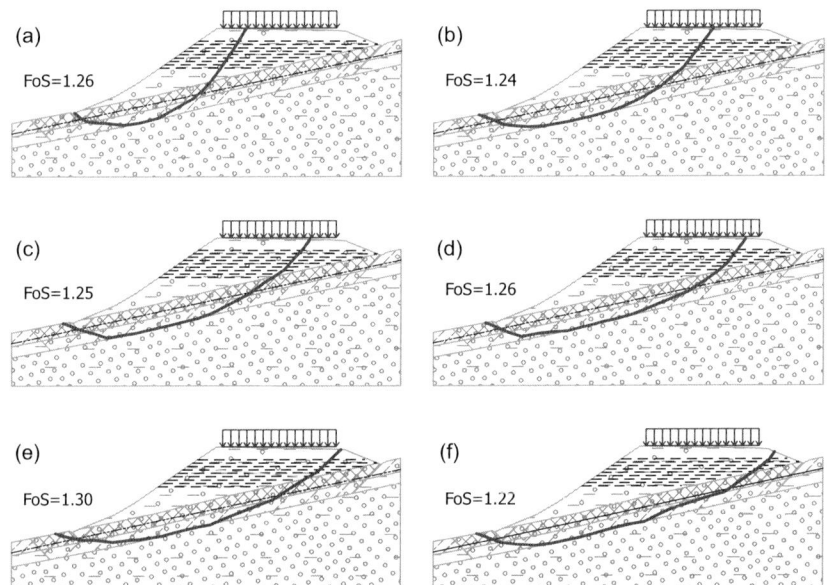

Fig. 6.8 Internal stability calculations for the motorway embankment; determination of reinforcement required for internal stability including their length

6.2 Transport Embankment – Fill

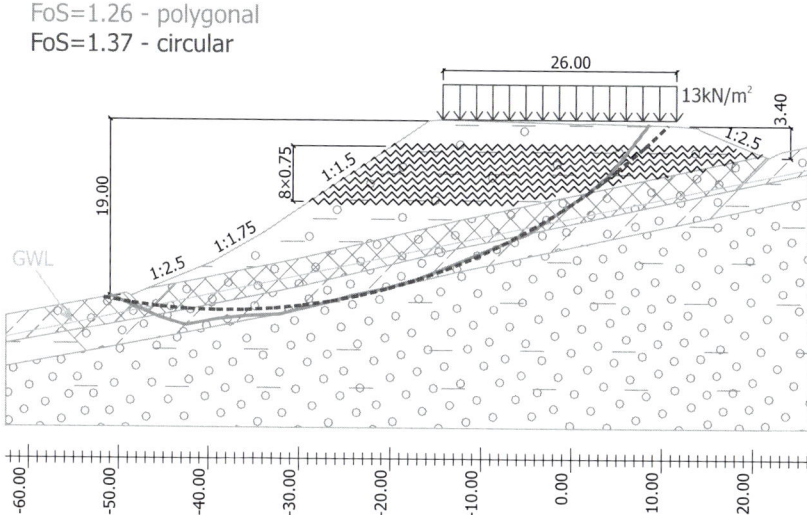

Fig. 6.9 Comparison of reinforced slope stability results for circular and general slip surfaces

6.2.1.2 Erosion Protection of Steep Green Reinforced Slopes

When applied to steep slopes the demands on erosion protection increase, especially in cases where a green slope is preferred. In Fig. 6.10 the basic cases of details of green reinforced slopes are represented, where it is possible to differentiate:

a) Straight reinforcement – reinforcement is laid down horizontally and straight, then soil is spread and compacted to the required height, levelling the face, preferably with face compaction. The method is used mainly for slopes with an inclination up to 50°. The face is covered by topsoil, again slightly compacted and seeded by grass. Topsoil can be situated also into different geomats or geocells in the manner which was described before – see Section 3.3.6. But different types of prefabricated sheets of geosynthetics with top soil and grass seeds can be used as well.

b) Reinforced wrapped around the face where the same geosynthetic used for reinforcement is also used for face protection. The wrap around installation procedure can be applied with or without formworks. Formworks are preferred where higher demands on smooth and uniform face finishing are defined. After laying reinforcing geosynthetics (which overlap the face by the length demanded for wrapping around) fill soil is compacted and smoothed into demanded shape and then the geosynthetic is wrapped around the face and fixed to the compacted soil layer. When using formworks the shape of the face can be continuously inclined or stepped. Details of the different types of patented formworks are described e.g. by Rimoldi and Jaecklin (1996). The distinguishing types consist of:

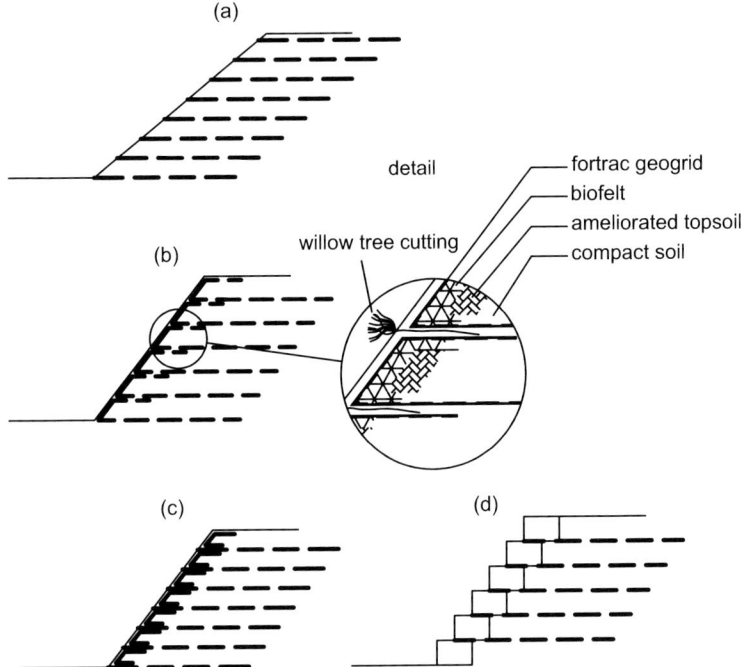

Fig. 6.10 Alternatives of green faced structures for reinforced soil. (**a**) Straight reinforcement. (**b**) Wrap-around reinforcement, plus possible detail. (**c**) Mixed scheme. (**d**) Face blocks with straight reinforcement

- Movable × sacrificial formworks,
- Formworks from wood or from steel mesh,
- With filling from top soil and subsequently hydraulically seeded or from vegetation mat (turf). Detail of facing can be also as described by Fantini and Roberti (1996), where willow tree cuttings are situated between individual layers.

c) Mixed scheme – where straight geosynthetic acts as reinforcement, while the second geosynthetic, wrapped around the face, acts as an anti-erosion element. The reinforcing element has high tensile strength and can be very stiff to bending. On the contrary the protection element is more flexible, lighter and is engineered to support the growing vegetation and to retain the soil, preventing wash out and erosion.

d) Front blocks tied back by straight reinforcement. These blocks, often referred to as gabions, big bags or just as containers, are usually prefabricated, filled with soil, on the face side often with topsoil or even with turf. The blocks are acting as a formwork for the fill soil and are mechanically connected to straight reinforcing geosynthetics.

6.2.2 Embankment on Soft Subsoil

This case is the most frequently discussed in the geotechnical literature because its solution generates on the one hand many difficulties, but on the other hand it offers a large number of variable solutions. Some basic principles are described in this respect for example by Bjerrum (1973), who also more closely specifies clayey materials considered as soft subsoil – soft clays – these are normally consolidated clays which he further specifies as young and old. Included in the category of soft subsoils also are soils with a higher content of organic substances, right up to turf-like soils. It is always necessary to specify the geological history of the deposit. Next these basic principles will be specified and there will be indicated methods of current technical solution for basic cases of improvement in soil and subsoil which have been already mentioned in previous chapters. The variant of the replacement of the soft subsoil by an another better material will not be further specified because its economic advantage exists only for a smaller thickness of soft subsoil and it can be inadmissible for ecological reasons and does not bring with it special technical problems.

The gist of the problem consists of the fact that due to the surcharge of the soft subsoil by the embankment, an increase of pore pressures emerges. Theoretically pore pressures reach their maximum at the time of completing the construction of the embankment. After, this results in their dissipation and decrease. Thus this leads to the increase of shear strength and thus also to overall stability. That is why the lowest stability is reached at the time of completing the construction. Bishop and Bjerrum (1960), Bjerrum (1973) clearly described this situation – see Fig. 6.11.

For the change of pore pressures Δu as a result of a change of main stresses $\Delta \sigma_1$, $\Delta \sigma_3$ Skempton and Bishop (1954) deduced the relation:

$$\Delta u = B(\Delta \sigma_3 + A(\Delta \sigma_1 - \Delta \sigma_3)) \quad (6.3)$$

Where A and B are coefficients of pore pressure. For the case of fully saturated soils the coefficient $B = 1$ and it is true for:

$$\Delta u = \Delta \sigma_3 + A(\Delta \sigma_1 - \Delta \sigma_3) \quad (6.4)$$

The size of the coefficient of the pore pressure A is different, depending above all on the factor of overconsolidation. In a simplified fashion it is possible to consider for dilatant (overconsolidated) soils that $A = 0$ (while it can also be negative) and for contractant soils (soft clays, normally consolidated soils) it is possible to consider that $A = 1$. For a monitored case of an embankment on soft subsoil it is therefore possible to consider that the coefficient of pore pressure $A = 1$. It is obvious that here the decisive state is short-term because with time – with ongoing consolidation – the situation is improving.

Note: The above mentioned development of the change of strength over time is valid for soft subsoil. For an embankment body the suction in partially saturated compacted soil can play a certain positive role during the building process. During

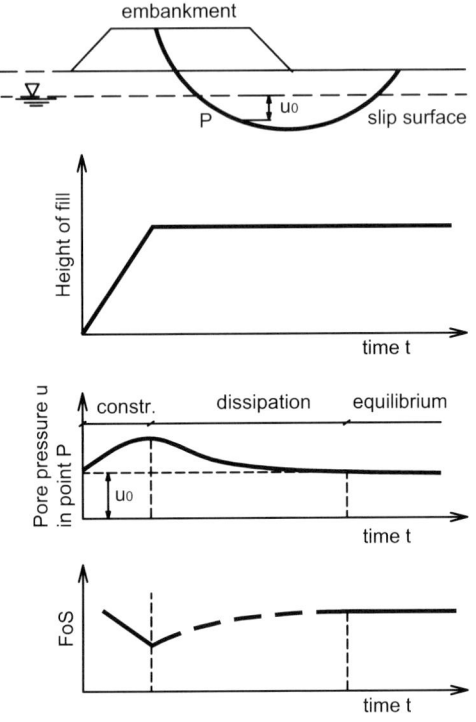

Fig. 6.11 Fill on soft subsoil; variation of pore pressure and factor of safety at point P with time (according to Bishop and Bjerrum)

a period of drought and heat suction pressure can play a significant role. However, after rains or in winter this advantage disappears. Thus this effect must be dealt with very cautiously.

From the viewpoint of pore pressures short-term slides represent undrained conditions. It is possible to solve their stability either with the help of effective parameters of shear strength, on condition that knowledge (in practical terms forecasts) of pore pressures, or with the help of total parameters of shear strength, where it is possible to consider saturated clays with $\varphi_u = 0$, $\tau_u = c_u$.

In the case of short-term stability it is a very often chosen solution to consider with total parameters of shear strength, the determination of which, especially in situ and with the help of the vane test, does not create special problems. Their correct assessment is, however, dependent on several factors especially velocity of loading and anisotropy of soil strength – orientation of the sample in the direction of the sliding surface.

In the laboratory the shear tests are carried out considerably faster compared with practical reality in situ. Bjerrum (1973) paid great attention to this reality in reaction to failure of embankments based on plastic clay. Back analysed in situ strength was considerably smaller than the strength measured in situ by the vane test. He also found out large influence when measuring undrained strength by the test of the type UU – undrained, unconsolidated. In Fig. 6.12 there are given results for plastic clay from Drammen carried out by different velocity. The fastest tests

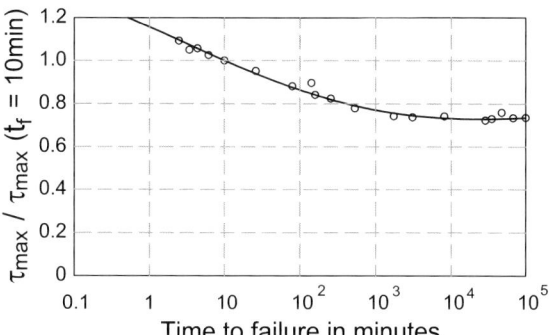

Fig. 6.12 Correlation between shear stress level and time to failure established from undrained triaxial compression tests on samples of plastic clay from Drammen (according to Bjerrum)

were executed with the velocity of the deformation about 35% of the height of the specimen per hour, which approximately corresponds with the velocity of standard field vane tests. The slowest used velocities were about 25 000 × smaller, which vice versa approximately corresponds with the velocities occurring in situ. From the dependence it is obvious that the strength falls with a lowering velocity of surcharge. The explanation for the decrease of strength over time is as a rule justified by the increase of moisture content in the central part of the tested specimen, either by suction during dilatation or by migration of the moisture with the influence of non-uniformly decomposed pore pressures, with laboratory specimens particularly at the bases of the specimens and in their central part.

Anisotropy of soil strength is induced by two mutually influencing aspects:

- By geometrical anisotropy, i.e. preferred decomposition of particles;
- By stress anisotropy.

Geometrical anisotropy is caused at deposition, when settled particles have the tendency to deposit themselves with the longitudinal axis horizontally. Stress anisotropy is caused by the combination of the history of loading and geometrical anisotropy (for example soils consolidated under K_0 conditions). The resulting effect is a strength anisotropy in relation to stress-strain behaviour. Skempton and Hutchinson (1969) state the influence of the specimen orientation on undrained strength – Fig. 6.13. The smallest strength is reached for the case when the sliding surface is approximately parallel with a horizontal level – level of sedimentation. It is necessary to take this reality into consideration because commonly the tested specimens are oriented vertically.

That is why Bjerrum (1973) or Menzies and Simons (1978) recommend the undrained test of shear strength given by the vane test in situ $(s_u)_{vane}$ to multiply by correction factors μ_A and μ_B for gaining the real undrained strength in situ $(s_u)_{field}$:

$$(s_u)_{field} = \mu_A \mu_B (s_u)_{vane} \tag{6.5}$$

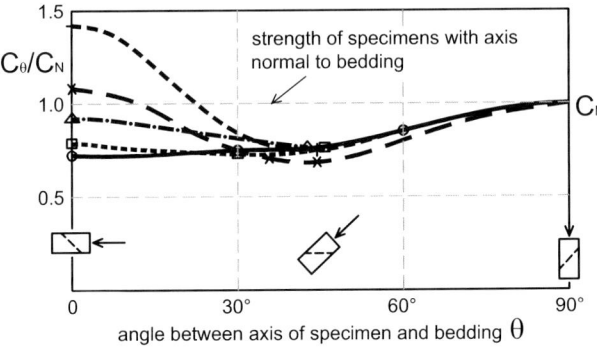

+ London Clay, Ashford (Wared 1965)
× London Clay, Wraysbury (Agarwal 1967)
△ Clay from Surte (Jakobson 1952)
□ San Francisco Bay clay (Duncon & Seed 1966)
○ Clay from Welland (Lo 1965)

Fig. 6.13 Effect of sample orientation on undrained strength for different clays (according to Skempton and Hutchinson)

where

μ_A is a correction factor expressing the influence of strength anisotropy – see Fig. 6.14a

μ_B – a correction factor expressing the influence of the velocity of the undrained test towards the real conditions – and depends according to Bjerrum on the index of plasticity – see Fig. 6.14b.

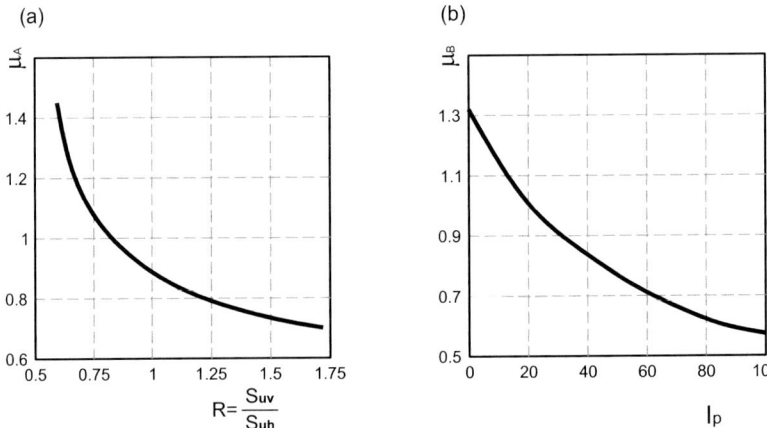

Fig. 6.14 Correction factors for undrained strength. (**a**) μ_A – as a function of anisotropy of undrained strength. (**b**) μ_B – as function of index plasticity

The solution of stability when considering effective parameters of shear strength has a certain advantage in the fact, that the effective parameters are more or less the characteristics of the soil and are not so dependent on secondary conditions, as is the case with total parameters of shear strength. Nevertheless, here also the shear test of the type CU – a consolidated undrained test with pore pressure measurement – showed interesting results from the anisotropy viewpoint – e.g. Mitchell (1976, 1993), Škrabal (1977) – Fig. 6.15. Anisotropy has a small influence on shear parameters φ', c' but a significant influence on the effective stress paths. In reality it means that for the variant of the orientation of a specimen 2 the specimen superimposed in the direction of geometrical anisotropy has a greater strength for a given starting state, given by consolidation pressure σ_c. It is obvious that for a varying geometric anisotropy the different development of pore pressures is given by anisotropy of specimen permeability.

A pore pressures forecast can cause a certain problem when using effective parameters of shear strength. The basic relation between the increment of pore pressures and the change of stresses is derived from the axisymmetric problem. For the 2D problem typical for embankments it is thus possible to use modified relationships (e.g. Henkel 1960). The increment of pore pressures can be also influenced by the historical origins of clayey subsoil or the occurrence of a certain stronger surface layer. However, the measurement of these pressures in situ when carrying out the embankment can vice versa give good feedback information about the accuracy of the forecast and so it is possible when constructing the embankment to apply the method of design with the help of the observation method. The speed of

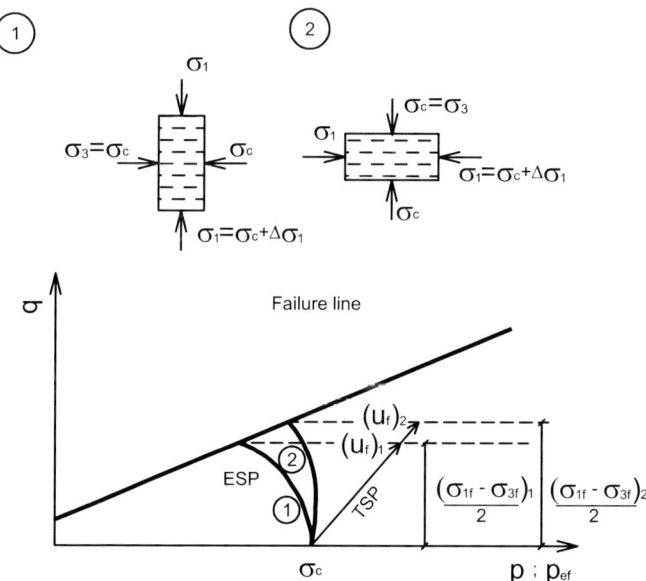

Fig. 6.15 The influence of sample anisotropy on effective stress path

embankment construction can be thus managed with regard to the recorded values of pore pressure.

Next it is assumed that the development of the degree of stability through time is to be made in comparison with the required degree of stability in accordance with Fig. 6.16a. It means that for a certain time period, when the growth of pore pressures was the greatest, stability is insufficient and it is necessary to react to the given situation. In principle 3 basic possibilities are offered:

a) To build the embankment in stages, by this the increase of pore pressures will reach only to the level where the stability is still at the required range and the embankment will be left at this level for a certain time when pore pressures are dissipating. This lowering of pore pressures will enable further filling, however, only in accordance with a maximum permissible increase of pore pressures, see Fig. 6.16b;

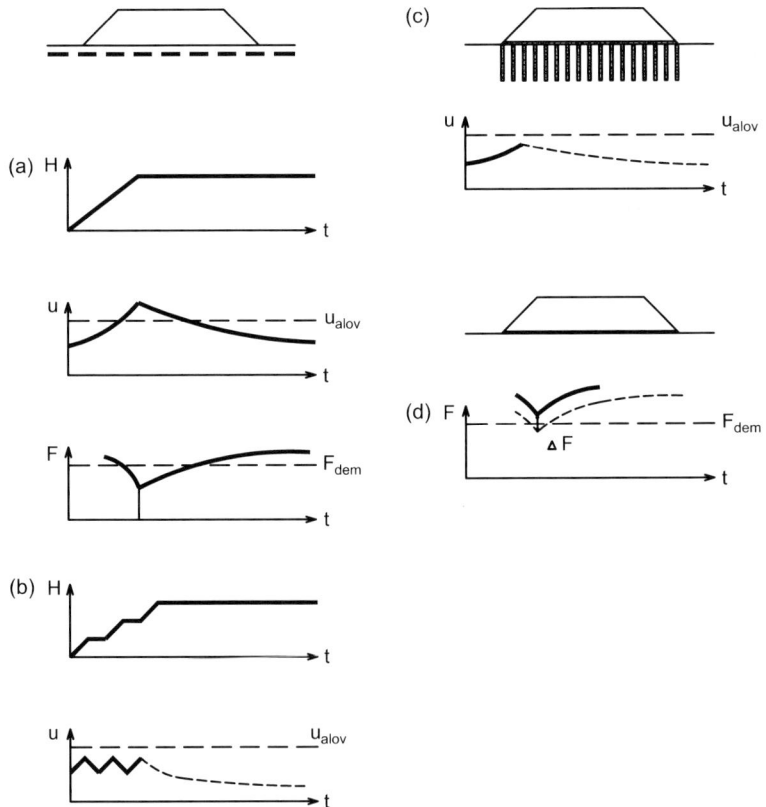

Fig. 6.16 Embankment on soft subsoil. (**a**) Possibilities how to solve instability for basic case. (**b**) Building up step by step. (**c**) Vertical drains. (**d**) Contact reinforcement

b) The application of vertical drains in the subsoil of the embankment for the speeding up of consolidation – the increase of pore pressures during the construction of the embankment will be smaller because a significant part of the theoretical increase will already dissipate during the filling of this embankment – see Fig. 6.16c, e.g. Suhonen and Tupala (1983);

c) To reinforce the subsoil contact with the embankment – see Fig. 6.16d, especially with help of geosynthetic reinforcement materials which will bring required additional stability into the solution. It is necessary to note here that in the instance of the basic case when long-term stability is greater than required (for the case without additional precautions) reinforcements in this case have an additional, temporary role. Reinforcements can, however, also be used in the case, when the soil, even after the completion of consolidation, is not itself able to assure the sufficient shear strength and thus also the stability. Then it, however, fulfils the long-term function – permanent, and by such a method of reinforcement it is necessary to view more strictly than for the basic case because the creep influence plays here a more significant role.

The above mentioned 3 basic possibilities are pointed towards the area of stability, a limiting state of failure. They also bring, however, certain positive features from the viewpoint of a limit state of deformation. In particular, the first two possibilities have a minimum impact on total deformation. However their contribution is that in fact, after the completion of the embankment, before the implementation of final adjustments on the surface (surfacing of road, rail lines, drainage systems etc.) they decrease further deformation. In this category we could also include the method referred to as preloading – its application, however, must be very cautious because some even higher embankment increases in principle the danger of failure – it is possible to avoid this danger if the preloading is applied with the help of suction (vacuum) . From the viewpoint of consolidation the first two possibilities ensure that the degree of consolidation after the completion of construction is significantly higher than without these precautions. The third possibility – the reinforcement of contact – has from the viewpoint of deformation an especially positive influence on the evenness of deformation, as has also been demonstrated on the physical model of a reinforced contact.

If, however, neither of these solutions are sufficient from the viewpoint of a limiting state of deformation, then it is possible to choose the following variants of a solution:

- The compaction of the surface of soft, strongly compressible subsoil. With any need for the compacting of subsoil to greater depths, even several meters deep, from the methods mentioned in Chapter 2, there comes into consideration compacting with the help of dynamic consolidation.
- The embankment construction on a group of piles. Attention here will not be focused on the type of piles (from the classical ones through to lime columns,

gravel piles, sand-gravel geosynthetic piles – geotextile encased columns – etc.), but on ensuring the load transmission from the embankment to the pile heads.
- The construction of the embankment from light weight materials – fill lightening.

All the above given possibilities or at least some combinations of them will be demonstrated in practical cases.

6.2.2.1 Speeding up of Consolidation

A very well described and documented example of speeding up of consolidation process was presented by Brenner and Pbebaharan (1983). In practice it was a large in situ model test in the scale 1:1 realized on 17 m thick stratum of soft clay over stiff clay at Pom Prachul, Bangkok. A test embankment, 33 m wide and 2.35 m high, was built in two stages with a total length of 90 m. The three test areas comprised a section with 1.5 m vertical drain spacing, one with drain spacing 2.5 m, and a section without drains. The drains were wicks of 50 mm diameter which were installed by the displacement method to the full depth of the soft clay. Such an example combines the two basic opportunities for speeding up consolidation:

- Construction in 2 stages, allowing the dissipation of the excess of pore water pressure in the interim period, 10 weeks long,
- Application of vertical drains.

The embankment was well instrumented with settlement gauges and piezometers. For the purpose of predicting settlements and pore pressure, as well as for the back-calculation of consolidation parameters, a finite difference model was developed which could take into account a layered soil profile, two-dimensional water flow, time-dependent loading, a stress distribution varying with depth, and undrained deformations.

All necessary soil parameters affecting compressibility and permeability were determined on samples obtained from 250 mm piston samples. Index plasticity was generally higher than 50%, natural water content closer to the liquid limit.

The first stage of the test embankment consisted of placing a sand blanket of thickness 0.35 m and then a fill of 1.1 m in five layers, reaching a total height of 1.45 m and width circa 33 m. Geosynthetics was used between the sand blanket and subsoil, fulfilling a filtration role. In the second stage, another five layers were placed to reach a final height of 2.35 m and width of about 18 m.

The drains were wicks of 50 mm diameter arranged in a square pattern (1.5×1.5 or 2.5×2.5 m) giving equivalent drain spacing (radius of drainage column) of 0.85 m and 1.41 respectively. Installation of the drains was by the displacement method using a casing (ID = 75 mm) with a disposable wooden plug.

For the wick stocking the same geosynthetics was used as for protection of the sand blanket. The sand in the drains had a mean diameter of 0.6 mm and a uniformity coefficient of 1.4. The drains penetrated the full length of the soft clay. A cross-section of the test embankment (section with 2.5 m spacing) is shown on Fig. 6.17.

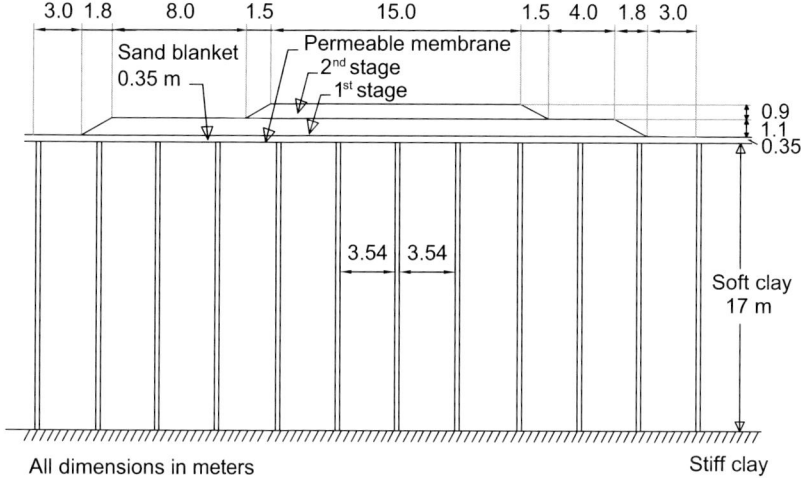

Fig. 6.17 Cross-section of test embankment – section with 2.5 m drain spacing (according to Brenner and Pbebaharan)

Numerical modelling used following differential equation for consolidation by sand drains:

$$\partial u/\partial t = c_r(\partial^2 u/\partial r^2 + 1/r\ \partial u/\partial r) + c_v\ \partial^2/\partial z^2 + R(z,t) \quad (6.6)$$

Where u is the excess pore pressure, r is radial distance, z is the depth, t is the time, and R is the rate of loading.

When measuring settlement, only initial and consolidation settlements were considered; the creep component was neglected. The initial settlement for both stages of loading was found to be 200 mm. In this calculation the increase in stiffness due to consolidation during the first stage of loading was not taken into account. The computed ultimate consolidation settlement was 971 mm for subsoil surface. Skempton's equation for predicting initial pore pressure was used.

Brenner and Pbebaharan (1983) presented many results of in situ measurement and these results compared with calculated (predicted ones). Figure 6.18a, shows observed settlements at the centres of the three test sections for the surface and four different depths. Figure 6.18b, shows observed excess pore water pressures at the centres of the three test sections also in different depths. Observed and predicted excess pore pressures at centre of a section with 2.5 m drain spacing are shown in Fig. 6.19. The observed values were used for back analysis, varying c_v and c_r in each layers until the best possible agreement was obtained at all depths where the pore pressures and the settlements had been measured. For these recalculated values the pore pressure dissipation around vertical drain was specified – see Fig. 6.20.

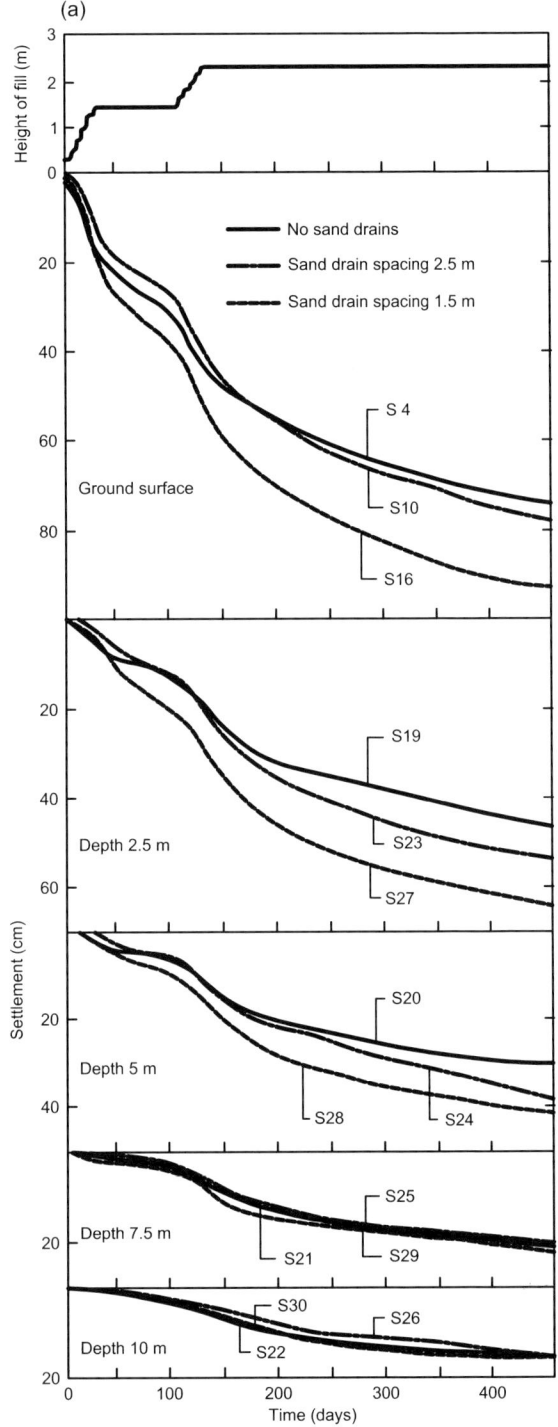

Fig. 6.18 Speeding up of consolidation with vertical drains – results at centres of the three test sections – no sand drains; sand drain spacing 2.5 m; sand drain spacing 1.5 m (according to Brenner and Pbebaharan). (**a**) Observed settlements. (**b**) Observed excess pore water pressures

6.2 Transport Embankment – Fill

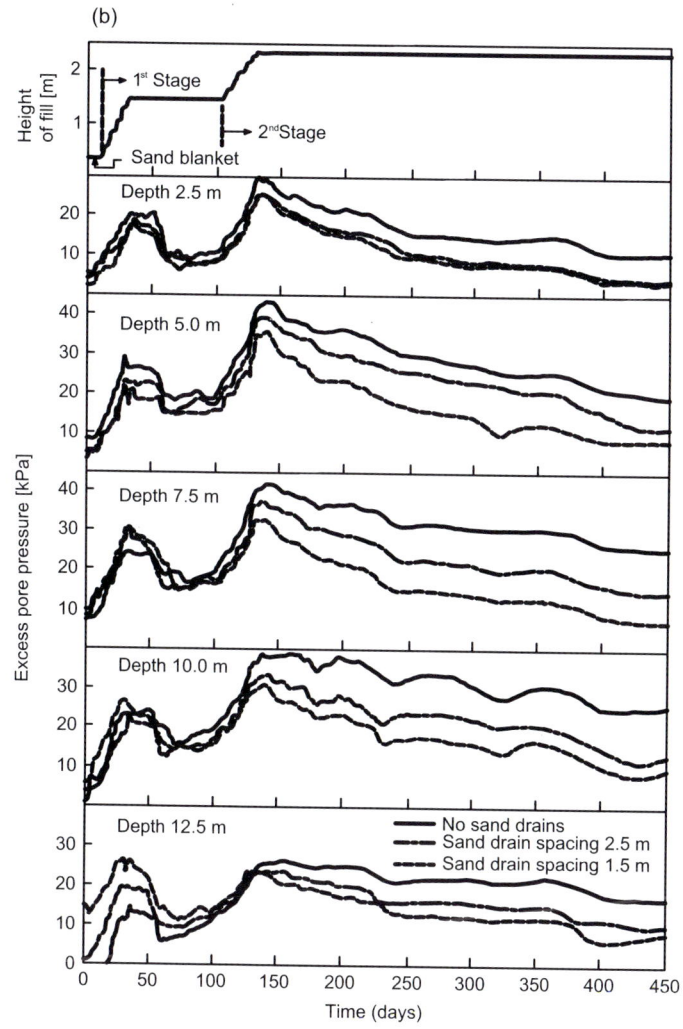

Fig. 6.18 (continued)

From this very well documented case some conclusions can be made as follows:

- The time gap between two stages of embankment construction had a positive effect on pore water dissipation;
- Vertical drains speeded up consolidation (settlements) under the embankment, but for higher depths the differences were rather small ones – Brenner and Pbebahan stated that below a depth of about 7 m sand drain installations is for this soil profile uneconomical;
- Vertical drains speeded up pore pressure dissipation, but for the phase of embankment construction the differences were rather small;

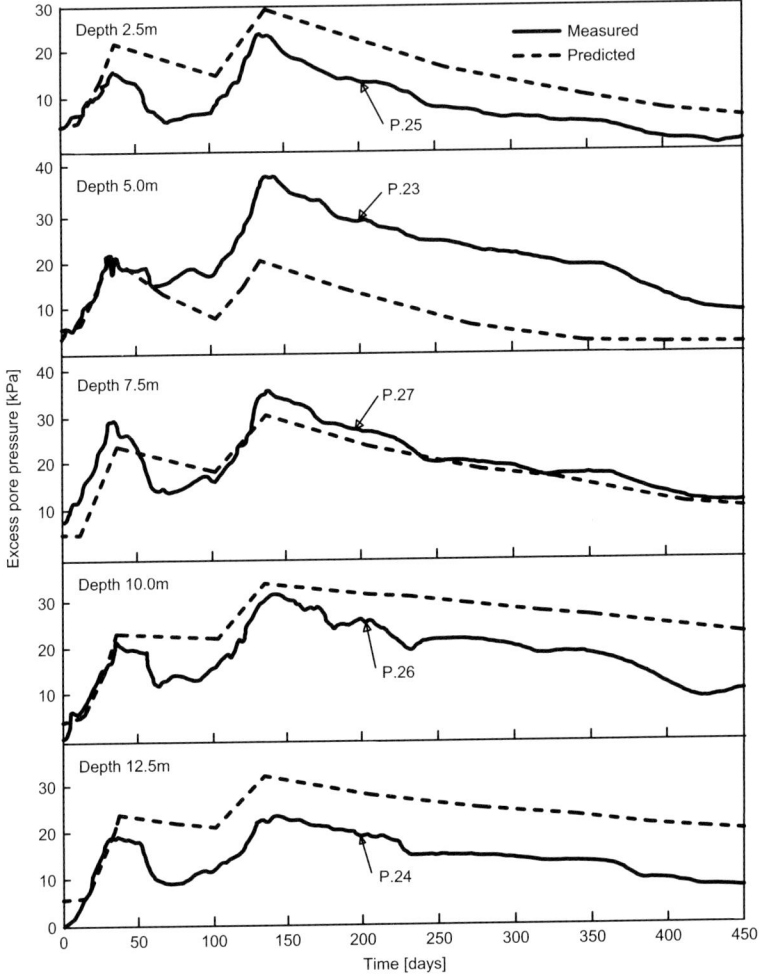

Fig. 6.19 Observed and predicted excess pore water pressures at centre of section with 2.5 m drain spacing (according to Brenner and Pbebaharan)

– The very often predicted excess of pore water pressure was lower than the measured one;
– Installation of sand drains generated excess pore pressure which at some piezometers locations had not yet fully dissipated when placement of the sand blanket began.

According to these authors attention should be paid to the following problems:

– Sensitivity in the collection of samples in very soft clays – there is a high probability that the samples would be a little bit compressed and measured values of deformation parameters higher than in situ;

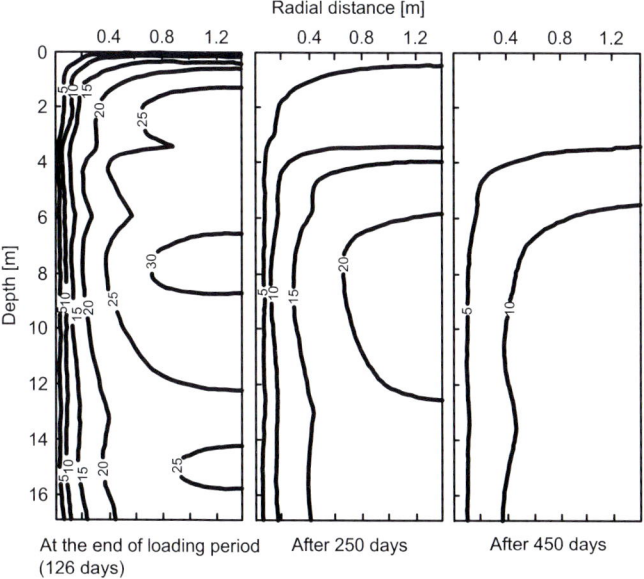

Fig. 6.20 Isobars of excess pore water pressure for section with 2.5 m drain spacing (units in kPa) (according to Brenner and Pbebaharan)

- The influence of drains installation – the displacement method can affect the behaviour of surrounding soil – lower permeability, lower compressibility;
- The combination of these two aspects can explain settlements behaviour – settlements without drains were higher than predicted, settlements with 1.5 m spacing were lower than predicted (therefore for 2.5 m spacing the best agreement was observed);
- The presence of small layers of more permeable soil such as fine sand and silt – this situation can significantly speed up the consolidation process for subsoil without vertical drains.

Park et al. (1997) described a case study of vacuum preloading with vertical drains. The vacuum preloading method does not require surcharge fill and has the advantage of avoiding shear failure when surcharge loading is applied quickly. The authors concentrated on a numerical model, during which they compared two cases of vacuum preloading. The conventional method is modelling vacuum preloading as surcharge loading in the consolidation analysis. A second one, proposed by Park et al., applies negative pore pressure at the surface of drains, see Fig. 4.16. The method was realized on area 5 000 m², vertical drains reaching the depth 26.7 m. Over soft subsoil 1 m thick sand layer was placed, after that vertical drain with spacing 1.15 m, supplemented later by additional drains to reach a spacing of 0.81 m. After the installation of an airtight membrane an additional surcharge of 0.5 m was added and the vacuum started, reaching 0.7 bars (70 kPa) after 2 weeks. Measured total settlement after 450 days was more than 3 m and measured values were in very good agreement with numerical models. One of the assumptions of this model was that a smeared zone around the drain is considered as twice the vertical drain diameter. But when comparing the measured excess pore pressure with the modelled one, good agreement was obtained only for the second model.

Puumalainen and Vepsalainen (1997) also describe a test field experiment observing the speeding up consolidation by vertical band-shape drains and the vacuum preloading. Vacuum preloading was used in the range 50–80 kPa and equals a surcharge of 2.5 m to 4 m height. According to previous calculations maximum preloading by classical method could be only 30 kPa (1.5 m) for given subsoil conditions. Even negative pore pressure was applied only for 3 months the water content had decreased and the shear strength increased significantly. The smear effect was not taken into consideration when determining the drain distance. Field measured data were compared with numerical modelling by FEM program Z–Soil. Vacuum suction was simulated with pore pressure boundary conditions. Good correspondence was found between the actual measured settlements and numerical modelling. During the vacuum pumping process the program was very useful since it allowed the use of a layered structure and variations of the under pressure load.

Vidmar et al. (1981) describe a successful highway construction on very soft soils. The highway Ljubljana-Postojna crosses the northern boundary part of the Ljubljana Marsh. Due to high compressibility of the upper peaty, silty and clayey layers of the subsoil with total thickness of about 12 m, settlements up to 1.6 m were foreseen. In order to reduce the post-construction settlements to tolerable values, vertical wick drains, preloading, light fill material (ash) and partial excavation of superficial peat were used.

A test embankment was built three years before the beginning of the construction of the motorway in this part. First of all a preloading of 15 kPa was applied for a period of roughly 1 year. Afterwards 160 m of test embankment was constructed, divided into 4 parts, and where for 3 of them different methods of speeding up consolidation were applied. After 2 years of observation and measurement, the rest of the motorway was constructed. To the calculated initial and consolidation settlements up to 0.25 m was added to eliminate secondary settlements for the first 20 years. The motorway is in a very good state now, roughly 30 years after construction.

At the end of this chapter section the conclusion can come very close to the one by Vidmar et al. (1981).

- Large and non-uniform settlements of very soft subsoil can be efficiently accelerated during highway construction by applying vertical drains and preloading.
- The non-uniformity of settlements due to varying thickness of soft deposits can successfully be overcome by using different construction methods in different sections of the highway.
- The excavation of the superficial peat and organic soil and the use of sandwich fill with prevailing light fly ash layers are suitable means for reduction of the total amount of settlements.

- The settlements and consolidation process may be reliably enough predicted by simplified computation procedures provided that the forecasted values are favourably checked by field measurements made on an appropriate test embankment.
- Preloading – loading in the first phase – can verify and enhance a subgrade model for the calculation of the settlement in line with the observational method.

6.2.2.2 Contact Reinforcement

In any application of the contact reinforcement it is in principle necessary to respect the same principles mentioned regarding the treatment of this contact for the basic case with the solid subsoil – especially ensuring its drainage.

When selecting the method of reinforcement it is possible to choose one continuous layer of reinforcement, combinations of more layers, ensuring a certain anchorage at the edges, or possibly the creation of a wrapped reinforced cushion when the space between the two reinforcements is filled with sandy gravel. It is necessary to control shear strength along the contact, especially that of the embankment and reinforcement, so that this contact would not be the weakest place and the slip surface of failure would not be coming through this contact. Several variants illustrate Fig. 4.24. The attractiveness of using reinforcements, in contrast to the previous methods, enabling the construction of embankments on soft subsoil, is relatively high. This consists above all in the faster technology of construction without time delays for partial pore pressure dissipation, with the possibility of filling more powerful layers or creating higher embankments. This situation led to the relatively rapid development of using geosynthetics for the contact reinforcement of the embankment and subsoil. In many cases plane surveying at a scale of 1:1 preceded these operations, especially in the period 1975–1995. In some cases there was an endeavour to get to a failure of the subsoil and so to monitor the case of a total collapse, and sometimes there was an attempt only to go to the phase, which can be classified as one of pre-failure, when the degree of stability based on classical methods showed the value F about 1.2. Finally a wide range of measurements were carried out on the particular constructions to give proof of good functionality. For individual cases it is possible to cite for example Brakel et al. (1982), Delmas et al. (1990), and Blume (1996). In the majority of cases there was a tendency for geosynthetic reinforcements of high strength 100–400 $kN.m^{-1}$, with elongation to a failure of about 10%. They tended to polyester reinforcements for the decreasing of creep influence. From several interesting conclusions we can extract the following:

- The evolution of the pore pressure during the construction confirms that the behaviour of the soft soil is not modified by the geotextile.
- During the first step of loading it seems to be confirmed that the lateral displacements of soils have a major action on the development of strains in the geotextile.
- The failure of the reinforced geotextile was caused by its rupture; the anchorage was sufficient and did not case pulling-out.

– When the failure occurs, the tensile force in the reinforcement was roughly in harmony on the basis of expectation, i.e. for relatively high tensile loading and at the strain, moving towards the limit value.

Blume (1996) describes the case in which The Federal Institution for Roads in Germany initiated and executed a short-term (1986–1990) and additionally a long-term (1990–1996) measurement program to verify the used calculation methods and to prove the short-term and long-term stability, deformations and serviceability of a road embankment reinforced with a high-strength low-creep polyester woven geotextile. The measurement program was applied to the national road B 211 near Grossenmeer in Northern Germany. The embankment height up to 4.5 m is overlaying saturated non-consolidated peat and organic clay with a thickness from 3 m to 5 m. Below this soft subsoil was sand. The short-term measurements were carried out on a trial embankment which was then integrated into the final road embankment due to the positive results of the first measurements, verifying the effect of the high-strength reinforcement and the assumptions made in the design stage.

This case is also important because of the fact that it also includes the preloading (prefilling) method. It is necessary to realize here that the reinforcement provides additional security from the viewpoint of stability but in practice does not speed up the settlement – consolidation. The combination of methods, preloading and reinforcement, is thus significant from this viewpoint – and it is just reinforcement that will enable the application of preloading. In the given case the short-term stability was calculated for the undrained shear strength – values of $c_u = 7$ and $9\,\text{kN.m}^{-2}$ and $c_u = 20\,\text{kN.m}^{-2}$ were used respectively for the peat and the upper soil layer. These values were determined by shear vane tests in the field. The magnitude of the preloading was established for the entire construction section as a sand mass of 4.5 m thickness. Because the subsoil consolidates due to the load acting upon it, and its shear resistance consequently increases, an initial factor of safety $F = 1.2$ at the time of filling was regarded as adequate. To guarantee this factor of safety a geotextile having the trade name Stabilenka 400 was used as reinforcement in the embankment base. This high-strength woven geotextile consisting of polyester yarns has a maximum tensile strength around $400\,\text{kN.m}^{-1}$ at a breaking strain around 10%. The force-strain curve behaviour of the fabric is approximately linear. In rough terms it was expected that the factor of safety could be assured when 50–60% of maximum tensile strength is activated. By field measurement these expectations were confirmed, Fig. 6.21. After unloading strain in the geosynthetic reinforcing element remained constant as well as in the embankment settlement even for a 3 year period of traffic. Because elongation was close to 6% (due to extreme settlements), activation of reinforcement was also close to 60%. This is rather a high value compared to the recommended value for polyester reinforcing element, as mentioned in Chapter 3. Nevertheless even for this activation the creep deformation for this material was satisfactory. Probably it is due to the fact that the soil begins to increase its strength as a result of the consolidation process.

The next illustrative example combines the reinforcement of the contact with vertical geodrains and staged construction. This again concerns the combination of

6.2 Transport Embankment – Fill 291

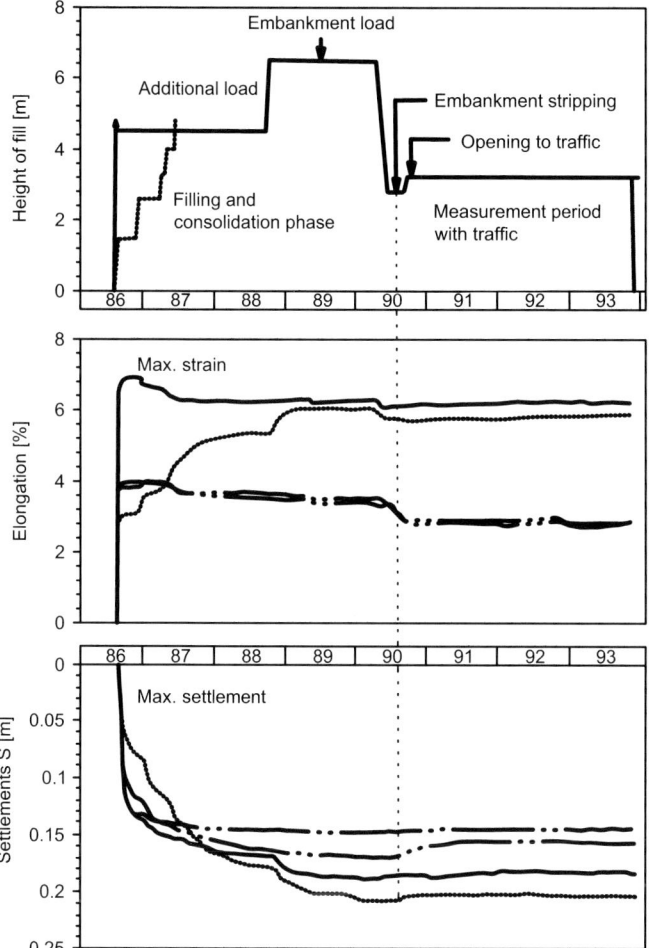

Fig. 6.21 Observed settlement in the centre of embankment and elongation in reinforcing geosynthetic observed there as a function of construction fills (according to Blume)

additional assurance of stability with the speeding up of deformation in the strongly compressible subsoil. The design proposed by the authors for the embankment leading to the motorway over bridge is depicted in the Fig. 6.22. In this case the subsoil from soft to firm clays is to the depth of about 12 m.

The construction scenario does not have enough time for the slow construction in order to build the embankment without speeding up of the consolidation. The design was optimised in order to reduce the required strength of the contact reinforcement and assumes construction in two stages that can be accommodated in the construction programme. During the first construction stage the embankment will be built to the maximum height of 6 m and waiting period of 7 month is specified. The waiting time is designed for dissipation of pore pressures up to the consolidation ratio of

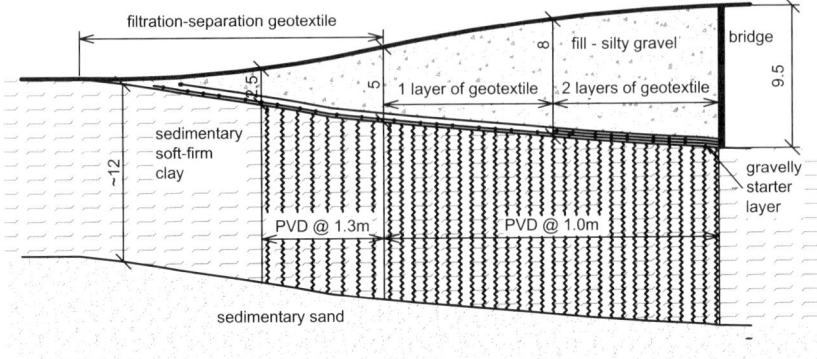

Fig. 6.22 Longitudinal section through bridge approach embankment on soft subsoil with indicated solution for speeding up the consolidation and maintaining the stability

about 97%, which also allows for the increase of undrained strength of the subsoil. In the second stage the embankment will be build to its final height of max 9.5 m and another waiting period of 7 months will be required to eliminate excessive post construction settlements (40 mm) required for bridge approach embankments before final road surfacing will be placed. For the embankment with the height between 2.5 and 5 m the proposed triangular grid of prefabricated vertical drains (Colbonddrain) is 1.3 m and for the higher embankment the grid is proposed to be 1.0 m.

In order to achieve short-term stability during the consolidation of the subsoil, contact reinforcement is required for the embankment height of more than 5 m. Up to the embankment height of 8 m high strength polyester geotextile Stabilenka 150/45 will be required and for the section of more than 8 m high embankment Stabilenka 300/45 will be required. Also for long-term stability of the embankment higher than 8 m contact reinforcement will be needed, in this case Stabilenka 150/45 will guarantee the required level of safety.

The reinforcement of the contact can be useful also in the case, where it is possible to expect significant, sudden deformations, for example in the collapse of underground spaces in the subsoil in mining or karstic areas. A very interesting case on this matter is described by Alexiew et al. (2002) and Leitner et al. (2002) for railway junction at Gröbers. This railway junction is situated in the old mining area in Germany, where coal was deep-mined but no precise records exist. The sinkholes that are developing in this area are usually about 4 m in diameter. This railway junction of Inter City Express (ICE) trains running with the speed of up to 300 km.h with high speed lines of 160 km.h and suburban trains could not be relocated due to the issues with not very flexible alignment design requirements. In this area up to nine railway tracks are running in parallel. The maximum allowable differential settlement between rails is for ICE trains 3 mm and this will have to be ensured also in this area prone to subsidence. In this case the decision was made by the railway authorities that warning and safety system have to be implemented in order to guarantee location of the developing sinkhole and protecting the railway track for at least one month, during which grouting of the newly developed cavern can be implemented.

That kind of system which was developed for this particular project is shown in Fig. 6.23 and consists of:

- Cement stabilised base layer (1);
- Geogrid reinforced gravel cushion including warning layer and extensionmeters (2–8);
- Cement stabilised bearing layer (9).

The system was designed according to German standards DIN 1054 (1996) and EBGEO (1997) using the calculation model presented in Fig. 6.24. During the design of this project, several new procedures for calculations were developed to guarantee very safe but effective design. The requirements for the properties of the cement stabilised (up to 4.5%) bearing layer were determined from one of the new calculation procedures and produced in onsite plant. The cement stabilised bearing layer was placed above the reinforced cushion to help with arching in this only 3.5 m thick layer above the developed sinkhole. The geosynthetic reinforcement in the gravel cushion was placed in two layers perpendicular to each other to have strong enough biaxial reinforcement above the circular sinkhole. The geogrid used in this project (Fortrac 1200/100–10 AM) was made from aramid in the longitudinal direction and polyvinylalcohol in the transverse direction with ultimate strengths of $1200\,kN.m^{-1}$ longitudinally and $100\,kN.m^{-1}$ transversally. The reason for using these special raw materials was their high modulus (see also Section 4.3.1.3). In order to achieve immediate mobilisation of the geogrids into action when sinkhole would open, these were installed on site with special prestressing device to a pre-stress force of at least $3\,kN.m^{-1}$, see Fig. 3.20. The preliminary design was also

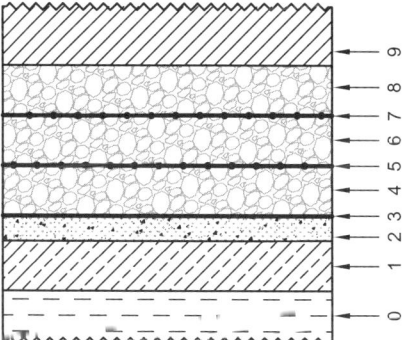

Fig. 6.23 Warning and safety system from reinforced and stabilised soil in mining area prone to subsidence at railway junction Gröbers (according to Leitner et al.). 0 – clayey subsoil; 1 – cement stabilised base layer; 2 – sand cushion; 3 – warning system with extensionmeters; 4 – gravel layer 0/32 mm; 5 – transversal layer of reinforcement (Fortrac 1200/100-10AM); 6 – gravel bearing layer 0/63 mm; 7 – longitudinal layer of reinforcement (Fortrac 1200/100-10AM); 8 – gravel bearing layer 0/63 mm; 9 – cement stabilised bearing layer

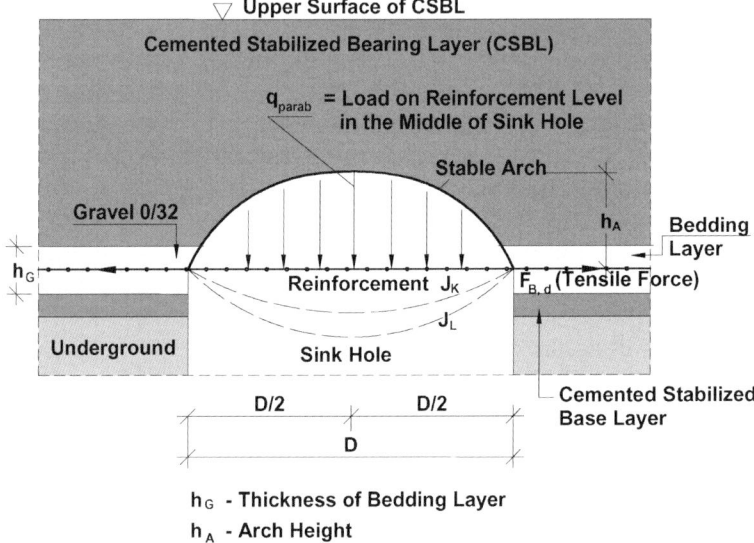

Fig. 6.24 Calculation model for design of reinforced soil platform for voids overbridging at Gröbers project (according to Alexiew et al.)

successfully tested in real scale on site, when creation of a sinkhole of 8×4 m (larger than realistic) was simulated. The results from this test were also used during the final design.

The warning layer is an integral part of the solution and can pinpoint the location of the newly developed sinkhole. The system consists of detection part from two nonwoven geotextiles with inserted electric resistance wires forming an orthogonal grid with a mesh size of 0.25×0.25 m and back-up part from fibre-optic extensionmeters. The change in the electric resistance of the wires or their rupture indicates the location of the sinkhole. In case of confusion the extensionmeters help with the localisation. All the cabling was securely taken up to the surface and in cable ducts to the control room with computer operated warning system that enables permanent control of the situation.

6.2.2.3 Pile Foundations

Pile foundation underneath transport infrastructure embankments is used to avoid both serviceability and stability problems associated with soft to very soft subsoils, especially when there is limited time available for the construction and other techniques describe above could not be used. Very often it is an issue of serviceability – of total and differential settlement. Piles as we understand them in the context of this chapter mean all systems that transfer vertical load from embankment to load bearing strata deeper in the subsoil and includes besides traditional piles among others vibro concrete columns, stone columns and geosynthetic encased sand columns, Raithel et al. (2004).

6.2 Transport Embankment – Fill

In some cases it is enough to install just piles to reduce settlements of the whole embankment, and then it is necessary to check if the piles are not propagating to the surface of the embankment. Most of the cases when the embankment is built on piles are with load transfer platform from geogrids. This system can be designed using several procedures, some of them are codified in different standards, and the rest are just from published papers. A nice overview of the methods that are routinely used in Europe is presented by Alexiew (2005). From this comparison it can be seen that the most advanced method is so called "New German Method", which will be available in EBGEO and the principles were presented by Kempfert et al. (2004). This method is applicable to the non-cohesive fill above geogrid reinforcement that should be in maximum two layers. It accounts for the strength of the fill material using "multi-shell arching" theory, and also for strain related counterpressure from subsoil in between the piles. The design schematic is shown in Fig. 6.25. The allowance for counterpressure is beneficial to reducing the requirement on the geosynthetic strength, but it has to be treated with caution because if overestimated it can lead to under design of the whole system.

As this design method is new, most of the up to date designs are based on one of the older methods, e.g. form BS 8006 or Hewlett and Randolf (1988).

One of the early designs of piled embankment over soft soil was presented by Brandl et al. (1997). It is a project of upgrading railway track between Magdeburg and Berlin to the train speeds up to $200 \, km.h^{-1}$. At this location the old track was built over deep deposits of soft peat and organic soils and within its 100 year history suffered considerable settlements. The proposal of constructing the upgrade using the technique of piled embankment with load transfer platform was preferred over conventional dig and replace method for economical and mainly environmental reasons. The 2 100 m long section of this structure was built in two stages between 1994 and 1995. The embankment height is between 2 and 3 m. The cast iron tubular piles 118 mm in diameter were driven to the depth of 10–20 m in the spacing of 1.9 m and covered with 1×1.25 m concrete pile caps. The piles were designed to have

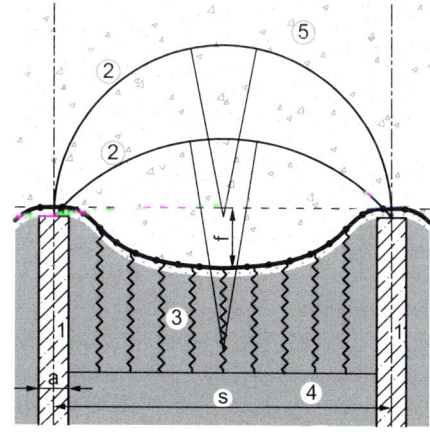

Fig. 6.25 Calculation model for the design of load transfer platform over piles using multi-arching approach (according to Kempfert et al.). 1 – piles; 2 – soil arches; 3 – support from subsoil; 4 – subsoil; 5 – embankment fill

an ultimate capacity of 1 000 kN. Within the three layers of geogrid well-graded gravely sand compacted to 97% Proctor standard (PS) test was placed and above it coarse sand compacted to 97–100% PS. The design of the reinforcing geogrid was done using several design methods and long term design strength of each layer of geogrid was determined to be at least $30\,\text{kN.m}^{-1}$ and in order to reach acceptable serviceability limits, the maximum allowable long-term strain was specified at 3%. From this specification was determined the required geogrid using the isochronous curves from serviceability limit state (SLS) and using all necessary partial reduction factors (creep, mechanical damage, environment, ...) for ultimate limit state (ULS). In this case the SLS was the leading requirement and finally geogrid Fortrac 150/150-30 was selected. The designed cross section can be seen on Fig. 6.26.

As this was one of the first piled embankments, it was well instrumented and monitored. Monitoring included settlements of the pile caps and different locations on the geogrids as well as strains in most important points on the geogrid. About one year of monitoring data was presented and it showed already levelling of the deformation processes in the whole system. The monitoring of strains installed was so sensitive that was possible to measure the response of the strain in geogrid on the passing train in full speed in the order of 0.001%. The monitoring also confirmed that the structure was designed with high level of safety.

6.2.2.4 Fill Lightening

As was seen in some previous examples the total settlements of the soft subsoil can reach even a few meters. When additional material is added to eliminate creep settlement, new settlement is accelerated and this negative tendency can continue for many years. Therefore there is a natural effort to use lighter fill material to decrease this process.

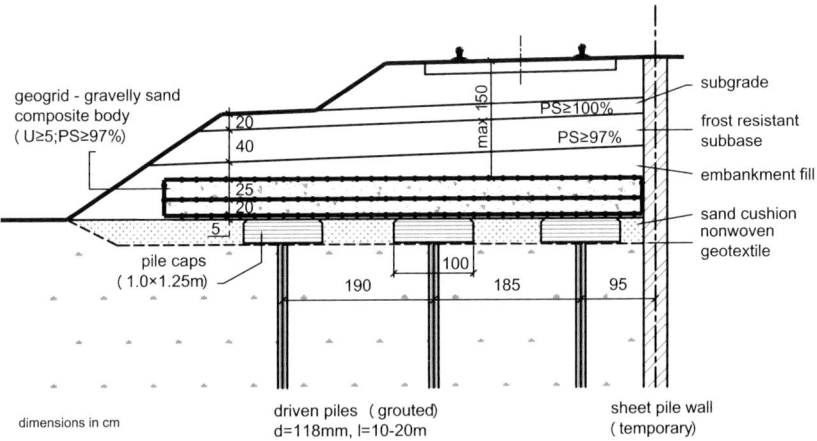

Fig. 6.26 Cross section of the railway embankment on piles between Magdeburg and Berlin (according to Brandl et al.)

6.2 Transport Embankment – Fill

The following three basic types of materials are now the most often used to fulfil this demand:

- Ash (flying ash, pulverized ash),
- Light weight aggregates from expanded clay,
- Expanded polystyrene (EPS).

Ash is a product of burning coal in thermal power stations. In many countries there is a significant surplus of such material and therefore different ways of utilization are sought for them. Earth structures of transport engineering are one of the serious opportunities, because the properties of ash are similar to sand (from angle of internal friction and modulus of deformation under 1D conditions.) Ash has usually a dry volume density lower than $1\,000\,kN.m^{-3}$, therefore is roughly 2 times lighter than natural soils. As utilization of ash is also connected with a potential environmental aspect, this matter will be described in more detail in the Section 6.8.1.

Light weight aggregates from expanded clay, known in the Czechia under the name Liapor, and which are produced in the kiln under a temperature of 1 100–1 200°C from tertiary clays which overlay brown coal (lignite) in the Sokolov brown coal basin. Liapor can be classified as ceramic material, which has generally good strength, resistance, low absorption rate, excellent thermal insulating properties and is also environmentally friendly. The shape of aggregates is nearly circular with an inner porous structure and with an encapsulated sintered surface.

The size of produced aggregates is 0–16 mm, which is assorted into narrower gradings, with different dry volume density:

Grading	8–16 mm	$\sim 290\,kg.m^{-3}$
Grading	4–8 mm	$\sim 350\,kg.m^{-3}$
Grading	0–4 mm	$\sim 525\,kg.m^{-3}$

The advantage is that Liapor is not hydroscopic. Individual aggregates or backfill has no capillary structures, with practically zero capillarity attraction. However when submerged, water slowly fills the inner porous structure. Nevertheless this porous no capillary structure allows expanding frozen water in aggregates. Therefore Liapor withstands cyclic freezing and gives excellent frost resistance to the products in which it is used.

During laboratory tests – Vaníček and Chamra (2001) proved that deformation characteristics are good, in the range for natural soils. For an expected range of stresses no structural collapse was observed due to aggregates break down. Deformation is more connected with small displacement between individual aggregates. Shear strength is close to the sand.

A small note to the two above mentioned lightweight materials should be added. Due to the granular structures and low volume density, the strength for low confining pressure is very low, as is well bearing capacity at the surface. Also sensitivity to surface erosion is high. Therefore this lightweight material should be covered by natural soil, top soil to allow grass growing, or some additives should be added, such as cement or lime to increase the unconfined strength. Lightweight concrete from Liapor aggregates is also a very well known product.

Expanded Polystyrene (EPS) is a super light fill material since it is about 100 times lighter than ordinary fill material and nearly 50 times lighter than other materials traditionally used as light fill (e.g. ash). Frydenlund and Aaboe (1994) describe 20 years experience with the utilization of EPS in road structures. The density of the material is close to $20\,kg.m^{-3}$ when delivered from the producer. But a value about $100\,kg.m^{-3}$ is recommended when counting with limit states of stability and deformation, allowing for some water content increase over its service life. If the EPS be submerged, buoyancy forces must be considered, and for that purpose a unit density of $20\,kg.m^{-3}$ is applied. This case was shown when an uplift limit state was discussed in Chapter 5, see Fig. 5.2a.

Frydenlund and Aaboe specified the following material requirements for road construction:

– The unconfined compressive strength measured on $50 \times 50 \times 50$ mm dry cubes should have a mean value not less than $100\,kN.m^{-2}$ at 5% strain,
– The block must reach specification from the dimensions point of view, minimum thickness 0.5 m, dimensions should be within 1%, and the evenness of the block surface measured with a 3 m straightedge should be within 5 mm.

EPS is a very stable compound chemically, and no decay of material is expected when placed in the ground and covered with soil. Also creep properties are good, cyclic loading tests kept below of 80% of the compressive strength showed that the material can take an unlimited number of load cycles. EPS is also resistant to biological destruction from bacteria and enzymes. Nevertheless a certain care should be taken, especially for the following three cases:

– When EPS is getting into direct contact with petrol,
– When potential for fire is high,
– When the probability of attack from small rodents is high – but up to now such case was not observed.

The first project completed in Norway is dated 1972, when with EPS the case was solved of continuous settlement of embankment, resting on a 3 m thick layer of peat above 10 m of soft sensitive clay and situated close to the bridge abutment founded on piles. Annually 0.2 m of refill was added to compensate this settlement. The problem was solved when 1 m of fill was substituted using EPS.

Another example is a temporary bridge based on EPS across Euroroad 6 at the Lokkeberg intersection near the Swedish border. Since the one lane Acrow steel bridge is temporary and possible deformations may be adjusted during the period of operation, it was decided to carry out the project as a full scale test. Design calculations showed that the bridge foundations could be supported by the EPS fill provided the EPS blocks ($0.5 \times 1.0 \times 3.0$ m) could sustain a load of $60\,kN.m^{-2}$. The bridge and adjoining embankments are located in an area with soft clay deposits. Below a drying crust of about 1.5 m the soil consists mainly of quick clay down to bedrock 6–16 m below the ground surface. A longitudinal profile is shown in Fig. 6.27. In addition to the concrete slab on top of the EPS, another slab was cast in the middle of the fill. Shotcrete was sprayed on the front slopes of the EPS while ordinary soil protection was provided on the side slopes. Deformations measured just after the structure was completed, were in the range of 50 mm (i.e. approx. 1% strain). No significant changes occurred after that and there were no signs of creep effects apart from in the lowest EPS layer. Here creep effects of the order of 1.7% (9 mm) were measured one and a half years after completion. After three years of usage the fill was operating well.

Beinbrech and Hohwiller (2000) show more examples of EPS utilization in geotechnical engineering. They write about rigid expanded polystyrene foam as a deforming and cushioning layer. In this context they pointed to curves of compressive stress versus strain for different bulk densities of EPS, see Fig. 6.28. Up to a strain of about 2% the expanded foam behaves in an approximately elastic manner. At a greater strain the expanded foam cells are deformed and the slopes of the pressure-strain curves become markedly flatter. The strains contain appreciable plastic components. Therefore the determination of strain to which EPS behave as an elastic material is very important, especially when Frydenlund and Aaboe (1994) were mentioning a higher strain −5%.

In the above sense lightweight materials were used to reduce vertical loading and deformations, but for the reduction of horizontal loading – as a backfill behind the retaining wall- they can be used as well. Smekal (2000) describes the application of EPS – expanded polystyrene and XPS – extruded polystyrene in railway track as thermal insulation in Sweden.

6.2.3 Embankment Widening

With growing demand for transport the requirements for transportation networks grow too, especially for road networks. That is why usage of the existing route through the enlarging of capacity is one of the interesting ways to increase transport

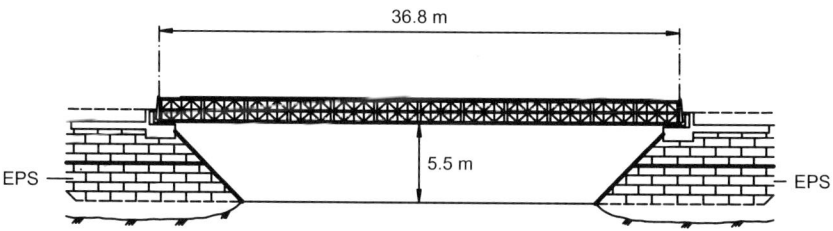

Fig. 6.27 Longitudinal section of bridge at Lokkeberg (according to Frydenlund and Aaboe), EPS – Expanded polystyrene blocks

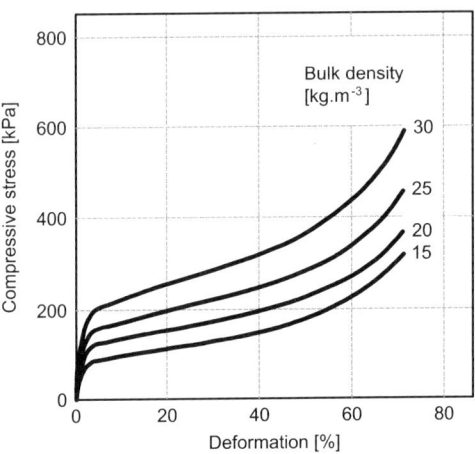

Fig. 6.28 Curves of compressive stress versus strain for different densities of EPS (according to Beinbrech and Hohwiller)

capacity. As an example serves a reconstruction of a 4 lane motorway to a 6 lane motorway. In the majority of cases this means greater intervention in the existing earth structure.

At the same time the reconstruction will also have to take into consideration non-technical problems and this will be above all requirement for new land adjacent to the present route. That is why where there is the possibility, we may expect that priority will be given to the technical solution which will not require any other appropriation of land.

With regards to any embankment and the viewpoint of technical solution the following problems will be involved:

– The elimination of the additional deformation of the existing embankment and its subsoil during widening of the embankment,
– The elimination of the danger of slope stability failure and load bearing capacity of the subsoil for a new part of the embankment, which in most cases will be implemented with a steeper slope,
– The elimination of the increased danger of surface erosion for steeper slopes.

It is possible in the same way as in the previous Section to divide the problem into two basic cases. In the first case the quality of the subsoil will be relatively good and will not require a special additional treatment. In the second case the opposite applies and it is again possible to address construction on the soft subsoil.

The advantage of a solution with zero requirements for additional land is not only connected with the process of its acquisition, but also with the possibility of preserving some parts of the existing construction, for example drainage channels along the route or fences. The solution without land acquisition can be carried out at certain heights of the existing embankment and for road constructions from the heights of the embankment around 3–4 m. For the case of relatively good subsoil the possible alternatives are illustrated in Fig. 6.29. From these results attention is then focused on:

6.2 Transport Embankment – Fill

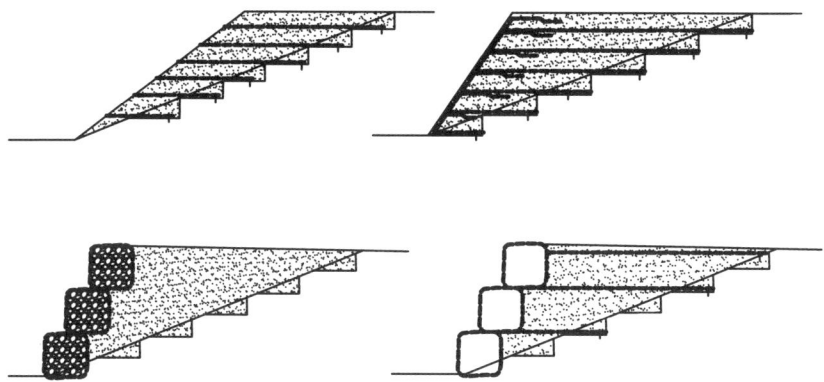

Fig. 6.29 Embankment widening alternatives

– A good interconnection of the old and new part of the embankment,
– Securing the stability of a steeper slope together with its anti-erosion protection.

The first condition is solved by the benching of the contact and with this by the choice of an appropriate material for the new part of the embankment, so that after the compaction, the deformation of the surface would be minimized and there would thus not appear longitudinal cracks on the carriageway at the point of widening. The second condition is solved by the reinforcement of the embankment using the calculation methods of the first limit state – limit state of failure. At the same time there must not be omitted the creep effect of the reinforcement so that the additional deformations do not influence the limit state of serviceability. The alternative of gabions (from wire cages) without reinforcement often recommended for this type of application should be carefully investigated. From a static viewpoint their effect on the increase of slope stability is very small. Anti-erosion protection has been already specified in the Section 6.2.1.2.

Note: The vertical face is also possible but its application will be dealt with in the chapter on retaining walls that is to come.

An interesting cross section when there was achieved not only the widening of a railway embankment but also its heightening is described by Mannsbart and Kropik (1996) – see Fig. 6.30. The increased level in the Danube after the construction of the hydro-power station Freudenau in Vienna required the raising of bridges to allow ships to pass through. The cross section pertains to the approach ramp of the bridge. The description is only general because Mannsbart and Kropik focus on the non-standard reinforcement of the embankment slope with non woven geotextiles with high elongation during failure. The application was possible only because the durability of this temporary construction was only 1 year.

The cases with soft subsoil are significantly more sensitive because the elimination of uneven deformations and the load bearing capacity of the subsoil will play a more significant role. That is why attention is focused on speeding up the consolidation of subsoil and its reinforcement and it is possible to choose various

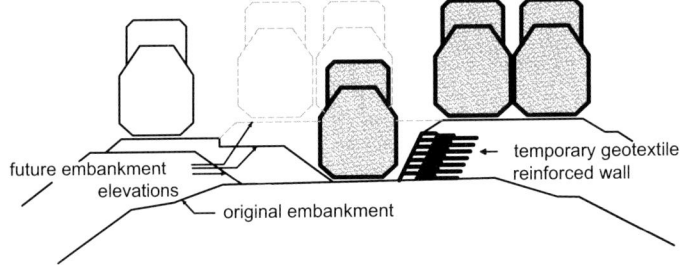

Fig. 6.30 Example of widening of railroad track in Vienna (according to Mannsbart and Kropik)

combinations which are given in the previous chapter. If the above stated precautions have been already applied in the first phase it will involve their connection and widening.

For the case of reinforcement it is necessary also to consider besides the reinforcement in horizontal direction the alternative of reinforcement in the vertical direction by the element which would sort out horizontal thrusts, eliminate horizontal deformations and thus ensure that deformations under the new embankment had rather taken the character of 1D. An interesting solution is the combination of vertical sand drains with their further reinforcement with the help of geosynthetic reinforcement. It involves the application of Kempfert (1996) or Kempfert et al. (1997): "Sand columns, where these columns are coated with a sewn geotextile composite of polyester threads and filter cloth, which guarantee the filter effect". At the same time the radial supporting of the sand columns is strengthened by the geotextile coating combined with the surrounding soft soil. Therefore the geotextile is subjected to ring tension forces. This method was first applied in widening a 6 m high railway embankment, which was built on peat and soft clay to a depth of 7 m. One of the constraining conditions of this construction was that the widening of the embankment should be finished within one month. The railway operation on the new embankment should be started again after a construction period of about 4 months. Therefore, it was demanded of the foundation method that only very small settlements of the new embankment were allowed after the completion of filling. Fig. 6.31 shows the existing and planned foundations with soil layers and parameters. The existing embankment has a very low stability. The soft ground conditions can be derived from the settlement basin of the existing embankment, which has a maximum settlement between 1.2 and 1.5 m. Sand columns were constructed in two phases, in the first one under the new embankment and in the second one under the old one from the new working level 4.00 m to prevent the stability failure of the existing railway embankment under dynamic loading resulting from the driving of casings. Measurements validated the proposed method, maximum settlement was roughly 0.2 m and a few days after the end of filling the rate of settlement was nearly zero. Also lateral deformation was much lower, reaching on the contact of the embankment with reinforced sand piles maximum value about of 70 mm.

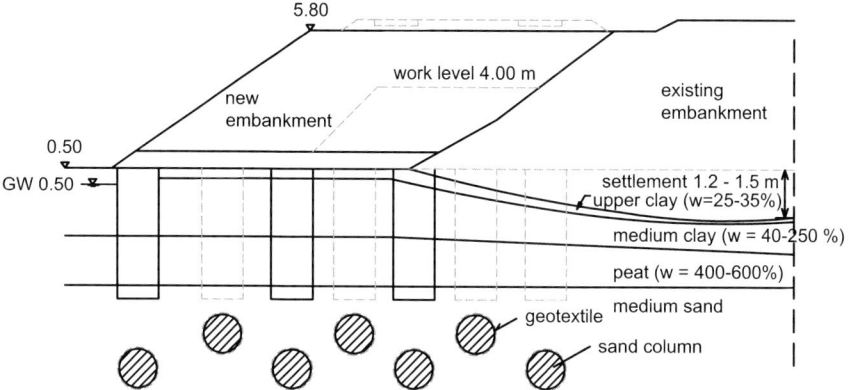

Fig. 6.31 Application of geotextile-coated sand columns in soft ground for railroad track widening (according to Kempfert)

6.3 Cutting

One of the basic questions in the case of cuttings of transport infrastructure is that of stability of slopes, how to design steep slopes and how long they will be stable. This question is sensitive for deeper cuttings. On the one side we want to ensure sufficient security from the viewpoint of the limit state of failure and on the other side it is necessary to minimize the financial costs of the cutting including considerations regarding the size of land appropriation for this structure. That is why every solution should reach a certain optimum between these two opposite views. The main attention here will therefore be focused on the limit state of failure. The cutting can, however, have a negative impact with regard to the ground water level. The lowering of this level will also be revealed away from the ground plan of the structure. That is why it is necessary to take into account the surrounding developments because with the lowering of the water level there will be an increase in effective stresses under these developments and thus through this also their settlement. In concluding this Section there will also be solved the case of the widening of a transport infrastructure route dealing with the extent to which it is possible to secure the steepest slope of the cutting with the new technical solutions so that the requirement for another land appropriation is minimized.

6.3.1 Classical Case – Limit States

As a basic case the slope of the excavation in clayey soils with the groundwater level near the surface will be examined. In practice this case is the most frequent. The occurrence of a deeper cutting for the route in non cohesive permeable soils with a high groundwater level is very improbable. If it exceptionally occurred it is more appropriate to avoid it because it would require resolving the long-term inflows into this cutting.

In the same way as for the case of the embankment as for the case of the cutting in clayey soils changes of pore pressures is brought about as a result of change of stress condition. Due to unloading the pore pressure is also decreasing. In the situation of the completing of the excavation, pore pressures thus reach their minimum and thus shear strength reaches its maximum. Over time, however, the pore pressure is increasing to the value determined from the flow net for the steady flow. By this the shear strength decreases and through this also the stability of the slope. The schematic illustration of these changes over time is stated in Fig. 6.32 – according to Bishop and Bjerrum (1960) – see also Skempton and Hutchinson (1969), where the change of the pore pressure on the general point of a potential sliding surface P is examined. From this a simple conclusion is derived – the situation compared to the embankment is exactly opposite. Stability shortly after the completion of the excavation is at its highest and decreases with time. Under stable conditions, from the viewpoint of pore pressures, the slope stability is at its lowest. This reality is generally recognized and often used for short-term excavations and has already been discussed with the case of in situ modelling in Chapter 5 in the example given by De Beer (1969). In Fig. 6.32 there are two basic possibilities given according to

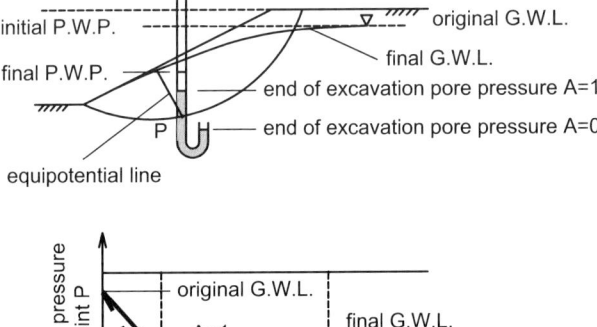

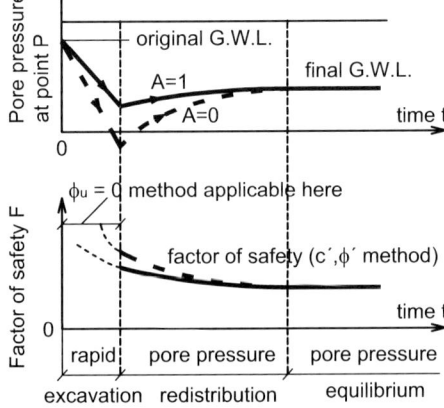

Fig. 6.32 The changes in pore pressure and factor of safety during and after the excavation of a cut in clay (according to Bishop and Bjerrum)

6.3 Cutting

the value of the coefficient of a pore pressure A, which can be in a simplified form for the contractant soils – soft clays and soils normally consolidated considered as $A = 1$, or for the dilatant soils (overconsolidated clays) considered as $A = 0$, (although we know that the coefficient A can also be negative).

$$\Delta u = \Delta\sigma_3 + A\,(\Delta\sigma_1 - \Delta\sigma_3) \tag{6.7}$$

It is obvious that certain differences exist for the basic types of soils, that is why there is an appropriate differentiation for:

– Soft clays, intact (normally consolidated),
– Stiff, up to firm clays, intact (slightly overconsolidated),
– Stiff, up to firm clays, fissured (strongly overconsolidated).

From this viewpoint it is obvious that it is possible to differentiate the problems of slope stability into several basic alternatives given by the type of soils, by short-term and long-term conditions and by the origin of the slope – embankment, cutting, natural slope – see Vaníček (1981).

As for embankments it is also possible here to solve the short-term stability under undrained conditions with the help of total parameters of shear strength (with the exception of fissured clays), but also with the help of effective parameters of shear strength, on condition of the knowledge – in practice a prediction – of pore pressures produced by a change of stress condition. When determining undrained strength it is necessary to take into consideration the influence of anisotropy, the rate of strain, and the size of the sample, as has already been discussed for the case of embankments.

The case of slope stability analysis under long-term conditions is a typical case of drained conditions and the use of a total analysis is wrong in this instance. That is why it is permissible to use only the effective parameters of shear strength. The credibility of peak parameters of shear strength φ', c' for non-fissured clays, whether of soft or firm consistency, was validated in many cases.

In the fissured clays the strength at the failure, gained by backanalysis, is lower than the peak strength. On the other hand it is higher than the residual strength, which can be applied only in the cases of slides on pre-existing slip surfaces. For example Chandler and Skempton (1974) or Skempton (1977) discovered that for two types of overconsolidated clays the failure happened at the peak angle φ', but practically at zero cohesion. The significance of this special behaviour is associated with the influence of the *progressive failure*. In principle what is meant by this is the spread of the failure from a certain point on potential slip surface into its surrounding.

With a conventional analysis of slope stability it is assumed that peak strength is mobilized simultaneously on the whole sliding area. This means that there is assumed an ideal elasto-plastic relation between shear strength and shear deformation. The real deformation curve, however, considerably differs from this assumption.

Generally the ratio of shear stress to shear strength along the slip surface is not uniform. There can exist a point where the strength has been exceeded and there is a question of how this failure will spread. Here it will depend on the nature of the soil behaviour. Let us assume that for soil that is elasto-plastic the failure started at point O and that the failure is spreading in both directions – see Fig. 6.33. Terminal points of the area of failure – plastic area AB – will lie as far from each other until there appears equilibrium between active and passive forces. In any case the slope will reach the indifferent equilibrium when the average shear stress on the slip surface equals the shear strength:

$$\tau_{aver} = \tau_f \tag{6.8}$$

A more complicated situation will prevail for a soil that is strongly overconsolidated – see Fig. 6.34. The failure originated at point S can spread to the sides. In time t_1 strength at point S decreases and the plastic area is limited by points A_1, B_1, where peak strength was reached. However, the failure can continue to spread as is represented by time t_2. As regards to this progressive failure, however, Bishop (1967) does not write only about the failure through time, but mainly in spatial terms. The failure of spreading can, however, stop because the points away from the plastic area have not yet reached their peak strength τ_f. Much will depend on the index of brittleness I_B according to Bishop, see Fig. 6.35.

$$I_B = (\tau_f - \tau_r)/\tau_f \tag{6.9}$$

From these considerations we can conclude that indifferent equilibrium for τ_{aver} is between the peak and the residual strength:

$$\tau_r < \tau_{aver} < \tau_f \tag{6.10}$$

Bjerrum (1967b) states 3 conditions of a progressive failure:

a) The soil shows brittle behaviour and a significant lowering of strength at deformations greater than those for the peak strength – a basic property of strongly overconsolidated clays.

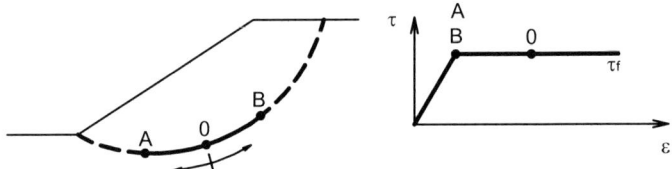

Fig. 6.33 Spreading of failure along slip surface for elasto-plastic soil

6.3 Cutting

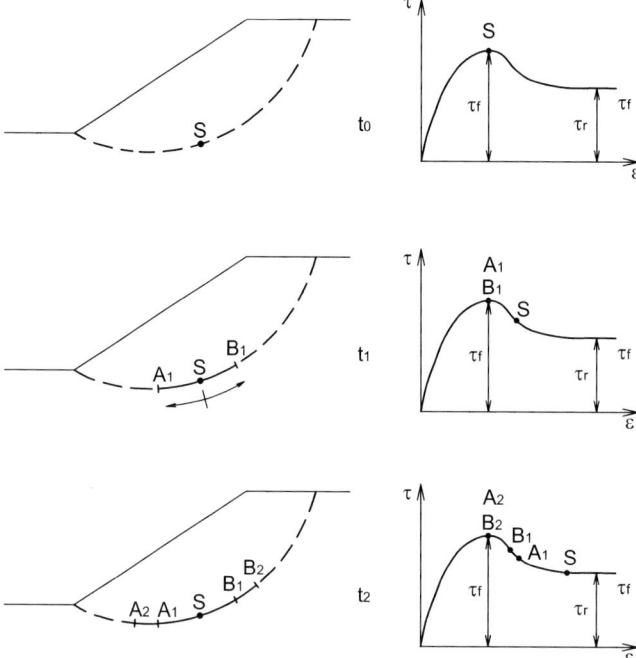

Fig. 6.34 Progressive failure along slip surface for overconsolidated fissured clay

b) There exist places of stress concentration – the simplest is with fissured clays at corners of micro-fissures where the strength can be exceeded locally and so a progressive failure can start.

c) Limit conditions are given by different deformations – for example by unloading at the toe of the excavation here various deformations are occurring which, in many cases, exceed the deformation necessary for reaching the peak strength. In this regard there are very instructive observations of the development of slip surface at the toe of the slope of overconsolidated fissured clays – soil above slip surface is visibly moving into the excavation while the soil under the slip surface is static – as mentioned by Havlíček (1981).

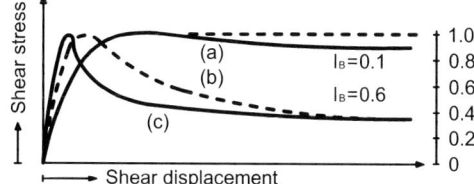

Fig. 6.35 Index of brittleness according to Bishop

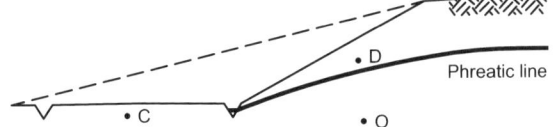

Fig. 6.36 Typical points of cuttings where different behaviour is studied

Up to now there has been a concentration on point O laying on the potential slip surface. But also the behaviour of other points such as C or D can be discussed in more detail, see schematic Fig. 6.36. Point D lying on the slope of the cuttings is under the influence of unloading. Due to this fact moisture content is increased as a result of suction. This unloading can also affect structural strength, e.g. loss of cementations bonds. For fissured clays this fact means also the opening of fissures connected with a significant decrease of shear strength there. This process, referred to as softening, is also connected with the process of weathering, where temperature changes (freezing, drying) are playing a most important role. These processes of softening can go to the depth of 6–8 meters. The final effect, connected with shear strength decrease, can lead to shallow slip failures. Therefore some countermeasures should be recommended do reduce this process there.

For point C the situation is similar and for us also important from the point of view of deformation or heave. To estimate this heave the samples should be reconsolidated in the laboratory very quickly after sampling and after that unloaded under the same stress path as expected for the given case. A certain care should be devoted to the possibility of shear failure. According to e.g. Lambe and Whitman (1969), there are four different main stress paths leading to failure – see also Section 5.2.2.2. This situation is shown on Fig. 6.37 – (starting from the point A representing anisotropic geostatic pressure). The discussed case represents one of them for which the following assumptions are specified:

$$\Delta\sigma_v = -; \Delta\sigma_h = 0 \qquad (6.11)$$

Now back to the Fig. 6.32; fundamental attention will be devoted to increasing long term stability. This can be done by two main basic measures:

- To decrease the underground water table for long term conditions, to decrease pore pressure along the potential slip surface,
- To reinforce the soil in the wider area of the potential slip surface.

These two measures will be discussed in the next sections.

6.3.2 Ground Water Lowering

Generally, the consequence of the excavation is a lowering of the groundwater level. In the period shortly after completing the excavation the lowering of pore pressures is the highest, but it concerns especially the area where there was brought about the release of pressure, i.e. in the nearest surrounding of the excavation. After the pore

Fig. 6.37 Four main stress paths leading to the failure

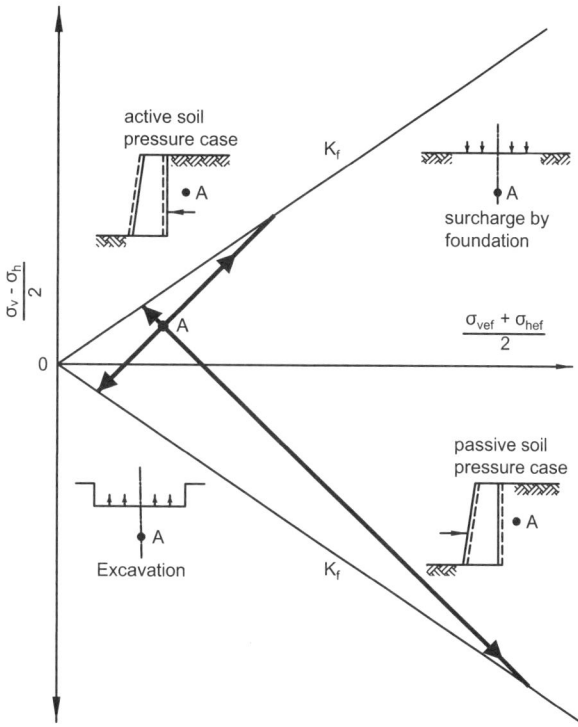

pressures stabilization from a long-term viewpoint the level is lower than before the excavation and the phreatic surface will intervene in the wider area. It can thus cause the lowering of the groundwater level also under the surrounding structural objects, which can have the consequence of their settlement with regard to moderate dip of this level, a non-uniform settlement. This danger does not have to be great because the level reduction for 1 m will cause an increment of effective stresses of $10\,\text{kN.m}^{-2}$. That is why by calculation it is recommended to verify the case, when the cutting is deep and the water level reduction significant.

For the basic case see also Fig. 6.36, where the stabilized groundwater level is influenced by the level of toe drain. This one ensures that water did not run out on the slope and did not start the process of internal erosion. The measures for increasing long-term stability paradoxically tend to the area of the lowering of the groundwater level under this basic level. Among the most frequent measures belong:

– Implementation of trench drains perpendicularly on the slope, Fig. 6.38a;
– Implementation of horizontal drainage boreholes Fig. 6.38b.

Trench drains can be constructed by current techniques very effectively. The ditch is narrow, approximately 200–400 mm wide. It is filled in with permeable sandy gravel which at the same time ensures filtration prevention against the washing out of soil

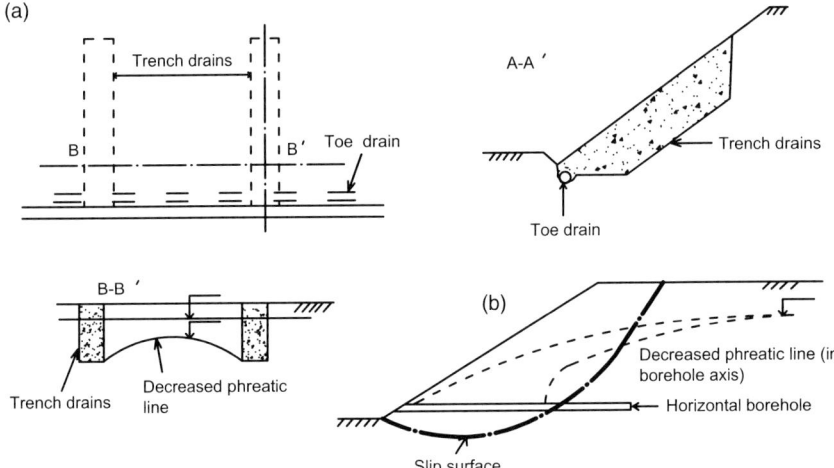

Fig. 6.38 Measures for phreatic surface decrease. (**a**) Trench drains. (**b**) Horizontal drainage boreholes

particles from the surrounding clayey soil. The surface of the trench drain is sealed by a low permeable soil. It is also possible to apply filtration geotextiles along the perimeter so that the drain can consist of coarser material. Using geosynthetics with a higher tensile strength can also play an additional supporting role.

Horizontal drainage boreholes are the most frequently implemented at a diameter of 80–150 mm. The borehole is equipped with perforated casing ensuring stability of the interface.

In both cases the design must take into consideration the 3D effect and how far from each other the trench drains or boreholes should be.

6.3.3 Cuts Widening

If the lowering of the groundwater level does not have the desired additional effect of increasing long-term stability, it is necessary to apply some of the elements of reinforcement. For the cutting, as has already been emphasised in Chapter 4, priority is given to nailing.

Subsequently three basic cases of nailing will be identified, according to the dip of the slope and according to the type of facing and the arrangement of the surface of the slope e.g. Phear et al. (2005).

For prevailing slopes of cuttings (~ 1 : 1.5), application of soil nailing with soft facing. These facings perform no long term role but provide stability until vegetation becomes established. Their primary role is to retain the vegetation layer (top soil) and prevent surface erosion. Materials commonly used for this purpose are erosion protection geo mattresses, geogrids, cellular geo-fabrics, geosynthetic sheet, light metallic mesh/fabric, or degradable geotextiles from natural materials like jute or coir matt. The long term effectiveness of a slope with a soft facing depends on the growth, and subsequent management, of vegetation. Good practice is to vegetate the face with grass, shrubs or with the help of turf. Secondary pins should be installed between the soil nails to control and restrict the movement of the soft facing until the vegetation becomes fully established. Soil nailing plays two basic roles. From the stability point of view, the potential slip surface must go behind the reinforced area or must cross the reinforced area. A second role is connected with the limitation of softening in the upper zone of the excavated slope.

For steeper slopes up to 60–70 degrees, soil nailing with flexible structural facings. These structural facings provide long-term stability to the face of the soil-nailed structure by transfer of the soil load from the soil nails to the nail heads. The facing materials allow greater soil movement and minor bulging between the head plates should be expected, although this is reduced with closer nail spacing. Materials used for flexible facings include geo grids and coated metallic meshes, in conjunction with head plates. These facing materials need to have sufficient strength and durability and will require either jointing or overlapping to provide structural continuity between the soil nails. The flexible facing should be applied after soil nail installation but prior to the nail heads application and also prior to the next excavation stage, to minimize degradation of the slope face. The application of pins between nail heads as well the application of a vegetation layer as in the previous case are needed to guarantee good system functioning. Also here, soil nailing is playing the above two mentioned roles.

What is also common to both these cases is a drainage system. Toe drain should guarantee that the water will not run on the slope surface, starting some problems with internal erosion. Nevertheless supplementary action to decrease the ground water level for steady flow, as in for example, horizontal drains, can be very useful.

A nice example of replacement of the mass concrete gravity wall by soil nailed slope to support widened cut for railway in the Czech Republic is shown in Fig. 6.39. In this case the soil nailed slope was designed as green with the facing consisting of organic (coir) mattress to improve the grass grow of shortly after construction and biaxial geogrid reinforced erosion protection mattress (Enkamat S55). The nail head plates were about 25×25 cm big and nails were spaced as such that 1 soil nail was placed in every 1 m^2. There were also intermediate pins fixing those two layers of geosynthetics to the surface of cut slope more rigorously. The Czech Railways as owner and investor have been very satisfied with this solution which nicely blends in the environment and was also built cheaper and quicker, therefore a very successful project.

For vertical or nearly vertical slopes, soil nailing with hard structural facings. This case was described in Chapter 4 as a basic case of soil nailing. Hard structural facings generally comprise sprayed concrete (shotcrete) reinforced with steel mesh. Other structural facing materials include conventional cast in situ concrete or precast concrete panels. This sort of facing helps to limit deformations. More problems are connected with the drainage system. Water pressure can readily build up behind the hard structural facing making it necessary to include weep holes within the facing, and/or a drainage system behind the facing. Behind the facing strips of geosynthetic drainage mattresses can be used.

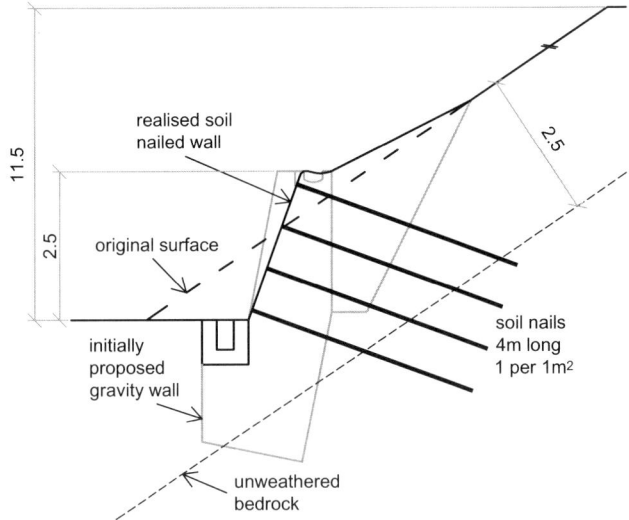

Fig. 6.39 Soil nailed wall replacing the gravity wall in railway cutting

From this short review it can be seen that during cuts widening either flexible or hard structural facings can be applied, but for less steep slopes flexible structural facing is preferred taking into account cost and aesthetic considerations, particularly when compared with sprayed concrete.

One example proposed by the authors for a narrow road in hilly conservation area near Czech and Slovak border is a kind of compromise between hard and flexible facings. The original design for this structure was soil nailed wall with shotcrete facing that was for aesthetic reasons buried behind independent gabion wall and the space in between backfilled with soil. The authors have been asked by the contractor to suggest a more economical solution but still keeping the aesthetics in mind. This challenge was solved keeping the idea of the original solution but using different layout. The facing gabions were used but this time we proposed to use slim panels (only 250 mm thick) that have incorporated steel reinforcing mesh, which can be embedded into the soil. For this reason the cut could be made smaller and savings were made. Soil nailing was still used but this time without shotcrete, hence another saving. In order to keep both structures stable, we proposed to use the reinforcing steel grids of gabions to reinforce the backfill in between the structures and more over to connect these grids to soil nails at their heads. Therefore the gabion wall was in fact acting as a structural facing for the soil nailed wall, which was not either hard or flexible, but something in between. The scheme showing the difference between original design and authors' proposal can be seen in Fig. 6.40.

6.3.3.1 Soil Nailing Design

There are many similarities with other types of soil reinforcement. The reinforced zone creates a quasi – homogenous block for which – from the *ultimate limit state*

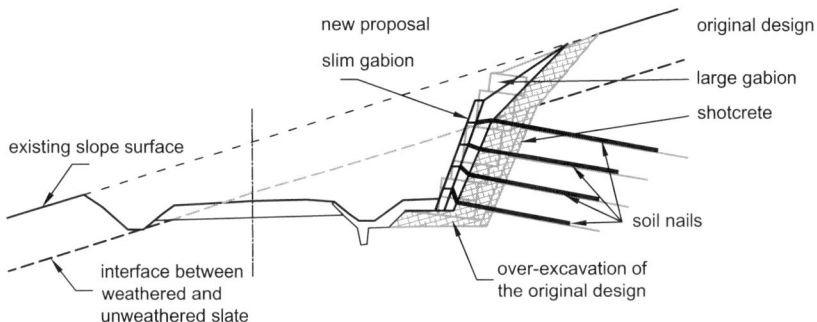

Fig. 6.40 Soil nailed wall with gabion facing in mountainous area

(ULS) point of view – internal and external stability should be controlled. From the external stability view the following possibilities of failure have to be checked – overturning, sliding, bearing capacity and finally overall slope stability along the slip surface going beyond the zone of reinforcement. From the internal stability view e.g. Plumelle and Schlosser (1991) consider three types of failure – Fig. 6.41:

– Failure of the nailed soil mass due to breakage of the bars (a),
– Failure of the nailed soil mass during the excavation phases (b),
– Failure of the nailed soil mass due to the lack of adherence on the bars (c).

Failure at (b) is strongly connected with the technology of construction. During the excavation phases the soil wall is unprotected but some soil arch develops there. But for a deeper excavation the soil arch becomes unstable and failure can occur there very rapidly. During the experimental program Plumelle and Schlosser also observed failure at (a), with the following conclusions:

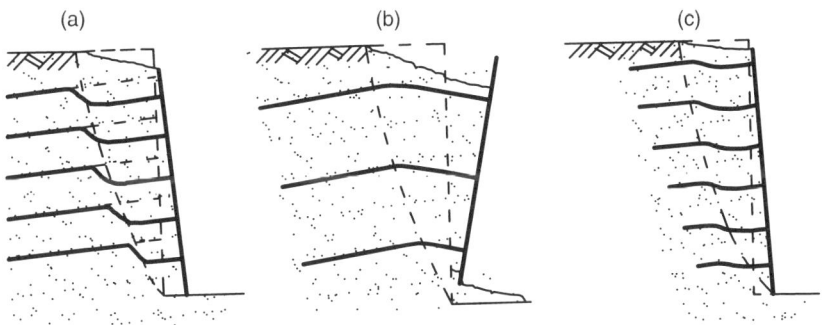

Fig. 6.41 Soil nailing – three types of failure from the internal stability view (according to Plumelle and Schlosser)

- The maximum tensile force (T_{max}), is not located at the front of the wall (T_0) but at some distance away from the facing, the ratio T_0/T_{max} decreases as the excavation progresses.
- The line of maximum tensile forces lies within the failure zone.
- The first resisting force mobilized is the tensile force in the nail. Prior to failure the bending stiffness gives an additional safety factor and prevents a quick collapse.

From these conclusions some recommendations can be summarized:

- The main resisting force mobilized in the nail is tensile force – before bending stiffness is giving additional support, serviceability limit states are exceeded,
- The limiting tensile force is therefore determined by friction at the soil/grout interface,
- The solution has to find the failure line along which the ratio between resistive forces against failure from nails and soil and disturbing forces is minimal.

Explained by the limit state approach, this ratio is expressed below and shown in Fig. 6.42, see also Phear et al. (2005):

$$\Sigma R_N/\gamma_N + R_S/\gamma_S \geq R_D \gamma_D \quad (6.12)$$

where

R_N and R_S – are characteristic values of resistive forces against failure for the nails and soil respectively,
R_D – is the characteristic value of the disturbing forces,
$\gamma_N, \gamma_S, \gamma_D$ – are partial factors to obtain design values (≥ 1.0).

The practical solution of this formula using the analytical solution of limit state equilibrium method can differ according to the different method of slope stability used, and according to the assumption as to how the resistive force from nails is included into slope stability solutions (see e.g. Bishop, Janbu, Sarma and

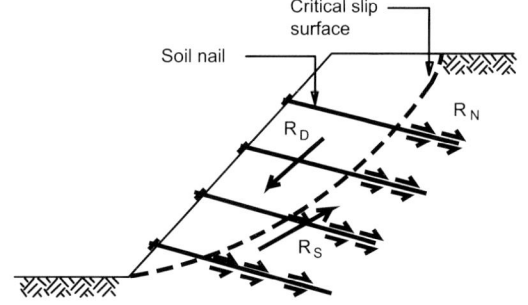

Fig. 6.42 Resisting and disturbing forces for soil-nailed slope stability (according to Phear et al.)

6.3 Cutting

others). More specifications are given in national recommendations as e.g. in France, Clouterre, or in the UK, Ciria Best Practice guidance.

The ultimate pull-out resistance of soil nails in cohesive soils can be estimated using the undrained shear strength c_u and an adhesion factor α by equation recommended by Ciria Best Practice guidance:

$$P_{ult} = \gamma_p \pi d L_b \alpha c_u \qquad (6.13)$$

where

> γ_p – partial factor for natural variation in pull out resistance
> d – diameter of grouted hole (or tendon diameter – for driven nails without grout annulus)
> L_b – bond length over which pull out resistance of soil nail can develop.

Note: This approach follows the suggestion by BS 8081:1989 for Type A anchors in cohesive soils (which are tremie grouted like soil nails). This method will tend to give an upper boundary estimate of ultimate pull-out resistance in cohesive soils. For reasonable confidence of long-term performance, it is probably most appropriate to calculate the pull-out resistance of soil nails using an effective stress design method, as is used for granular soils:

$$P_{ult} = \gamma_p \pi d (K \sigma_v' \tan \varphi_d' + c') L_b \qquad (6.14)$$

Recommended values of coefficient of earth pressure K are specified only for granular soils, taking into account the influence of constrained dilation in the surrounding granular soil.

- Medium dense to dense sandy gravels $K = 1.4$ to 2.3
- Dense sands $K = 1.4$
- Fine sands and silts of high relative density $K = 1.0$
- Fine sands and silts of low relative density $K = 0.5$

In some approaches vertical effective stress σ_v' multiplied by K is substituted by effective radial stress σ_r':

$$\sigma_r' = 1/4 \, (3 + K_a) \, \sigma_v' \qquad (6.15)$$

This approach tends to give a lower boundary assessment of pull-out resistance. One of the reasons for this is that K_a gives a lower boundary value for the lateral stress.

Matejčeková (2005) made an overview of different methods of soil nailing design and for LEM – internal stability calculation – collected 22 computer codes, but for practical comparison selected four of them – Czech GEO 4, France Talren, German GGU a US Snail. In most cases an individual code has some modification. For the general case the variability of results were inside 5%, with exceptions up to 12%.

Besides the LEM approach other methods can be used such as Earth pressure method and FEM. The earth pressure method – see Fig. 6.43 seems to be very

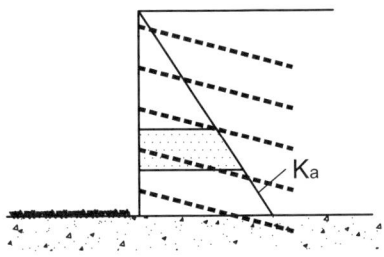

Fig. 6.43 Earth pressure acting on soil nail

simple; calculating that an individual nail will be loaded by calculated portion of earth pressure, but does not explicitly examine all possible failure modes of soil nails themselves and the overall structure. Additional stability checks have to be adopted. In spite of these limitations, the valuation of earth pressure in soil nail wall design is applied usually in the wall facing design.

Numerical modelling by Finite element or Finite difference methods are preferred where precise deformation analysis checking *Serviceability limit state* has to be performed.

At the stage of design, analysis of deformations of soil nail wall is a difficult task, because displacement at the head of the facing is influenced by various factors such as – rate of construction, height of excavation phases and spacing between nails, properties of nails, facing, bearing capacity of the foundation soil etc.

Full scale experiments on nailed walls have shown that only small displacements are necessary to mobilize nail forces. Generally, horizontal δ_h and vertical displacement δ_v at the head of the facing seem to be roughly of equal value – see Fig. 4.40. They are usually put at between $H/1000$ and $3H/1000$ (French Clouterre). But when a nailed wall has a low global factor of safety or the ratio of H/L is high, the values of displacements tend to be higher.

When applying numerical models – the following problems are a matter of discussion:

- 3D versus 2D model – modelling in 3D is very demanding on element mesh. In the case of a simple and regular geometrical arrangement of reinforced wall especially, 2D modelling under plain strain conditions is preferred.
- Modelling of soil nail – e.g. software Plaxis and CESAR-LCPC offers: a defined body, structural elements with axial stiffness only, structural elements with axial and bending stiffness, (the deformation characteristics are governed by isotropic linear elasticity).
- Modelling of contact problems soil × nail – interface elements are of virtual thickness and usually have elastic-plastic behaviour.
- Modelling of contact problems soil × facing – where different approaches are applied for hard and flexible facings.

The advantage of FEM is the ability to simulate the soil nail wall behaviour during the individual steps of the construction. The predicted values, above all displacements or tensile forces in instrumented nails, can be compared very quickly with measured ones and in this way can correct the design when necessary.

6.4 Specific Cases of Interaction

In this chapter two basic examples of interaction of soil with other structural elements will be specified:

- Interaction with retaining structures, applied to a retaining wall and bridge abutment,
- Interaction with other structural elements which can be part of an earth structure such as different pipes, different subways etc.

From the first point of view Jones (1991) did a very useful overview of different earth retaining systems. He distinguishes between:

- Externally stabilized systems, which use an external structural wall, against which stabilizing forces are mobilized. This system can be divided into in situ walls such as sheet piles, soldier piles, cast in situ slurry wall, or into gravity walls as masonry, concrete, cantilever, gabion, crib, bin, cellular cofferdam.
- Internally stabilized systems, which involve reinforcements installed within and extending beyond the potential failure mass. Within this system shear transfer to mobilize the tensile capacity of closely spaced reinforcing elements has removed the need for a structural wall. A facing serves here to prevent local ravelling and deterioration rather than to provide primary structural support. Typically representative is geosynthetic reinforced soil for fill reinforcement and soil nailing for in situ reinforcement.
- Hybrid systems, which combine elements of both internally and externally supported soil. Typically representative here are gabions or concrete wall blocks with reinforcing elements.

Hereafter in accordance with the specified aim of this book, attention will be focused mostly on internally stabilized systems supplemented with some examples from hybrid systems. The possibility of reducing the pressure on the retaining system by applying lighter fill materials will also be a matter of interest.

6.4.1 Retaining Walls

Before specifying different types of retaining walls based on internally stabilized systems, some essential distinctions can be made according to Fig. 6.44. From the construction processes:

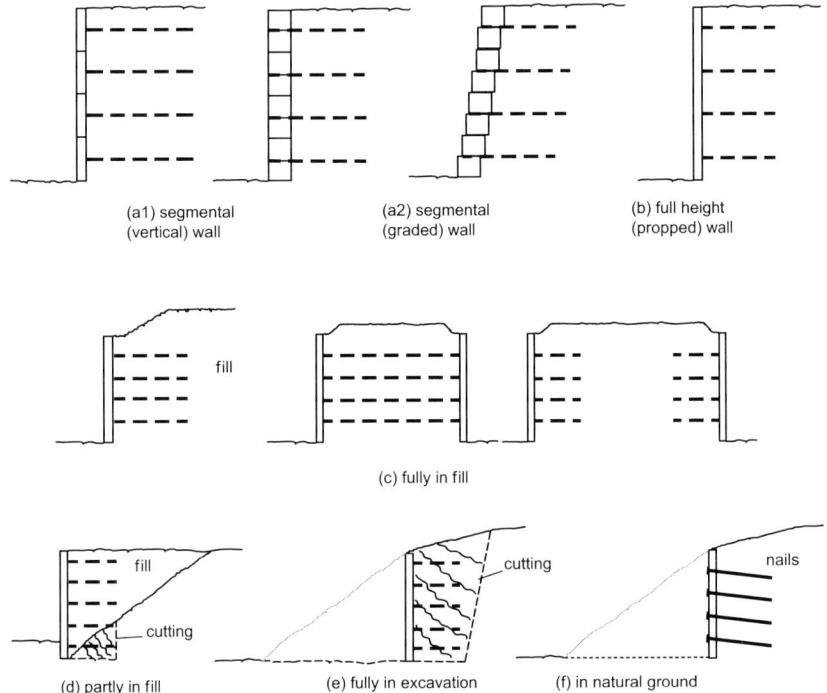

Fig. 6.44 Different types of retaining walls based on internally stabilized systems

- Segmental wall – constructed step by step – not only fill but also facing element (a1),
- Full height wall – facing is composed from full height panel and connection with the backfill via reinforcing element is going step by step – reinforcement is located between compacted fill layers (b). This system needs support during the initial steps, and so is often called the propped wall.

Note: Up to now the segmental wall is strongly preferred.
From the shape of facing:

- Vertical walls,
- Graded wall – with respect to the previous distinction it is possible only for the segmental walls (a2).

In relation to the surrounding soil:

- Reinforcement is realized fully in fill (c),
- Reinforcement is realized partly in excavation (d),
- Reinforcement is realized mostly in excavation (e).

6.4 Specific Cases of Interaction

For the last case making a decision between reinforced soils and in situ reinforcement (soil nailing) will depend not only on the technical aspect, the possibility to excavate more materials so decreasing slope stability, but also on an economic view. In most cases soil nailing is preferred, because this situation is typical for this sort of reinforcement.

6.4.1.1 Segmental Concrete Block

Small prefabricated blocks have different shapes as patented by different producers, but are mostly similar to hollow brick. Weight is around 20 kg, so that the block is easily transported by one man. Connection between individual blocks and the reinforcing element (geogrid sheet) is realized by way of friction. Detail of such a connection is shown in Fig. 6.45 where gravel grains filling the inner space in the block are partly sunk between a mesh of grids, so the size should be in a certain ratio to meshes. Facing can be vertical or graded (stepped) and for the first case the connection can be improved e.g. by a vertical steel bar. For the graded wall, a small buttress ensures stability of the blocks during the compaction of a new layer and also after completing the whole wall, increases inner stability, especially against bulging. Typical examples of external and internal stability which have to be checked are shown on Fig. 6.46, M Vaníček (1998). View of the finished bridge wing wall for the project of major road I/13 – bypass of Ostrov in the Czech Republic is in Fig. 6.47. To improve the aesthetic aspect producers can create a face to look like natural stone. Special blocks, called concrete planter boxes, are a little bit wider and the front part is filled by top soil for better planting.

6.4.1.2 Partial Height Concrete Panel

Big concrete panels with an area at least 1 m^2 need to be transported on place by crane. Fair-face concrete has a hexagonal shape or the shape of briquette, generally a shape where there is no straight interstice. The castings of one row thus make possible provisional gripping of the second row. This second row can in this way

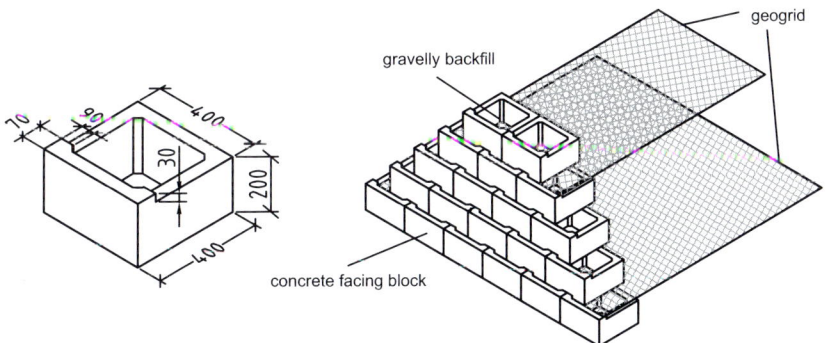

Fig. 6.45 Wall from small prefabricated blocks with a detail of typical block

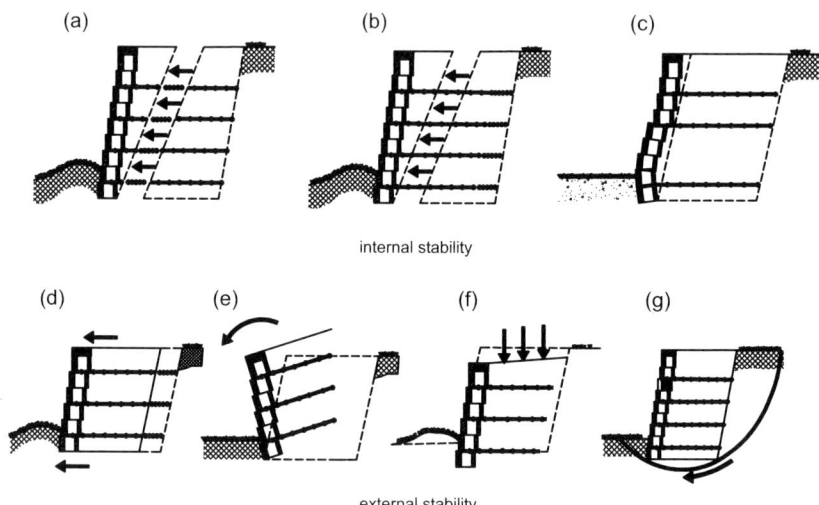

Fig. 6.46 Typical examples of external and internal stability for wall from small prefabricated blocks
Internal stability. (**a**) Breakage of reinforcing element. (**b**) Extraction of reinforcing element from soil (too short anchoring length). (**c**) Buckling of wall (too great distance between reinforcing elements)
External stability. (**d**) Sliding. (**e**) Overturning. (**f**) Bearing capacity of subsoil. (**g**) Overall stability – slip surface behind of zone of reinforcement

Fig. 6.47 Reinforced soil wall, facing from small prefabricates

perform a role of formwork, behind which the soil layer can be compacted. A panel is connected to the reinforcing element, which is usually a reinforcing strip or a sheet the width of the panel, Figs. 6.48 and 6.49.

6.4.1.3 L – Shaped Facing

A relatively problem-free technology of construction can be satisfied by a facing element with an L-shape cross section, see Fig. 6.50. On one side it works as a formwork, on the other, a vertical load on the horizontal part of the cross section can significantly help to improve total stability. L-shaped facing is in fact a cantilever wall which belongs to externally stabilized systems. L-shaped (or turned T) at full height can create classical concrete facing but can work also as support for a moveable formwork situated between the facing and the reinforced soil construction, Fig. 6.51. Reinforced wrapping around the face system as mentioned in Section 6.2.1.2 can be used here. Therefore after the completion of construction work, the geosynthetic reinforced fill has no contact with this wall. This principle can be used when an old masonry retaining wall has to be protected. Soil behind is excavated and reinforced and thus lateral pressure on a historical wall is fully eliminated. Alternatively, light fill material such as e.g. expanded clay can be used here.

6.4.1.4 Gabion Facing

Gabion wall is a classical gravity wall. When combined with reinforcement it falls into the category of hybrid systems. Until reinforced the benefit to stability is insignificant. In fact it is a steeper version of the face protection referred to in Section 6.2.1.2 as front blocks tied back by straight reinforcement, Fig. 6.52.

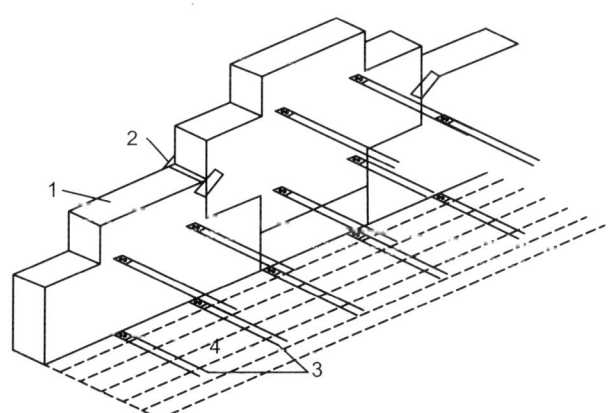

Fig. 6.48 Wall from large prefabricated concrete panels. 1 – face panel; 2 – provisional gripping; 3 – reinforcing strips; 4 – backfill

Fig. 6.49 Photos of retaining wall from large prefabricated panels. (**a**) Front view. (**b**) Connection detail

We can show here an example of such a wall at a car racetrack in Most in the Czech Republic, for the design of which the author was involved. Due to fire issues, the facing was made in this case from gabions. The slope is up to 6 m high with inclination of 78° and the gabions are embedded 2 m below the surface.

The interesting part of this project was also the use of crashed concrete as backfill material of the reinforced zone and hence the reinforcing elements had to be resistant to alkaline environment. In this case polyvinylalcohol based Fortrac® geogrids have been used.

6.4 Specific Cases of Interaction

Fig. 6.50 Retaining wall from L shaped facings

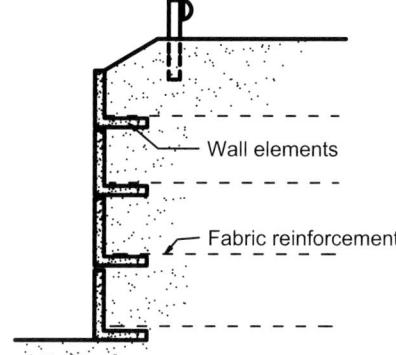

Fig. 6.51 L-shaped facing on full height

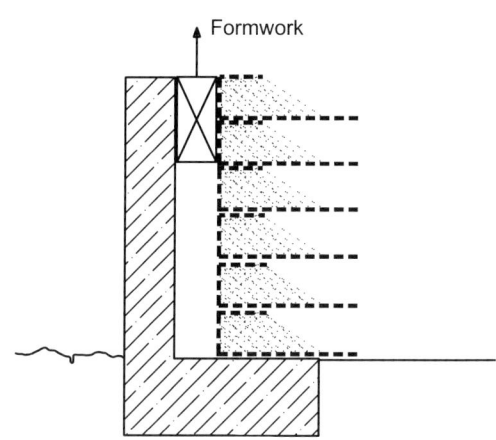

Even when the subsoil was of a good quality, better than backfill, still overall stability was an issue and therefore reinforced cushion underneath the gabions was used. The reinforced cushion was used for two reasons, one was already mentioned overall stability, and the other was improved bearing capacity for the gabions. In this case the cushion was reinforced by steel mesh with the same protection as gabions to guarantee durability.

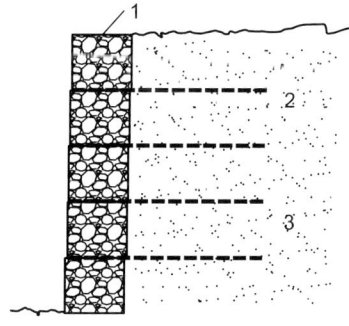

Fig. 6.52 Tailed gabions. 1 – gabions or geotextile bags; 2 – reinforcing element; 3 – backfill

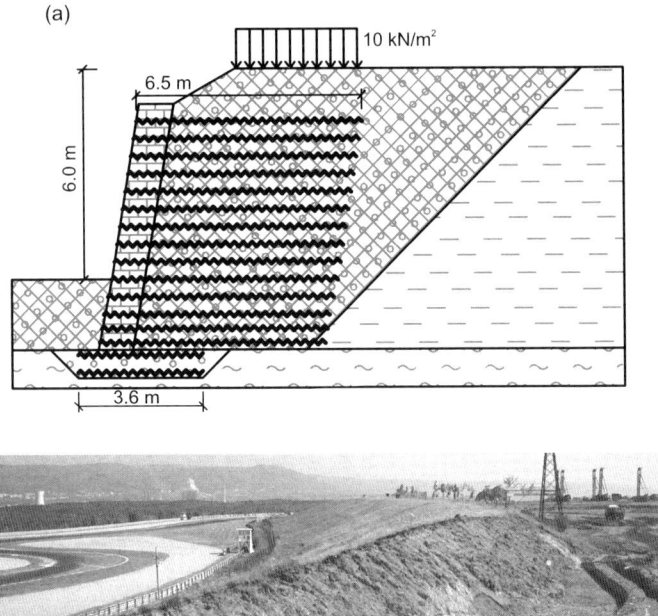

Fig. 6.53 Soil reinforced wall with gabion facing. (**a**) Design cross-section. (**b**) Photo from construction

Schematic design cross section of the highest embankment and photo from construction are presented in Fig. 6.53.

A similar function can be fulfilled by big bags made from synthetic materials filled with sand. But the problems with container damage limit their usage. The next example is an exception to this statement.

6.4.1.5 Permanent Geosynthetic-reinforced Soil Retaining Wall (GRS-RW System)

This system is now widely used in Japan, especially when constructed to support bullet train tracks. The main features can be stated in line with e.g. Tatsuoka et al. (1996) as follows:

a) The use of a full-height rigid facing cast-in-place by staged construction procedures – see Fig. 6.54;
b) The use of a polymer grid reinforcement for cohesionless soils to enhance good contact with soil and of a composite of non-woven and woven geotextiles for

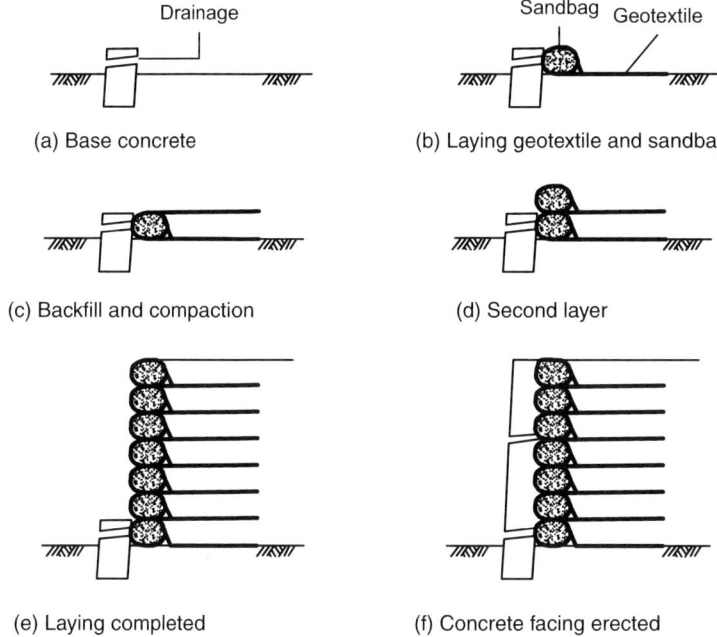

Fig. 6.54 Standard staged construction procedures for the GRS-RW system (according to Tatsuoka et al.)

nearly saturated cohesive soils to facilitate both drainage and tensile-reinforcing of the backfill;
c) The use of relatively short reinforcement;
d) The use of low-quality on-site soil as backfill, if necessary.

The staged construction is clear from the figure. The last step is connected with a cast-in-place construction of thin lightly steel-reinforced concrete facing directly on the wall face after the deformation of the backfill and subsoil layer beneath the wall has taken place. After that a good connection between the facing and the main body of the wall is assured.

6.4.1.6 Others Application

It would be difficult to describe all the systems which have been used up to the present time. Roughly two directions can be mentioned – the application of more flexible facings and the application of an anchoring element, where the anchor is helping to transfer tensile force in the reinforcing element not counting only with the friction between element and soil. A retaining wall made from old tyres can be the example of such an application – see Fig. 6.55.

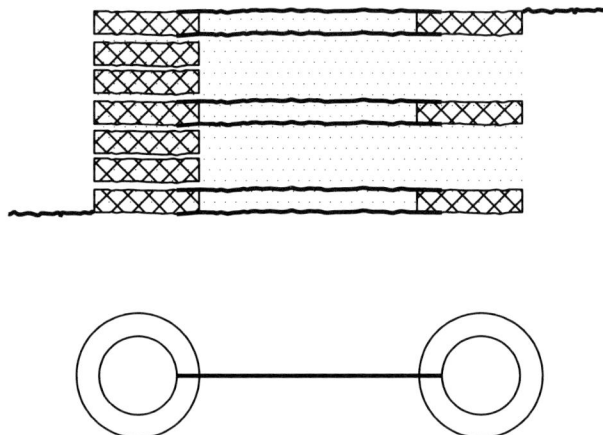

Fig. 6.55 Retaining wall from old tyres

6.4.1.7 Design of Geosynthetic-Reinforced Soil Retaining Wall

For all types there are close similarities to the design of wall from small segmental concrete blocks, which were mentioned at the beginning. Unlike external stability, internal stability is controlled along the failure surface, for which the probability of rupture of the reinforcing element or pull-out due to insufficient anchoring length (anchoring resistance) is highest. When designing the connection between facing and reinforcing elements the distribution of earth pressure can be used (see Fig. 6.43) as a first estimation. But the exact distribution of earth pressure along the height of the retaining wall will depend also on the rigidity of the spacing. There is also a connection to the horizontal deformation, as discussed by McGown et al. (1987) – see e.g. Jones (1991) – Fig. 6.56.

For better displacement evaluation, structure modelling is preferred, and above all numerical modelling. Good results for numerical modelling (application of FEM methods) are obtained when some recommendations discussed in Section 5.3.1 are respected.

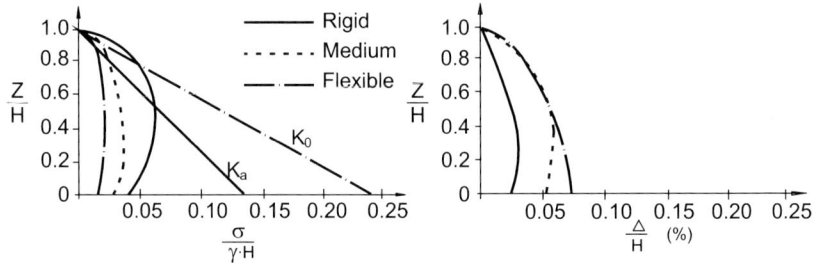

Fig. 6.56 Variation in lateral facing pressures and displacement for different facing stiffness when using high quality backfill material – $\varphi = 50°$ (according to Jones; McGown et al.)

6.4.2 Bridge Abutments

Relatively high demands are generally placed on bridge abutments. From the viewpoint of the actual bridge construction, the demand is involved of a relatively high bearing load on the foundations of bridge abutments and the effort to minimize deformations, especially non-uniform ones, between the individual bridge abutments or the pillars. This first requirement increases the problem of interaction of the bridge abutment with the approach embankment. The difference between the settlement of the actual abutment and the approach embankment can reach unacceptable values and the detail which expresses itself by the bump on the surface of the carriageway must be solved.

A classic reinforced concrete bridge abutment has its advantages and disadvantages. It is possible to solve the problem of the lightly bearing subsoil by the foundation of the abutment on piles with a minimum settlement – which increases the demands on the approach embankment. The second significant disadvantage, especially during reconstructions, can be the time factor – the necessary time from the starting of the reconstruction until the possibility of a full load on a bridge abutment during the wet process of its construction.

An interesting solution presents here the application of the principle of a retaining wall from the reinforced soil, see Fig. 4.27. The requirements are in principle greater:

– Under the constraint of vertical deformations there is used soil with a high modulus of deformation, especially gravely sands,
– Under the constraint of horizontal deformations a rigid facing is applied,
– Under the constraint of horizontal strains after the completion of the construction process, because of creep reinforcements with a low creep coefficient b such as polyester, aramid, polyvinylalcohol are chosen.

Bridge abutment from reinforced soil can thus bring the following advantages – I Vaníček and M Vaníček (1997):

– In reconstructing older structural objects a significant role is played by time – the construction time. By using the reinforced earth abutment it is possible to significantly shorten the time of reconstruction, generally also of the new construction.
– With lower demands on the quality of the subsoil, the foundation of bridge abutments on piles in practice could be avoided.
– By using bridge precast units in the form of simple girders the demands on lateral deformations are moderate. By distributing the loads on a wider area at the bridge abutment from the reinforced soil, it is possible legitimately to expect that the difference in the settlement of abutments will be minimized. Moreover, the deformation can also be influenced by the type of soil used.
– Last but not least, a significant role is played by price – the most frequently asserted evaluation gives a saving of 12–15% – see e.g. Brady and Masterton (1991).

Examples of particular implementations of bridge abutments from reinforced soil are described for example by Tateyama and Murata (1994), or Tatsuoka et al. (1996) in application to Japanese railways, or Kirschner and Hermansen (1994), respectively by J Vaníček and M Vaníček (1999) for a case implemented in Denmark. In the first case in one place at the Nagoya site and in four places at the Amagasaki site, the GRS-RWs were constructed as bridge abutments, supporting a simple beam bridge girder through RC blocks placed on the crest of the wall near the wall face – see Fig. 6.57. The length of bridge girder was 13.2 m. Considering the importance of the structure, a relatively strong grid with a rupture strength of 60 kN.m^{-1} was used. In addition, high quality especially selected well-graded gravel was carefully compacted to a dry density as high as about 2200 kg.m^{-3}. And the facing was constructed using a larger amount of steel reinforcement than other ordinary portions of the GRS-RW. All applications were very closely monitored, using earth pressure cells, strain gauges, stress transducers, suction meters, differential settlement gauges, inclinometers and acceleration meters. Loading tests were divided into static load and the measurement of dynamic response. A static load of 300 kN (30 tons) which was the total weight of a loaded dump truck and a compaction tired roller, was applied on the girder immediately above the concrete block having a base area of 2 m by 10 m. This static load was equivalent to the design train load. From the settlement point of view it was proved that the settlement of the block was less than 0.1 mm and almost recoverable. Whereas the largest allowable vertical displacement of the track during the passing of the bullet train should be 1 mm or less according to the Japanese Railway Authority.

The success of this system in Japan is also demonstrated by the fact that Tatsuoka et al. (1996) have already referred to about 17 applications. Tatsuoka nevertheless states that the case represented in Fig. 6.57 is still the case with longest bridge girder so far supported by GRS (geosynthetic reinforced soil) bridge abutment with FHR (full height rigid) facings. To support a longer and heavier bridge girder, GRS bridge abutment should be stiffer than those constructed so far. As reinforcement works only after the surrounding soil extends sufficiently in the horizontal direction, it

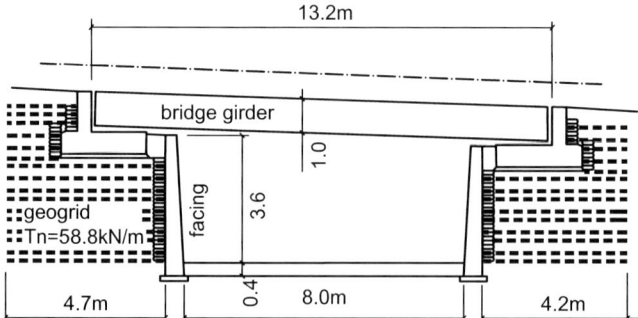

Fig. 6.57 Cross-section of geosynthetic-reinforced soil bridge abutments at Shinkansen Yard, Nagoya (according to Tateyama and Murata)

is very difficult to substantially increase the vertical stiffness of a reinforced soil mass against a relatively small vertical load even when reinforced with densely spaced long inextensible reinforcement. Therefore they recommended "Preloaded and prestressed GRS-RW" as mentioned earlier in Chapter 4 – Fig. 4.34.

In the second case Danish State Railways (DSB) opted for reinforced soil abutments, when a road bridge was built over the Nyborg-Fredericia main railway line on the island of Funen. Due to the subsoil for this bridge at Ullerslev being glacial clay, considerable settlements of the abutment had to be accommodated. The conventional pile foundation was ruled out for financial reasons, and instead, the DSB chose a flexible supporting structure of earth and geogrids. The road bridge, with a span of 15.5 m, consists of an 11 m wide steel structure resting on 8 m high abutments. Each abutment has to carry 2 000 kN of dead weight and 1 700 kN of traffic loading. A gravel-sand compacted to a Proctor density of 100% was employed as fill material for the abutments. The soil was reinforced with polyester geogrid type Fortrac 110/30-20 at a spacing of 0.5 m for each layer, which enabled the construction of an abutment with a frontal inclination of 81°, almost vertical, see Fig. 6.58, e.g. Kirschner and Hermansen (1994) or Alexiew (2007). Final facing is composed from 150 mm thick reinforced concrete panel.

The abutment was built between December 1991 and January 1992. The bridge itself was erected in August 1992. Since December 1991 the deformations of the abutment facings have been measured at 40 measuring points. The results show that almost all of the abutment deformations are caused by the subsoil. Up to the summer of 1993 the settlements amounted to max 40–50 mm, of which approx.

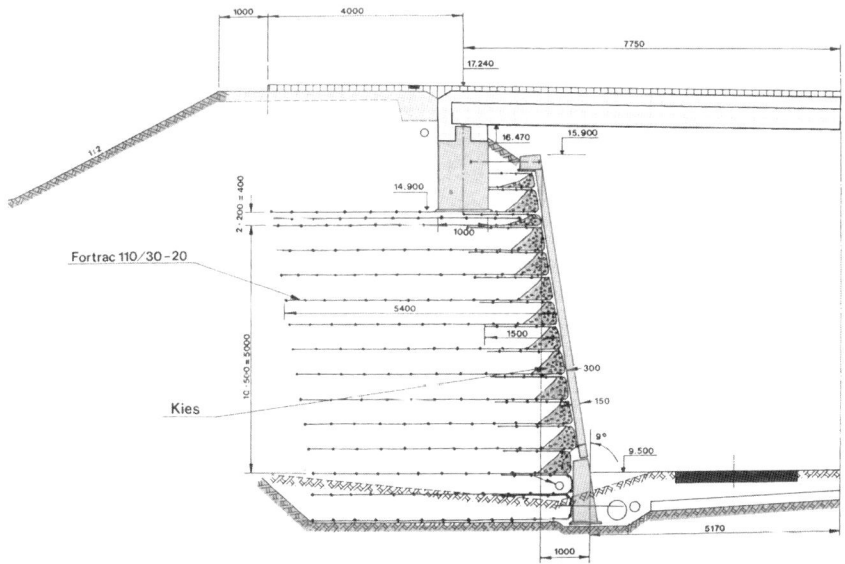

Fig. 6.58 Cross-section of the geosynthetic-reinforced soil bridge abutment at Ullerslev/Funen Denmark

10 mm occurred after the erection of the bridge superstructure. The reinforced soil abutments had themselves only settled approx. 2 mm. DBS's reaction to this project, in both technical and financial aspects, was positive.

Note: The combination of flexible facing during compaction and completion of rigid supplemented facing after the horizontal movement has nearly stopped can also help to solve one problem connected with stiff concrete bridge abutment. There is a tendency to lower vertical deformation faces due to heavy compaction of fill material. This heavy compaction can significantly increase earth pressure acting on the rigid wall, exceeding even pressure at rest – see e.g. Duncan et al. (1991). If not counting with this higher value, this negative impact can cause some problems, e.g. development of cracks in the wing-wall. On the contrary the flexible facing can help to eliminate this aspect before the additional stiff facing is added.

Geosynthetic reinforced soil bridge abutment is a very interesting alternative solution for bridge abutments. It is possible to expect applications also to bigger spans of bridges and also on greater loads. It will be necessary to overcome a certain conservative approach of not only designers but also of investors/owners – railway or motorway authorities. That is why in Japan, where support comes from the side of the Japanese railways themselves, it is so significant. Development can be expected towards Preloaded and prestressed GRS-RW. The possibility of an application to a lighter embankment made from expanded polystyrene blocks as bridge abutments, described in Section 6.2.2.4 in Fig. 6.27 has so far been used exclusively for temporary constructions.

6.4.3 Interaction of Fill with Different Underpasses

In this chapter there will be a very concise depiction of cases of interactions of the earth structure, most frequently the embankment, with the structures which go through this embankment and cross it. These structures include different underpasses both for the mobility of people and animals (eco-ducts). In the latter case it concerns at least partial interconnection of a natural area for animal mobility that is divided by the new structures. However, the most frequent case is various culverts for diverting surface waters. Different piping and pipelines etc., from those for drainage to those for gas, are distinguished by the smaller cross section.

In any case it concerns a different type of structure with different rigidity, which naturally leads to significant interaction. This interaction will in principle have a different character according to whether it concerns:

– Ductile structures such as e.g. polyethylene pipes, flexible culverts, corrugated metal conduits,
– Very rigid structures, either concrete or earthenware.

For the basic case of the interaction of piping with the embankment – see Fig. 6.59 it is possible to give consideration to the following – see e.g. Jesenák (1993). At the

6.4 Specific Cases of Interaction

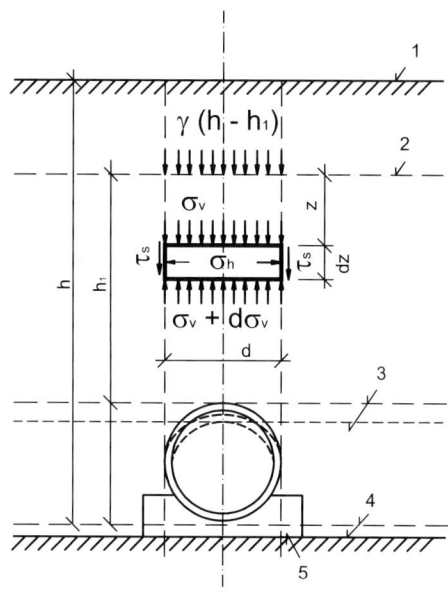

Fig. 6.59 Pipeline under fill.
1 – surface of fill; 2 – level with same deformation; 3 – critical level; 4 – stiff subsoil; 5 – stiff bed

level of the apex of the pipeline there can be defined the critical plane. If the apex of the pipeline settles, due to its rigidity, less than the critical plane – i.e. less than the surrounding soil – the embankment surcharges the pipeline above the value of geostatic pressure. In simplified form it is possible to imagine that this surcharge is transmitted by shear forces in the vertical planes parallel with the sides of the pipeline. So if the pipeline settles, due to the flexibility, more than the critical plane, certain unloading is occurring, because part of the geostatic pressure is transmitted by shear forces into the surrounding soil.

For a low embankment the shear forces reach up to the surface. For higher embankments the deformations of the soil are levelling at height h_1 over the apex of the pipeline and hence the shear stresses in vertical planes above this height are zero. That is why besides the rigidity of the pipeline it is necessary to distinguish the cases of low and high embankments. For high embankments in practice the non-uniform deformations on the surface will be negligible.

In case of a very rigid structure, the vertical loading therefore will be higher than the geostatic one – i.e. higher than $\gamma \times h$. It will grow with the rigidity ratio of this structure towards the rigidity of the surrounding soil. In principle therefore two solutions present themselves:

- The increase in rigidity of the surrounding soil, for example by the choice of material or through a greater compaction effort.
- The application of a more flexible element above this structure.

Beinbrech and Hohwiller (2000) in their pronouncement on the use of expanded polystyrene blocks also give the solution for the examined case. They specify that by arranging a deforming layer from EPS over the inflexible pipe a rearrangement of the lines of stress around the pipeline is achieved. The strains arising in the deformed layer mobilize the internal shear forces of the overlaying soil and in this way affecting the load transfer to the sides and relieve the pressure on the pipe. With reference to Vaslestad et al. (1994) they stated that the highest load on the pipe is considerably reduced by the presence of this compressible inclusion. Recommendation for the arrangement is specified in Fig. 6.60.

In the instance of a flexible structure there can exist yet more basic cases, for example the pipeline of the circular section can be produced from plastic materials which are brittle or from plastic materials which are more ductile, e.g. the pipeline from HDPE – high density polyethylene. The deformation of such a pipeline depends on the symmetry of loading. One of the basic precautions is therefore the increase of the rigidity of the surrounding soil at the sides of the pipeline. This can be achieved for smaller diameters by special compacting divided rollers when at the same time compaction is taking place along both sides of the pipeline. It is even possible and appropriate that the pipeline would after the compaction indicate side squeeze and vertical expansion. By the consequent compacting of the embankment the shape of the pipeline returns back to its original circular profile.

With big profiles, such as corrugated metal conduits it is necessary to pay great attention to compacting the soil circumferentially especially from the viewpoint of symmetry. The rigidity can be increased also by the application of reinforced soil. Just as the interaction for the rigid pipeline helped the deforming plate from EPS, the situation for the ductile pipeline would be improved by the application of the rigid plate. It is possible also to replace this one with the help of reinforced soil. The schematic picture for corrugated metal conduit is shown in Fig. 6.61. Detailed

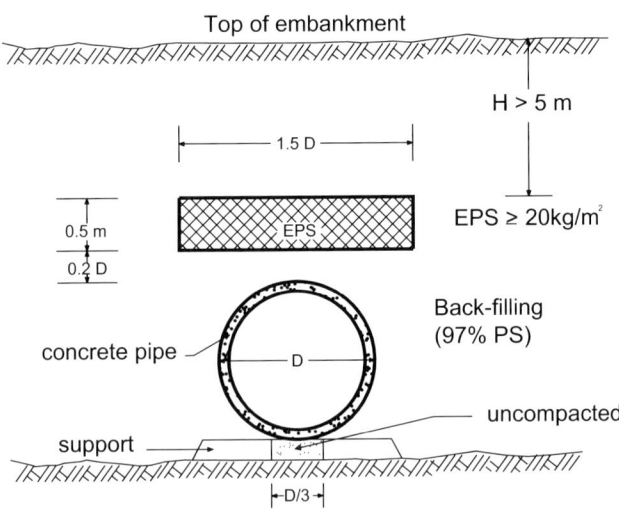

Fig. 6.60 Reduction of soil pressure on rigid pipeline using deforming layer over pipes (according to Beinbrech and Hohwiller)

6.4 Specific Cases of Interaction

Fig. 6.61 Schematic picture of reinforced soil around corrugated metal conduits to eliminate asymmetrical loading

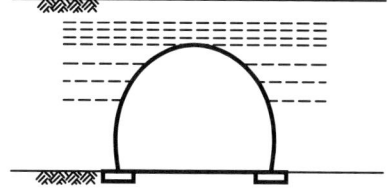

specifications for the design of different types of pipelines or corrugated metal conduits are given in special publications e.g. Selig (1985) or Moser (2001).

Sargand and Nasada (2000) describe the performance of large-diameter honeycomb-design HDPE pipe under a highway embankment. The wall configuration of this type of product consists of a series of inner circular tubes formed between two flat sheets to increase longitudinal stiffness. Pipe with diameter of 1.07 m was situated into an excavated trench with width of 1.6 m. Backfill around the pipe up to the shoulder level was river gravel, the upper part from compacted sand cover. The pipe was instrumented and soil pressure measured as function of highway embankment fill height – with maximum 15.9 m. Changes in soil pressure around the test pipe over time are shown in Fig. 6.62. Pressure at the crown is much lower than the geostatic pressure, and a small relaxation over time was observed. The vertical and horizontal deflections of the test pipe stabilized at -10% and $+3\%$, respectively. Nevertheless Sargand and Nasada stated, using some other results, that the performance is strongly dependent upon quality of compaction and quality of installation. The obtained results were also compared with numerical modelling, using both elastic solutions and finite element analysis. Both of these methods were partially successful in predicting the actual field pipe performance.

For all above mentioned cases numerical modelling is a very good tool for interaction simulation. Nevertheless practical results are very often strongly influenced by quality of performance on the site.

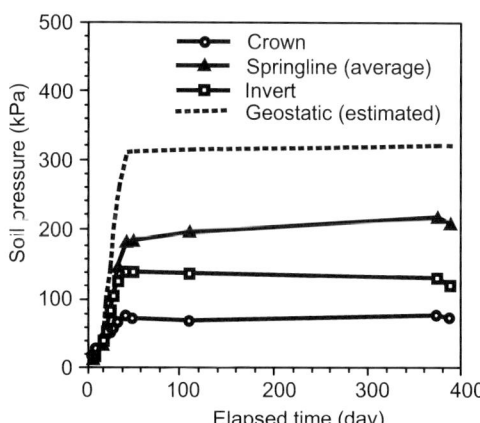

Fig. 6.62 Changes in soil pressure around the test pipe over time (according to Sargand and Masada)

6.5 Specificity for Roadways

In Fig. 6.63 a typical cross-section through the structure of a pavement is shown. Up to now most attention has been devoted to the foundation up to the formation level. There is a small exception to this when speaking about soil stabilization, because the soil stabilized with lime or cement can create not only subgrade but also capping layer or the subbase. The structure of the pavement can have a different arrangement, according to national recommendations e.g. Gschwendt et al. (2001) and he distinguishes four main structures:

– Asphalt roadways with non-cemented base, where different types of asphalt mixtures of the type AB (asphalt concrete) are used as well as aggregates covered by asphalt (OK) and dry-bound macadam (ŠV,ŠD), mechanically stabilized aggregates (MSK), or gravel-sand (ŠP) – see Fig. 6.64a;
– Asphalt roadways with a cemented base where under the asphalt layers are layers stabilized by cement such as aggregates stabilized by cement (KSC) or cement stabilization (SC) or mechanically stabilized soil (MSZ) or ballast chipping and gravel-sand – see Fig. 6.64b;
– Concrete roadways with concrete pavement processed in one or two layers (CB), or from continuously reinforced concrete (CBA); below this is usually porous concrete (MCB) or aggregates covered by asphalt and the subbase is composed from mechanically stabilized aggregates or from cement stabilization – see Fig. 6.64c;
– Composite roadways with a wearing course from asphalt situated on a concrete or reinforced concrete road base – Fig. 6.64d.

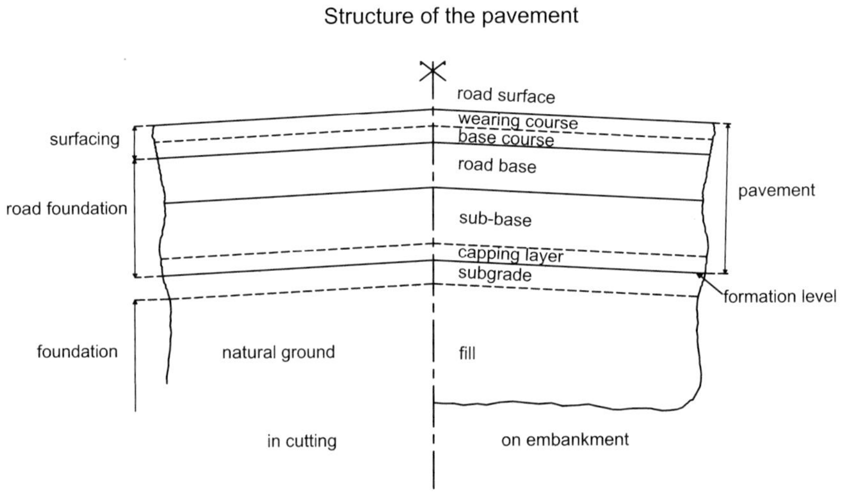

Fig. 6.63 Schematic cross-section of road structure on embankment and in cutting

6.5 Specificity for Roadways

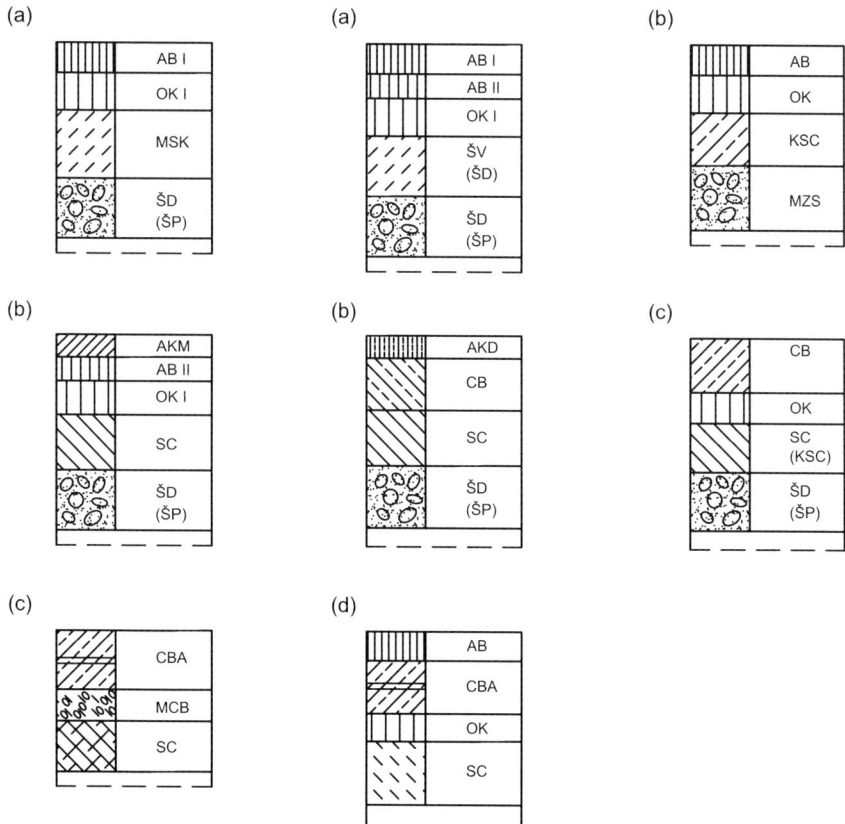

Fig. 6.64 Typical composition of road structure (according to Gschwendt)

Detailed composition of the structure of the pavement and the thickness of individual layers depends on the type of roadway and degree of transport loading. It is not the intention of this chapter to go into any detail on the pavement structure specification or specification of materials used there. From the geotechnical point of view the following aspects deserve specific attention:

– Transfer of loads to the surface of formation level or to the level of soft subsoil in relation to the bearing capacity and settlement. This problem is more sensitive when the ratio of the total load (dead load plus traffic load) to the dead load is a high one – which is the case with low embankments.
– Frequency of dynamic loading by traffic and the behaviour of subsoil layers under such loading.
– The behaviour and degradation of ballast, chippings under loading – crushing on individual contacts.
– Protection against freezing, which is strongly dependent on climatic conditions, having a direct negative impact mostly on the material of fill or natural ground, especially when the underground water level is close to the surface. This direct

impact is therefore very sensitive for low embankment and unpaved roads. For paved roads the thickness of the pavement is usually greater than is the depth where freezing can occur at what is about 0.8–1.0 m for Central European countries. Nevertheless it is necessary to check the resistivity to freezing-thawing cycles for upper layers – for ballast (chippings) and for stabilized soils.
- Service life of the pavement structure – potential risk of cracks developing on the pavement surface or the development of ruts on an asphalt surface.

The last point deserves more attention because cracks and ruts are the main reason for the defects observed on motorways and 1st class roads. Cracks on the pavement surface can develop not only as a result of differential settlement but also due to propagation of cracks from lower layers. Very sensitive to critical reflective cracking and their propagation into the surfacing are cement stabilized soils, which are used in road base courses. These cracks result from the hydration process, the drying up of the layer, or a temperature change. The higher the tensile strength is, the larger is the spacing between the cracks and, furthermore, the width of the cracks. Brandl (1999) describes the experiences gained during the construction of the Austrian-Hungarian motorway and some fundamental points are therefore described below. The following composition of pavement structure was used there – from the top:

- 40 mm asphalt concrete,
- 2 × 110 mm binder course,
- 360 mm – cement stabilized base course constructed by a mixed-in-place technique,
- 200 mm – unbound granular sub-base (frost blanket course).

Brandl specifies that commonly, cement stabilized base courses of highways under heavy traffic loads are constructed by mixing-in-plant, while a mixed-in-place technique was widely limited to sub-bases and subgrades until a few years ago. The main advantage of the mixed-in-plant technique is the high homogeneity of the mixture and a low sensitivity to construction weather. On the other hand, it is more expensive than the mixed-in-place technique (roughly about 30%). Moreover, the latter also impacts less on the environment due to the shorter transport distance of the construction materials. Nevertheless to reach the same technical quality results some recommendations are specified starting from the increase of the thickness of the cement-stabilized layer by at least 10% in relation to mixed-in-plant, additional stabilization of the topmost 30 mm of the granular sub-base – in one process together with the base and ending with a recommendation to increase volume of testing in order to detect local difficulties with in homogeneities or weak points over time. Final design of the cement stabilized base course consisted of:

- Sorting and homogenizing of the soil from the cuttings along the highway section;
- Placing of the soil (sandy gravel, locally silty) on the precompacted sub-base;

6.5 Specificity for Roadways

- Light compaction to close surface;
- Scarification of the prepared gravel surface with the teeth of a caterpillar;
- Adding of water (adjustment of water content);
- Distributing of cement with a spreader;
- Mixing-in-place with a rotovator;
- Compacting with a vibratory roller;
- Levelling with a laser grader;
- Final compaction pass with a vibratory roller – including primary micro-crack rolling;
- Sprinkling of bitumen emulsion as a protective coat against evaporation;
- Secondary micro-crack rolling 24 hours later.

Control testing was divided into three phases. *Suitability tests* comprised primarily the determination of the grain size curves, the standard and modified Proctor values, and the portion of weak, weathered grains. The maximum allowable grain size of the cement stabilization was limited to 63 mm in order to provide a homogeneous mixture and to minimize the demand on the rotovator. To achieve a sufficient matrix of the mixture, a portion of at least 35% (even better 40%) of grains smaller than 4 mm were required. The fines (lower than 0.063 mm) should vary between 8–15%. Furthermore, the unconfined compressive strength and freezing-thawing resistance were determined by cylindrical specimens with different cement and water contents. The allowable minimum of the unconfined compressive strength after seven days was $q_{7D} = 2.5$ MPa (single value); the mean value of three specimens should be $q_{7D} = 3.0$ MPa. The investigation of the frost resistance was necessary if the percentage of the fines (< 0.063 mm) exceeded 20% or in case of a high portion of weathered particles. The suitability tests resulted in required cement content within a range of 85–60 kg.m^{-3}. But soils which needed more than 125 kg.m^{-3} of cement (PC 275 H) were not used.

The *control tests* tests on the construction site referred to the following parameters:

- Mixing factor (at the beginning of the construction works);
- Soil quality and cement content (daily);
- Water content (continuously).

And finally the *acceptance tests* comprised the following data:

- Evenness (continuously);
- Required level and thickness of layer (daily);
- Degree of compaction (8 times per 5 000 m^2 if the nuclear gauge test was used);
- Unconfined compression strength (2 tests per 5 000 m^2) usually after 7 days.

The determination of the compressive strength after 7 days, as is usual for soil stabilization with cement, proved to be insufficient in several sections. The time period between mixing/compaction and acceptance testing was too long, especially during hot weather. Consequently, the 2-days and 3-days compressive strengths were also determined which facilitated a proper adaptation of the mixed-in-place technique to locally differing soil properties and weather conditions.

From the order of the operations in producing a stabilized layer it is noticeable that for elimination of reflective cracks there was a method selected referred to as relaxation rolling or micro-crack rolling with continuous compaction control. A slight over-compaction creates a fine meshed network of numerous surface-near thin cracks which do not propagate into the bituminous top layer. The formation of wide reflective cracks at larger spacing can be prevented by this slight over-compaction. Conventional micro-crack rolling is performed only by experience without any quantitative background. However, the continuous compaction control (CCC) – described in Chapter 2 – represents an essential improvement which is based on the measurement of the dynamic interaction between roller and (stabilised) soil. The operator of the roller is informed about this interaction, about soil response to the vibratory roller compaction by way of the so called dynamic compaction value. A proper relaxation rolling, which causes a specific re-loosening of the surface-near zone of the compacted stabilised layer, is then achieved by exceeding the maximum at response curve with one to three additional roller passes. This methodology has proved very suitable and also practical because an absolute, conventional value of the degree of compaction (e.g. modulus E_{v1} and/or E_{v2}; percentage of Proctor density) is not needed during this phase. Absolute values require a calibration of the dynamic compaction values by means of conventional tests. The measuring depth of dynamic continuous compaction control is significantly greater than that of load plate tests. Hence, weak zones beneath the topmost layer can also be detected. But if acceptance tests in this specific case disclosed that the unconfined compressive strength after two days exceeded $q = 5$ MPa, micro-crack rolling was intensively repeated on the third day and/or on the 4th day at the latest.

Brandl also discusses some other methods which can eliminate crack development such as:

- Reduction of the initial strength by adding less cement (but this can reduce freezing-thawing resistance),
- Cutting of notched joints into the cement stabilised layer (prolonged construction period and increased costs),
- Installation of a "bitumen-geotextile" on the top of the stabilised course as a reinforcing and sealing element within the pavement structure,
- Increase of the thickness of the bituminous structure,
- Concrete pavement instead of the bituminous structure with asphalt surfacing.

But especially promising are the below mentioned two possibilities:

- To perform the stabilisation with cement and bitumen, – especially adding foamed bitumen together with water during the soil milling and cement mixing process of the rotavator – which significantly reduces the brittleness of the cement stabilisation and the crack tendency.

- To perform the stabilisation with cement and plastics – adding small pieces of plastic sheeting, bottles, old tyres etc., which should not be too soft in relation to the stiffness of the cement-treated soil. The technology is a challenge for the recycling industry, but this needs additional research, especially field tests, before a large-scale application can be recommended.

6.5.1 Unpaved Roads

Unpaved roads represent the type of road that is most frequently used in Europe in the countryside as e.g. in forest roads or roads in mountainous areas. They also can be utilized as provisional roads and on many occasions as building site roads. Their outstanding feature is that they follow the surface of the existing ground, which is why their embankments or cuttings are very limited. Their construction is very straightforward, on flat terrain, even possibly a low embankment there is placed a layer of coarse gravel. For permanent unpaved roads the surface of gravel layer is filled by finer material and compacted – the final result is referred to as dry-bound macadam. Not only so that the surface is smooth, but also that rain water is more easily drained from the surface to the lateral drains. Dry-bound macadam should not be susceptible to freezing.

Most problems for this type of road are connected with soft subsoil, e.g. Gourc et al. (1983). The first condition of the design of such a structure is connected with the bearing capacity of this soft subsoil, mostly soft clay. Houlsby and Burd (1999) present an approximate calculation, based on bearing capacity theory. They assume that the problem of the design of an unpaved road over soft clay can be idealized as in Fig. 6.65a. Soft clay is characterized by an undrained strength s_u and granular fill with a thickness D by an angle of friction φ'. The loading is assumed to be uniform, of intensity q and acting over a width B. The uniform loading of the surface of the granular fill is assumed to spread at a load-spread angle β, so that the effective width of the loaded area on the clay is $B' = B + 2D \tan \beta$ – see Fig. 6.65b. Houlsby and Burd recommended an empirical expression to calculate load-spread angle β:

$$\beta = \varphi' - 10(1 + \ln(s_u/\gamma D)) \qquad (6.16)$$

To design an appropriate thickness of granular fill D, uniform load on the surface of soft clay should be compared with its bearing capacity expressed by the well known equation:

$$\sigma = (2 + \pi) s_u + \gamma D \qquad (6.17)$$

Note 1: Bearing capacity of soft clay for this type of loading can be calculated also as a classical stability problem along the slip surface, as was mentioned in Chapter 5. Numerical methods can be used as well.

Fig. 6.65 Unpaved road (according to Houlsby and Burd). (**a**) Schematic cross-section. (**b**) Assumptions for bearing capacity calculation

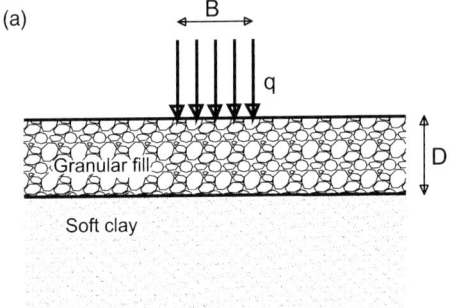

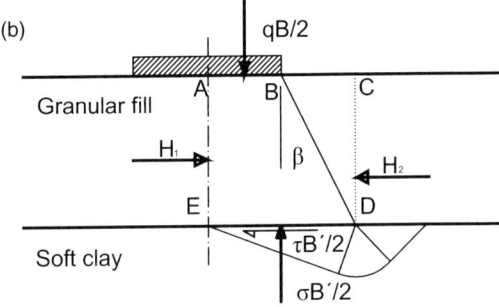

Note 2: There is high potential that the soft clay will penetrate into the pores of granular fill. Therefore the application of some separator as e.g. geosynthetics is very useful in this case.

In most cases the final solution leads to the application of reinforced unpaved road when a layer of reinforcement is included between the clay and granular fill. This reinforcement is able to fulfil the following functions:

– It can act as a separator between coarse fill and fine clay, as was just mentioned in Note 2;
– It can support the shear stress from the granular fill, causing tension in the reinforcement, and transferring purely vertical loading to the clay below, thus increasing the bearing capacity of the clay;
– At large rut depths the combination of tension in the reinforcement with its curvature results in additional bearing capacity (the so called "membrane effect").

To solve the problem of bearing capacity two approaches can be applied. If there are no strict demands from the beginning, the third function – membrane effect – can be taken into account. This effect was described e.g. by Giroud and Noiry (1981) and can be schematically illustrated by Fig. 6.66. Due to surface loading the reinforcement is curved concave upwards below the loaded area. The tensioned membrane effect depends both on the curvature of the reinforcement and on the tension. Because both are linearly related to the rut depth, it is calculated that the tensioned

Fig. 6.66 Principle of tensioned membrane effect

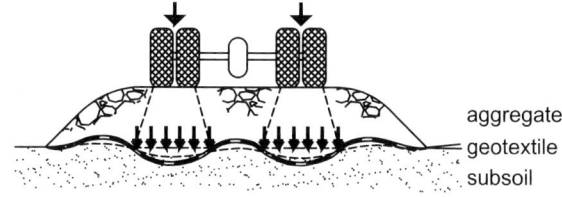

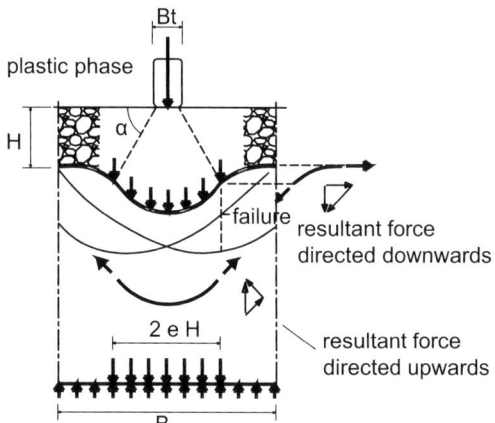

membrane effect varies approximately with the square of the rut depth, and linearly with reinforcement stiffness. Calculation assumed that the reinforcement is firmly anchored at some point outside the loaded area. After mobilization of this membrane effect and the refilling of rut no another significant rut development is expected.

Houlsby and Burd concentrate on a stricter case, where the quality of surface has to be good from the beginning and therefore the second function is taken into account. The following equation is recommended for the capacity of a two-layer system:

$$Q = (2+\pi)s_u(B'/B) = (2+\pi)s_u(1+2(D/B))\tan\beta \qquad (6.18)$$

For this case Houlsby and Jewell (1990) propose design charts which allow the necessary depth of granular fill, and the required reinforcement tension, to be determined.

Beyond this approach other calculation methods can also be applied, such as using a stability calculation where enhanced bearing capacity from the reinforcement can be calculated as an additional horizontal force, whose amplitude depends on maximum tensile strength and allowable elongation. The finite element approach to the second function is described by Burd and Brocklehurst (1990). Because the main aim of the reinforcement in this case is to transfer horizontal stresses that develop in the fill layer when a vertical load is applied, the reinforcement system can be composed not only from single sheet reinforcement but from reinforced sand cushions or from geocells – see Fig. 6.67.

Most problems are connected with the prediction of settlement. The first step usually consists of the calculation of settlement for total static loading. No special problems exist there compared to the settlement of classical fill. Because the ratio of traffic loading to the total loading is very significant in this case, more problems are associated with prognosis of additional settlement as a function of the number and intensity of traffic loads. Due to the difficulties in solving this problem in most cases this prognosis is done on the basis of previous experience. On the basis of an extensive set of test data Delmas et al. (1986) propose a hyperbolic curve to simulate this function. Sawicki and Swidzinski (1990) use a numerical model to observe the deformation below traffic loading as a function of number of loading cycles N. A numerical algorithm is based on their own compaction theory – on the model of compaction of sand under cycling loading. They suppose flexible pavement composed from bituminous concrete under which is a layer of standardized

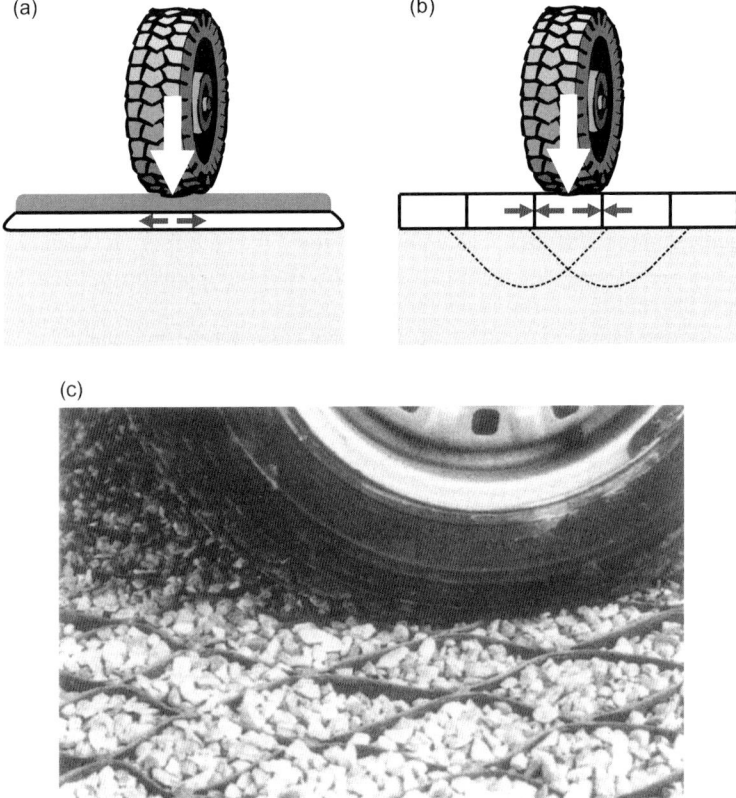

Fig. 6.67 Reinforced unpaved roads. (**a**) Cushion reinforcement. (**b**) Application of geocells. (**c**) Photo of geocell road

sand followed by natural sand. For all layers elastic constants are defined. The compaction Φ is measured as an irreversible porosity change due to the cyclic shearing, and described by the following constitutive equation:

$$d\Phi/dN = D_1 J \exp(-D_2 \Phi) \qquad (6.19)$$

Where D_1 and D_2 are coefficients determined in laboratory for natural sand with the help of a cyclic simple shear apparatus. J is the second invariant of the strain amplitude deviator. In spite of the elastic constants used for modelling some conclusions are worth mentioning:

- Pavement settlements of the order of a few centimetres due to traffic can be expected for dense and medium dense sands in subsoil.
- Subsoil settlements are strongly influenced by the compaction characteristics of a sand D_1 and D_2.
- The compaction occurs mainly in the subsoil layer adjacent to the pavement. The depth of this layer is approximately 0.5–1.0 m. In order to prevent excessive settlements due to traffic this layer should be precompacted down to 1 m depth.
- The influence of the pavement rigidity (Young's modulus and thickness) on settlements is practically negligible.

A last note should be directed to the quality of top granular fill. This fill should be well compacted to prevent the bearing capacity being determined by the strength of the granular fill. This type of failure deserves attention when loading from tyres of lorries are close to the edge of the road. Therefore from the practical point of view it is useful to define a certain boundary beyond which it is forbidden to pass.

6.5.2 Interaction of Construction Layer with Subsoil

Generally all what was mentioned for the unpaved roads is true for the paved roads. The pavement in this case is there to redistribute the loading from the passing vehicles to the subsoil in such a way that the surface of paved road would not be affected by any deformation as it is the case for unpaved roads. Therefore if everything would be fine there would not be surface deformations on the paved roads. Unfortunately it is not the case on large amount of roads. This is because of different reasons and to mention some – the traffic loads were increased above estimated levels, or the design life expired, or the subsoil properties were poorer or became poorer with time. So when the road surface is not properly fulfilling its purpose, it is necessary to repair the road. Of course, there are different ways of road repair, from simple resurfacing to the complete rebuilt, depending on the level of damage.

In the case when the problem of unsatisfactory road does not lie in the subsoil, the resurfacing option should be sufficient. In this situation, we can extend the life of the new asphalt overlay by inclusion of reinforcing geogrid, which is specially adapted for this application. This special geogrid should be impregnated by asphalt, in order to have a good bond with the asphalt road layers and do not act on contrary as separator. The inclusion of asphalt reinforcing geogrid can lower the requirement for the overlay thickness, or prolong the maintenance cycle or combination of both.

If the reason for road surface deficiency is the weak subsoil (usually defined as $E_{def2} < 45$ MPa), the most rigorous solution is to rebuild the whole road from its base, where the treatment of the subsoil is necessary. This subsoil treatment can be done by several ways as mentioned in Chapter 4 – soil stabilization and soil reinforcement or more classical way – by soil replacement or by addition of higher quality soil layer. But again there is an option just for resurfacing and when this is done with the use of geogrids for asphalt reinforcement, it can be seen as an economic solution. But then the authors should be very careful with the selection of the right geogrid, as we know that some perform very well under such circumstances and some do not perform as reinforcement at all and make the condition even worse than without that reinforcing product.

One good example of road resurfacing using reinforcing geogrid, even when the rest of the pavement structure was significantly affected by the cracking, is from Ochtrup, Germany. This road was significantly affected by heavy loading from lorries and went through resurfacing in 1996, when it was found during the works (after milling off the cracked wearing course) that the whole pavement structure was cracked (Fig. 6.68a). Due to both financial and time reasons, reinforcing geogrid for asphalt reinforcement HaTelit 30/13 was used, see Elsing (2006). The condition of the road surface in 2003 was still without any cracks (Fig. 6.68b).

Fig. 6.68 Photos of road before treatment with asphalt reinforcement and after 7 years of use (according to Elsing)

(b)

Fig. 6.68 (continued)

6.5.3 Parking Place

The number of parking lots grows sharply with the development of road transport. It concerns above all mass parking lots near supermarkets, airports, big firms as a part of the system P+R within the radius of big cities, parking lots near big petrol stations, and near customs facilities etc. It is possible in a simplified way to summarize their specificities into the following points:

a) The need for a relatively large area,
b) The need for a good drainage system for conducting rainwater away,
c) The need to control of the quality of water intake from the parking lot surface.

The first point in many cases requires the partial positioning of the parking lot surface into the cutting and partially into the embankment. This fact can significantly influence differential settlement of the parking lot surface area. Sensitivity regarding deformation is strongly tied to point (b) and is moreover intensified by the demand for the minimum transfer of materials, which is connected with the effort to use local soils and which may be also less appropriate ones. The effort to use the area as much as possible leads to the proposal for steep slopes of the cutting and the embankment.

The surface of the area of the parking lot can be dealt within a relatively short time including the proposed inclination for conducting rainwater away. A sensitive question is that of estimating not only consolidation settlement for a part in the embankment but also deformations caused by traffic loading, which can lead to a change of designed and implemented inclination ratios. Here again it is valid that the traffic loading is relatively high regarding the overall loading. The drainage system itself causes a significant interaction with the surrounding soil. It involves:

- Different deformations of ductile and rigid drainage pipes,
- Different deformations of the surface of the backfill of ditches and the surrounding natural ground,
- Different deformations between the vertical manhole and the surrounding ground, which if close to the manhole cannot be easily compacted,
- The question whether to construct the drainage network in the area of the embankment after creating the formation level or to built it at the same time with the construction of the embankment.

It is obvious that all these interactions can additionally influence gradient ratios and thus also the functionality of the area from the viewpoint of its dewatering. It is also clear that the forecast of the impact of these interactions is not simple. In many regards the resulting effect will be influenced by the quality of work carried out.

The third point deals with the environmental protection. There are two basic problems here:

- The size of the parking lot area (and ultimately also the adjacent structural objects) influences the subsoil, because this surface area prevents direct infiltration of rainwater – this matter is connected with the long-term influence on the groundwater level. That is why it is more appropriate to check the quality of water before enabling infiltration into the subsoil. Therefore, some sort of water infiltration is usually recommended, e.g. by infiltration wells.
- The second one deals with the quality of surface water intake, when together with the rainwater there also gets into the drainage network wastes from the whole surface area. Here this does not concern only solid particles but moreover various foreshots – of oil and petrol from parked cars. Thus it is necessary before releasing waters from the drainage network or before its infiltration to test their quality, for example in sealed sedimentary tanks and in line with any need to carry out additional protective measures. This point also deals with wastes from larger surface areas of highways and therefore it will be also mentioned in Section 6.8.2.

6.6 Specificity for Railways

Over recent decades the problems connected with railway systems can be divided into three main areas:

- Maintenance of existing old lines,
- Track reconstruction for trains with higher speed,
- Construction of new tracks for high speed trains.

6.6 Specificity for Railways

Some basic principles and specificities for railways on old lines will be briefly shown.

Very often these old lines were constructed more than a century ago. In those days the compaction equipment was rather different from the present day one and also frequency of traffic, loading and speeds were much lower than is prevailing today. A typical cross-section is shown on Fig. 6.69. A basic division is between substructure, (where the top layer termed subgrade, sub-ballast, base course, structural layer is situated on the earth plain – on natural soil for cuttings or on fill for embankments) and superstructure (composed from rails, sleepers, track bed, ballast). The rails connected to wooden sleepers are laid down on the track bed, composed mostly from ballast, good quality aggregates – crushed rock. However, in some countries with shortage of rock, as in The Netherlands, sand was used as well. The ballast layer is a critical link in the load absorption and transfer chain. It must distribute the loads to prevent the track being displaced and it must also provide good drainage. The thickness of ballast is usually close to 0.5 m with a minimum of 0.3 m. Thickness of ballast plus subgrade is not only a function of the quality of the earth plain in order to be able to carry loading from the moving trains, but also a function of climatic conditions, in order to protect the soils in foundation against freezing (if they are composed from soils sensitive to freezing). The direct contact of ballast with the earth plain is possible where the quality of foundation is good – with relatively high deformation modulus, with coarse grains, and it is not so sensitive to freezing. The main reason for subgrade is to protect the migration of fine particles from foundation to ballast. Subgrade has to fulfil filtration criteria not only on the contact with the foundation but also with the ballast. Another reason for subgrade is to guarantee a continuous increase of stiffness in any multi-layered system. The final cross-section should meet the following criteria:

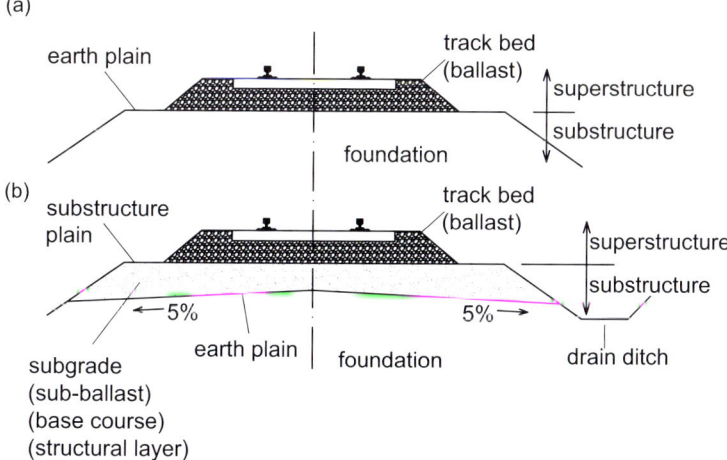

Fig. 6.69 Typical cross-sections for single-track line. (**a**) Direct contact between ballast and foundation. (**b**) With a subgrade situated between ballast and foundation

- To guarantee that the foundation will be sufficiently safe against main limit states – the ultimate limit state (e.g. as a result of insufficient bearing capacity) and the limit state of serviceability (deformations).
- To limit infiltration of water from superstructure to foundation, to limit worsening of the quality of soil there. This can be achieved by very good compaction of the foundation, by a drainage layer situated above with an inclination roughly about 5% to the side ditch. In this case more than 80% of infiltrated rainfall is drained by the side ditch.
- To protect migration of fine particles from the foundation to the pores of the above coarse layer, very often in the form of mud, as a result of traffic loading and water found there. This effect is often called pumping. Restriction of the migration of fine particles upwards can be achieved by applying a filtration layer. This filtration layer can be a classical granular filter or a geotextile filter (e.g. Tyc and Vaníček (1981), Faure and Imbert (1996)). Migration can be also stopped with the help of a geomembrane.

Nevertheless with time it is necessary to count with some rail track deterioration. Moving trains apply dynamic loads and vibrations which cause not only wear in the rails and sleepers but also deformation and ageing of the ballast and other layers below it. Let us discuss only ballast itself. Even if from the beginning ballast contains a practically negligible amount of fine particles (usually 3% of finer than 0.02 mm are allowed), this amount is increasing as a result of:

- Crushing process during compaction, dynamic loading and as a result of freezing – thawing effect.
- Falling small particles from the surface of rail track to the pores of ballast, especially in areas where freight trains are carrying some fine substances such as e.g. flying ash, coal etc.

An increasing number of fine particles have a negative influence on freezing-thawing behaviour and drainage; even a pumping effect can start there. All these negative influences affect deformation of rail track. The solution to these problems is usually tamping with an addition of new ballast or in more problematic cases replacement of the ballast.

6.6.1 Track Reconstruction for Higher Speed

Old tracks were not constructed for nowadays speeds, higher thrust loads and higher intensity of traffic operation. Maintenance, improvement or reconstruction of old tracks was mostly connected with superstructure, not with foundation. Therefore the reconstruction of old tracks for higher speed is directed to an improvement of the earth plain and the construction of subgrade. The requirement to improve the quality of substructure is supported by field measurement. Sunaga (2001) presented

6.6 Specificity for Railways

the results of measurement of vertical acceleration as well as vertical and horizontal displacements measured by an accelerometer situated on the top of subgrade and wheel loads measured by strain gauges installed at the rail web, Fig. 6.70. Measurement was performed on conventional line but with high-speed express trains. Embankment height was not higher than 3 m. Fig. 6.71 shows an example of the relation between vertical subgrade displacement and train speed. Results indicate that the vertical displacement is:

– Increasing with train speed,
– Strongly depends on quality of ground.

Note that CPT test value (q_c) was 260 kPa for soft ground and 15 MPa for hard ground.

But not only train speed but also quality of track have an influence on maximum dynamic track forces which are used for track dimensioning. German Railway Authorities – see e.g. Brandl (2001e) give a simple correlation between the maximum dynamic force Q_{max} and the mean force (quasi-static axle load) Q_{mean}:

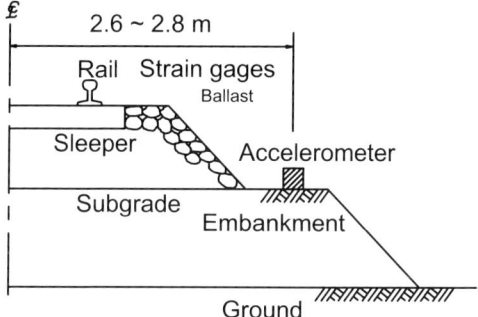

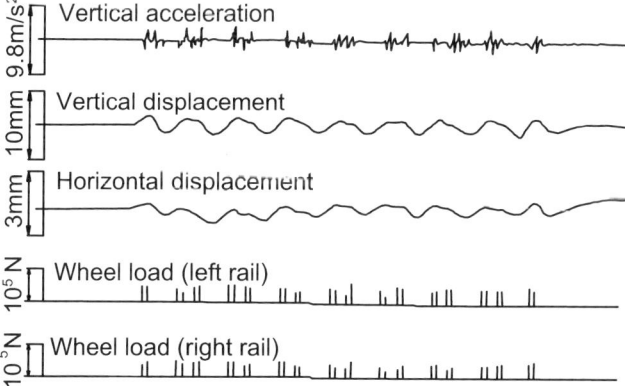

Fig. 6.70 Results of vibration measurement on conventional line (according to Sunaga)

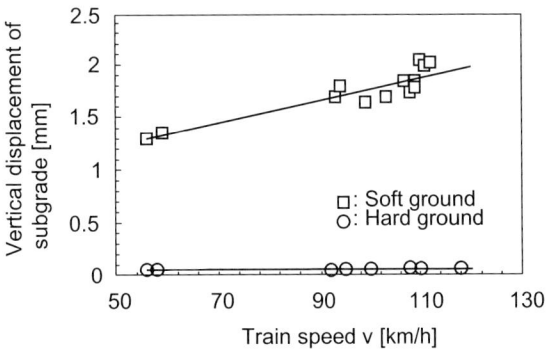

Fig. 6.71 Relation between train speed and vertical displacement of subgrade on conventional line (according to Sunaga)

$$Q_{max} = Q_{mean}(1 + t.n.\varphi) \tag{6.20}$$

where

t describes the statistic safety $t : t = 1$ for 67%; $t = 3$ for 99.7%.
$n = 0.1$–0.3, depending on track quality and type of railway line.
$n = 0.1$ for excellent track; 0.2 for average track and $n = 0.3$ for poor track.

$\varphi = 1$ for train speeds of $v < 60$ km.h^{-1}
$\varphi = 1 + 0.5(v - 60)/190$ v for passenger trains and $60 < v < 300$ km.h^{-1}
$\varphi = 1 + 0.5(v - 60)/80$ v for freight trains and $60 < v < 140$ km.h^{-1}

Figure 6.72 illustrates the increase of additional loading above the quasi-static axle load as a function of train speed and track quality. The last recommendation for coefficient n expressing the influence of track quality is based on statistical mean standard deviations observed for different lines in Germany: $n = 0.15$ for high-speed lines, main lines; $n = 0.2$ for secondary main lines and $n = 0.25$ for other tracks. Rail design is based on $t = 3$, hence on a statistical probability of 99.7% because a rupture of rails would have severe consequences. On the other hand, base pressures on the track bed (ballast $\rightarrow$ subgrade) are calculated with $t = 1$ (i.e. statistic safety

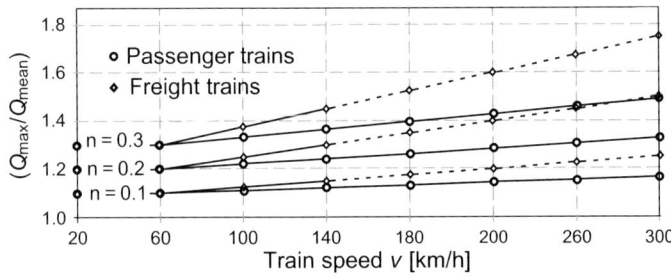

Fig. 6.72 Maximum dynamic track forces Q_{max} versus speed of passenger and freight trains; n – track quality factor (according to Brandl)

6.6 Specificity for Railways

of 67%) because this part involves less risk. This corresponds with the usual scatter of geotechnical parameters.

Briefly stated, the possibility to increase the speed of trains on existing railway lines needs an improvement in the quality of track in order to decrease additional loading above quasi-static axle load and limit vertical displacement which can not only cause technical damage but is also uncomfortable to passengers.

Therefore hereafter the experience with the reconstruction of the main railway lines in the Czech Republic will be described below. In 1994 the reconstruction of 4 main corridors started with a total length of 1400 km (15% of the whole railway network) with a projected completion in 2010 for proposed speed of 160 km.h^{-1} (with possibility to go up to 200 km.h^{-1}). These 4 corridors, which are part of the Trans-European Networks include:

- North – southeast direction – boundary with Germany (Dresden, Berlin) – Děčín – Praha – Brno – Břeclav – boundary with Austria (Vienna) and Slovakia (Bratislava);
- North – south direction – boundary with Poland – Ostrava – Přerov – Břeclav – boundary with Austria (Vienna) and Slovakia (Bratislava);
- West – east direction – boundary with Germany (Nurnberg) – Plzeň – Praha – Olomouc – Ostrava – connection to Poland (Krakow) and Slovakia (Žilina);
- North – south direction – boundary with Germany (Dresden, Berlin) – Děčín – Praha – České Budějovice – boundary with Austria (Linz, Vienna).

The reconstruction was performed in accordance with the amended directive of the Czech Railway ČD S4 – Substructure. The main demand for new substructure is connected with modulus of deformation on the earth plane $E_0 = 30$ MPa and on the substructure plain $E_{pl} = 50$ MPa, measured by static load test with plate diameter of 0.3 m. This directive also precisely defined the application of geosynthetics in subgrade or on the contact of subgrade with the foundation. There are four main possibilities of application:

- As a separation, filtration and drainage element – main requirement - nonwoven geotextile with unit weight ≥ 250 g.m^{-2};
- As a reinforcing element – main requirement – woven geotextile – applied on the contact of subgrade with foundation – minimal ultimate tensile strength of 30 kN.m^{-1} and strength of 10 kN.m^{-1} at 3% elongation, for geogrids it is recommended to be applied in subgrade layer to guarantee good interlock of coarse grains in mesh of geogrid; Fig. 6.73;
- As a reinforcing element fulfilling also the additional separation and filtration functions – main requirement – geocomposite fulfilling all previous criteria;
- As a sealing element – geomembrane protecting infiltration of rainfall into the foundation, has to be protected from both sides by nonwoven geotextiles or by fine sand. After a positive experience with the first application of geomembrane in 1957, this possibility was for the first time implemented into ČD directive in 1960 – see Tyc (2000). Also Imbert et al. (1996) describe 20 years of positive

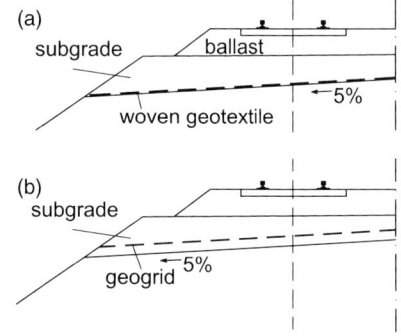

Fig. 6.73 Application of reinforcing geosynthetics. (**a**) Woven geotextile applied on the contact can also fulfil the filtration role. (**b**) Geogrids inside of subgrade layer

experience with thick bituminous geomembrane, preventing not only upward migration of fines but also protecting the foundation against rainfall. But geomembrane can guarantee success only in the case that the maximum underground water table is at least 1–2 m below this geomembrane. Geomembrane can also protect underground water against chemical contamination – e.g. from foreshots from railcars. In this case the geomembrane should be resistant to these chemicals. More details will be given in Section 6.8.

A brief summary of improvement in earth plain and subgrade is as follows (see for more detail e.g. Minář (2000)).

6.6.1.1 Earth Plain

If not fulfilling basic demands on the modulus of deformation $E_0 \geq 30$ MPa the following methods were applied:

- Replacement of top layer was applied only in the case of high plastic clays with very soft consistency – used material for mechanical stabilisation – crushed aggregates grading 0–250 mm or grading 0–63 from recycled ballast;
- Stabilization with lime, cement and mixture of lime with cement – results were very good – measured modulus much higher than demanded – often in the range of 85 MPa–240 MPa, even high plasticity clays ($I_p > 21$) of soft consistency (with initial modulus $E_0 \leq 10$ MPa) had after lime stabilisation $E_0 = 43$ MPa;
- In many cases it was too much, so that after that subgrade with thickness of 0.2 m was only symbolic – but protection against freezing after that was poor;
- Reinforcement with geosynthetics – geogrids were applied below the layer of mechanical stabilisation – but this tendency was stopped because S4 directive later on allowed the application of geosynthetic reinforcement on the contact of subgrade with earth plain if measured modulus was higher than 60% of the demanded one and it is 18 MPa;
- When the foundation consisted of sandy soils, reinforcement was applied directly on the earth plain, reinforcing contact between sand and ballast – thickness of ballast was increased up to 0.4–0.45 m, but subgrade layer was omitted;

6.6 Specificity for Railways

– For a sandy foundation with a close level of underground water, mechanical stabilisation was used with gravel grading 0–63 mm and the surface was closed by emulsified asphalt;
– In some cases we learned that our predecessors also used some sort of improvement – the top of the earth plain was constructed from hand-packed rockfill – discussion whether to leave it or remove it was inconclusive or ambiguous.

6.6.1.2 Subgrade

– Demanded value $E_{pl} \geq 50$ MPa was not reached when applied sandy gravel grading 0–63 mm ($C_u \geq 18$);
– Demanded values were guaranteed when crusher-run material grading 0–32 or 0–22 mm was used (E_{pl} in average $= 92$ MPa) or when recycled material from ballast, grading 0–32 resp. 32–63 was used (E_{pl} in average $= 94$ MPa);
– When subgrade was applied on the top of stabilised soil the final modulus was a little bit lower but still fulfilling the demanded value;
– Reinforcement with geosynthetics was also successful, woven geotextiles were applied directly on the contact where geogrids were inside subgrade layer;
– Minimum thickness should be at least 0.25 m.

6.6.1.3 Brief Summary

Briefly summarizing, three main possibilities proved to be successful in fulfilling the main required criteria:

– To increase the thickness of coarse materials, or partly substitute soft material by this coarse layer;
– To apply soil reinforcement by geosynthetics – this reinforcement can decrease the thickness of coarse material at least by 0.1 m; because of relatively strict demands on coarse material, which can become expensive, more precise comparison of these two approaches deserve attention – on Fig. 6.74 such comparison is presented from Swiss guidance SVG;
– To apply soil stabilisation process.

Any selection between these three main possibilities will depend not only on the quality of the earth plain, but also on the possibility to change the rail level, on comparative economics and also on machinery used. In recent times the technological method of track reconstruction has been able to perform all individual steps in one sequence without removing the rail grid. So this means removal of ballast, its cleaning, crushing, grading and back deposition into the structural layer, its compaction (by vibrating plate compactor), and in addition also reinforcement or soil stabilisation application, see photo Fig. 6.75.
Note: Before using recycled ballast it is compulsory to check whether or not it is not contaminated.

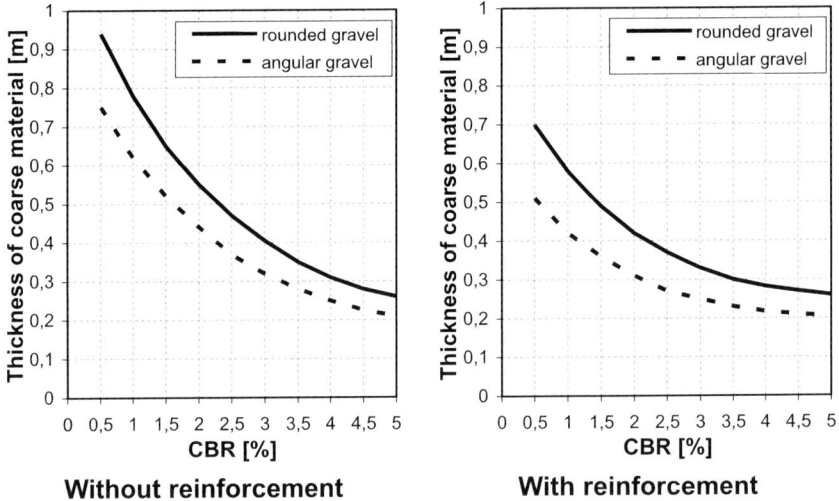

| Without reinforcement | With reinforcement |

Fig. 6.74 Diagram showing the possibility of subgrade reduction due to reinforcement by geosynthetics (according to SVG)

Fig. 6.75 Photo of reconstruction technology without removing rail grid; application of reinforcing geosynthetics

Final discussion can be connected with the main criteria, which are based on modulus of deformation measured by static load test with plate diameter of 0.3 m. The main points of the discussion can be:

– Small diameter is giving information only about the surface part of plain;
– Test is performed relatively quickly – for saturated clay the obtained modulus (practically, under undrained conditions) has to be lowered by a reduction factor;

- Results obtained at the bottom of the narrow trial pit during geotechnical investigation are usually higher than that measured after on excavated plain.

But on the other hand the test is relatively simple, quick; many of such tests can be performed before and after reconstruction and the results can be compared with each other.

6.6.2 Construction of New Tracks for High Speed Trains

At the end of the last century some countries began the construction of a new generation of the railway networks, as in Japan with "Bullet – Shinkansen" trains, TGV lines in France and Belgium, ICE trains in Germany, the AVE lines in Spain, the Channel Tunnel Rail Link in Great Britain. These are proposed in many other countries. Proposed speed is most often between 250 and 300 km.h^{-1}. These new networks aimed at passenger traffic offer some advantages compared to road transport or air travel. This is primarily due to the direct connection between centres of large cities and greater comfort for passengers.

Nevertheless, the high speeds proposed bring new problems to geotechnical engineering.

Fujii (1977) when describing geotechnical aspects of construction of the Shinkansen express train system directed his attention to problems of substructure (embankments and cuts) not on superstructure or on interaction of superstructure with substructure. He admitted some relatively big problems for the first line – Tokyo – Shin-Osaka (Tokaido Shinkansen):

- Frequent occurrence of mud pumping;
- Failure of slope faces due to rain;
- Settlement of banks in areas of poor ground;
- Settlement of bridge approach embankments.

It is worthwhile to mention two problems. To prevent problem of mud pumping the following standards were specified for 300 mm thick top layer of foundation: $d_{10} \leq 0.074$ mm, $d_{70} \leq 0.42$ mm, $w_L \leq 35$, $I_P \leq 9$. The allowable amount of settlement was decided from the track maintenance limits and economic considerations. The annual permissible amount of settlement was specified by 100 mm in areas of straight track and 50 mm in areas with horizontal or vertical curvature. This value started from the assumption that a track correction is made once a week to correct the permissible settlement 2 mm – mainly from the viewpoint of passenger comfort. The amount of residual settlement which continues after opening of the line has been set at a maximum value of 500 mm in 10 years. Where these limits were not attainable, rail track was situated on viaducts. The Tokaido Shinkansen line was opened in October 1964. The experiences gained there were immediately applied on the other such lines not only in Japan but also in other parts of the world. That is confirmation of the fact that geotechnical engineering is very quickly able to absorb not only new theoretical findings but pioneering experience gained on completed structures.

After two or three decades concentration is much more directed to the interaction of individual parts of the railway track which need close co-operation between railway and geotechnical engineering. Lord (1999) mentions two main problems with which we have to deal:

- High speeds demand tighter tolerances on track alignment for the purposes of safety and passenger comfort. Consequently much more attention needs to be paid to the design of the railway track formation and subgrade, particularly in the light of higher axle loadings.
- Where the new railway line is carried on shallow embankment across soft alluvial deposits, the passage of a high speed train, with a live load as much as, if not more than the dead load of the embankment, has been predicted to generate pressure waves in the alluvium travelling ahead of, and at the same speed as the train; the impact of such waves on the ground beneath the embankment, particularly when passing trains meet, has been the subject of considerable speculation.

Before going deeper into each problem a different track model can be mentioned. A brief overview is given by e.g. Kopf and Adam (2005). In railway engineering the track structure is prevalently modelled as an infinite, flexible beam resting on continuous springs loaded by a single non-moving force. This static model is improved by a dynamic model where a single load moves at constant speed generating dynamic effects. Furthermore, the springs are replaced by spring-dashpot elements in order to take into account damping effects. Kopf and Adam are proposing a solution, which is equivalent to the previous one but more easily worked out – non-moving load on a moving beam. Nevertheless Brandl (2004) specifies that more realistic results can be obtained from the multi-layered elastic half-space model. This facilitates, for instance, the assessment of the effects of different stiffness of the particular layers, of bond between the interfaces, and of an out-of-roundness of wheels. In Fig. 6.76a is a sketch of a flexible beam resting on continuous spring-dashpot elements loaded by a moving force. In Fig. 6.76b is a similar model for two coupled beams with a moving load. Brandl or Kopf and Adam specify a differential equation for the deflection line and all input data, which were also tested experimentally.

From these models the following three aspects are very important:

- The deflection line can specify vertical displacement – and compare obtained results with allowable displacement on one side or in return to modify the track composition;
- The deflection line can specify the pressure wave and emphasises the needs to study the impact of these waves on the ground beneath the embankment;
- Need to perform more laboratory and in situ tests to check the relevance of theoretical models to reality or to study not only single moving load but also multiple moving loads simulating individual axle loading.

The first point is strongly connected with the optimum design of a multi-layered rail track which involves a gradual increase in stiffness from the bottom to the top – from foundation (natural ground or fill) via subgrade and ballast layers to the rail

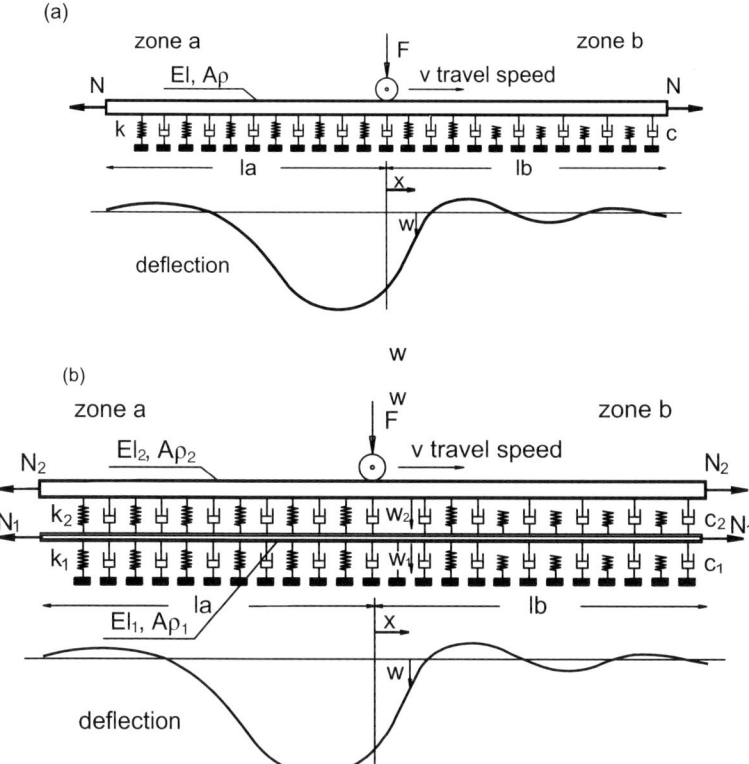

Fig. 6.76 Analytical rail track model (according to Brandl). (**a**) Sketch of a flexible beam resting on continuous spring-dashpot elements loaded by a moving force. (**b**) Two coupled beams with a moving load

grid. The achievable stiffness of each of the respective following layers depends not only on their material properties, thickness, and compaction degree but also on the stiffness of the underlying layer. Therefore, the adverse effect of soft subsoil cannot be compensated by excessive compaction of the next layer. In the end, the entire system should exhibit an overall stiffness that allows minimum rail displacements on the one hand, but limits maximum rail displacement on the other. Brandl states that up to now experience has shown that the elastic rail deflection Δz under a passing wheel load of about 200 kN should be in the range of

$1.0 \text{ mm} \leq \Delta z \geq 2.2 \text{ mm}$ for train speed of $v \leq 160 \text{ km.h}^{-1}$
$1.5 \text{ mm} \leq \Delta z \geq 2.0 \text{ mm}$ for high-speed trains

In Fig. 6.77 are two cross sections of ballasted rail track for high speed railway lines. Both these cross sections referred to as conventional and modified, were used for monitoring the scope of the construction of the high-speed railway line between Vienna and Salzburg – Kopf and Adam (2005). In the modified cross-section the unbound sub-ballast (stabilized layer) was substituted by a thin asphalt layer. For

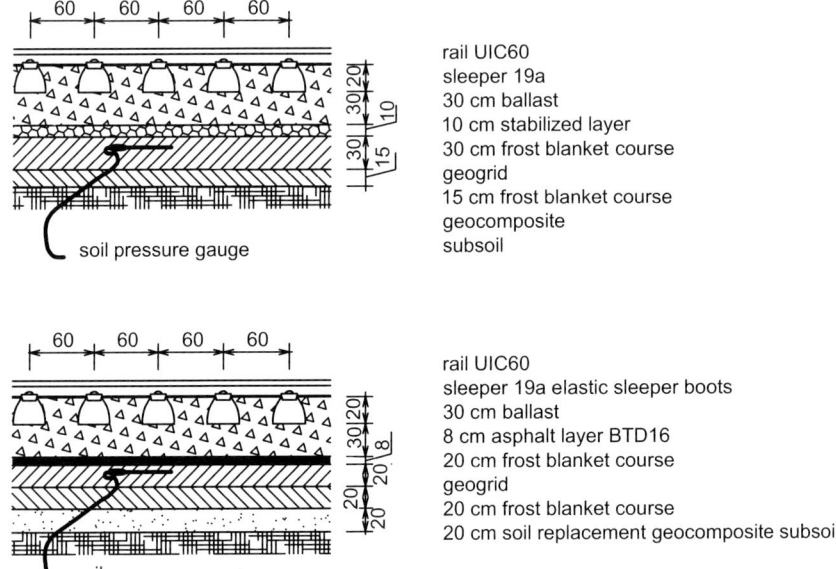

Fig. 6.77 Two cross sections of ballasted rail track for high speed railway lines (according to Kopf and Adam)

the modified structure the sleepers rest on synthetic pads in order to guarantee the minimum required track deflection under a train wheel. To achieve uniformity in stiffness of the individual layers the Continuous Compaction Control (CCC) system can play a very significant role.

More authors described the problem connected with the stress wave propagation for the foundation of high speed train lines on soft subsoil, e.g. Madshus and Kaynia (1999), Adolfsson et al. (1999). A precise summary is given by Madshus (2001). Soft soils pose challenges both for the static performance of the embankment with respect to bearing capacity and settlements, and for dynamic performance. Broadly speaking, when counting with the limit states of foundation, dynamic effect should be taken into account. Moving axle loads themselves only lead to quasi-static deformations in the track-embankment-soil-system. At higher speeds, the axle-load-induced-deformations will gain dynamic amplification, and reach excessive values as a certain speed, called "critical speed" is approached. Such high deformations may pose a threat regarding train de-railing, rail fatigue, distortion of the ballast, excessive settlements in the embankment and ground materials and even degradation of the bearing capacity of the supporting soil. It can also lead to unacceptably high vibrations in buildings along the track. The critical speed phenomenon is thus a match both in speed and wave length between the train and the site. For those cases where the bogie spacing of the train does match the natural wave, the dynamic amplification around critical speed will be particularly high. Provision of a stronger, stiffer track bed is one way of limiting deflections and increasing the critical speed at which the rising wave propagates. Under conditions of soft subsoil the design leads to paved (non-ballasted) rail tracks. Madshus gives two proposals coming from Norway and Holland, see Fig. 6.78.

One of the practical examples of how to solve the problem of soft subsoil has been stated by Lord (1999) and O'Riordan (2004). The Channel Tunnel Rail Link crosses the Rainham Marshes, a 7 km long section where soft alluvial deposits 3–9 m thick overlay river gravels. Alignment constraints restricted the maximum embankment height to about 2 m and hence there was a strong probability that a rising wave would propagate ahead of the high speed train. As a robust solution was required for the track bed, it was decided to construct the entire length on a concrete slab cast on a shallow embankment. The slab is designed in units of 80 m in length to accommodate thermal movement of the rails, supported on piles about 10 m long driven into the gravel at 10 m centres. The horizontal braking forces of the train are designed to be absorbed by two rows of barrettes in the centre of the slab which provide longitudinal stiffness and are very efficient in transferring these horizontal loads.

The construction costs for non-ballasted paved rail tracks are higher than for rail tracks with ballast. The economic advantage of paved systems lies in reduced maintenance costs. On the other side the non-ballasted, paved rail tracks can suffer only very small total and differential settlements because relevelling is extremely costly.

Brandl states that the allowable residual settlement should be:

$s \leq 60\,\text{mm}$ for train speed of $v \leq 160\,\text{km.h}^{-1}$

$s \leq 30\,\text{mm}\ (s \leq 15\,\text{mm})$ for high-speed trains (for $v \geq 300\,\text{km.h}^{-1}$)

And allowable differential settlements between the rails should be:

$$\Delta s_{transversal} \leq 2-3\,\text{mm}$$

To decrease sensitiveness to the differential settlements new sophisticated trains are proposed by railway engineers.

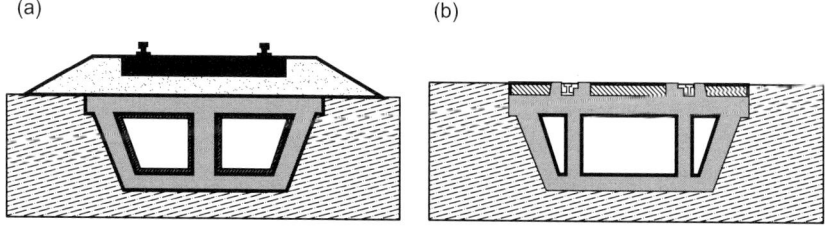

Fig. 6.78 Two proposals of countermeasures based on longitudinal concrete box girder (according to Madshus). (**a**) Using ballast on top (NGI proposal). (**b**) Completely without ballast (Holland Railconsult proposal)

Fig. 6.79 Photos of testing equipment for high speed train track in Geotechnical Laboratory Madrid

Due to the very complicated problems especially connected with vibration and due to different types of subsoil, in many countries they started with experimental modelling of these processes as was also mentioned before. Measurements of actual response of railway embankments during high speed train passage is essential for monitoring its conditions, for planning countermeasures and for validation of nu-

(c)

Fig. 6.79 (continued)

merical models. On Fig. 6.79 there is a measuring centre in the Laboratorio de Geotecnia of CEDEX Madrid where they are able to perform the test in the scale 1:1.

6.7 Specificity for Airfields

The development of civil aviation has undergone unprecedented development since the second half of the last century. There has been not only a growth in the intensity of air traffic but at the same time there has been a sharp increase in the total size and mass of aircraft themselves. If around the year 1960 total mass tended to be about 100–200 tons, in the year 2000 the maximum mass of a passenger aircraft approached 500 tons and there are even heavier ones planned. This brings about higher demands on the quality of the structural system of landing runway, including demands on the quality of subsoil.

On the one hand, together with this growth, new lands for airfields are sought, not only with regard to the growth in their numbers but also on the other hand with regard to the growing demand for longer runways. Obtaining such land close to big cities is very difficult, not only for environmental reasons but also because such land is simply not available. Therefore there comes to the fore in the development of new international airfields reliance on places with very difficult geotechnical conditions. From the viewpoint of the unfavourable quality of subsoil it is possible to state three basic conditions:

- The construction of airfields on very soft foundation soils;
- The construction of airfields on artificial islands, where part of the runway or also the whole airfield is situated on embankments put into place on the seashore or in the sea bay;
- The construction of airfields in places which were contaminated by previous activities, for example on the sites of previous military airfields.

Next there will be given some applications especially from the recent past and then a summary of some experience gained with the implementation of such airfields.

6.7.1 Practical Examples

The extension of runway No. 1 at the *Marseilles/Marignane International Airport* was carried out from 1974 to 1976 by means of fill materials on the Mediterranean Sea bed – Simon and Perfetti (1983). Soft clay having very poor properties is located less than 5 m below sea level. The thickness of the compressible layer varies, with the prolonged runway area, from 2.5 m at the beginning of extension to 6 m at the end. Total volume of the fill material, which was applied between peripheral dykes, reached 1.6 mil.m^3. An inner section was divided into 3 parts, 2 of them under the axis of runway and taxiway using nonwoven geotextile to separate compressible clay layer from first finer layer (0–4 mm) of fill (3 × 0.3 m), Fig. 6.80. The rest of the fill was composed from rocky material, grading 0–250 mm, deposited in regular layers of 0.5 m thickness from floating pontoons. The settlement of compressible layer was roughly 25% of its thickness (1.3 m). The insertion of geotextiles prevented in particular the penetration of fill material into the clay layer. This penetration in part without geotextiles was evaluated at 0.7 m. The reduction of amount of volume of fill material was estimated to be 5–10%. But the main advantage was the reduction of mud ridges as a result of extrusion of soft material due to non uniform loading by deposited fill, leading to the reduction of differential settlement.

New Bangkok International Airport is a typical example of construction of runways on soft clays. This new airport, which is scheduled to open in 2006, called also Suvarnabhumi Airport, is situated at Nong Ngu Hao in Samut Prakan Province, 30 km east of downtown Bangkok. This new airport situated on an area of 3220 hectares can accommodate 45 million passengers and 3 million tons of cargo annually. Bangkok and its surroundings are situated on a flood plain underlain with soft marine clay deposits known as "soft Bangkok clay". This clay poses particular problems for construction because of its low strength, highly compressible behaviour and low permeability. Consolidation by preloading was applied there which utilized geodrains – prefabricated vertical drains with the application of overburden pressure (i.e. preload). Some experiences with the application of geodrains there are described in more details by Bergado et al. (2002). The degree of consolidation estimated from the pore-pressure dissipation measurements agreed with those obtained from settlement measurements. The full-scale study made there confirmed that the

6.7 Specificity for Airfields

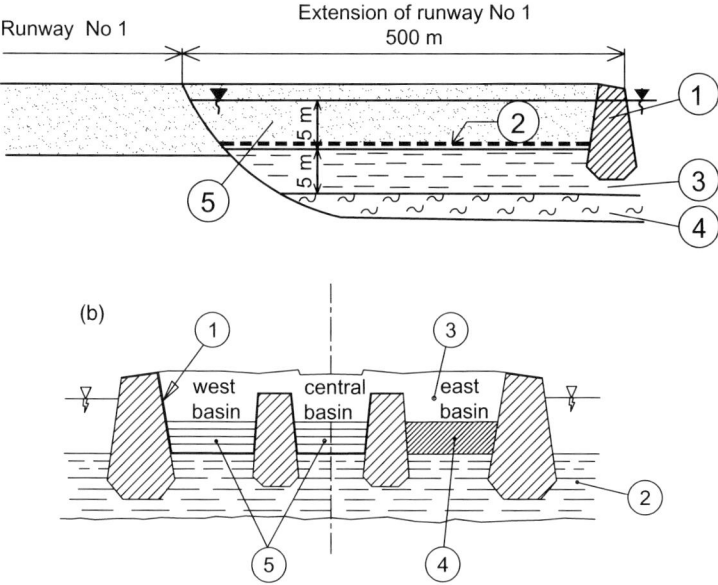

Fig. 6.80 Schematic section through runway extension in Marseilles (according to Simon and Perfetti). (**a**) Section along runway axis. 1 – dyke; 2 – geotextile; 3 – soft clay; 4 – marly substratum; 5 – embankment. (**b**) Cross-section. 1 – geotextile; 2 – soft clay; 3 – ordinary fill; 4 – fill layers poured from floating pontoon; 5 – fill performed with grab bucket

magnitudes of consolidation settlements increased with the corresponding decrease of PVD spacing at a particular time period.

Hong Kong International Airport has been built on a largely man-made island at Chek Lap Kok and was opened in July 1998. The airport is a phased development and as such opened with a capacity of 35 m passengers and 3 Mt of cargo annually and will ultimately have a capacity of 87 m passengers and 9 Mt of cargo, Plant (1999). The total proposed airport area is 1248 hectares of land, of which 310 hectares of the total area is two islands – Chek Lap Kok (302 ha) and Lam Chau (8 ha). The average water depth prior to reclamation was typically 5 m. The airport platform was situated at level +6.5 PD (Principal Datum), when the mean sea level at site is +1.3 PD with a typical tidal range of 2 m. The site of the reclamation is underlain by a succession of deposits comprising a very soft marine clay that overlies a mixed alluvial, a marine sequence of clays, silts, sands and gravels. The average thickness of the recent soft marine mud was 8 m. All this mud was dredged (69 Mcm – million cubic metres) and taken to a marine dump some 40 km away. The total fill requirement was 197 Mcm, of which 108 Mcm was obtained from the granitic islands of Chek Lap Kok and Lam Chau, 6.6 Mcm from the levelling of the two Brothers Islands to the east of the airport and 7.3 Mcm from other contracts.

The remainder was sand brought from marine borrow areas within Hong Kong waters. A total of 1.7 km of vertical seawalls were built using precast concrete blocks and 13 km of sloping seawalls were constructed using rock products from the quarrying of Chek Lap Kok. Limited areas of the reclamation have been treated generally either by surcharging or vibro compaction to improve the creep settlement of the fills. As the result of compressible mud replacement, settlement of a substantial magnitude occurred in the early days of the reclamation works. Therefore main attention was devoted to the residual settlement in the individual parts to estimate the expected differential settlement which will have the main influence on the quality of runways.

Also other airfields such as Changi Airport in Singapore or Incheon International Airport in South Korea were partly constructed on natural ground and partly on fill situated on the sea floor. But in these cases more different methods of soft ground improvement were applied.

Ground improvement methods, which were applied at Changi, included soil replacement, preloading, surcharge with vertical sand drains (displacement and non-displacement), and surcharge with flexible prefabricated geodrains, and dynamic compaction for hydraulic sand-fill, Yang and Chiam (1999).

In the case of Incheon International Airport the vertical drain methods were applied for their economical efficiency and the capability of the construction included sand drains, plastic board drain, pack drain, and sand compaction. These methods were evaluated along with the preloading method – Chae et al. (2003).

Kansai International Airport (KIA) in Japan was opened in September 1994. The airport was built about 5 km offshore at the south shore of Osaka Bay by reclaiming 511 ha of island (4.37 km long and 1.25 km wide) in about 18 meters of water depth in the sea, see Akai and Tanaka (1999).The geological profile of the seabed at the Airport site is slightly inclined toward offshore. The uppermost alluvial (Holocene) layer is the normally consolidated clay layer, the thickness of which is about 20 m. The underlain Pleistocene deposits are over-consolidated clay layers with OCR of about 1.3 and sand and gravel thin layers. These layers alternately deposit to a total thickness of several hundreds meters. The surface elevation of the Airport is specified to be not lower than HHWL (CDL+3.2 m), 50 years after opening the airport, Matsui et al. (2003). The total amount of settlement of the alluvial and Pleistocene layers was estimated at 11.5 m after 50 years, including the compression of the fill material of 0.5 m. As a result, soil reclamation of about 33 m in thickness and 180 million m^3 in volume was required to form the island. Holocene clay layer was improved mainly by sand drain method in order to promote settlement and to increase soil strength at an early stage. This was followed by construction of an 11 km long seawall around the island. Then reclamation inside the seawall was carried out.

In 1999 Kansai International Airport launched its second phase construction to complement its current facilities with an additional 4000 m runway. The construction is even more complicated, depth of water is about 2 m higher, reclamation area 545 ha, volume of fill 250 Mm^3, seawall length 13 km, Furudoi (2005). Between these two artificial islands is a gap, decreasing the influence of new fill on the

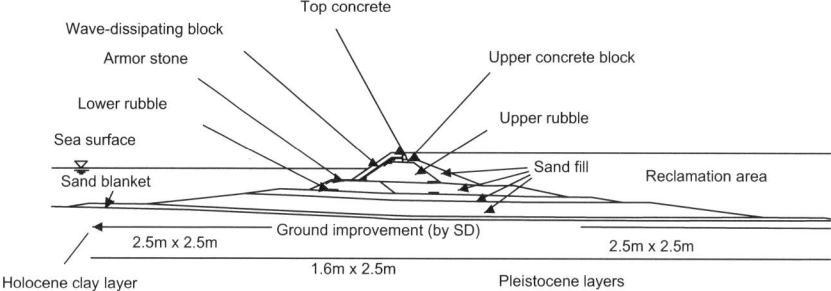

Fig. 6.81 Kansai International Airport; typical cross-section (according to Matsui et al.)

stresses under the fill of the first phase. Construction of the second phase started with an application of sand blanket on the sea floor followed by sand drains. Great attention was devoted to the seawall, especially from the wave erosion point of view. A typical cross section is shown in Fig. 6.81. During the construction of the seawall the increase of strength of Holocene clay was checked to prevent loss of stability (bearing capacity of subsoil). After that the filling of the inner area started. The reclamation work consists of three stages: soil dumping by hopper barges, soil heaping by reclaimer barges and multi-layered construction by bulldozers and rollers – see Fig. 6.82. While the settlement of Holocene clay layer will be completed during the reclamation work by ground improvement, Pleistocene clay layers will continue to settle after the airport opens. This large and long-term settlement was reflected in the setting of thickness of reclamation land. Furthermore, to minimize the differential settlement, which will occur after the completion of the airport island and may cause damage to airport facilities, reclamation work was conducted as evenly and uniformly as possible. The second phase construction project is now in progress for opening of the new runway in 2007.

Positive experiences with the construction of Kansai International Airport initiated some others, such as Central Japan International Airport in Nagoya – Mitarai et al. (2003), or Kobe Airport, both constructed on artificial islands.

6.7.2 Experience Gained from the Construction of Airfields in Challenging Geotechnical Conditions

Practical examples, especially from the construction of airfields on embankments implemented on the sea bottom, show what progress has been reached in the construction of earth structures in the last decades. There is not only involved their demanding nature with regard to the technology of execution and to the huge amount of earth work carried out, but also especially the ability to forecast the behaviour of such a big earth structure, especially with regard for sensitivity to the evenness of the runway surface.

Fig. 6.82 Photos of second phase construction of Kansai International Airport

6.7 Specificity for Airfields

The sea bottom is in the majority of cases covered by very young clayey sediments. Their compressibility is high and shear strength is low. Solutions direct us to:

- The extraction of these clays and their replacement by a new embankment which in practice come into consideration if they are found at a low depth and their thickness is about a metre;
- Their gradual strengthening by the created dune, when in the majority of cases vertical drains are used for their faster consolidation;
- The overlap of the filtration geotextile to prevent the stirring up of these clays with deposited material. The experience from the construction of the airfield KIA shows, that the even pouring of sands on this layer can also significantly prevent this stirring up;
- The extraction of these clayey materials and their redeposition after admixing the stabilizing agent (cement) – see Mitarai et al. (2003) on the application at the airfield in Nagoya. Tsuchida and Kang (2003) state an additional alternative, when the slurry of dredged soil is mixed with cement and air foam, or cement and expanded polystyrol beads with a diameter of 1–3 mm. The adjusted wet density of lightweight soils is usually $1000 \, kg.m^{-3}$ above sea level and $1100–1200 \, kg.m^{-3}$ below sea level. The conventional strength ranges from 200 kPa to 400 kPa. One of the practical examples was applied to Tokyo International Airport (Haneda Airport).

The next technical problem is connected with the construction of objects around perimeters such as seawalls and dykes. In principle this deals with the construction of a relatively high embankment on extremely low- bearing capacity subsoil. This leads on the one hand to the construction of vertical drains. Along with their help, if it is necessary not only to speed up the consolidation but also to increase the strength of subsoil, priority is given to sand vertical drains, especially to displacement sand drains. The actual surplus loading must be undertaken slowly and uniformly in order to monitor the development of pore pressures together with evidence of the growth of drained strength and with the help of this information, to manage the speed of the construction process. In principle, this involves the application of the observation method to the proposal. Also, in some cases smaller lentils of sandier material which are drained by vertical drains contribute to faster consolidation. Generally they thus contribute to greater permeability in a horizontal direction and to a significant speeding up of the consolidation. This phenomenon has also significance for the anisotropy of deposited clayey sediments – (a higher horizontal permeability than the permeability in a vertical direction significantly speeds up the consolidation and the draining away from pores of clayey sediments and thus decreases their porosity).

Objects along the artificial island perimeter, near the sea level, must also fulfil an anti-erosion defending function. On the outside face there are deposited wave dissipating blocks. Moreover, the granular composition has to prevent the washout of finer particles deposited on the inside face of the seawall or from the internal space, where there are most frequently deposited sands extracted from the sea bottom. The tidal levels have to taken into consideration when judging the granular composition of the deposited materials.

The filling of the internal space between seawalls and dykes must be done evenly and closer to the surface there are deposited coarser and less compressible materials. At Changi airport dynamic compaction was used extensively for the densification of hydraulic fill of between 4 and 8 m thicknesses. The compaction above the sea level is done in the classical way, by compacting in layers and significant attention is paid to control of the quality of this compaction. Estimation of the settlement over time is approximately in metres and for the second phase of the airfield KIA construction the already measured settlement exceeded 10 m. The prediction uses the 1D theory of consolidation. In applying the methods of speeding up the consolidation its primary component will happen within a framework of completion for earth works, which is approximately 4–6 years. Great attention is therefore paid to the component of secondary consolidation, the creep component of deformation after the dissipation of pore pressures as a consequence of the surplus load. Furudoi (2005) specifies more closely the calculation models. Their continuous verification depends on the measuring of the settlement in situ. In accordance with development over time (both predicted and also measured) it is possible consequently also to carry through a forecast into the future and to compare the results with the limit values. At the same time the installed monitoring units will help to separate the development of the deformation for individual layers from the subsoil ones across soft clayey sediments to the actual fill. Plant (1999) mentions the criteria of ICAO – International Civil Aviation Organisation for runways, which are as follows: "The transition from one slope to another should be accomplished by a curved surface with a rate of change not exceeding 0.1% per 30 m" (minimum radius of curvature of 30 000 m) and "Undulations or appreciable changes in slopes located close together along a runway, should be avoided. The distance between the points of intersection of two successive curves shall not be less than the sum of the absolute numerical values of the corresponding slope changes multiplied by 30 000 m or 45 m which ever is the greater". ICAO gives guidance on the application of these requirements and in particular notes that "The operation of aircraft and different settlement of surface foundations will eventually lead to an increases in surface irregularities. Small deviations in the above tolerances will not seriously hamper aircraft operations. In general, isolated irregularities of the order of 25–30 mm over a distance of 45 m are tolerable".

The expected settlement has had a significant impact on airfield pavements. In the case of Hong Kong Airport, concrete pavement was designed on an area of original island where there were no problems with differential settlement. The main reason is that generally high quality concrete pavements are preferred for busy airports as they resist deterioration from fuel spillage, do not deform under channelled traffic and have a durable surface resulting in relatively low maintenance and long life. Nevertheless flexible asphalt pavements were selected for the runways and taxiways where they are situated on the reclaimed land undergoing settlement.

Yang and Chiam (1999) give more detail for aircraft pavement design on reclaimed terrain at Changi Airport. Pavements for both runways I and II are of the flexible type consisting of a 750 mm thick granite stone base course supported on a

6.7 Specificity for Airfields

(a)

(b)

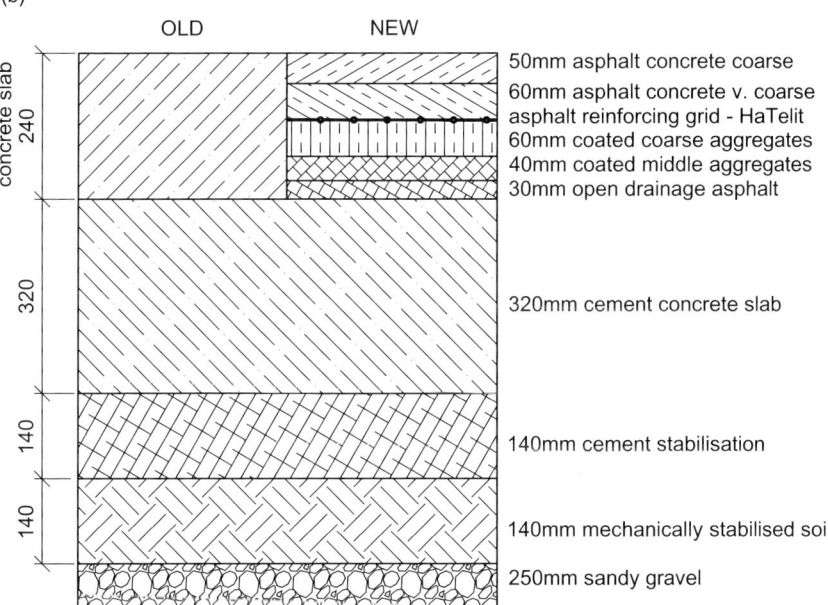

Fig. 6.83 Asphalt reinforcement at Prague's airport Ruzyně. (**a**) Photo from grid installation. (**b**) Cross-section of the old pavement and the repaired one

prepared earth subgrade compacted to 98% modified AASHTO maximum dry density. The base course was overlain with a 150 mm thick asphalt concrete surface. A system of subsoil drains has been installed on both edges of the structure pavement to lower and control the groundwater table. Pavements for the aircraft parking bays are of a concrete type, consisting of a 360 mm thick concrete slab (flexural strength 4.15 MN.mm^{-2}) and a 150 mm thick base slab (flexural strength 2.5 MN.mm^{-2}). The two slabs were designed as direct overlay of partially bonded concrete.

Generally great attention is devoted to the prevention of crack propagation. Propagation of cracks can be limited not only for reconstructed pavement but also for new ones by applying a geosynthetic reinforcement layer integrated into the top layer. The authors proposed this solution not only for reconstruction of the military airfield in Čáslav, but also for the reconstruction of the main international airport in Prague – Ruzyně – see Fig. 6.83.

6.8 Relation to Environments

When there is a new construction of transport infrastructure, there is legal requirement in most of the world countries to prepare the Environment Impact Assessment, in order to determine how the structure to be built will impact the environment. In here we can already distinguish what structures are having positive and which have negative impact. Hereafter we will discuss three main earth structures that are affecting the environment, structures being built using waste or by-products, structures that are separating surface water from underground water and noise protection bunds. The reader can easily recognise that the first one can have a negative impact and the other two have a positive impact on our environment. Therefore for the utilisation of waste materials in earth structures it is necessary to investigate the possibility of contaminant spreading.

6.8.1 Impact of Waste Material Used in Earth Structures on Environment

With the increasing volume of waste being produced every day, we are obliged to next generations to do something about it, to start to behave more sustainable. And in here we will try to indicate that such possibilities exist and we as geotechnical engineers can help to promote them. On one hand the efforts to minimize the volume of wastes in general, production wastes, construction wastes, wastes related to building reconstruction or demolition and on the other hand to maximize reuse of the secondary resources – are important elements of the sustainable construction. The waste materials that can be potentially used in construction of embankments are those that are being produced in large volumes and have similar mechanical properties as soils, e.g. Havukainen (1983), Tsonis et al. (1983).

The embankments of transport engineering projects offer a huge opportunity to store and to use large volumes of waste materials, such as waste from mining activity, the coal-burning residues, extracted material from underground construction, demolition of old structures etc. In comparison with the standard construction material, where the main focus is on limit states design, the use of waste materials also introduces an environmental aspect, such as leakage of contamination from the used waste materials.

6.8 Relation to Environments 371

The advantage of using the "waste" large volume materials in the embankment construction is that it is not necessary to store it in other special purpose built storages like landfills, tailing dams and etc.

This can be of benefit also for the selection of the road/railway alignment design, which does not have to try to get the balance of volumes of embankments and cuttings to equilibrium, but can be designed, with higher volumes of embankments.

As the environmental issue is the most important one for this type of structure and the other design issues are covered elsewhere for general earth structures, in this chapter we will concentrate on the modelling of contaminant spreading for the use in risk assessment of the environmental impacts of the embankments built fully or partially from waste materials.

The use of the waste materials mentioned above in the infrastructure embankments can be arranged in different configurations. But they should not be arranged like a landfill type of construction, as that type would require a permit and conditions that are the same as for regular landfills under the relevant not only European directives. The other environmental constraints by European directives are on leachates, from which is necessary to distinguish between substances on so called "List 1" and "List 2". The substances on "List 1" cannot be allowed (no leachate at all) into the "Controlled waters" – sources of potable water. Therefore the infrastructure embankments in water sensitive areas cannot be built using the waste materials that are leaching dangerous substances.

The specification of the properties of waste materials has to be divided into mechanical-physical properties as for ordinary soils and on chemical properties. From the contaminant transport point of view we need to gather the information about the waste leachate composition and parameters for contaminant transport modelling. The leachate composition is done on source by source basis; of course for the same type of waste, we can assume certain general properties. And again for modelling parameters those should be determined on project basis (see Section 5.3.2.5), but here more general parameters or range of possible values are available from other researchers, details are available in Section 8.3.2.2.

6.8.1.1 Embankments from Ash

As this subject is too wide, we will concentrate on the waste that produces leachate and is available in huge quantities in the Czech Republic. This waste material is flying ash that is a by-product from coal burning in power and heating stations. The mechanical-physical properties of which are presented in Section 8.2.2.1. The leachate composition will be shown later in the example on risk assessment.

Generally there are two options for the use of ash in embankments, either the full internal body of the embankment is built from ash (see Fig. 6.84a) or the embankment core is built in layers inter-bedding the waste material with "normal" soil layers (see Fig. 6.84b). Usually the second type is called sandwich construction. In this type of construction the "normal" soil is usually represented by fine grained locally won soils that do not have good shear strength properties because the waste material can have them much better. Another advantage of using clayey soil in the

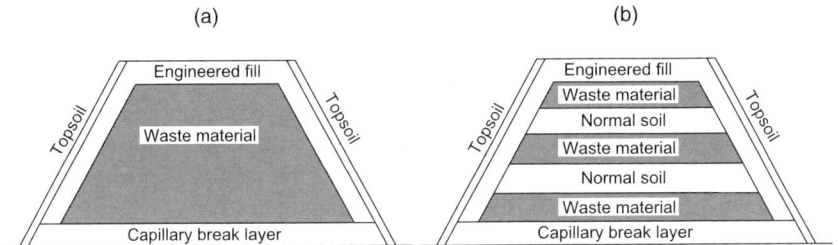

Fig. 6.84 Embankments from waste materials – cross-section alternatives. (**a**) Body only from waste material. (**b**) Core from layers of waste material and normal soil

sandwich embankments is their capability of sorption of some of the substances in the leachate.

With respect to limit state design (overall stability, settlement) ash does not have negative influence, because its properties (shear strength, modulus of deformation) are in the range of natural soils used in such structures. On the other hand its higher permeability can speed up in some cases consolidation process of the embankment itself. However more sensitive for ash is the limit state of surface erosion as the material is composed mostly from fine grains. Therefore it is recommended not to use it directly on slopes. Materials should be under the protective layer of surrounding materials and grassed topsoil.

6.8.1.2 Risk Assessment Example

The ash from the burning of coal is usually considered as waste material as it contains contaminants in much higher concentrations than it is typical for standard construction materials like soils.

The data presented in this chapter were collected during a preparation of recent motorway project in England (M Vaníček 2003), but the proposed and assessed embankment was not at the end built, because source of other non-waste material was found.

The chemical properties of the ash leachate that will be discussed hereafter are taken from three power plants in the UK.

Chemical Properties of Ash

Table 6.1 presents the chemical composition of the leachate for the selected hazardous substances. In order to provide a benchmark for comparison, the results of the leachate analyses have been compared to the UK Environmental Agency's "Contamination Classification Thresholds for Disposal" (EA, Version 3 April 2001).

The leachate analyses for the twenty four samples from all three sources (in different conditions) show that there are elevated levels of Arsenic, Cadmium, Chromium, Mercury, Selenium, Sulphate, pH and PAH when compared to the Leachate Quality Threshold Limits (see Table 6.1).

6.8 Relation to Environments

Table 6.1 Chemical composition of results for 3 sources of ash

Source	Determinant	Arsenic (soluble)	Boron (soluble)	Cadmium (soluble)	Chromium (soluble)	Mercury (soluble)	Nickel (soluble)	PAH (total)	pH	Phenols (total)	Selenium (soluble)	Sulphate as SO_4	Zinc (soluble)
Units		µg/l	mg/l	µg/l	µg/l	µg/l	µg/l	µg/l	pH units	µg/l	µg/l	mg/l	µg/l
1		< 10	0.46	< 0.50	220*	< 1.0	< 10	0.31*	11.2	< 0.50	9.0	73	< 10
1		16*	1.5	< 0.50	45	< 1.0	< 10	0.20	9.3	< 0.50	10	320*	290*
1		< 10	0.31	< 0.50	250*	< 1.0	< 10	0.29*	11.0	< 0.50	9.0	150	140
1		< 10	0.16	< 0.50	150*	< 1.0	< 10	0.39*	10.7	< 0.50	2.0	300*	320*
1		< 10	0.98	< 0.50	< 10	< 1.0	< 10	0.45*	10.8	< 0.50	8.0	68	60
2		< 10	0.52	0.89	72*	0.23	< 10	0.20	10*	< 0.50	13.0*	180*	< 10
2		< 10	0.46	1.18*	110*	0.23	< 10	0.20	9.9*	< 0.50	14*	270*	< 10
2		< 10	1.2	< 0.50	32	0.3	< 10	1.2*	9.5	< 0.50	6.1	400*	< 10
2		< 10	1.2	< 0.50	27	0.3	< 10	0.48*	9.4	< 0.50	5	400*	43
2		14*	0.84	4.63*	100*	1.3*	< 10	0.59*	9.3	< 0.50	14.0*	780*	< 10
2		18*	0.78	2.52*	79*	0.85	< 10		9	< 0.50	9.9		< 10
3		18*	1.1	< 0.50	31	0.75	< 10	0.43*	10.8*	< 0.50	3.1		10
3		< 10	0.43	< 0.50	25	0.32	< 10	0.56*	11.1*	< 0.50	4.3	< 10	< 10
3		< 10	0.1	< 0.50	< 10	< 0.20	< 10	¡0.20	11.8*	< 0.50	2.0	< 10	< 10
1		< 10	1.2	< 0.50	11	0.28	< 10	¡0.20	10.5*	< 0.50	7.0	70	< 10
1		< 10	1.5	0.65	< 10	0.38	< 10		9.5	< 0.50	8.5	360*	< 10
1		< 10	0.95	0.91	52*	0.25	< 10	0.23*	10.4*	< 0.50	9.8	180*	< 10
1		< 10	0.71	0.65	32	0.3	< 10	0.20	10.6*	< 0.50	7.1	100	< 10
1		< 10	1.1	< 0.50	11	< 0.20	< 10	0.20	10.2*	< 0.50	8.5	160*	< 10

*Value exceeds the LIMIT

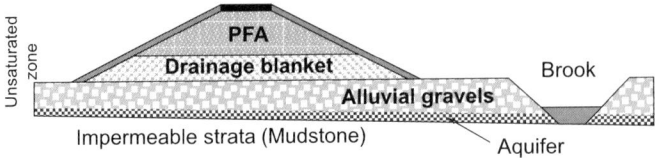

Fig. 6.85 Schematic cross section of PFA embankment used for risk analysis

Risk Analysis

The site in question covers the proposed embankments (Fig. 6.85) in the immediate vicinity of the brook that passes under the motorway. Existing ground level varies between 91 mAoD and 105 mAoD.

The geology of the area is alluvium deposits from the brook, comprising mainly clayey silty sandy gravel, overlaying mudstone, whose upper layers are weathered.

At the site, the alluvium gravel layer has thickness varying 1.5 and 2 m. The upper weathered layers of mudstone formation comprise silty clays of between 2 and 10 m thick. The interface between these two strata varies between 89 mAoD and 100 mAoD.

The groundwater table is situated in the weathered mudstone layer at about 90–95 mAoD. A flood level of 93.04 mAoD is indicated by the Environment Agency (EA). As a consequence, any leachate derived from ash in the embankment is likely to flow to the brook on the contact between alluvium gravel and weathered mudstone.

The potential source of contamination here is the ash material being proposed for use in the construction of the embankments. Given that the ash will be isolated from groundwater rises by the inclusion of a 500 mm drainage layer situated above the 100 year flood level the only potential risk of leachate generation would be via precipitation onto the embankment and subsequent percolation through it.

For the risk assessment analysis, we assumed for this site the following ground model.

At this location the source of contaminants (ash) will be situated on top of an unsaturated granular layer comprising a 500 mm thick drainage layer and alluvial deposits of between 1.5 and 2 m thick. The Minor Aquifer at this location comprises the basal section of the alluvial gravels and is about 0.1 m thick with permeability of about 1×10^{-3} m.s^{-1}. The hydraulic gradient of the aquifer tends to be the same as the inclination of the surface.

For the risk analysis the results of the leachate testing have been taken and the impact of those determinants showing elevated results with respect to the EA threshold levels have been examined. The model assumes that immediate leachate of contaminants into the unsaturated gravel layer (alluvial gravels) beneath the embankment occurs over the total area of the structure. The only receptor in the area of interest is the brook.

The ash used in the embankments will be covered by clean imported topsoils to promote the growth of vegetation. The slope of the embankment, the vegetation and soil cover and road surface will reduce infiltration into the material but may

6.8 Relation to Environments

not prevent all of it particularly in the very long term. The possibility that rainwater penetrating the embankment could leach contaminants from material used in its construction and this leachate could be transported to the brook as part of the natural groundwater flow has therefore been further examined.

A probabilistic analysis using computer program CONSIM (recommended by the UK Environmental Agency) has been used for the risk assessment. The stochastic part of the solution in CONSIM is based upon the Monte Carlo method, which means that the model is analysed several thousand times with randomly generated input parameters that satisfies the entered probability distributions. The output shows probability of exceeding given trigger values at given times. CONSIM divides the problem analysis into two phases. The first phase models vertical movement of contaminants from the source within unsaturated zone. The second analyses horizontal spreading of contaminants within saturated aquifer. As both phases are utilizing one-dimensional analytical approach, concentrations of contaminants are calculated in given target locations (interface between saturated and unsaturated zone and surface water course). The model to evaluate contaminant movement within unsaturated zone would not be mentioned here as the concentration of contaminant at the base of unsaturated zone remains the same as the concentration of the leachate only shifted in time.

For contaminant transport modelling in the saturated zone (aquifer) CONSIM uses modification of analytical solution based on Ogata's equation (see Section 8.3.2.2) allowing for lateral spreading, which is:

$$\frac{c}{c_0} = \frac{1}{2} \left[\mathrm{erfc} \left(\frac{R \cdot z - v \cdot t}{2 \cdot \sqrt{D_z \cdot R \cdot t}} \right) + \exp \left(\frac{v \cdot z}{D_z} \right) \cdot \mathrm{erfc} \left(\frac{R \cdot z + v \cdot t}{2 \cdot \sqrt{D_z \cdot R \cdot t}} \right) \right] \cdot \mathrm{erf} \left(\frac{S_x}{4 \cdot \sqrt{D_x \cdot z}} \right) \tag{6.21}$$

where

$c(z, 0) = 0 \quad z \geq 0$
$c(0, t) = c_0 \quad t \geq 0; c_0 -$ source concentration of contaminant
$c(\infty, t) = 0 \quad t \geq 0$
erfc (x) – complementary error function
D_z – hydrodynamic dispersion coefficient
t – time of interest
v – groundwater velocity
R – retardation factor
S_x – width of contaminant source (plume)
D_x – lateral dispersivity coefficient in saturated zone

Key inputs to the analyses are as follows:

- The distribution concentration of each of the most common contaminants was modelled as a normal or uniform distribution based upon the minimum, mode and maximum values for the leachate test results.
- An infiltration rainfall of 180 mm/yr has been applied which is a very conservative value.
- No retardation factors have been applied.
- The groundwater level in this area lies close to the surface (at the contact between the sand and gravel (alluvium) and the underlying clay layers (mudstone).
- Transport of the leachate takes place in the unsaturated sand and gravel layer and assumed 100 mm thick aquifer just on top of the weathered mudstone – this is a conservative assumption as it neglects the possibility of vertical dispersion/diffusion and therefore dilution and sorption of the leachate.
- The permeability of the sand and gravel layer is assumed to be 1×10^{-3} m.s^{-1}.

The results of the CONSIM program are expressed as function of probability in which the contaminant with certain concentration can reach the aquifer at given time. The analysis was performed for all determinants, but only for Chromium will be presented here (Table 6.2) just to show the principle.

Examination of the results of this analysis suggests the following, M Vaníček (2006), I Vaníček and M Vaníček (2006):

- The results of the CONSIM analysis indicate that the concentrations of Cadmium, Chromium, PAH and Sulphates entering the brook would exceed the EA Leachate Quality Threshold Values (LQTV) with only certain probability. From the Table 6.2 it is visible that for Chromium it is 20%.
- If a very conservative dilution factor is applied to the CONSIM output (maximum of ×5) then all determinants fall below the identified LQTV indicating that actual levels in the brook are extremely unlikely to be significant (For Chromium it is lower that 1×10^{-4}). Based on the identified area of the embankment and infiltration rates a maximum leachate flow of 2.7×10^{-4} m^3.s^{-1}(0.27 l/s) is calculated which gives a dilution of over 7 000 times based on flows in the brook of about 2 m^3.s^{-1}. For this actually derived degree of dilution the probability for Chromium would be less that 1×10^{-7}.
- Leachate tests on the source ash materials give pH values ranging between 9.0 and 13. Again, the significant dilutions that will occur will ensure that there will be no adverse effects of the leachate on reaching the brook.

In summary the analyses undertaken have used very conservative or "worse case" parameters. Even based on these parameters, only extremely minor dilution in the

Table 6.2 Concentration of Chromium reaching the brook after 10 years as a function of probability

Concentration [mg/l]	0.05	0.10	0.15	0.20
Probability [–]	0.20	0.015	0.001	1×10^{-4}

Note: For chromium the limit value is 0.05 mg/l

brook is required to ensure that the LQTV are observed. Therefore, it can be concluded that the CONSIM analysis demonstrates that there is no significant risk of contaminants arising from the ash impacting upon surface water.

Concluding Remarks

The risk analysis is a very powerful tool to show Environmental Impacts of the proposed infrastructure projects generally, but as it was intended by this Chapter to also use this tool for small parts of big projects. The embankments of transport infrastructure offer good possibility, where to use some large volumes of waste materials and therefore to decrease significant amount of these materials that are to be stored as classical waste in spoil heaps, landfills or tailing dams. Risk of potential contaminant spreading is of significant importance and there are several options how to deal with those issues. In the example shown, the risk associated with contaminant spreading was checked by stochastic approach and it was shown that the probability to contaminate surface waters for the given case is very low. Nevertheless the significance of dilution was also important.

6.8.2 Collection of Surface Run-off and Subsoil Sealing

In order to protect our environment from the spills on the transport infrastructure embankments, it is necessary to collect the surface run-off water into collection ponds, where the water will be temporarily stored and cleaned if necessary. Not only are the collection ponds in need of a sealing system, but also the whole road, if it is passing through the area of potable water protection zone.

The run-off surface water from roads is being collected by the drainage system and before releasing into surface water courses, this water is passed through interceptors. These interceptors can be either mechanical, working on sedimentation theory basis or biological, principally special plants and their roots are extracting contamination from water. The sedimentation ponds have to be sealed off from the subsoil in order not to leak dirty water into the environment. These ponds are constructed nearby the road once the volume of collected water demands it. The sealing system is usually made of natural clay, geosynthetic clay liners (GCL) or geomembranes. Each of them has their advantages and disadvantages and those are discussed in more detail in Section 8.3.2.1 about landfill sealing systems. The collection ponds do not act only as sedimentary ponds but also during heavy rainfalls as attenuation ponds that are holding the excess water that is not allowed to be discharged into watercourse during such event. Example of sedimentation pond sealed with GCL next to the motorway is shown in Fig. 6.86 . And this way is helping to reduce the problems of localised flooding.

In the case of railways or roads passing through water protection zones, the requirements are usually such that the whole road or railway structure has to be isolated from the environment to be sure that the protected water would not get polluted. This is again done using the same systems of natural clays, GCLs or geomembranes. Example of such protection with GCL underneath the road is shown in Fig. 6.87 a and for railways embankment surface in Fig. 6.87b.

(a)

(b)

Fig. 6.86 Photo of settling pond near motorway. (**a**) During construction. (**b**) In use

6.8.3 Noise Protection Barriers

With the increase of traffic volumes on our roads and construction of new manufacturing facilities that run 24 hrs a day on both greenfields and brownfields we are more and more exposed to the increased levels of noise. In order to protect ourselves, very often noise protection barriers are being built. There are several options for noise protection barriers, but as we are describing earth structures, we will discuss hereafter only those that are built as earth structures. These barriers, known also as noise bunds can be used to store surplus material available from the excavations for other structures (cuttings, tunnels, underground spaces in building construction) or even waste materials as was already mentioned in Section 6.8.1. If there is enough available space around the bund to be erected, it is possible to make it with very flat slopes as landscaping and those can be reused for example for farming purposes. On

6.8 Relation to Environments

Fig. 6.87 Photo of GCL installation to protect environment from spillage from. (**a**) Road. (**b**) Railway

the other hand, if there is lack of space available, they are built as reinforced earth structures, as it is shown on Fig. 6.88.

One example of a recently built noise protection barrier from reinforced soil designed by the author is near the TPC (Toyota, Peugeot and Citroen) factory in

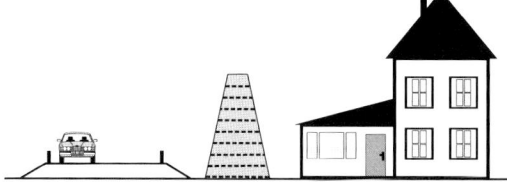

Fig. 6.88 Schematic section of soil reinforced noise protection bund

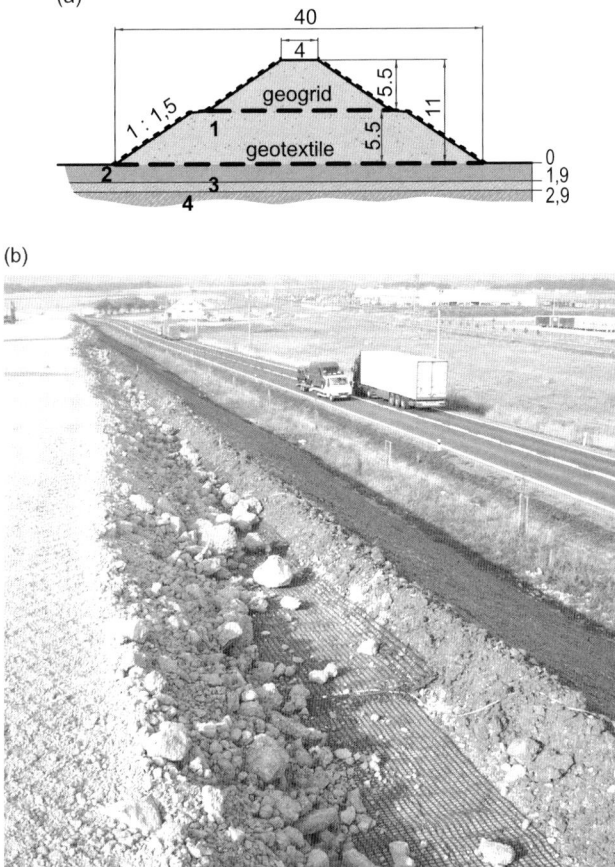

Fig. 6.89 Noise protection bund at Ovčáry near TPC factory. (**a**) Cross-section. (**b**) Photo from construction

the Czech Republic. Here the noise bund protects a nearby village from the noise produced by heavy transport on nearby road. This bund was due to the limited space available proposed with steeper slopes (1:1.5). The material used for the bund construction (1) was reasonably good so the reinforcement within the bund body was needed only for the stretches with the height of more than 7 m (up to 11 m). But the approximately 1.9 m top layer of the subsoil (2) has lower bearing capacity and therefore contact reinforcement was required for bund higher than 4.5 m. Underneath this soft layer was 1.0 m of weathered siltstone (3) underlaid by siltstone (4). The slopes were also protected against surface erosion using erosion protection mattresses. The cross-section of the bund and photo from its construction are presented in Fig. 6.89.

Chapter 7
Earth Structures in Water Engineering

Earth structures in water engineering are one of the oldest constructions, certainly of these, which are now called civil engineering. Some of them are very old, about 5 000–5 500 years, as for example the small dam Mokhrablur in Armenia (close to the town Kasak), or Jawa in Jordan (close to the town Mafrag). Their earliest role in providing storage for irrigation water formed a major contribution to the development of our civilization. The height of such dams steadily increased but still was only around 20 m in 1800. It is worth mentioning the dam Horná Hodruša in Slovakia with a height of 21.4 m from 1614 or the system of small (fish pond) dams in South Bohemia close to the town Třeboň from the end of the sixteenth century.

The height of embankment dams is of considerable geotechnical interest in consideration to the stresses and water pressures created. The world's highest embankment dams exceeded 100 m in 1926, 200 m in 1968 and 300 m in 1980 (Nurek, Rogunski) – e.g. Penman (1986), Lukáč and Bednárová (2001), Votruba et al. (1978).

The main reason for dam construction was water storage when there was a surplus and its utilization when there was limitation for both for irrigation and drinking water. Then also, water retention during extreme surplus – during floods – to limit the negative impact of floods on property below the dam profile. Last but not least, this aim is connected with the utilization of water energy, firstly for powering different machinery, and today mostly for electric energy production.

During the interaction of an earth structure with retained water our attention is devoted to the observation of seepage of water through the soil. In Section 5.3.1.3 resp. 5.3.2.3 the attention was devoted to steady flow, when the outflow from the soil element is equal to the inflow for the unit of time. The best way to demonstrate steady flow is a flow net, where a set of flow lines represent lines along which water is passing and a set of equipotentials are lines connecting places with the same water pressure, e.g. Fig. 5.53. With the help of a flow net we can determine the:

- Amount of water seeping through the soil body (e.g. sealing layer),
- Magnitude of pore water pressure in individual points of earth structure,
- Magnitude of hydraulic gradient for the individual element in the earth structure,
- Magnitude of seepage pressure applied on this individual element.

With the help of this information, the safe design of an embankment dam can be made reflecting all relevant limit states of this earth structure as limit states of type GEO, HYD and UPL. With a change of water level in a water reservoir the character of seepage changes from steady flow to unsteady flow. It is necessary to control the limit states also for the unsteady flow, and the problem is more difficult. Therefore the finite element method can be used, e.g. Kazda (1990).

Individual group of problems are connected with the influence of embankment dams on the environment. Already a great impact is connected with dam construction. Water reservoirs flood previous existing areas of fauna and flora (and very often with historical buildings), punctuate sediment transport, transport aquatic animals, mainly fish, significantly change underground water seepage in the immediate environment and can lead to waterside erosion. We can even speak about the aesthetics of dam construction, about their integration into the surrounding environment.

Last but not least, we have to speak about risk assessment, about the risk of the failure of the dam on the area below the dam profile, e.g. Stewart (2000). The design, construction and monitoring of the dam have to take this risk into account, e.g. De Mello (1977).

It is not the aim of this chapter to specify the details of all the problems associated with all types of earth structures in water engineering, but to focus only on some specific examples, which can be decisive for the safe design and maintenance of these earth structures. Specific attention will be devoted to the potential risk of crack creation in the core of earth and rock-fill dams, Vaníček (1988). The solution of this problem bears many new opinions on the construction of earth and rock-fill dams as from a theoretical view (tensile strength of soils, inter-particle forces) to a practical view on how to adjust the design of new fill dams to these new findings. At the end of the chapter some specific problems of small old earth fill dams as well as anti-floods dams will be presented.

7.1 Basic Cases of Earth Structures in Water Engineering and Their Features

The range of earth structures used in water engineering is very wide. Instead of classical fill-dams, which were briefly mentioned in the introduction and will be dealt with in more detail later on, we can distinguish between:

- Dams for storing reservoirs of drinking water or water used for different utilization such as for fire-fighting – fluctuation of water level is rather small, in a very limited range.
- Dams for lakes (ponds) – once during a one or two year period the lake is completely drained to harvest all fish bred there.
- Dams for hydroelectric power plant reservoirs (pumping station) – specificity is given by great fluctuation of water level, practically once a day the reservoir is emptied and filled again. The direction of seepage in the earth structures changes all day.

- Flood protection dams – can have a different character – from dams along the river stream to dams on little rivers, which do not retain water with the exception of floods to dams of polders, which only retain water diverted from the main stream during floods. Unsteady flow prevails with change of water saturation – a very complicated problem of unsteady flow in unsaturated soils.
- Sea canals, connecting sea waterways (Suez, Panama).
- Floating channels on river streams, also connecting two river-basins, – mostly cuts, partly embankments, affected by waves produced by river boats.
- Feed ditch to the hydro electric power station (e.g. Gabčíkovo on Danube river) – cuts but also embankments, small fluctuations of water level.
- Protective dams along sea shore, protective dams of sea ports – fluctuation of sea level, great waves.
- Cofferdams – fill dams ensuring the construction of dams, spillway dams, ... – so that these structures can be founded in dry foundation pits. Cofferdam design height is determined by floods, so that the risk of overflowing during floods greater than the design value is very high. The construction of a cofferdam can be included to the dam profile. When protecting the construction of large hydro structures (dams) the cofferdam alone is a very challenging earth structure (e.g. cofferdam of Three Gorges Dam) – a unique feature that the construction has to dam up the running stream.

7.2 Earth and Rock-Fill Dams, Typical Cross-Section

Schematic typical cross-sections for a dam body are presented in Fig. 7.1 e.g. Votruba et al. (1978). Dams are constructed as:

- Homogeneous earth fill dams, constructed from soil ensuring the design from in consideration of permeability – the subsoil should be also sufficiently impermeable; if this impermeable subsoil is in low depth, the cut-off is used for connection of the dam body with this low permeable subsoil. To protect the outflow of water on the downstream side, different drains are constructed, toe drain, planar drain or chimney drain, see Fig. 7.2. The drain is protected by a filter.
- Earth-fill or rock-fill dams with earth (clay) core, when the upstream and downstream stabilization zones are created by more permeable soil (earth-fill dam) or rock (rock-fill dam). Earth core can be situated close to the upstream face or in the central part with a vertical axis or with an inclined axis. Protective layers around the earth (clay) core guarantee filtration stability on contact with different layers (filters) or if needed lead away permeate through water (drains).
- Earth-fill or rock-fill dams with a sealing layer from artificial materials (asphalt concrete, concrete, steel, plastics) situated on the upstream face or in the central part with a vertical axis – protective layers protect sealing against loading by deformation of stabilization zones and enabling to drain permeated water away.

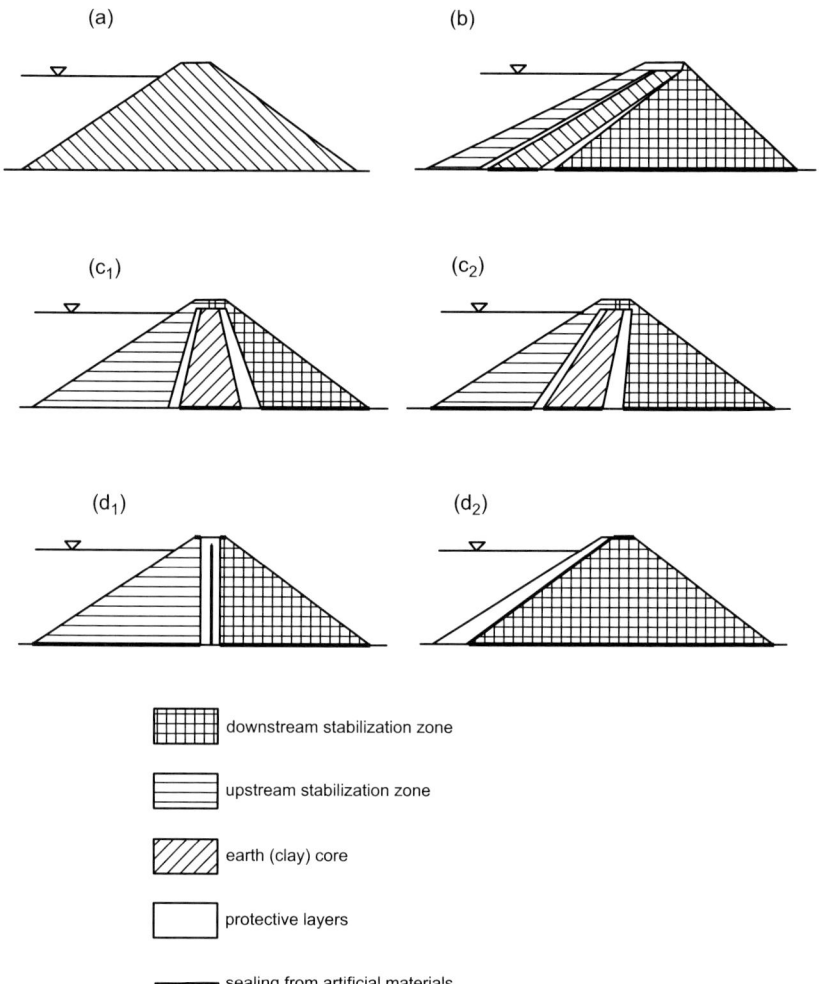

Fig. 7.1 Schematic typical cross-sections for fill dams. (**a**) Homogeneous. (**b**) With upstream clay core. (**c**) With central clay core. (c_1) with vertical axis; (c_2) with inclined axis. (**d**) With artificial sealing element. (d_1) central sealing; (d_2) upstream sealing

Schematic cross-sections did not specify the character of the subsoil. Usually the sealing element passing through the dam body joins on a vertical sealing wall passing through the subsoil (sealing slurry walls, injection walls, sheet-pile wall).

The presented cross-sections represent the earth and rock-fill dams constructed in layers, which were compacted in place. The dam construction can be performed also by hydraulic fill, by technology, which mixes the soil and water and then deposits the sediment there. This construction technology was popular in the USA roughly from the end of nineteenth century up to the 1940s and usually represented the largest dams – e.g. Fort Peck 73.8 m (1940). In the former Soviet Union, the peak

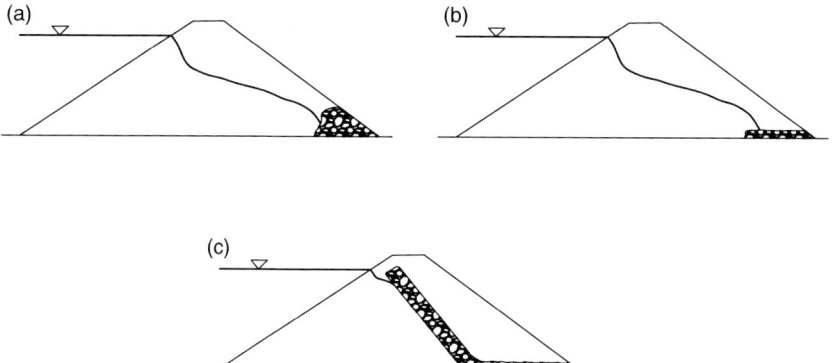

Fig. 7.2 Homogeneous dam. (**a**) Toe drain. (**b**) Planar drain. (**c**) Chimney drain

of this technology was in the 1950s – biggest such dam – Mingecaur has a height of 81 m, total volume 15 mil. m^3 (1954). But with the improvement of construction technology earth and rock-fill dams prevail. Hydraulic filling technology was very limited after monitoring extremely low pore pressure dissipation in the central part of the dam body. Deposited material by hydraulic filling is very sensitive to liquefaction, especially during dynamic loading. The San Fernando earthquake led to serious damage of both the Upper and Lower San Fernando dams in 1971. Subsequent verification of the dynamic stability by finite element method for 36 dams in California showed that most of them must reduce their functionality or at least essentially be reconstructed, e.g. Votruba et al. (1978).

7.3 An Analysis of the Most Frequent Failures of Earth and Rockfill Dams

A standard approach to the project of earth-fill dams is based on the choice of appropriate local materials for component parts of the dam, the sealing, the stabilization part, the filters, respectively the transitional zone, and on the determination of cross-section profile arrangement which can be considered safe on the basis of slope stability calculation. And as we show later, in most cases the project of slopes made in a classical way will do. Cases of dam failure due to slippage of its downstream or upstream face are not too many, and above all most of them happened before modern methods of soil mechanics have been adopted. In the majority of cases if there were some problems during construction of earth-fill dams, they were mainly related to the internal erosion, a crest overflowing and to undetected parts of subsoil weakness and the like. It concerns the causes which are not related directly to the degree of safety defined on the basis of stability solution.

The above-mentioned conclusions are confirmed by many various papers on analysis of failures of earth-fill dams. One of the first research studies was written by Reinius (1948) who made an analysis of 109 dams which failed in the first half of the last century. Causes of failures are divided into eight groups, the most important are:

1) A crest overflowing with subsequent surface erosion (36 failures)
2) A seepage along or seepage from outlets and spillways (25 failures)
3) A seepage through dam body or subsoil (25 failures)
4) An insufficient slope stability – a slope failure (7 failures)

The remaining failures are due to the deficient compaction, the subsoil shearing, the increased seepage caused by animals etc.

Since that time the number of failures caused by crest overflowing reduced because an increased monitoring of flow rate leads to a higher exactness of prediction. So failures occur only when there is a lack of hydrological data, or when there were some considerable changes, e.g. deforestation. The technology and control of compaction are getting better so that the shear strength and deformation characteristics are better, too.

This development was confirmed by the survey made in 1965–1972 carried by the International dam commission – ICOLD, Paris in 1974. It evaluated 466 failures and accidents in all, and from those 271 cases were the dams which were higher than 15 m, dams built after the year 1900; the cause of their failure was determined. With a view to 7 757 dams built in the years 1900 and 1965 in 43 countries it means that on the whole there was 3.5% of defects from which 1.16% cases were failures. From the above-mentioned number of problems are 206 earth-fill dams (64 failures and 142 defects) and 4 316 built ones. The fault liability rate is higher than average it is 4.77%, while the failure rate is 1.48%. Rockfill dams have worse results in comparison with earth dams. Low dams lower than 15–20 m, have the highest number of failures. The fact that an internal effect of seeping water has caused the most number of failures was confirmed again. Votruba et al. (1978) made a similar analysis for the former Czechoslovakia. They found out that one 14 m high earth-fill dam in Bílá Desná failed in 1916, which reflects approx. 1% of failure rate. The most probable cause of the failure of the Bílá Desná dam was the differential settlement of subsoil under the embankment and the bottom outlet founded on timber piles, and the following internal erosion along a developed crack.

The author's research work, which was realised between 1970 and 1985, was based on their own studies; Vaníček (1975, 1977c, 1988). The research of all questions connected with an increased seepage through the body of a dam caused by tension cracks in earth sealing is the main subject of our following chapters. The development of tension cracks is very dangerous because it can cause a full destruction of a dam. Among cracks which occur in a general direction in a body of fill dams, the most dangerous are those going in the direction of flowing water, they penetrate the sealing and they are not visible on the surface of a dam. They are so-called internal transversal cracks. Generally the development of internal cracks can result in an exceeding of allowed quantity of seepage so that the basic function of a dam is damaged. The next two aspects of increased crack seepage could affect the safety on the whole. They are:

- Increasing internal erosion, especially in the case of a large susceptibility to erosion,
- An exceeding of the drain capacity followed by an unpredictable rise of seepage and pore pressures.

Knowledge of the conditions and reasons of the development of tension cracks in a dam body and consequently their behaviour is very important for a correct assessment of safety of fill dams. An analysis of conditions under which cracks occur, shows that total and effective tensile strength of sealing soil is its limiting criterion, or it is their deformation properties under the tensile loading. But there are some known cases of cracks being in a dam body which can increase seepage but this does not exceed an allowed limit by itself and it does not rise up with time. We can conclude that the study of soil behaviour during water flowing through the crack is very useful. It is necessary to pay attention to this aspect especially in the connection with soil dispersivity, where dispersive soils are extremely susceptible to the internal erosion. If it is not possible to prevent the outwash of the grains from cracks, the stability of the layer interface is important. There is a close linkage with the problem of filtration criteria. The requirement of the complex approach to it is justified and emphasized by the occurrence of failures of fill dams registered since 1965. It concerns e.g. Hyttejuvet dam in Norway, Balderhead dam in England, and a group of high dams in Mexico and Teton dam in the USA in 1976. Numerous accidents of these dams were not caused by the lack of knowledge of designers or contractors, but by the relatively, more risky approach to the possible height of the dam, to the maximal use of local less suitable materials and at the same time to the questionable location of these dams on places with a less convenient cross section profile. A lot of dams were well-equipped with measuring devices for monitoring the behaviour of the dams during their construction and operation; received data serve as a very good source of useful information. Experience gained during the construction, during the first filling the reservoir and operation of Czech dams, e.g. Dalešice dam, is important without question, Vaníček (1982b). Other publications in the last period, e.g. Höeg (2001, 2003) or Foster et al. (2000a) confirm that the proposed way of problem investigation was the right one and in some directions are helping to specify it in more detail, e.g. Foster et al. (2000b).

7.4 Conditions and Reasons of the Tensile Cracks Development

Some simple prerequisites given by Wilson (1977) allow us to imagine how the dam body is changing during its construction, the first filling of the reservoir and its operation. In Fig. 7.3 a simplified dam section is shown where the whole body is homogenous from the point of view of deformation modulus, it consists of two basic areas with different permeability: a permeable stabilization zone on upstream and downstream side and a low permeable, relatively narrow central sealing. Another prerequisite is a linear relation between vertical load and vertical deformation; this deformation does not depend on time, i.e. it is realized at the moment of loading.

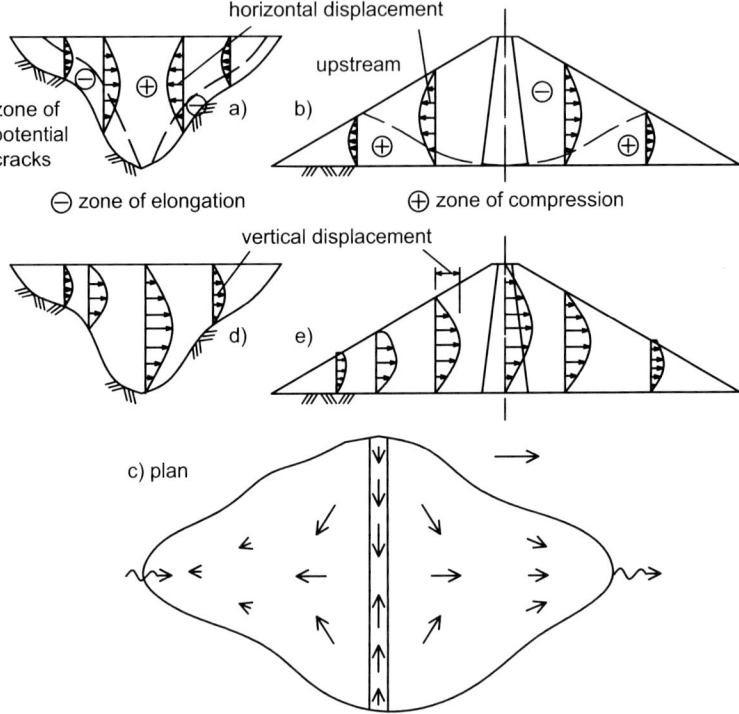

Fig. 7.3 Vertical and horizontal displacements of dam during construction period (according to Wilson)

During the construction the individual layers are put out and compacted horizontally. The settlement of the surface of each layer is in proportion to the number of layers under this level and to the number of layers made above it. For example we can suppose ten layers. Under the surface of the first layer is one layer, and above that there are nine layers, it means that settlements will be in proportion to number 9:

$$s = 9.h/E \times \Delta\sigma_z \qquad (7.1)$$

where

h – a thickness of one layer
E – a deformation modulus
$\Delta\sigma_z$ – an imposed load carried by one layer

The highest settlement is for the surface of the 5th layer (five lower layers are compressed 5 times – 5×5 = 25. Distribution of vertical settlement is parabolic; maximum is in the middle of the height of the dam. These relative settlements are shown in Fig. 7.3d,e. Generally the settlement takes place under the triaxial

7.4 Conditions and Reasons of the Tensile Cracks Development

conditions, it means that each component during a vertical settlement can spread around (if it is possible). It is just relevant to the direction of main section – upstream and downstream faces of a dam are bulging by the effect of this lateral enlargement, Fig. 7.3b. The situation is different in the middle profile, Fig. 7.3a. Generally dams with trapezoid profile of valley in the middle part have practically no horizontal displacements in the direction of dam axis. In the case of a triangle profile the trajectory of point displacements, which are situated lower in the direction to the centre of the valley, is influenced by building of separate layers and by slope inclination of valley. Horizontal displacements in vertical sections are shown in Fig. 7.3a, resultants of displacements are schematically shown in a dam plan in Fig. 7.3c.

In accordance with the previous prerequisite the crest of a dam did not sag (there was no surcharge load), but the lower layers sagged. A completed layer is compressed in the middle part more than by edges because under this part there are more layers. It means that at this point an extension takes place (in accordance with displacements illustrated above); it is along the sides of valley. At the same time we must realize that the tensile stress does not have to act necessarily in the area of extension. The lateral extension always takes place when an increase of lateral stress σ_x in relation to vertical surcharge load $\Delta\sigma_z$ is smaller than the value of coefficient of earth pressure at rest K_0.

The size of vertical deformations depends mainly on the compressibility of used material. When deformation modulus is lower the vertical deformations are larger and an extension is larger, too. Horizontal displacements do not depend on the shape and steepness of the sides of valley only, but they are also proportional to the size of vertical displacement. Wilson (1977) presents a relation of horizontal displacements to vertical ones from zero for the middle part up to 0.5 nearby sides of valley.

Similarly he gives presumable range of stresses for the middle part of sealing:

– vertical stress: $\qquad \sigma_z = \gamma.h.K_z$

where $K_z \approx 0.75$ for a triangle profile,

– horizontal stress in the direction of water flow: $\qquad \sigma_x = \gamma.h.K_x$

where $K_0 > K_x > K_a$, i.e. app. in the range of 0.3–0.5,

– horizontal stress in the direction of dam axis: $\qquad \sigma_y = \gamma.h.K_y$

where

> $K_0 > K_y > K_p$, i.e. app. in the range of 0.5–0.75.
> γ – volume weight of soil; h – depth of considered point under surface,
> K_a, K_p, K_0 – coefficients of an active and a passive earth pressure and a pressure at rest.

Ratio of these stresses is changing in accordance with placement of considered points. Value of σ_y close to the dam valley will be different because $K_y < K_0$.

During the first reservoir filling the saturation of grained material in the stabilization upstream zone takes place and at the same time further deformations occur. These deformations are caused by moistening of contact surface of coarse grain where a relatively high intergranular stress acts. The type of vertical and horizontal displacements differs from the type of displacement which takes place during the construction, Fig. 7.4. Maximum of vertical and horizontal displacement appears along the crest and face of a dam. In crest the tensile deformations appear nearby the valley and pressure deformations are in the middle zone, Fig. 7.4a. As far as the development of transversal cracks, the situation is the most critical because changes take place very quickly with respect to speed of filling the reservoir, and the sealing is not able to adapt to the change of stresses. From the point of view of effective stresses there are undrained conditions. An absolute size of relative extensions along the crest is not only a question of the properties of used material, but it is also significantly affected by shape and slope of the upper part of the sides of valley.

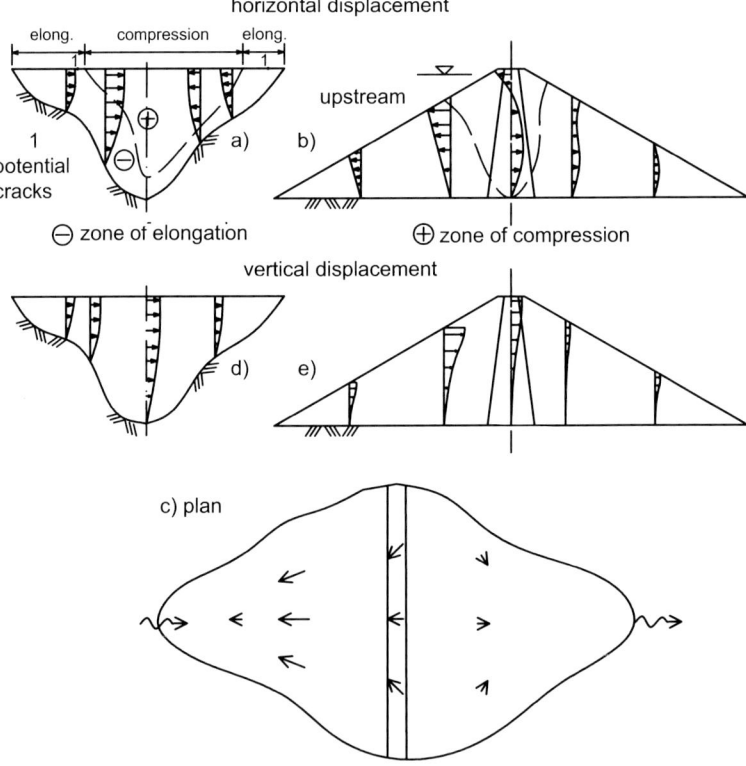

Fig. 7.4 Vertical and horizontal displacements of dam during first reservoir filling (according to Wilson)

7.4 Conditions and Reasons of the Tensile Cracks Development

Deformations found in cross section, Fig. 7.4b,e are different in comparison with ones formed in construction period. The upstream stabilization zone is the most deformed while the sealing and downstream stabilization zone are substantially affected less by the first filling of the reservoir. Final trajectories of displacements of the upstream stabilization zone, Fig. 7.4c, are in opposite direction to water and they cause a certain lightening of sealing above the water level in reservoir. A horizontal earth pressure acting on the sealing is reduced, possibly up to an active earth pressure. At the same time the total load on sealing of the low part is growing up, it particularly depends on the rising water level in reservoir. An earth pressure is also getting down there however the decrease is lower than hydrostatic pressure increase. The result of these different situations above and under the water level in reservoir can be an upstream shift of the sealing in the upper zone and a downstream shift of the sealing in the lower zone, Fig. 7.4b.

A relatively sudden compression of the stabilization zone in comparison with the sealing causes a big shearing stress and deformation along the upstream edge of the sealing. In view of time trajectories of the displacement, this area, where big shearing deformations happen, is not limited very much, but its speed is continually falling. In the first place a consolidation of less permeable soil, particularly the core, is taken into account here. Then a secondary deformation happens – a continuing crushing of contact surface of coarse grain, a water seepage through the sealing, a moistening of downstream stabilization zone (e.g. by rain, by cyclic lightening and surcharging, by water movement in reservoir, by rise of water level above the level of the first filling the reservoir, or by earthquake).

In line with the taken elementary model, we can say that deformation of the dam body is a very complicated phenomenon. The largest problem is non-homogeneity of the dam body with respect to modulus of deformation which makes the situation more difficult. Subsoil cannot be also taken as an incompressible material, and if it is non-homogenous in a plan, then this phenomenon makes a situation more challenging. In practice sometimes the first two parts cannot be separated because filling the reservoir often runs together with its construction. An assumption of deformation which does not depend on time is also simplified. This is often manifested by sagging of the dam crest even in the phase after filling the dam body.

Today there are two basic ways which help us to understand the complicated processes occurring in the body of a dam. In the phase of prediction it is an analysis of the state of stresses and deformation of the dam body and the use of finite element method. In the phase of checking the behaviour modern measuring devices are used. They are put into the dam body during its construction. They are: extensometers, inclinometers, pressure indicators, indicators of water pressure – piezometric indicators, precise levelling of the surface points etc.

An instrumentation and a dam monitoring are used not only for checking of the supposed behaviour of a dam, but their role is more universal because they have to ensure both technical and safety supervision. In some countries this supervision is integrated into the laws. In the Czech Republic it is the Water Act (Law). Supervision is the duty of water course managers. Reports worked out on the basis of their measuring of some hydraulic structures are a very useful source of information.

Numerical methods of modelling of dam behaviour are improving. In some cases they were very successful, especially in prediction of deformation; in others they were less effective. One of the main reasons of these differences (limitation of prediction) is a practical limitation of possibilities of laboratory simulation of loading of the individual elements under such conditions which can occur in reality. As it concerns the tensile zones, it is necessary to take more tensile tests and to provide good initial data for the behaviour of these tensile zones. This is one of the main goals of the following specification of the tensile tests. The use of FEM is a great advantage because it allows making fast parametric studies. An initial value is varied in a range that will probably exist, and an effect of this variability on calculated behaviour is observed. The result of these parametric numerical studies is the recommendation for appropriate building materials (a choice of the borrow pit, a possibility of special ground treatment etc.), and the recommendation for placing measuring devices on areas with specific behaviour etc.

7.4.1 Conditions of the Tensile Cracks Development

In principle we recognize two types of conditions under which tensile cracks can develop, Vaughan (1976), Vaníček (1978a):

- Under dry conditions – cracks are created above the water level in the reservoir and their filling is air;
- Under wet conditions – cracks are created under the water level in the reservoir and their filling is water.

In both cases the situation differs a little, it depends on the place which can be near the dam crest, where a vertical pressure is very small here, or on the place which is inside of the dam.

In the first case stresses can be compared to the loading during simple tensile test or during bending test and calculated or measured elongation can be compared with the results from these tensile tests. In the second case the vertical main pressure – vertical pressure induced by weight – plays a much more important role – and therefore the conditions there are closer to the triaxial tensile test. The results from the triaxial tensile test are considered to be a more important condition for cracks development. Calculated or measured elongation is then better to compare with results of extension triaxial test.

A dry crack is created under the condition of the smallest main stress drops in negative tensile area and it is absolutely lower than tensile strength. This condition is demonstrated in total stresses because a dry crack is developed under undrained conditions:

$$\sigma_3 < -\sigma_t \tag{7.2}$$

7.4 Conditions and Reasons of the Tensile Cracks Development

where σ_3 is the smallest main stress

$$\sigma_t \geq 0 \tag{7.3}$$

where $-\sigma_t$ a tensile strength in total stresses, determined under undrained conditions

A wet crack is created under the condition of the rising water level in reservoir, water pressure can go through the permeable stabilization zone up to the surface of sealing (with a certain irregularity – Fig. 7.5), it exceeds the smallest main stress acting here, which is enlarged by tensile strength. If filling the reservoir is fast enough, we can take into account undrained conditions and this condition looks as follows:

$$p_w > \sigma_3 + \sigma_t \tag{7.4}$$

where

$p_w = \gamma_w . h_w$ – a hydrostatic pressure
σ_3 – the smallest main stress at the end of construction period
σ_t – a total tensile strength determined under undrained conditions

If filling is slow enough so that the pore pressures in the sealing could be balanced with a water level in reservoir, we could consider the drained conditions. After filling the reservoir there is a pore pressure increase Δu which is in accordance with p_w on condition that there were no losses of pressure height. This assumption looks as follows:

$$p_w > \sigma_3 + \sigma_t' \tag{7.5}$$

$$\text{resp. } \sigma_3' < -\sigma_t' \tag{7.6}$$

σ_3 is the smallest main total stress at the end of construction stage

$$\sigma_3' = \sigma_3 - p_w' \tag{7.7}$$

σ_t' is an effective tensile strength, generally $\sigma_t' \geq 0$

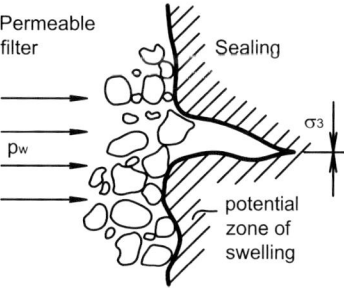

Fig. 7.5 Scheme explaining the development of wet crack under undrained conditions

Process of wet crack development is generally called hydraulic fracturing. All mentioned conditions under which both dry and wet cracks are created, are always necessary conditions, but they are not sufficient. From this standpoint we must take into account a general state of stresses of an element, particularly, the highest main stress σ_1 and changes which could occur since the end of construction phase for possible development of wet crack. For example Ohne and Narita (1977) formulate an additional condition for creation of a dry crack:

$$\sigma_1 \leq q_u \tag{7.8}$$

where q_u – a simple compression strength (unconfined compression strength). This condition allows the transfer of the smallest main stress into tensile zone.

Vaughan (1976) or Alberro (1977) describe the conditions of wet crack development. Vaughan (1976) explains the development of wet cracks under drained conditions and he uses stress paths for showing the state of stress at point A and also for the changes of the state of stresses – see Fig. 7.6. He supposes that point A is not affected by pressure of seeping water until the reservoir is full. Initial stresses occurring before filling are marked a, or a' for an effective state of stresses. State of stresses is changed from a to b, or from a' to b' by water load (water pressure acts on impervious sealing of reservoir). Pore pressures then will rise up as a result of seepage and an effective state of stresses will change from b' to c'. Total state of stresses is changed a little, let say from b to c. By an increasing water pressure the point c' (which is on the limiting line of shear strength) is reached; the failure will be caused by shear. But the crack development can happen after reaching the point d' only and for this a considerable shearing and a redistribution of stresses are necessary. Though this redistribution of stresses has not been modelled yet,

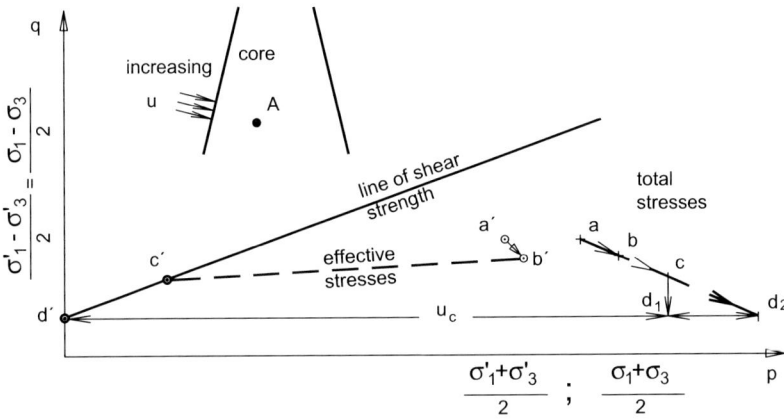

Fig. 7.6 Changes in state of stresses in point A in clay core during wet crack development under the drained conditions (according to Vaughan). a,a' – initial state of stresses before watering; b,b' – after watering; c' – shear failure; d' – origination of tensile failure; cd$_1$ – soil with zero swelling potential; cd$_2$ – soil with swelling potential

7.4 Conditions and Reasons of the Tensile Cracks Development

we can expect a slowing down of process of cracking. Tensile triaxial tests under drained conditions, which will be described later, can serve as an explanation of soil behaviour in this area.

Difference between d and d' is a value of pore pressure needed for crack development. The larger the difference, the smaller the risk of crack development. On this occasion Vaughan points out that a big swelling capacity of sealing soil can have a significant influence on an increase of total state of stress. In the case of zero swelling capacity the total state of stresses during the failure is given by point d_1, while the total state of stresses is growing by swelling and is given e.g. by point d_2. This last aspect is very important because:

- A large swelling capacity of sealing soil decreases the danger of crack development under drained conditions,
- A criterion, based on the condition of hydraulic fracturing formation, requires that an average total stresses $(\sigma_1 + \sigma_2)/2$ – including the growth of total stresses by water loading on upstream face – should be smaller than maximum hydraulic pressure of seeping water. But this aspect is considered to be conservative from the above-mentioned point of view.

The previous failure by shear before the potential failure by tension is of great importance. The failure by shear leads to large transfer of stresses from the zone of high concentration to the zone of its low concentration. Cracks will be probably created in limited zone of low stresses where the shearing failure will happen and the cracks will be developed earlier than the shearing stress will spread. This situation is important especially from the point of view of real danger of crack development by hydraulic fracturing in the dam clay core in consequence of drilling works.

Making drill holes into finished clay core for an installation of piezometers, an injection of subsoil or an injection used for some other purposes is usual nowadays. Full or partial loss of drilling fluid does not have to lead necessarily to the conclusion that cracks appeared in the core because of the way of drilling which could cause them. Here is a more unfavourable situation in comparison with the danger of crack development in the clay core during increasing water level in reservoir, because a water pressure is acting locally, so the redistribution of stresses is not taken into account.

If we compare undrained and drained conditions of crack development we find that a larger risk is in the case of undrained conditions

$$u \geq \sigma_3 + \sigma_t, \text{ resp. } u \geq (\sigma_1 + \sigma_3)/2 + \sigma_t \tag{7.9}$$

A risk is getting smaller when the filling is slow enough so that pore pressures in clay core could level with water surface in reservoir.

A rough estimation of potential danger of wet crack development a conservative condition is:

$$\sigma_3 < p_w \tag{7.10}$$

and the critical condition:

$$\sigma_1 < p_w \tag{7.11}$$

It is obvious that stress redistribution is necessary for wet crack development. In an ideal case for a homogeneous dam and plane problem the vertical pressure in dam axis:

$$\sigma_1 = \gamma.h \tag{7.12}$$

where h – depth of point A under the crest of a dam. Water pressure is on the level of point A – maximally:

$$p_w = \gamma_w.h_w \tag{7.13}$$

where h_w – height of water in reservoir.

Generally: $h > h_w$ and $\gamma \sim 20\,\text{kN.m}^{-3}$ and $\gamma_w \sim 10\,\text{kN.m}^{-3}$, and for this ideal case p_w/σ_1 is less than 0.5. As mentioned above in Fig. 7.3, as a consequence of the three-dimensional problem, when there is an interaction between a dam body and sides of valley, the value σ_1 is reduced. For triangle profile of valley:

$$\sigma_1 \sim 0.75.\gamma.h \tag{7.14}$$

In this case p_w/σ_1 is less than 0.67. For horizontal stress σ_x the first more conservative condition can be fulfilled, as for $\sigma_x = \sigma_3 = (0.3–0.5)\gamma.h$ ratio p_w/σ_3 is about 1.5–0.9.

In real case, besides the interaction between a dam and sides of a valley, there are also interactions:

– Between a dam and subsoil.
– Between individual parts of a dam such as clay core, filters, transitional and stabilization zones.

As a result of these interactions in some areas we find a decrease of the state of stresses in comparison with geostatic state of stresses which can lead to the development of tensile zone in an extreme case. Zones with reduced stresses are the potential zones where the development of wet cracks can occur by hydraulic fracturing.

On the other hand in some zones there is an increase of the state of stresses which can cause the development of plastic zones. It means that an increasing state of stresses reached a limiting state of failure. Though our main attention will be paid to a danger of tensile crack development, our following note concerns the plastic zones and their secondary manifestation:

– Plastic zones can affect the position of the most dangerous slip surface during assessing of slope stability by limit equilibrium methods,

7.4 Conditions and Reasons of the Tensile Cracks Development

– A plasticity leads to a reduction of deformation characteristics and as a consequence to a further increase of difference in deformations of individual zones in a dam.

Plasticity in cohesionless materials leads to an increase of porosity caused by dilatancy. As such it could be critical, so a risk of seismic effect is more probable.

7.4.2 Reasons of the Tensile Cracks Development

Cracks in clay core of a fill dam can have a general direction. In accordance with Fig. 7.7 we classify them as follows:

– Transversal external cracks
– Longitudinal external cracks
– Transversal internal cracks
– Longitudinal internal cracks

External cracks are visible, they are on dam surface, and internal cracks are invisible as they are covered by non-cohesive soil of the stabilization zone. For reasons given above internal cracks could not be identified and repaired easily. With a view of possible spreading of cracks by internal erosion the transversal cracks are more dangerous because they are in the direction of water flow.

Sherard et al. (1963, 1973) describes various possibility of crack development, he points out an irregularity of bedrock, a various toughness of dam subsoil, an influence of different stiffnesses of individual dam zones. Some specific examples will be shown in detail later. With respect to immediate causes of crack development in the dam body, we can determine their order and show them in the list below:

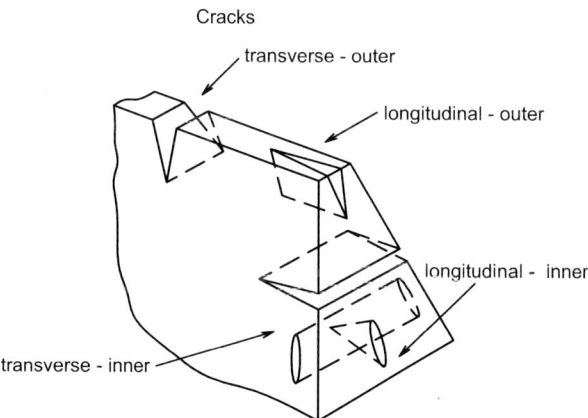

Fig. 7.7 Different orientation of cracks in clay core

- Differential settlement
- Hydraulic fracturing
- Desiccation
- Seismic effects

The last two causes have a local effect for both seismic areas and flood banks, which most of the time do not retain water. We will focus on the first two causes.

7.4.2.1 Cracks Initiated by Differential Settlement

In 1950 Casagrande pointed out that crack development as a result of differential settlement is very dangerous. He described a case of the 25 m high homogenous dam where after the first water filling a very concentrated seepage had occurred. He expressed his concern about the clay which is compacted on optimal moisture content but it does not have to be able to follow even small differences during settlement without crack development.

This fact was accepted and understood after a delay because of existing opinion about the ability of embankment dams to take over considerable deformations. By comparison with construction of older fill dams, a present-day practice is different; it can influence the number of cracks to some extent. For example dams are higher, their construction is faster, less suitable local materials are used, a dam profile is situated on more problematical places, and compaction efforts are more effective.

There is a specific tendency to compact a drier soil; it is possible because of using more powerful rollers allowing a more effective compaction. And at the same time there is a fear of developing too large pore pressures inside of the dam. However processing of a drier soil considerably reduces a flexibility of an embankment dam. In 1969 Casagrande expressed his view on an increasing height of dams. He thinks that in the case of high rock dams in valleys with steep sides it is impossible to avoid the tensile zones and transversal cracks in the crest of a dam which is near the sides of valley. It does not depend on the kind of building material. So it is necessary to protect the dam against the effects of cracks.

Differential Settlements – Practical Examples

Transversal Cracks in a Dam Crest

El Infernillo a Mexican dam was taken as an example of transversal cracks, in spite of fact that cracks had a rather limited extent. But an investigation of this dam has been carried out very carefully and in great detail, and in this respect obtained data became a reference data, e.g. Marsal and de Arellano (1967), Wilson and Squier (1969), Wilson (1973), Marsal (1977). El Infernillo dam is a rock-fill dam with a narrow central clay core, it is 148 m high, the crest is 344 m long and the width in the dam is 570 m, Fig. 7.8a. After construction of cofferdams and the river devolution, the filling has started in August 1962. The dam was finished on 7th of December 1964 and filling the reservoir started on 15th of June 1964. A narrow core was made of clayey soil ($w_L = 49\%$, $w_P = 24\%$) the compaction was done in layers of 0.15 m thickness under an average moisture content $w = 23\%$ (according to Proctor modified compaction test the optimal moisture content is $w = 19.3\%$).

Sealing was protected by filter layer from alluvial sand, by transitional zone and by stabilization zone from rock which was got from a quarry. Figure 7.8b presents vertical and horizontal (in the direction of longitudinal section) core deformations divided into the stages of construction and the first filling. It is obvious that there is a good concordance with theoretical assumption, see Fig. 7.3 and 7.4. Figure 7.8c shows horizontal displacements of the centre of clay core in the direction of water flow which developed after the end of construction phase. In the first two days of filling the sealing deflected against upstream direction. Vertical deformations of the dam crest and time of their development are shown in Fig. 7.8d. It is obvious that speed of settlement together with time is falling and a maximum value of settlement is 1.0 m after 10.5 years. Horizontal displacements of the whole dam crest in the direction of water flow are presented by Fig. 7.8e. A sharp bend in displacement curve λ_x by banks lasts here in spite of hydrostatic pressure on sealing occurring here for more than 10 years. Figure 7.8f shows values of longitudinal relative deformations of a dam crest and time. In central part a compression of 0.2–0.4% was found, by sides there is an extension approx. 0.3% on the right-hand side and 0.7% on the left-side of a bank. On the third day after the beginning of filling the reservoir the crack of 10 mm in width occurred by the left bank and a series of narrow, 1 mm wide, transversal cracks occurred by the right bank. At this moment the extension was smaller, it was about 0.1–0.2%. These cracks were repaired, and new ones have not occurred, though the crest elongation continued for the time being and in 1975 they reached a maximum – approx. 1.2%. A deformation of upstream stabilization zone caused by saturation was declared as the main cause of cracks development.

Longitudinal Cracks in a Dam Crest

In consequence of filling the reservoir the trajectories of upstream points of a dam crest move in upstream direction, the trajectory of downstream points move vertically or in downstream direction. Due to these different trajectories the central part of a dam crest elongates and longitudinal cracks occur. The final effect is strengthened if there is a larger compressibility of stabilization zone in comparison with a transitional zone, a filter or even sealing, Alberro (1977).

Gepatsch dam is a rock-fill dam, it is 153 m high, with central clay core and steep slopes 1:1.5 with berms. In the highest part the slope was 1:1, Schober (1967, 1977). Figure 7.9 shows trajectories of displacements of points A, B, C, D in relation to time and fluctuation of water in the reservoir. Measuring points move nearly horizontally in the direction of water flow during the filling the reservoir and sloping down in opposite direction of water during emptying. Absolute values of displacements are decreasing in connection with time and a number of filling and emptying cycles. A horizontal distance of the points A and B is getting larger and after 10 years it reached 0.488 m, and a horizontal distance of the point C is already 0.998 m. A maximum settlement of the crest is 1.22 m, i.e. approx. 0.8% of a dam height. The

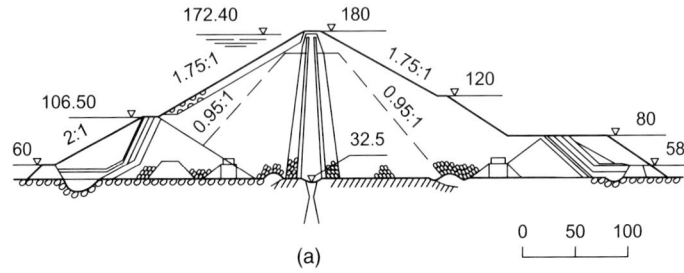

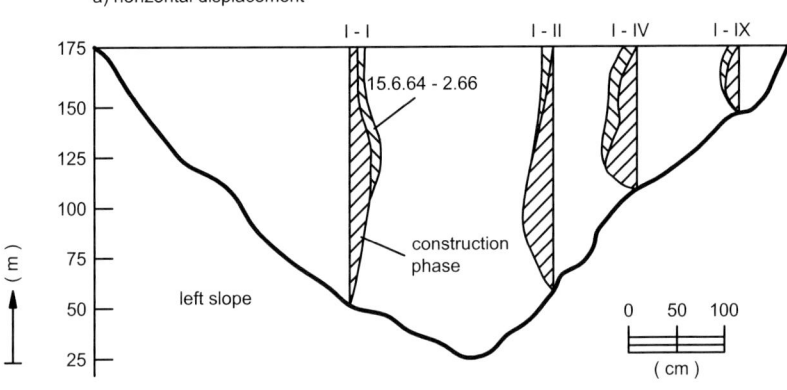

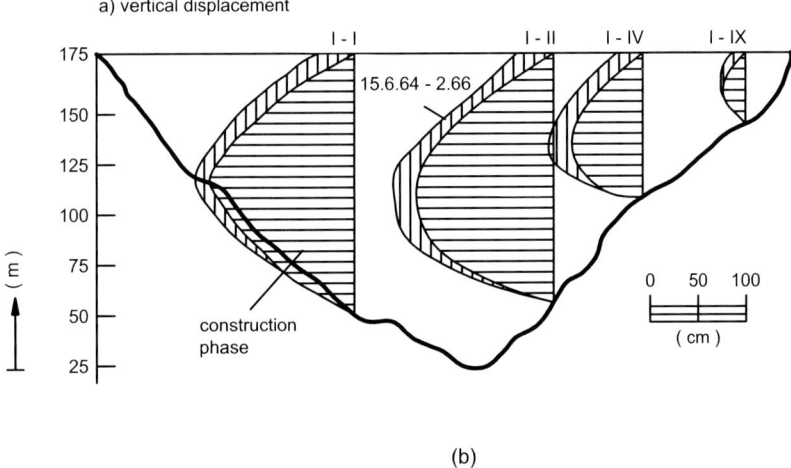

Fig. 7.8 El Infiernillo dam (according to Wilson). (**a**) Cross section. (**b**) Horizontal and vertical displacement of clay core during construction and first watering. (**c**) Core horizontal displacement along its height. (**d**) Vertical displacement of dam crest with time. λ_z – settlement (in centimeters). (**e**) Horizontal displacement of dam crest with time. λ_x – horizontal movement (in centimeters); M_i – measuring blocks. (**f**) Longitudinal strains of dam crest as function of time

7.4 Conditions and Reasons of the Tensile Cracks Development

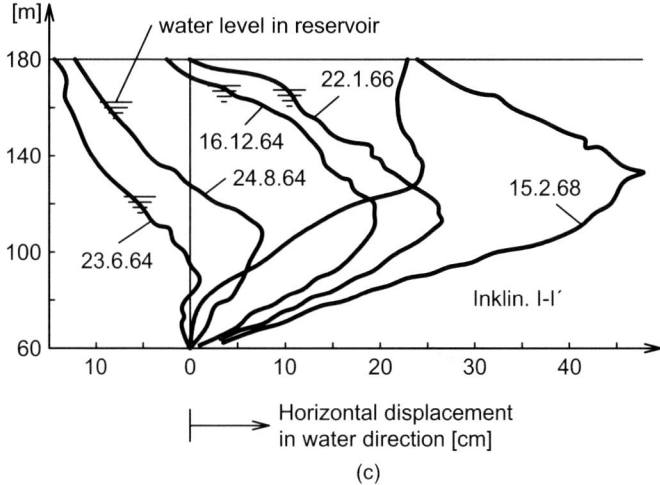

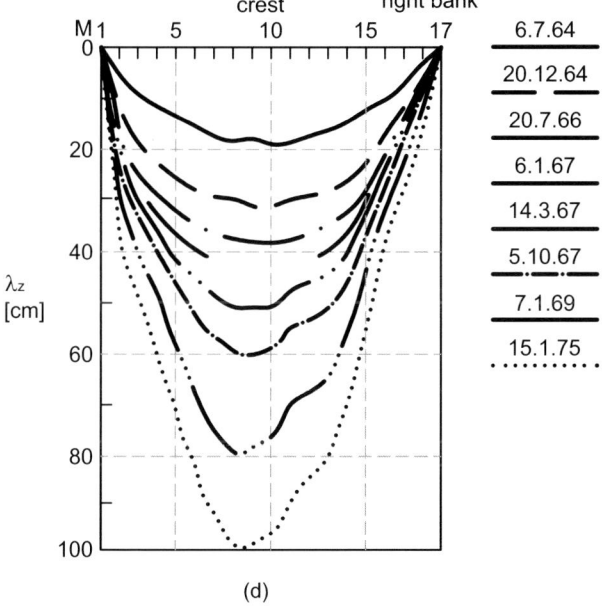

Fig. 7.8 (continued)

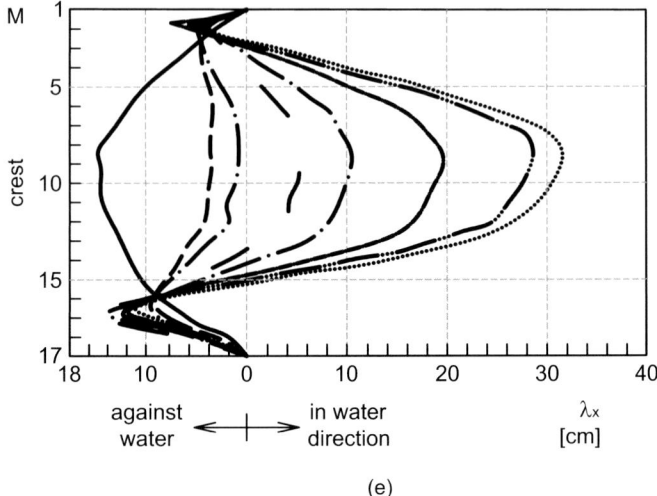

(e)

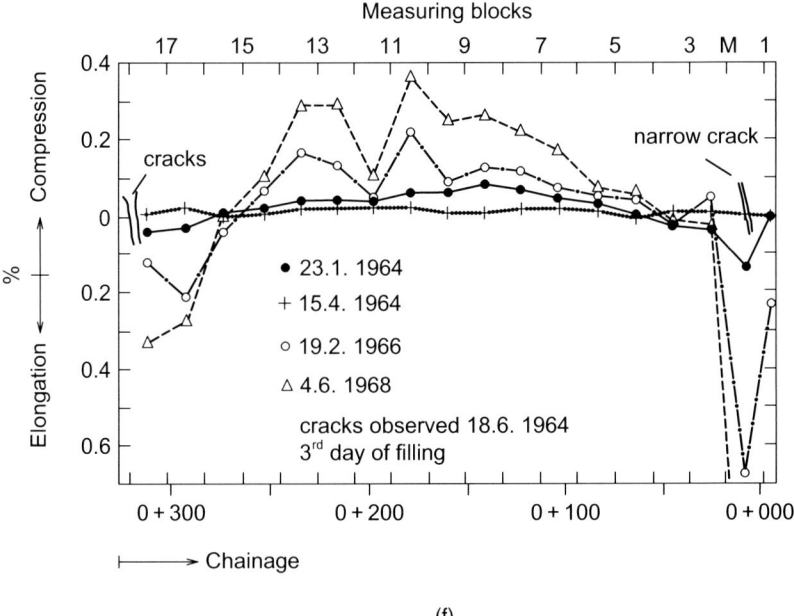

(f)

Fig. 7.8 (continued)

7.4 Conditions and Reasons of the Tensile Cracks Development

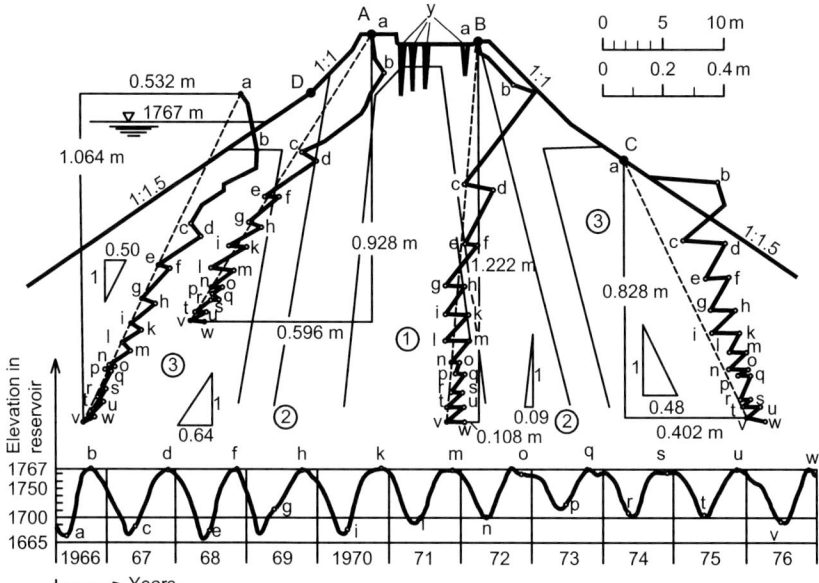

Fig. 7.9 Gepatsch dam – Crest movements (according to Schöber). A,B,C,D – measuring points; a–w – time measuring points; Y – longitudinal cracks on dam crest. 1 – clay core; 2 – transition zone; 3 – stabilization zone

depth of the cracks is relatively small, in 1967 the cracks were about 2 m deep and they reached up to the sealing.

Longitudinal Internal Cracks

Alberro et al. (1976) describe problems of the Guadalupe dam, which had to be reconstructed twice. The main problem was a variable thickness of compressible alluvial deposits of subsoil, Fig. 7.10a. The first dam was a rock-fill dam, it was 28.5 m high, with an upstream concrete sealing which was connected with an underground concrete wall 26 m deep. In 1947 after the first filling there was a high seepage loss – 4 m^3.s^{-1}; the reservoir was drained off and repaired. The concrete sealing was perforated, covered with 1 m thick filter and upstream soil sealing, Fig. 7.10b. Soil sealing was from clayey sand and it was relatively stiff. In 1949 after reconstruction the reservoir was filled with water again. A higher seepage (0.5 m^3.s^{-1}) was registered in September 1952 and the reservoir was drained off once more. On the surface of upstream face there was a depression in rock-fill cover. After exposure of material there was found a longitudinal crack connected with a cave 2 m high, 3 m wide and 5 m long going towards the place of connection with an underground wall. It was determined that the cause of the development of this crack was a differential settlement between downstream and upstream zones. The explanation may

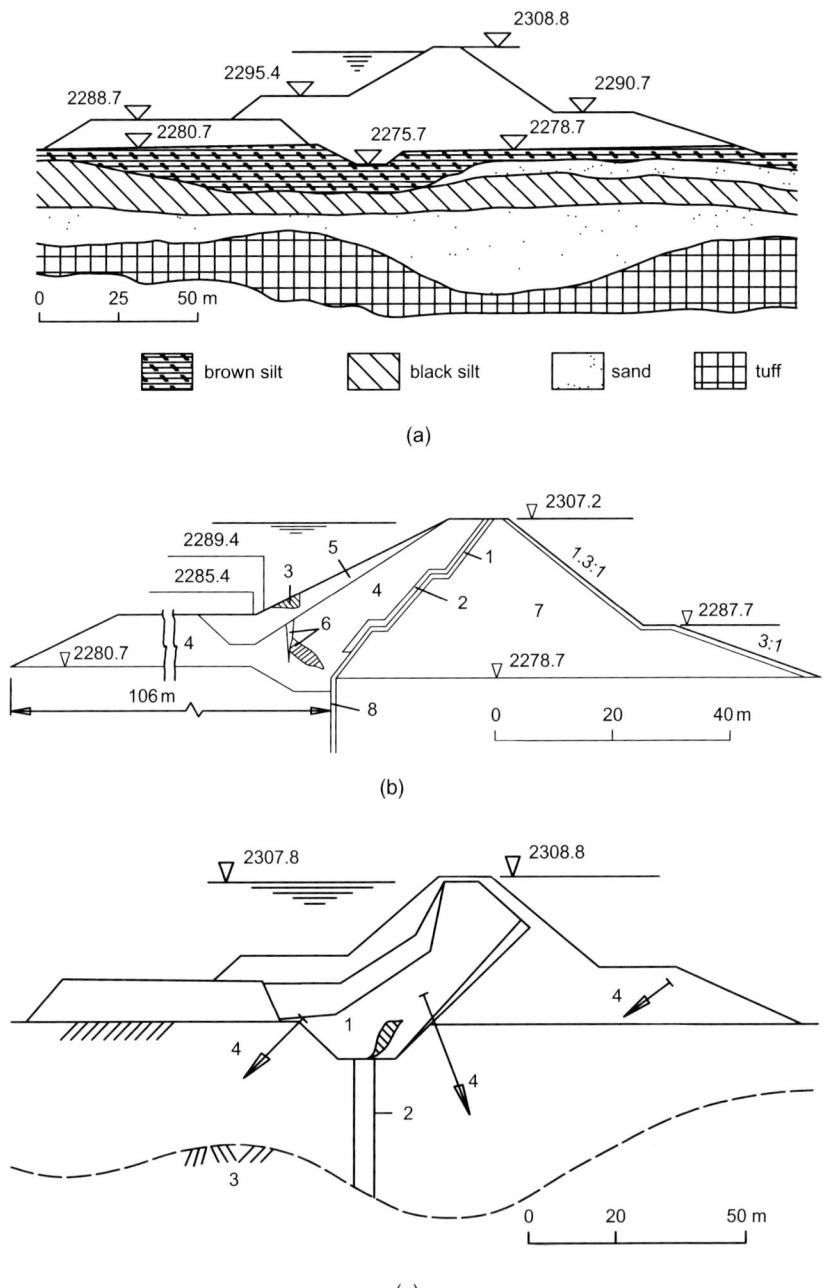

Fig. 7.10 Guadalupe dam (according to Alberro et al.). (**a**) Schematic cross-section and geological profile of subsoil. (**b**) Dam II. 1 – perforated old concrete sealing; 2 – filter; 3 – sifting on upstream face; 4 – clay core; 5 – rock-fill; 6 – crack and cavern; 7 – stabilization zone; 8 – sealing wall. (**c**) Dam III. 1 – tensile crack; 2 – sealing wall; 3 – rock subsoil; 4 – vector of movements with scale

7.4 Conditions and Reasons of the Tensile Cracks Development

be connected with fracturing above rigid element of concrete wall, when both zones settled more.

Remedial actions connected with making a dam higher by about 1.6 m – began at the end of 1967. Material of the upstream zone was removed up to the rock-fill under the initial concrete sealing. A cut-off trench was made along the underground wall and a new upstream sealing from plastic clay, Fig. 7.10c. A new built part of the dam was well-equipped with instruments. Above-mentioned processes of a differential settlement were noticed just in the case of the second dam, but to a smaller extent. Monitoring devices situated in the place of connection with an underground wall indicate a tensile stress and probable development of longitudinal crack in an upstream zone connection. A good function of the dam, which started in 1975, is due to the use of plastic clay and to the reduced differential settlement of subsoil.

7.4.2.2 Cracks Initiated by Hydraulic Fracturing

Bjerrum (1967a) paid attention to the risk of the tensile cracks in a core of the dams caused by hydraulic fracturing. A narrow core with high water content is much more compressible than the supporting members of the dam. The result is that the weight of the core will hang on the filter material and the vertical pressure to which the core is subjected will never be comparable to the nominal overburden pressure. It means that such a core would never consolidate fully. This would not be a serious thing in itself, but what is worse is that when a core is standing with a low vertical pressure and is a subjected to a horizontal water pressure exceeding the existing vertical pressure, we have the risk of hydraulic fracturing.

Bjerrum et al. (1972) also deal with hydraulic fracturing but from a different point of view. He describes its development during the field test of soil permeability (an infiltration test during which the overpressure is applied). He derives an equation defining the ratio of pore pressure to the vertical pressure under which the hydraulic fracturing will start. This case is closely linked with an artificial crack development in the sealing of a dam during drilling. Vaughan (1972) defines the ratio between the pore pressure and the minor main stress. Results of the hydraulic fracturing tests conducted using Geonor piezometers installed in the core of Alvito dam are presented by Seco e Pinto and Das Neves (1985).

Hydraulic Fracturing – Practical Examples

Hyttejuvet Dam

Kjaernsli and Torblaa (1968) describe their experience with the Hyttejuvet dam. This rockfill dam, 90 m high, and a 400 m long crest, was built in Norway in 1964–1965. A cross-section profile, Fig. 7.11a is characterised by sudden change of the width of sealing core. Change of core width was a certain reaction on high pore pressures in the low part of sealing, which have been measured by the end of 1964. Grain size distribution curves of materials used for construction of a dam are shown in Fig. 7.11b. In the case of the core there is a soil of glacial moraine with a

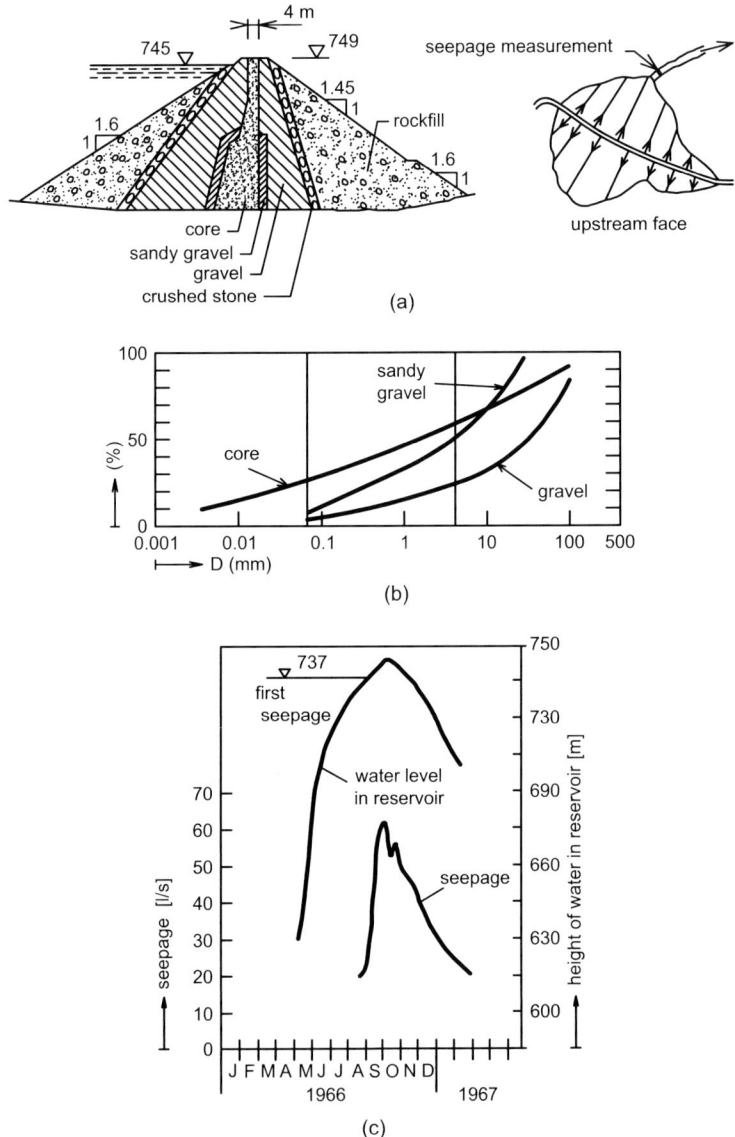

Fig. 7.11 Hyttejuvet dam (according to Kjaernsli and Torblaa). (**a**) Cross section and plan. (**b**) Grain size distribution of used soils. (**c**) Relation between reservoir water level and measured seepage

continuous curve. A liquid limit is 21%, and plasticity index $I_p = 6$. Compaction of sealing was done when moisture content was $w = w_{opt} - 1\%$ up to $w_{opt} + 2\%$.

During the first reservoir filling, when water reached nearly the maximum level (elevation 737) a concentrated seepage of turbid water had occurred, Fig. 7.11c.

7.4 Conditions and Reasons of the Tensile Cracks Development

A maximum seepage loss of 62 litres per second had decreased a little after the reservoir was filled up to a maximum level, then seepage substantially decreased when water in reservoir fell down. A pressure transducer which is placed in the sealing core 21 m under the crest, i.e. 17 m under the maximum of water level gave really useful information. In October 1965 a vertical pressure 170 kN.m^{-2} was recorded, in June 1966 before filling the reservoir, pressure reached even lower value of 140 kN.m^{-2}, but in October 1966 after filling the reservoir it was 230 kN.m^{-2}. These measurements indicate that more than 50% of the weight of soil above the place of measurement was transferred to a filtration or stabilization zone.

A sudden increase of seepage can be explained by the presence of an opened crack at the elevation 737 or by sudden development of the crack bellow this level. As seepage continued (to a lesser degree) during the falling of water level under this elevation, the crack had to be lower and closed otherwise the crack must not have existed before the reservoir filling. A small decrease of seepage before reaching the maximum level could be explained by the partial collapse of the crack in sealing. The sealing was grouted in summer 1967 and in spite of this seepage 15–20 l.s^{-1} was recorded after the repeated filling the reservoir.

Balderhead Dam

A very interesting case of an internal failure which was observed even on the crest of an embankment dam is described by Vaughan et al. (1970). It concerns a rockfill dam Balderhead, 48 m high, built in England in 1961–1965, Fig. 7.12a,b. A sudden increase of seepage of nearly full reservoir, was observed, Fig. 7.12c. It was found that during an increased seepage the place of sealing failure was not on the present water level, but it was deeply below it. The conclusion was that the cracks were created suddenly by hydraulic fracturing. This fact was indicated by a piezometer. Data of the piezometric height are in good accordance with a water level of the reservoir up to a sudden increase of seepage (point *A* on a time scale in Fig. 7.12c). We can see that the conditions have been seemingly nearly stabilized for 6 months (segment *A–B*). A few observations of turbid seepage water have shown that only a small amount of solid particles was washed out. In the next period (segment *B–C*) the seepage had been increasing, a piezometric level had decreased and conditions had changed into unstable. Turbidity of seepage water became greater because of a larger quantity of fine particles. At the end of this second 6 month long period the first, and after a short time, the second hole appeared in the crest of the dam. Therefore a water level was lowered in the reservoir by 7 m. After this lowering all measured values returned to normal limits, though the water level remained above the place of failure. From this point of view we can conclude that a water pressure was not so high to keep the cracks open.

The following investigation found out internal failure zones besides the two main failure zones, which propagated to the dam crest. Main failure zones contained only an insignificant amount of clay, but washed sand and gravel particles. In another place a thick layer of sand washed from clayey gravel was observed. Speed of seeping water was at the moment sufficient for taking the coarse particles to the filter where

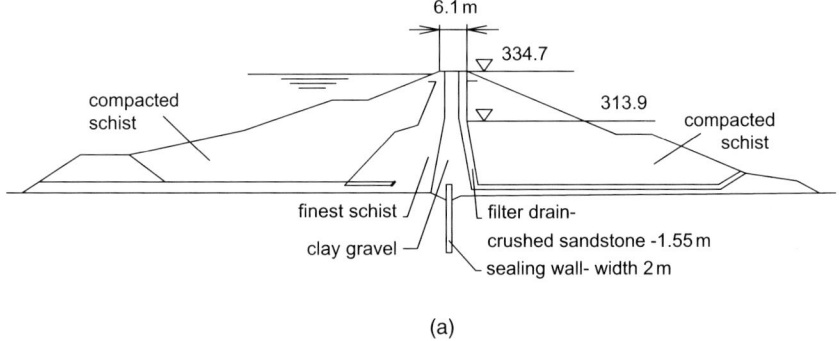

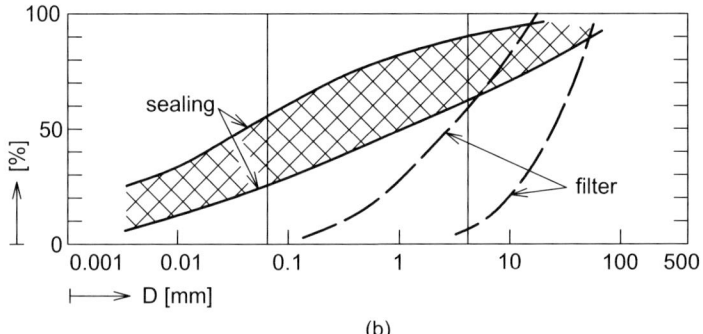

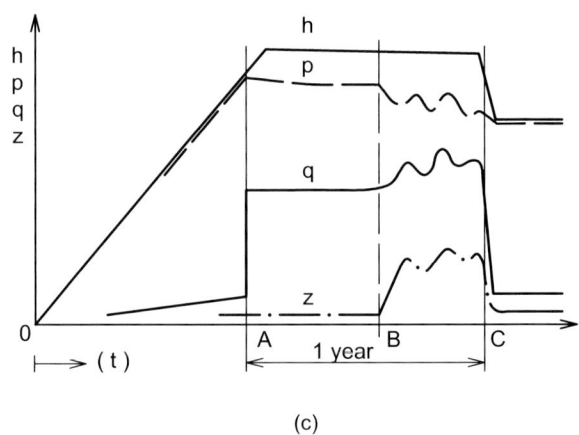

Fig. 7.12 Balderhead dam (according to Vaughan et al.). (**a**) Cross-section. (**b**) Grain size distribution of used soils. (**c**) Results of monitoring. h – reservoir water level; p – level in piezometer P; q – seepage; z – amount of washed-up fine particles. (**d**) Cross-section through failure zone. 1 – clay core; 2 – crater at surface; 3 – investigation probe; 4 – vertical filtration drain; 5 – zone affected by erosion; 6 – concrete sealing wall; P – location of piezometer P

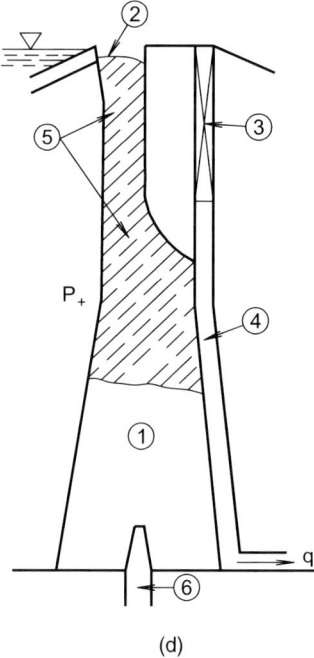

(d)

Fig. 7.12 (continued)

they partially helped to seal the cracks. But in main failure zones this phenomenon has not occurred, Fig. 7.12d. After the repair of the sealing by injection and by concrete diaphragm wall the increased seepage was not noticed. The application of sensitive and useful piezometers (69 pieces) put near the upstream face of sealing was very practical because the allowed monitoring of possible seepage paths.

7.4.2.3 Susceptibility of Dams to Cracking

In view of discussed points and described practical examples we can conclude that a tendency to develop cracks in fill dams is determined by more factors. Some of them are:

- Geometric factors – a height of a dam, a steepness of valley slopes, a width of clay core, its location,
- Material factors – a moisture content of the clay core, a ratio between modulus of deformation of clay core and stabilization zones, non-homogeneity of subsoil,
- Technological factors – filling the reservoir during the construction of the dam.

Numerical parametric studies, e.g. Doležalová (1970), Eisenstein et al. (1972), Kulhawy and Gurtowski (1976) as well as practical experiences obtained during the reservoir filling – Ničiporovič (1969) proved to be a very useful tool for studying of the individual factors. These factors will be more specified at the end of chapter.

7.5 Stress-Strain Behaviour of Soils in Tension

The significance of observation stress-strain behaviour of soils in tension was stressed by the author, Vaníček (1977c). Reasons for studying this phenomenon are different but basically we can divide them into two groups:

- Practical reasons – connected with the potential risk of tensile cracks – mostly in dam engineering, e.g. Vaníček (1971), Leonards and Narain (1963), Krishnayya et al. (1974), but recently also in landfill engineering,
- Theoretical reasons – connected with the behaviour of the soil at small, resp. tensile stresses, e.g. Rosenquist (1959), Bishop and Garga (1969), Parry and Nadarajan (1974).

But the division is not as strong as we can see in the example of earth dams. The risk of damage is connected not only with tension cracks but also with erosion (susceptibility to erosion) of the material. And the susceptibility to erosion is a function of the results of attractive and repulsive forces between soil particles.

There are much more practical examples where a tension zone can be observed as slopes, retaining walls, and tunnels. In classical solutions, this reality is usually neglected, but numerical methods such as FEM can take this fact into consideration. High potential risk exists for protection dams with a sealing carpet, see Fig. 7.13 or during compaction when the modulus of deformation decreases with depth, Fig. 7.14.

If we do not take into account the dimensions of tested samples, their preparation or time effect, the principal classification can be done by

- Principle of loading,
- Drainage conditions,
- Opportunity to measure elongation.

Furthermore, a brief description is provided about the principle of loading with remarks about two other aspects, see Fig. 7.15.

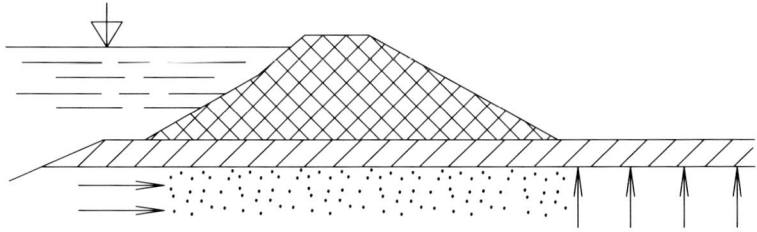

Fig. 7.13 Example of potential risk of tensile loading for protection dam with sealing carpet

7.5 Stress-Strain Behaviour of Soils in Tension

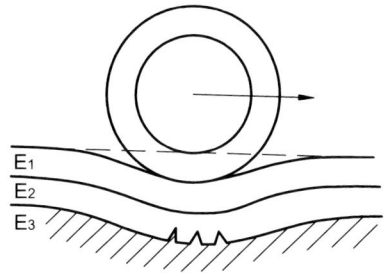

Fig. 7.14 Possibility of tensile cracks creation during compaction: $E_3 < E_2 < E_1$

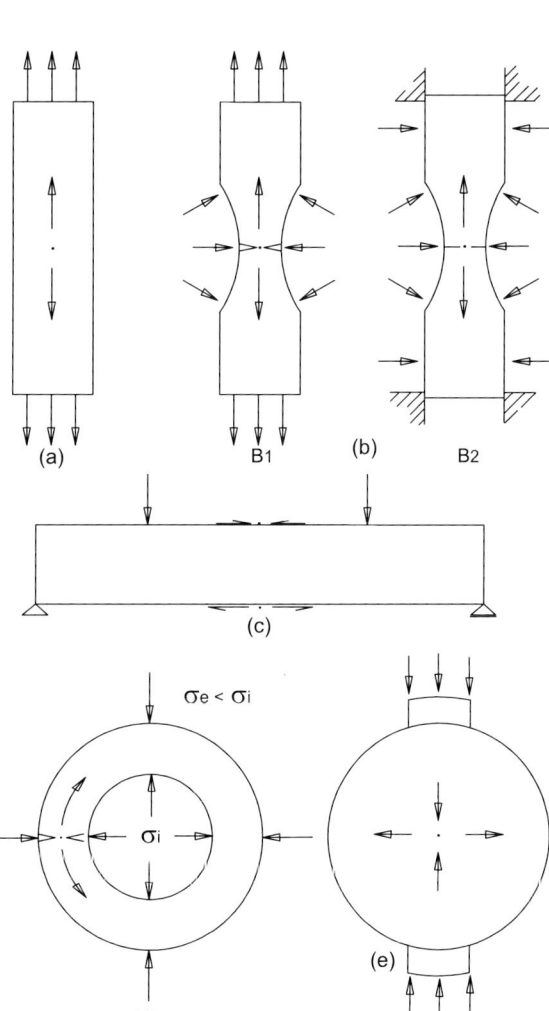

Fig. 7.15 Division of tensile tests according to principle of loading. (**a**) Axial tensile test (direct tension test). (**b**) Triaxial tensile test. (**c**) Bending test. (**d**) Test on hollow cylinder. (**e**) Indirect (Brazilian) tensile test

A. Axial tensile test (direct tension test)
B. Triaxial tensile test
C. Bending test
D. Test on hollow cylinder
E. Indirect (Brazilian) tensile test.

7.5.1 Tensile Tests Under Undrained Conditions

These tests are beneficial when checking the potential risk of dry cracks or wet cracks under undrained conditions. Cracks created on the dam crest, where vertical stress is very small, correspond to the loading conditions valid for an axial tension test and a bending test. The bending test especially represents the deflection of the dam crest. Therefore this test will be described in more detail. Other tests are performed under a more general state of loading, which better corresponds to inner crack creation. Tests on a hollow cylinder or the indirect tension test are rather the exception, see e.g. Šuklje and Drnovšek (1965) or Symes et al. (1984) for hollow cylinder and Narain and Rawat (1970) or Krisnayya and Eisenstein (1972) for the indirect, (Brazilian), tension test.

7.5.1.1 Axial Tensile Test

The transfer of the tensile stresses on the sample is probably the biggest problem for this type of test, because of the risk of adjoining (uncontrollable) stresses. Different proposals were connected with freezing, with clueing or just with friction between sleeve and sample. The experiment described by Tschebotarioff et al (1953) is very well known. The shape of the soil mould is similar to the shape of briquette. Tensile force is applied in a horizontal direction. One end section of the soil mould is fixed and remains stationary during the entire test. Ball bearing rollers permitting frictionless movement support the rest of the mould. The strain measurements were made on the central part.

7.5.1.2 Triaxial Tensile Test

The type of test, $B1$ will be described in more detail in the next chapter – tests performed under drained conditions. The type of test $B2$ was described by Ter-Martirosjan (1973, 1977), Zaretsky et al. (1977). The scheme is shown in Fig. 7.16a. A sample has a reduced central section. Independence on the ratio of the cross section areas of the central and end parts creates a complex state of stresses when the sample is partly under pressure and partly under tension, see Fig. 7.16b. Tensile stress in the central cross section is given by equation:

7.5 Stress-Strain Behaviour of Soils in Tension

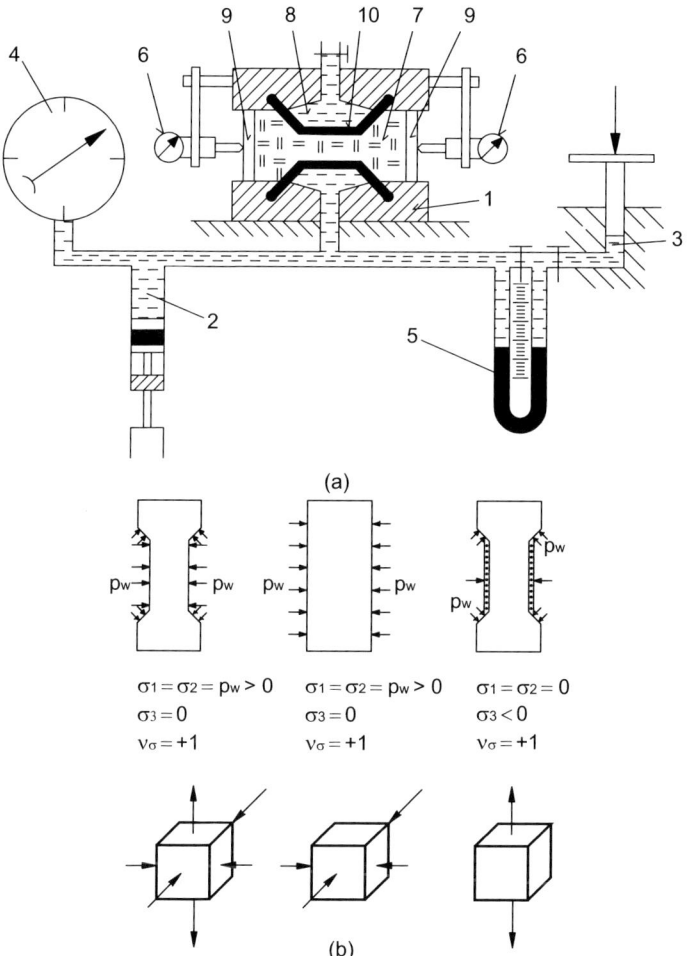

Fig. 7.16 Triaxial test type B2 (according to Ter-Martirosjan). (**a**) Scheme. 1 – sample socket; 2 – source of water pressure; 3 – loading stabilizer; 4 – pressure gauge; 5 – mercury volume gauge; 6 – strain gauge; 7 – tested sample; 8 – destilled pressure water; 9 – frontal base; 10 – rubber membrane.(**b**) Different arrangement of the loading state

$$\sigma_3 = p_w(1-A_F/A_c) \tag{7.15}$$

where p_w is the hydrostatic pressure acting on the sample via a flexible membrane.

For a cylindrical sample, the classical test at elongation (extension) can be modelled. Again when the reduced central part is covered by a stiff socket the unconfined tensile test is modelled ($\sigma_1 = \sigma_2$ and $\sigma_3 < 0$). Deformation is measured at the ends of the sample only.

7.5.1.3 Bending Test

The principle of the test consists in loading a soil beam sample by a pair of forces in the central part. Loading by such a pair of forces has such advantage that between these two forces, the shearing force is zero and the bending moment is constant. It is pure bending. The results of the bending test can be evaluated according to different theories of theoretical mechanics:

- Elastic theory,
- Navier's hypothesis (direct method),
- Differential method.

Fundamental assumptions of these theories are shown in Fig. 7.17. Elastic theory assumes that deformation and stresses at extreme tensile and compression fibres are equal. This theory was used by Leonards and Narain (1963). For the evaluation of measured data we need only the bending moment and the deflection curve of the beam.

Navier's hypothesis assumes that plane sections remain plane after bending and that the stress is linearly proportional to strain. This assumption was used by Vaníček (1971, 1977a,c).

The differential method assumes that plane sections remain plane and no creep occurs during bending. This method is not based on any preferred stress-strain law and is considered to be superior to the other two. This method was used by Ajaz and Parry (1975a,b).

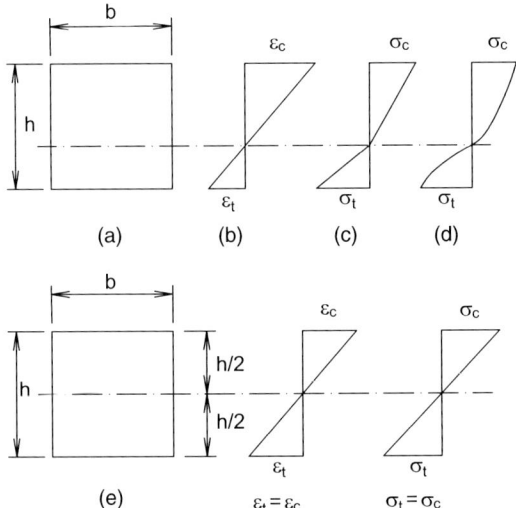

Fig. 7.17 Bending test – fundamental assumptions. (**a**) Beam cross section. (**b**) Strain diagram. (**c**) Stress diagram for Navier's hypothesis (direct method). (**d**) Stress diagram for differential method. (**e**) Elastic bending theory

7.5 Stress-Strain Behaviour of Soils in Tension

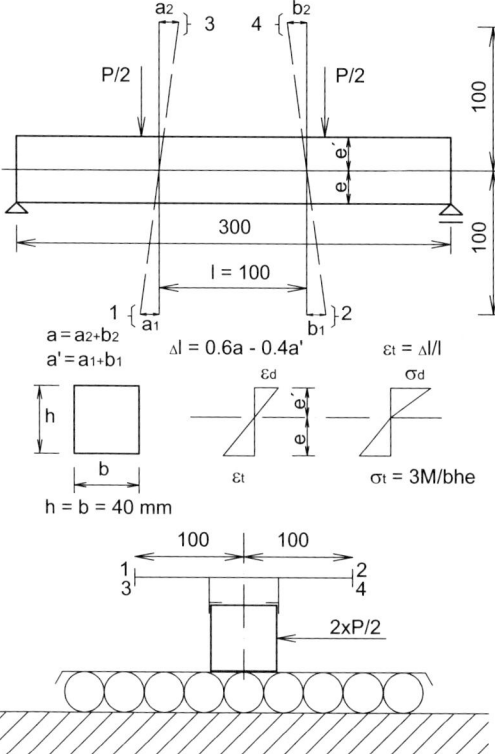

Fig. 7.18 The layout of the arrangement and evaluation of bending test proposed by author

The layout of the measuring device used by the author is presented in Fig. 7.18. The beam was 300 mm long and 40 ×40 mm in cross-section and was loaded in a horizontal direction. The weight of the beam and friction were eliminated by means of a horizontal ball-bearing. In the central part of the beam, which deforms under a constant bending movement, are two installed detectors connected with the beam. Deformation is measured at their ends with a magnification of five times. The tested soil is pressed in four layers in a special form in such a way for selected moisture content so that the demanded dry density is reached; e.g. determined by the Proctor standard test. After that the beam was coated with a hot mixture of oil and paraffin to prevent desiccation. Loading was increased at a constant rate in the direction of the compacted layers.

7.5.2 Results of Undrained Tensile Tests

Although the tests described above differ in many directions, some results and general conclusions are identical and provide the opportunity to forecast the behaviour of soils during tension loading.

During the investigation of the behaviour of clay cores of earth and rock-fill dams it is important to observe the aspects which can have some influence on the decision making process connected with soil modification – especially any modifications which assist in improving the adaptability to tensile deformations. Brittleness or ductility of the clay core is not only the function of initial moisture content but also the function of the degree of compaction. Even time factors play a certain role, which can be positive in relation to the higher ductility during slower deformation or negative as a result of the process that increases the structural strength by cementation bonds between individual particles. The investigation of different factors is based on the results from the core of Dalešice dam, which was finished in the Czech Republic at the end of 1970s. However the results will be compared with other published ones.

7.5.2.1 The Influence of Moisture Content on Tensile Characteristics

The Influence of Moisture Content on Tensile Strain at Failure

Tensile strain at failure increases and very often a linear interpolation is applied for rather nominal changes of moisture content than an optimal average. A wider range of moisture content is shown in Fig. 7.19 for the proposed relationship. The increase of the tensile strain at failure is relatively steep for a small change in moisture content from the optimum. The gradient decreases at the end of the observed range and for $w = w_{opt} + 6\%$ first marked occurrence of the decreasing maximum tensile strain, Vaníček (1977d).

The possibility to evaluate the influence the variation of moisture content (from an optimum one) on tensile strain at failure – the percentage of tensile strain ratio has been proposed by Ajaz and Parry (1975a), Fig. 7.20.

The percentage tensile strain ratio is defined as $100 \cdot (\varepsilon_t - \varepsilon_{t(OP)})/(\varepsilon_{t(OP)})$ where ε_t is the maximum tensile strain at any moisture content and $\varepsilon_{t(OP)}$ is the maximum

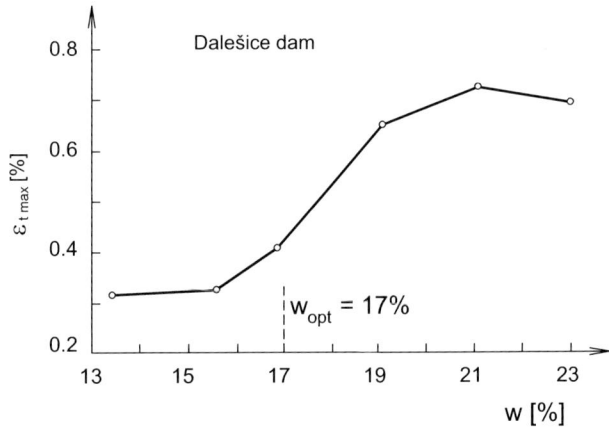

Fig. 7.19 Influence of moisture content on tensile strain at failure

7.5 Stress-Strain Behaviour of Soils in Tension

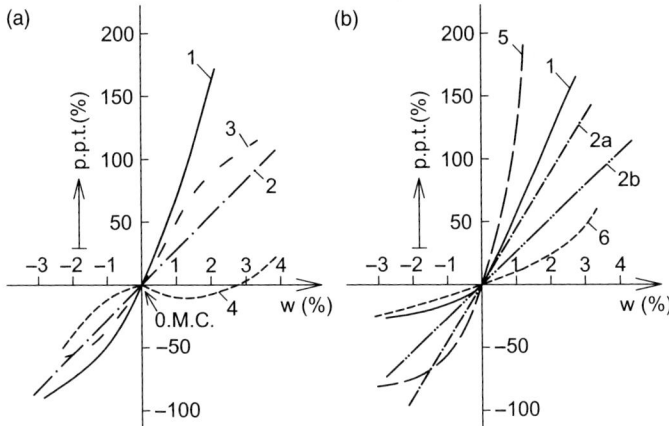

Fig. 7.20 Percentage tensile strain ratio (p.p.t.) as function of change of moisture content against optimal one (according to Ajaz and Parry) (O.M.C: – optimum moisture content according to Proctor standard test). (**a**) Bending tests. (**b**) Axial tensile tests and Brazilian test. 1 – balderhead clay; 2 – gault clay; 2a – controlled strain; 2b – controlled stress; 3 – silty clay; 4 – limestone clay; 5 – Mica till; 6 – Obaradai clay

tensile strain at the optimum moisture content for a given soil. It can be seen that, except for limestone clay, the maximum tensile strain increases significantly with the increase of moisture content above the optimum, irrespective of the type of tension test. The maximum increase in the tensile strain at failure is observed to be for Balderhead clay where compacting 2% wet of optimum increases the tensile failure strain by 172% on the basis of bending tests and by 101% on the basis of direct tension tests. For Mica till (Krishnayya et al. 1974) where the Brazilian test has been used; the increase in strain becomes disproportionately high at water contents greater than optimum.

To compare the results from different authors and different soils the index of plasticity I_p was used, see Fig. 7.21 – Vaníček (1977). Individual points represent tests on compacted samples by energy using the Proctor standard under optimum moisture content. It can be seen that the values of maximum tensile strains are mostly in the range of 0.1–0.5% and increase with the value of index plasticity. But this increase is not so significant to unilaterally lead to an application on only plastic materials. The changes in maximum tensile strain are more affected by moisture content. It is shown in the same picture for material from Dalešice dam ($I_p = 15.8$) by a dotted line.

The Influence of Moisture Content on the Maximum Tensile Strength

Generally the tensile strength of the tested soils decreased with an increase in moisture content. For small changes around the optimum moisture content this relationship was nearly linear. Figure 7.22 shows the influence of moisture content for a wider range. The maximum value of tensile strength is reached for $w = w_{opt} - 3.5\%$

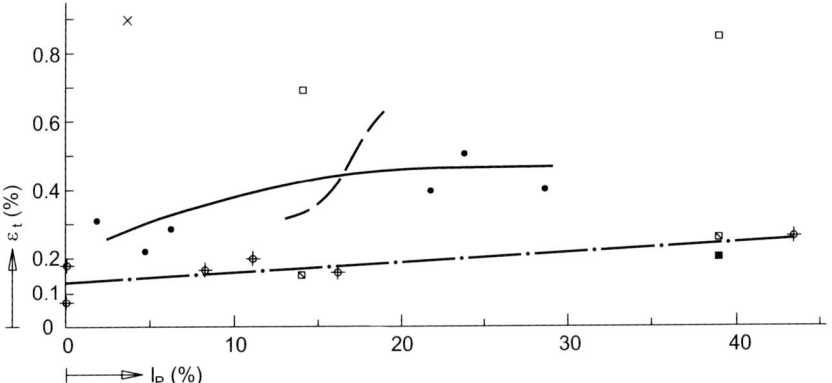

Fig. 7.21 Results of tensile strain at failure for different tests and soils compacted by Proctor standard at optimum moisture content

for the Dalešice clay core. We can expect a further decrease of the moisture content in the value of tensile strength due to lower dry density and a loss of capillary forces for very dry samples compacted from powder. With an increase in the moisture content the tensile strength falls.

Again the index of plasticity was used for the comparison of different tests and soils compacted at optimum moisture content, see Fig. 7.23. The increase in tensile strength with the index of plasticity is not convincing. For most samples with an index of plasticity lower than 30, the maximum tensile strength is in the range of 30–$80\,\mathrm{kN.m^{-2}}$.

The Influence of Moisture Content on the Tension Modulus of Deformation

The influence of moisture content is very important from the point of view of earth core cracking. The question is how much we can decrease the stiffness of the soil in

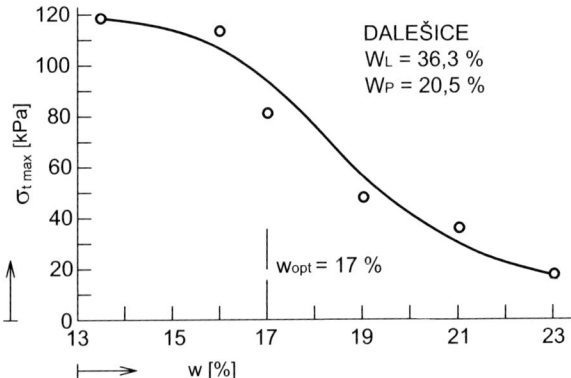

Fig. 7.22 Tensile strength as function of moisture content for clay core from Dalešice dam

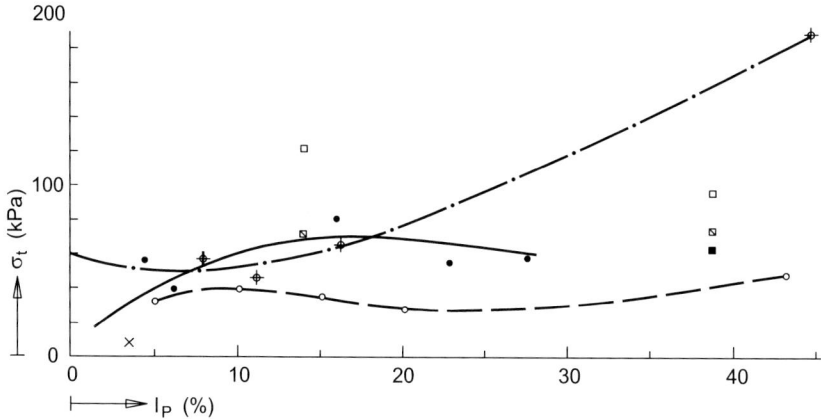

Fig. 7.23 Results of tensile strength for different tests and soils compacted by Proctor standard at optimum moisture content

the earth core by an acceptable increase in moisture content. The increasing flexibility and decreasing strength are partly limited to the differences in stresses, which have a tendency to create in the earth structure. Simultaneously, they can be a reason of higher relative and total deformations, which on the other side can help to create tensile cracks. Therefore, for each given case there is an optimum value of flexibility and strength, which will minimize the potential risk of crack development.

Two different tension moduli can be taken into account:

– Initial tangent modulus,
– Secant modulus at failure.

In all cases the tensile modulus decreases with moisture content increase. This decrease is again very significant around the optimum moisture content, for lower moisture content another decrease was not so significant. Results for the Dalešice clay core for secant modulus at failure are shown in Fig. 7.24. The obtained relationship is in good agreement with relation to $\sigma_t = f(w)$ – only slightly steeper. For the Dalešice clay core it seems to be useful to increase the moisture content above optimum about 1.5–2.0% – from 17% to 19%, to increase the flexibility but not to decrease tensile strength so much.

The Influence of Compaction Effort on the Tensile Characteristics

There are in principle two ways to study this influence:

– To study the differences for optimal moisture content for two basic compaction energies – e.g. the Proctor standard or Proctor modified. Here Leonards and Narain (1963) pointed out that an increase in the compaction effort substantially reduced the flexibility.

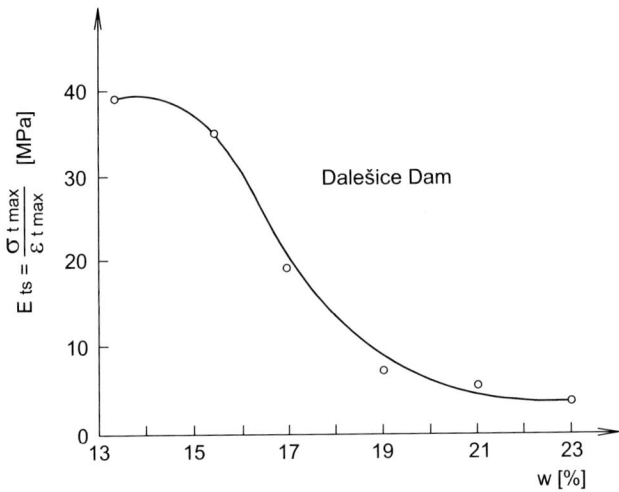

Fig. 7.24 Influence of moisture content on tensile secant modulus

– To study small changes of compaction energy e.g. against the Proctor standard for the same moisture content.

The second method was used – Vaníček (1975). For the certain moisture content the initial dry density is about 2.5% resp. 10% lower than from the standard Proctor test. One line represents a compaction effort for which the degree of saturation $S_r = 0.975$, Fig. 7.25a. The results are presented in Fig. 7.25b–d for tensile strength, maximum elongation and secant modulus. From the comparison of the results it is possible to summarize:

– The change in the compaction effort is more obvious for tensile strains, especially for wetter samples; lower elongation is for a lower compaction effort,
– The differences in tensile strength are not so visible, nevertheless for 90% PST yes,
– A sample compacted only at 90% PST always gave the lowest values.

Generally we can say that tensile strength and tensile strain at failure increase with compaction effort, so that the influence on the secant modulus at failure is negligible. This statement is very important because when the results of compaction are better from the view of maximum tensile strain, and so after that also shear strength and compression are better – shear parameters and modulus of deformation are higher.

7.5 Stress-Strain Behaviour of Soils in Tension

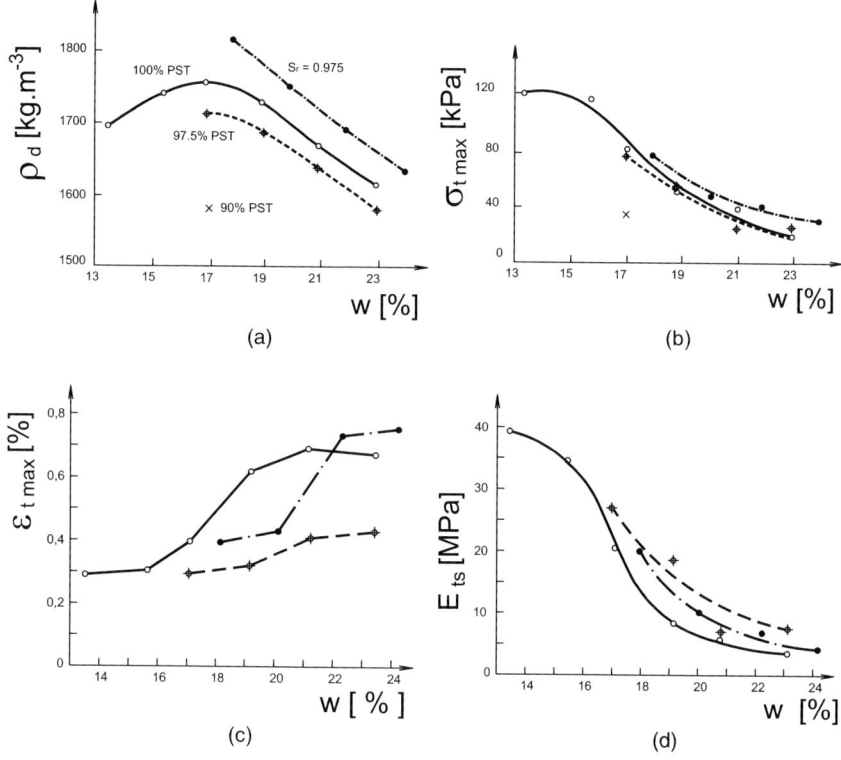

Fig. 7.25 Influence of compaction effort on tensile characteristics. (**a**) Different compaction effort. (**b**) Results of tensile strength. (**c**) Results of tensile strain at failure. (**d**) Results of tensile secant modulus

The Results for Beams Sampled from the Clay Core

The undisturbed samples were taken from the two sites from the surface of the clay core of Dalešice dam after a removal of a 100 mm thick upper layer, Vaníček (1977b). The size of block sample was approximately 150 × 150 × 400 mm. The sampling of block sample in situ was rather difficult, especially in the direction of the longitudinal axis of dam – i.e. in the direction of roller passes. In the laboratory, the tested beam was cut out from this block using a special knife. But only two beams were prepared to a full length of 300 mm and sampled in a perpendicular direction to the longitudinal axis without any visible cracks. Two other beams had very small cracks but not continuous ones. The other three beams had a length of only 200 mm. This problem displayed considerable extension of internal fissures with preferable orientation – which is perpendicular to the longitudinal axis of the dam. We can take the following reasons into account for the explanation of this fact:

– Tensile fissures after compaction by roller,
– The fissures were created by a crawler-mounted tractor,

- Non homogeneity of soils (the blocks of virgin soil with undisturbed structure connected with remoulded soil).

The last point was obvious for beam No. 7, which had a slightly different colour, consistency and was without the typical white calcium lattice of the original sample of loess material from the borrow pit. Good agreement with laboratory prepared samples was obtained only for beam No. 7. All others differ in some values. Extremely high values were measured for tensile strength – including samples, which had small fissures but not continuous ones. Their tensile strength was higher than for laboratory prepared samples, even the elongation was smaller and beams failed in the place where these small fissures were observed.

The problems connected with sampling of block samples and with the preparation of the beams together with the differences in the measured results are the evidence of fissures and non-homogeneity in the core material. The explanation of extremely high values of tensile strength for in situ taken samples is only due to the cementation bonds, in case of loess soil due to the strong undisturbed structure containing calcium lattice. This structure was remoulded during excavation and compactions only for beam No. 7 where good agreement with laboratory prepared samples was obtained. Nevertheless, the results show higher stiffness of the core material than was expected on the base of the laboratory prepared samples.

Results of Tensile Test Performed In Situ

A rather unique example is described by Romero and Rodrígues (1981). They give the following reasons to test tensile characteristics directly in situ:

- A rather great gradient of stresses and deformation in the cross section of laboratory samples with small dimensions, which have some influence on the strength in the tensile zone,
- In many cases the laboratory tests are performed only on finer fraction of clay core,
- The structure of soil particles in laboratory can be different from the structure after the deposition into the dam body, due to different methods used for compaction.

Romero and Rodrígues describe one in situ test completed during the construction of the Canales dam in Spain. The basic scheme is presented in Fig. 7.26. Experimental fill with a width of 8.5 m and high 2.5 m was compacted in the same way as other parts of the clay core. However, below this fill, a flat press constructed from flexible material was constructed. The flat press with a cross section of 300 × 3 mm was protected by a PVC plate and overlapped by a 3 mm thick steel slab. This slab transferred the loading from the flat press on the fill. The surface of the fill moved upward and its deflection was measured. A crack in the central part of the fill appeared when the elongation was of 0.24%. The duration of test was 9 hours. However, the used soil ($w_L = 65\%, w_p = 40\%$) was compacted on the dry side of the Proctor curve, with moisture content lower from optimum about 3–4%.

7.5 Stress-Strain Behaviour of Soils in Tension

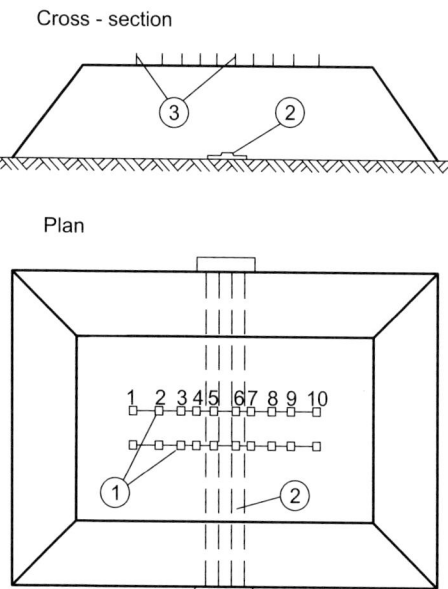

Fig. 7.26 Layout of tensile in situ test (according to Romero and Rodrígues). 1 – row of extensometers; 2 – flat press; 3 – measuring points

The Influence of Time

Time effect can have two dimensions:

– Firstly it is a time of test duration, where very quick and rather slow loading can be modelled. This aspect is important for the explanation of the El Infiernillo dam behaviour, where due to quick loading, the first cracks were observed even for rather small elongations of 0.1–0.3%. After repair of this place and for continued but much slower elongation new cracks were not observed.
– Secondly, it is time delay at which tests started after beam preparation.

For the first aspect there were different results presented by Tschebotarioff et al. (1953) resp. by Leonards and Narain (1963), but they performed tests under a different time of test duration. Author, Vaníček and Pauli (1973), therefore tested this aspect for the clay core of the Bulgarian dam Rosino. The results are shown in Fig. 7.27. The tensile strain at failure slowly increases for slower loading, supporting the observation from the El Infiernillo dam. Tensile strength is little bit higher for very quick and also very slow tests. The explanation is most likely in pore pressure distribution in tested sample under undrained conditions.

The second aspect is also interesting from a practical view. Most of the performed tests started in the range of 24 hours after the sample was prepared. But the core in the dam is not loaded by tensile stresses immediately after compaction. And finally, we know that during beam preparation, usually high pressures are used

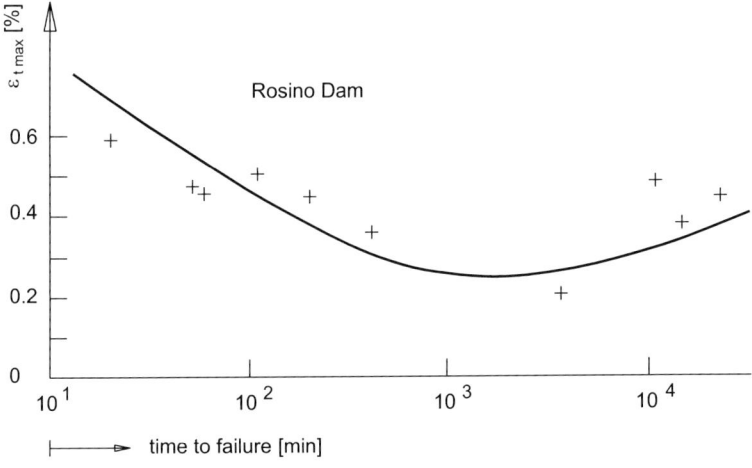

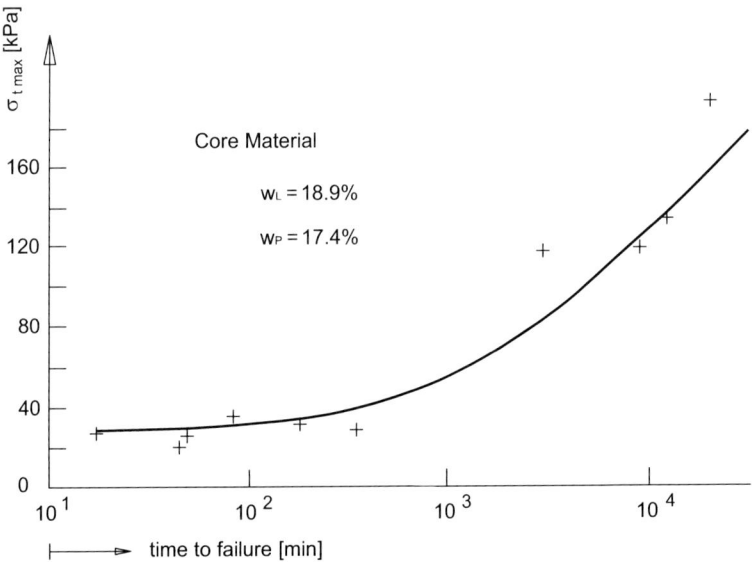

Fig. 7.27 Tensile strength and tensile strain at failure as function of test duration

during beam formation. After unloading negative pore water pressure can reach high values, which can increase tensile undrained strength. Series of tests really proved this expectation. We received not only lower tensile strength for tests which started with a delay 11, 31 and 107 days but also maximum tensile strain compared to the tests which started within 24 hours after beam preparation.

From the presented results it is obvious that the influence of time is a complicated problem and therefore the direct transmission of laboratory tests on the real

7.5 Stress-Strain Behaviour of Soils in Tension

behaviour of a clay core is not so easy. Different factors are blended together as local moisture content migration and thixotropic strengthening with time.

Comparison of Tension and Compression Modulus

When performing numerical analysis of the stress-strain behaviour of an earth structure there is a very important question connected with the differences between both tension and compression modulus. When the values are assumed equal the calculation is easier. On the contrary, not respecting the differences between them can influence the solution's authenticity. Therefore different authors have tried to compare obtained modulus; however the conclusions of these comparisons are multivalent. Leonards and Narain (1963) stated that these moduli are comparable; Lushnikov et al. (1973) compared direct tension and compression tests and concluded that the compression modulus is roughly 2–4 times higher and increases with moisture content increase. On the contrary, Ajaz and Parry (1975a) when comparing also both direct tests concluded that the tension modulus is higher.

The advantage of the bending test proposed is the possibility to compare these modulus during one test when the deformation of marginal fibres is measured and both tension and compression modulus are calculated from the obtained results with the help of Navier's hypothesis. In most cases the tension modulus was lower than the compression modulus, with a small exception for the long-term test.

Preferably this statement is possible to prove by tests performed for the Dalešice dam core material. All tests, not only tension, but also compression characteristics were calculated from the measured data. The initial phase of loading the compression modulus was only slightly higher than tension modulus, roughly about 10–30%. But when the secant moduli for state of failure were compared, the compression modulus was roughly in the range of 1.8–3.4 times higher than the tension modulus under the same conditions, see Fig. 7.28.

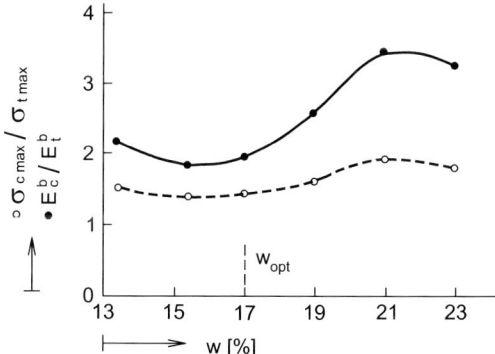

Fig. 7.28 Bending test: Ratio of compression strength to tensile strength and ratio of compression modulus to tension modulus as a function of moisture content

Summary of Undrained Tensile Test

To briefly summarize the results for undrained tensile tests the following points can be emphasized:

- Maximum tensile strength is in the range of $30-80$ kN.m^{-2} and is in the expected range of capillary forces (is roughly in the same range as negative neutral stresses in the soil after compaction),
- Maximum tensile strain at failure is roughly in the range of 0.1–0.5%,
- For the first phase of localization of the tensile zone by numerical modelling an equality for both compression and extension (tension) modulus can be reflected, but for detailed specification and localization of tension zone failures, it is necessary to distinguished between both modulus,
- Flexibility to tensile loading can be improved by small moisture content increase or by compaction effort, an additional small amount of bentonite presents also feasible solution,
- Time effect influencing the results of tensile characteristics have to be more precise for each individual case,
- Laboratory tests on laboratory prepared samples are very useful for parametric study, however the final check either on samples sampled from the original core or directly performed there seems to be helpful.

7.5.3 Results of Drained Tensile Tests

Bishop and Garga (1969) described the first triaxial drained tensile test without the use of end clamps, type $B1$ on Fig. 7.15. A sample with a reduced centre section is enclosed by a rubber membrane. The end caps will only become detached from the ends under the action of an axial tensile force T when the average effective stress at the ends of sample drops to zero. The axial effective stress throughout the centre section will at this point be negative (i.e. in tension), the magnitude of this tensile stress is dependent on the ratio of the end section and mid-section areas.

A controlled rate of the strain tension test with constant cell pressure will thus be a test with $\sigma'_1 = \sigma'_2 = constant$, and with σ'_3 decreasing (until the peak stress difference $\sigma'_1 - \sigma'_3$ is reached).

Bishop and Garga tested London blue clay ($w_L = 75\%$, $w_p = 29\%$) either on undisturbed samples, carefully sampled in situ, or on a remoulded sample. Differences in results are prominent. For the undisturbed samples, the measured effective tensile strength was in the range of 26.3–33.3 kN.m^{-2}, whereas for remoulded samples practically zero. Time to failure for the undisturbed samples was in the range of 6.7–55.2 hours and tensile strain at failure in the range of 2.19–16.7%. Some fundamental findings can be briefly summarized as:

- Failure of the sample has a character of a brittle material,
- Tensile stress at failure is almost independent of the value of σ_1' in the range examined,
- The variability of the maximum strain is rather large.

7.5 Stress-Strain Behaviour of Soils in Tension

There are some signals that the rate of loading for this type of clay was still insufficient to be full drained. The authors completed two series of drained triaxial tests performed at the Imperial College and at the Czech Technical University, using hydraulic triaxial apparatus described by Bishop and Wesley (1975), see Fig. 7.29.

During the first series instead of testing of Chalk Marl from the Channel Tunnel, the clay from the valley slopes downstream of Cod Beck Dam was used to study the behaviour of plastic clay material in a range of small compression and tension stresses and to study the influence of salinity of pore water on this behaviour, Vaníček (1977c). This clay was deposited in a fresh water lake and minimum of salts were supposed to be in the pore water.

The undisturbed sample with a moisture content $w_n = 28\%$ was remoulded and after mixing with a distilled water, the slurry was sieved through a British Standard No. 36 sieve (0.42 mm). Liquidity limit $w_L = 44.1\%$, plasticity limit $w_p = 18.5\%$, plasticity index $I_p = 25.6\%$.

The samples with a different salinity of pore water were prepared from the clay slurry. Samples No. 1, 2, 3, 7 and 8 (without salinity) were prepared by air drying to the moisture content near to natural (= 28%). Samples No. 4, 5 and 6 were mixed with brine (NaCl and KCl – dissolved in distilled water) so that samples No. 4 and 5 contained 2.5 g of salts per litre of pore water and the ratio of the cations of calcium (K) and sodium (Na) was K/Na = 0.2 for sample No. 4 and K/Na = 0.8 for sample No. 5. Pore water for sample No. 6 contained 5 g of salts per litre with a

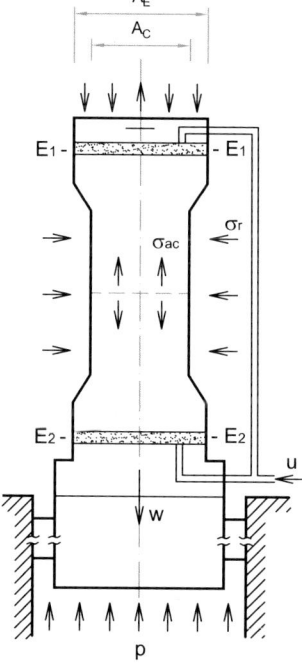

Fig. 7.29 Layout of the drained triaxial tension test performed in the hydraulic triaxial apparatus. σ_r – radial stress; p – stress in loading cell; u – back pressure applied to the drainage connection; W – weight of piston; σ_{ac} – axial stress on the centre section; A_E – area of end section; A_C – area of centre section

ration K/Na = 0.2. The clay slurry after was mixed with distilled water only, as the samples dried. Afterwards the clay-water mix was compacted by a wood stick in a special brass mould. French powder was used to improve removing the specimen from this mould. The format of sample corresponds with $B1$ on Fig. 7.15, with a diameter of 38.1 mm and a height of 76.2 mm.

7.5.3.1 Testing Procedure

An adapted hydraulic triaxial apparatus was used. Each sample was consolidated under small cell pressure, $10\,kN.m^{-2}$ or $20\,kN.m^{-2}$ against zero pressure (with an open air trap) to remove the air between the rubber membrane and the specimen. After that cell pressure (σ_r) and back pressure (u) were slowly increased to the value of $\sigma_r = 1\,000\,kN.m^{-2}$ and $u = 990(980)\,kN.m^{-2}$. Then the phase of consolidation began. Consolidation pressure $\sigma'_c = 600\,kN.m^{-2}$ ($\sigma_r = 1\,000\,kN.m^{-2}$, $u = 400\,kN.m^{-2}$) was used with an exception to test No. 1 ($\sigma'_c = 400\,kN.m^{-2}$).

After loading was completed, back pressure increased again so that the difference between the cell pressure and the back pressure was a low (from 5 to $25\,kN.m^{-2}$). The load cell was connected to the top cap and the test was started by applying a change in the pressure in the lower pressure cell.

7.5.3.2 Measured Results

Except for test No. 3, which was done as a drained compression test, all the others were performed as drained tension tests with drainage from both sides. The cross section area before the load cell was connected was calculated on the base of a measured volume change after loading and unloading or with the help of a cathetometer. The stress-strain curve was calculated for the narrowest cross-section area. The strains were calculated as average values – which means that the measured deformation was divided by the initial height of the sample.

Because the tests were performed as consolidated drained tests, great attention was devoted to the determination rate of loading, time to failure, and to ensure that the excess (change) of pore pressures had a chance to dissipate (to equalize). The coefficient of the consolidation was calculated from the consolidation stage. An average value from the three lowest results is $0.07\,cm^2.min^{-1}$ ($0.01\,m^2$ per day). Calculated required time to failure was in all cases slightly lower than real time to failure, which was between 30 and 40 hours.

7.5.3.3 Stress-Strain Behaviour

The character of the stress-strain curve is very interesting. The deviator of stresses increases after the first drop practically in all cases of the tension tests. This means that the character of stress-strain work softening behaviour of the clay material typical (and also obtained) for a drained compression test changes into stress-strain work hardening for a drained tensile test after the first drop. Summaries of the results

7.5 Stress-Strain Behaviour of Soils in Tension

are in Table 7.1. A typical stress-strain curve for a tension test (for sample No. 2) is shown in Fig. 7.30. Photos of samples after the end of the test are shown in Fig. 7.31.

The first peak on the stress-strain curve was observed when tensile stresses were in the range of 3–6 kN.m^{-2}. Slightly higher values (up to 7.6 kN.m^{-2}) were obtained for samples where salt was added. Elongation at this point was between 1.5% and 5.8%. The maximum measured values of tensile stresses were less than 9 kN.m^{-2} for samples without salt and between 12 and 24 kN.m^{-2} for samples with salt. Elongation for maximum tensile strength varied in wide range from 3.4% to 21.9%.

The maximum drained tensile strength was probably not overcome because of any detachment of the two parts of the sample, as it was typical for undrained tests and even for undisturbed London blue clay. Perhaps the exception is test No. 7 where the central cross section was reduced to the diameter of 19 mm (0.75"). A small crack was visible after removing of the rubber membrane there.

The central section area becomes smaller with an increase in loading and deformation. In the weakest point, a small neck developed and this development coincided with the first peak failure. For example in test No. 2, which is presented in Fig. 7.30, the first peak in deviator of stresses occurred for tensile stress in the central part of the sample equal to 3.2 kN.m^2 and for the elongation, 3.6%. After this first peak, a small neck in the central part was observed but the deviator of stresses rose again and a second neck was observed. This untypical behaviour – when the failure did not continue in the first neck with the highest concentration of stresses – was observed in soils for the first time.

It is rather difficult to explain this special behaviour. But with a high probability, the first failure is due to shear strength. This failure is accompanied by a fall in the deviator of stresses. After that – probably after rearrangement of the clay particles in this zone – the tensile loading began and the stress-strain work hardening behaviour

Table 7.1 Results of triaxial tensile drained tests of clays from fresh water sediments

Sample no.	Pore water salinity	Initial values			First peak				Max measured value		
		σ'_c	σ_r	σ'_r	$\sigma_1 - \sigma_3$	ε_1	t_1	σ'_{t1}	$\sigma_1 - \sigma_3$	ε_2	σ'_{t2}
		kPa			kPa	%	hour	kPa	kPa	%	kPa
1	0	300	1000	22.9	25.9	2.27	38	3.0	31.7	3.4	8.8
2	0	600	1000	24.4	27.6	3.6	27	3.2	29.3	6.6	4.9
3	0	600	1000	10.2	39.7	4.1	41				
4	2.5 g/l K/Na = 0.2	600	1000	19.0	26.66	3.12	27.7	7.66	40.2	18.2	21.2
5	2.5 g/l K/Na = 0.8	600	1000	18.7	24.6	4.17	30	5.9	42.9	21.9	24.2
6	5.0 g/l K/Na = 0.2	600	1000	19.9	25.4	5.21	36	5.5	32.4	11.9	12.5
7	0	600	1000	5.0	9.5	5.86	48	4.5	13.8	14.6	8.8
8	0	600	400	5.2	10.9	1.51	24	5.76			

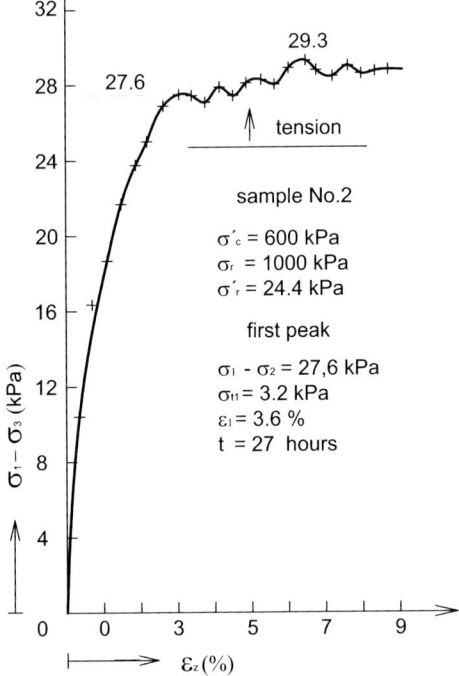

Fig. 7.30 The typical result of drained triaxial tension test

of plastic clay was valid for this loading. It makes it possible to develop another shear failure at a different point.

Mohr's circles for the first failure (tests No. 1, 2, 7 and 8) and shear failure for compression test (No. 3) are in good agreement – Fig. 7.32. The envelope of the Mohr's circles has a character of a parabolic curve with a cohesion of $c' = 9$ kN.m^{-2}. Mohr's circles of the samples with chemical additives (No. 4–6) are a little bit higher

Fig. 7.31 Triaxial drained tests No. 1, 2 and 7 – tension tests; No. 3 – compression test – shape of samples at the end of tests

7.5 Stress-Strain Behaviour of Soils in Tension

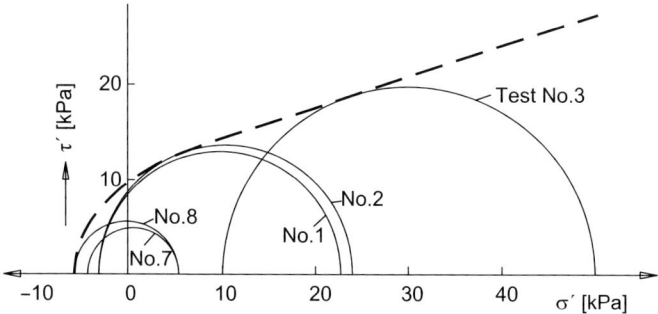

Fig. 7.32 Mohr's circles for triaxial compression and tension tests (for first failure)

with a cohesion of $c' = 10.5 - 12.5\,\text{kN.m}^{-2}$. It indicates a certain influence of the pore water salinity.

A second more extensive series of tests were performed on Most clay ($w_L = 53.1\%$, $w_p = 25.4\%$, $I_p = 26.7\%$). Dry pulverized clay was mixed with water to clay slurry and this clay slurry was consolidated in a large oedometer with a diameter of 250 mm and a height of 150 mm under effective stresses of 100 or 300 kN.m^{-2}. Afterwards, at the end of consolidation, cylindrical samples were cut out with a diameter of 38.1 mm (1.5″). From this cylinder the samples with neck in the central part, similar to the first series, were prepared, but now with the help of hand turning. Such prepared samples were tested under similar conditions as for the first series. Results were comparable, deformation curves were similar and Mohr's circles and envelopes as well. In Fig. 7.33 two necks are visible in the central part of the tested clay. Only slightly higher strength was observed for samples consolidated under higher pressure:

$$\sigma'_c = 100\,\text{kN.m}^{-2} \quad c' = 4\,\text{kN.m}^{-2} \quad \varphi' = 23.2°$$
$$\sigma'_c = 300\,\text{kN.m}^{-2} \quad c' = 5\,\text{kN.m}^{-2} \quad \varphi' = 23.6°$$

Value of effective tensile strength was even smaller than in the first series: from 0.34 to 3.08 kN.m^{-2}.

7.5.3.4 Brief Summary

The described triaxial drained tests generated a first larger set of such tests and helped us to make a first conclusion. Information about rather low values of effective tensile strength mobilized under large elongation is above all an untypical characteristic of the stress-strain curve, which to some extent, is similar to the stress-strain curve of steel in tension. The development of tensile crack under drained conditions in tested soils is very improbable, even if the tensile strength is very small. The reason is large tensile strain at failure and also the possibility of plastic failure spreading due to the ability of stress redistribution. Tension tests approved this

Fig. 7.33 Photo of tested Most clay with visible two necks

theoretical ability of stress redistribution. Tests also helped prove the character of Mohr's circles in the range of small positive (small compression) and negative (tension) stresses and showed the significance of internal structures elicited by chemical additives.

7.6 Soil Behaviour During Seepage Through Crack

As was mentioned in Section 7.3 most of dam failures and accidents are connected with local concentration seepage accompanied by progressive erosion. Today, when compaction methods are much better than at the beginning of fill dam construction, predestined seepage paths are connected with:

– Tensile cracks in clay core,
– Poor connection of soil on contact with a stiff element inside the dam body,
– Poor connection of clay core with subsoil, particularly if this subsoil is fissured and water passes through,
– Seepage of water from an outlet to the surrounding soil.

7.6 Soil Behaviour During Seepage Through Crack

Water seeping along a predestined path can initiate the process of progressive erosion of soil particles known as internal erosion. If the progressive erosion starts at the downstream face piping happens as well. In Russian a more general term known as filtration deformation is used, in the strict sense of the word also suffusion, scouring or washing ("vypor" or "razmyv").

Therefore it is very useful to know the sensitivity of different soils to this sort of erosion, whereas generally better resistance to erosion is valid for well-compacted soils. This resistance depends not only on particle size but also on inter particle forces. For frictional, non-cohesive soil, resistance depends mostly of grain size, but also on the shape of the grains. Very coarse grains have very good resistance, such as gravel, well graded and fine sands particles or even silt particles with no plasticity, as SP or ML soils have very poor resistance. On the contrary for cohesive soils the resistance increases with the increase of plasticity, usually connected with lower grain – clay – particles. Therefore clay with high plasticity (CH) and well compacted has very good resistance to erosion. But there exist a special group of soils, which behave rather different from most of "classical" soils. These soils are called dispersive soils and can be identified by the high percentage of sodium ion Na^+ in the pore water.

When the water starts to seep through a crack two contradictory processes usually start and therefore will be treated in this chapter in more detail:

– Internal erosion – piping – a process during which preferential path of seepage increases its volume,
– Swelling – a process during which there is a tendency to close the cracks.

7.6.1 Model of Seepage Cracks for Cohesive and Frictional Soils

Regardless of crack creation under dry or wet conditions, a crack is a potential risk from the view of internal erosion at the moment when water starts to seep through. Nevertheless there is a fundamental difference between crack behaviour created in cohesive or frictional soils and therefore these two cases will be discussed separately.

7.6.1.1 Frictional Soils

In the case of frictional soils, a dry crack can develop and can be stable as a result of negative pore pressure. But under wet conditions this negative pore pressure is eliminated – on crack walls effective stresses are zero. Because frictional soil has no effective tensile strength the walls and roof of such crack are unstable, a collapse starts very quickly and spreads upwards. The spreading of this collapse is very quick therefore significant erosion is not possible. Collapse results in filling in the crack by soil, although with a looser form. After that, a new equilibrium state is established without significant erosion. Such behaviour of frictional soils is one of reasons for strict demands on filter layers, so that a crack cannot be continuous, from the core up to the drainage layer. At the same time this fact warns us. The creation of a crack along the contact with the concrete structure, which is above the frictional soil, is very dangerous.

7.6.1.2 Cohesive Soils

As described in the previous chapter, cohesive soils have a certain, although rather small effective tensile strength. Because effective loading along the periphery of a crack is zero for wet conditions, the crack can be stable as long as the crack size does not exceed a certain critical dimension. Because the crack is at the beginning smaller, it does not reach this critical dimension, walls are stable, but, however, are under the influence of seeping water.

The speed of erosion is uncertain. Critical water velocity exist for open channels, below this critical velocity erosion is practically zero. Whether such similar a rule is also valid for seepage through a crack is not verified up to now, although it seems to be true. Elevated values of seepage were monitored for some dams indicating the seepage through a preferential path, however with time no changes observed there. A certain irregularity in the crack shape and changes in water velocity can initiate internal erosion even if the average water velocity is low. Therefore the estimation of this water velocity is very important especially in the case of laboratory simulation of this process.

Under the assumption that erosion started, the important question is connected with the degree of it spreading. Again, water velocity is very important from the view of possibility of segregation of eroded soil particles; whether the coarser particle will sink to the bottom of the crack and fine particles will be transported away. This process of segregation of particles can have a significant influence on the filter design. If fine particles can pass through the filter the total volume of clay core material decreases. The crack size is increasing dependant on erosion velocity. When reaching the critical size the collapse of the crack roof and walls starts. This collapse corresponds to the drained mechanism so that the speed by which the size of crack rises upwards depends on the degree of soil swelling along the crack periphery. The speed of crack spreading is much slower than for frictional soils and this is a larger opportunity for progressive erosion.

From the above-mentioned results, that water velocity seeping through a crack can manifest itself in two directions. For small water velocity, the speed of erosion can decrease; even stop, as a result of the swelling potential of the clay core. But if the swelling potential is small the erosion and segregation of particles can continue. On the contrary, for high water velocity the seeping water is transported with higher speed also along with coarse particles or clusters of fine particles (floccule). But after that there is a higher chance that the coarser particles will create a reverse filter and can install stability on the boundary of clay core with a filter, slowing down the erosion activity up to zero. From the view of segregation of particles probably the most dangerous are soils with continuous grain size distribution curve as is the case for boulder clay. Relatively low cohesion and low potential to swelling of boulder clay can allow upwards progressive erosion, which is accompanied by the segregation of particles.

7.6.2 Swelling Potential

Swelling potential can be determined in two different ways. Firstly by expressing this swelling potential with the help of volume changes which start after the contact of soil with water. For 1D case vertical deformation is measured. In connection to swelling potential of the clay core the sample is compacted by the Proctor standard energy for optimum moisture content and vertically loaded with very small pressure, in the range of 5–10 kN.m^{-2}. The second approach defines the swelling potential with the help of pressure, which is needed to protect any swelling deformation of the sample. This pressure is denoted as swelling pressure – Myslivec and Eisenstein (1965).

The main factors influencing the swelling potential are:

- The amount and character of clay minerals; the highest swelling potential has the most active clay mineral montmorillonite (bentonite); qualitatively of this potential can be controlled via index of plasticity I_P and index of colloidal activity I_A,
- Dry density,
- Initial moisture content.

Mitchell (1993) alludes to the works of Seed et al. (1962) and Chen (1975) where swelling potential is expressed as % of swelling and is observed as a function of the plasticity index I_P and colloidal activity I_A. Two equations are recommended:

$$S = 3.6 \cdot 10^{-5} \cdot I_A^{2.44} \cdot C^{3.44} \tag{7.16}$$

Where C is the percent of clay particles (% < 0.002 mm), or

$$S = 2.16 \cdot 10^{-3} \cdot I_P^{2.44} \tag{7.17}$$

The second equation was recommended on the base of compacted natural soils. An accuracy of ±35% is mentioned. Nevertheless, Mitchell states that while these recommended relations illustrate the influences of compositional factors and provide preliminary guidance about the potential magnitude of swelling, reliable quantification of swell and swell pressure in any case should be based on the results of tests on representative undisturbed samples tested under appropriate conditions of confinement and water chemistry.

Kassif and Baker (1971) pointed out the influence of ageing of compacted samples – on time delay between the sample compaction to sample testing, to time when the swelling potential is measured after direct contact of the sample with water.

Some results are presented in Fig. 7.34a–c. Clayey soil was tested, containing a high percentage of clay minerals from which roughly 60–70% is montmorillonite and with $w_L = 75.3\%$, $w_P = 29.1\%$. Swelling pressure increased with dry density and fell with the growth of moisture content. Figure 7.34b shows the influence of ageing of compacted samples. During first few days (2–8 days) swelling pressure increased, followed a small drop and steady conditions were established roughly after 20 days. For practical application for the core of dams this steady state will be decisive. From Fig. 7.34c it is obvious that steady values are not so sensitive to the initial moisture content and that the sensitivity decreases with the decreasing value of dry density.

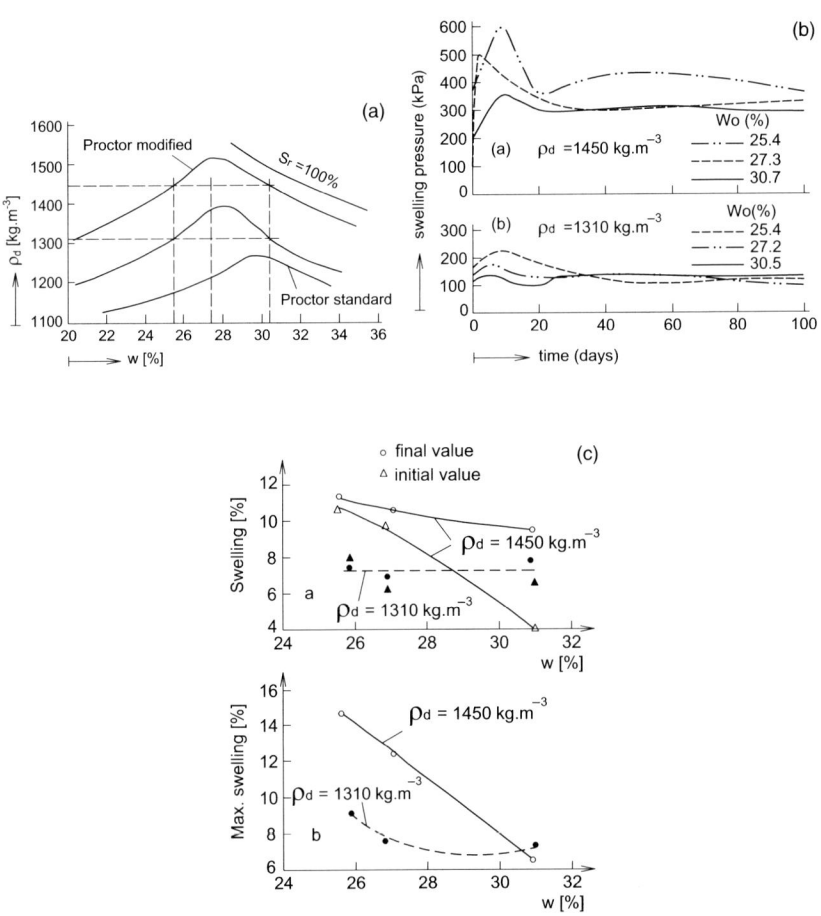

Fig. 7.34 Swelling potential –Results for compacted clay (according to Kassif and Baker). (**a**) Compaction curves with marked tested points. (**b**) Swelling pressure as function of time delay for different dry density and moisture content. (**c**) Swelling potential as function of initial moisture content

7.6 Soil Behaviour During Seepage Through Crack

Kassif and Baker deduce that immediately after compaction the arrangement of individual particles corresponds with the dispersive structure, namely with an unstable energetic level as the result of energy increase which the system absorbed. In the attempt to reach equilibrium there is an accompanied energy release, which is connected with water molecule reorientation inside of an electric field of the individual particles. Such a reorientation can invoke

- An increase of suction,
- A creation of bonds between individual clay particles.

These two aspects react differently from the view of swelling potential. Suction increases also the swelling potential increase, while the creation of bonds have a tendency to decrease the swelling potential either by an increase of strength or by limiting the surface area of particles, which is able to absorb more water.

7.6.3 Forces Between Individual Particles

Between individual particles there are forces of different types and character, attractive and repulsive. Interaction can be met via direct contact of mineral particles or through the medium of water or air phases. Individual forces have different radii; and we can distinguish between forces of high radius, acting on the distance of up to a hundred angströms (1 Ä = 1.10^{-10} m) and forces of low radius, acting only on the distance of a few angströms.

Forces of high radius are critical for fabric and orientation of individual clay particles in a water solution. Once the particles are closer to each other as due to the increase of effective stresses to values typical in engineering practice, afterwards mechanical properties are controlled by conditions on inter particle contacts that are created by forces of low radius.

To help imagine the creation of individual particles orientation better a model of soil deposited in a water solution can be used. The gravitational effect predominates during sedimentation on grains up to a certain size. Individual grains do not affect each other. Mutual interaction starts at the moment when forces between individual particles prevail gravitational forces. This case is valid only under the condition that mutual forces (electric charge) are acting on a large area with respect to the total weight of the particles. This boundary is valid for particles with a diameter around 0.001 mm (so called colloidal particles) or for particles having a specific surface higher than $25 m^2.g^{-1}$. Only clay minerals are able to fulfil these demands, though kaolinite is just below this boundary with a specific surface around $10 - 20 m^2.g^{-1}$. It means that for all silt, sand and higher particles gravitational forces prevail and the structure will have the characteristics of individual grains.

For clay particles Lambe (1953, 1958) presented three fundamental possibilities of internal structure based on the theory of a diffusive double layer, Fig. 7.35. He

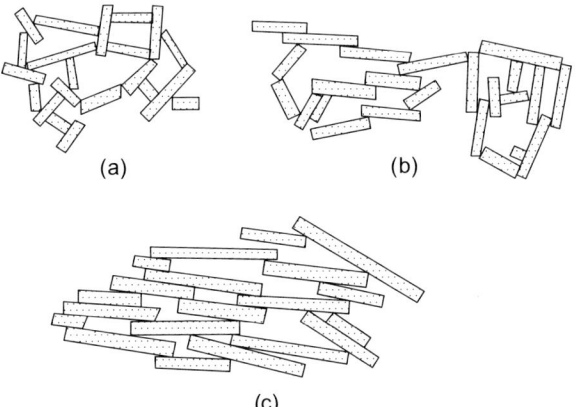

Fig. 7.35 Internal structure of clay particles (according to Lambe). (**a**) Nonsalt flocculation. (**b**) Salt flocculation ((according to Lambe). (**c**) Dispersive structure

stated that the sedimentation under very low concentration of electrolyte concentration leads to the arrangement of clay particles presented in Fig. 7.35a, which is called nonsalt flocculation. A relatively small change in particle arrangement should be valid for high electrolyte concentration, called salt flocculation. These arrangements are typical for contacts edge × edge or edge × surface, where there is high attraction due to different electrical charges. Lambe pondered that for the middle electrolyte concentration, which corresponds to brackish water, the final structure corresponds to the arrangement shown in Fig. 7.35c, which is called dispersive arrangement. Clay particles are relatively parallel, but without direct contact surface × surface, because a diffusive double layer separates individual particles.

The term flocculation, therefore, corresponds to the situation when the result of all forces is positive; particles attract each other. Term dispersion corresponds to the situation where result of forces is negative; between particles are repulsive forces as the result of a diffusive double layer.

Dispersion can be induced in the laboratory for example during a hydrometer grain size distribution test. Dispersive and flocculation structures were proved e.g. by Sides and Barden (1971) with the help of a scanning microscope, Fig. 7.36.

Nevertheless, real soils do not contain only clay minerals. Therefore real particle arrangement is more complicated. According to the observation of Collins and McGown (1974) coarser particles are not in direct contact, either they are encased by finer particles or are connected through bridges from finer particles. The stability of these bridges strongly influences the behaviour of soil as an aggregate, Fig. 7.37.

The walls of a crack in the clay core are not in any case smooth, internal erosion for dispersive soils will proceed very quickly, because seeping water even for low velocity will carry away the smallest particles easily detachable from others. For flocculated soils the process of internal erosion is much more limited. Needed water

7.6 Soil Behaviour During Seepage Through Crack

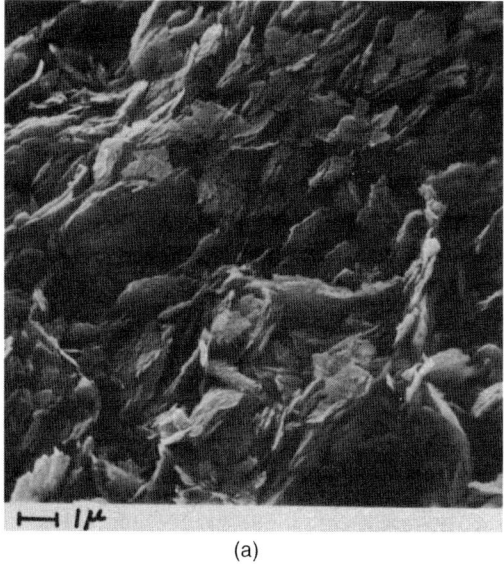
(a)

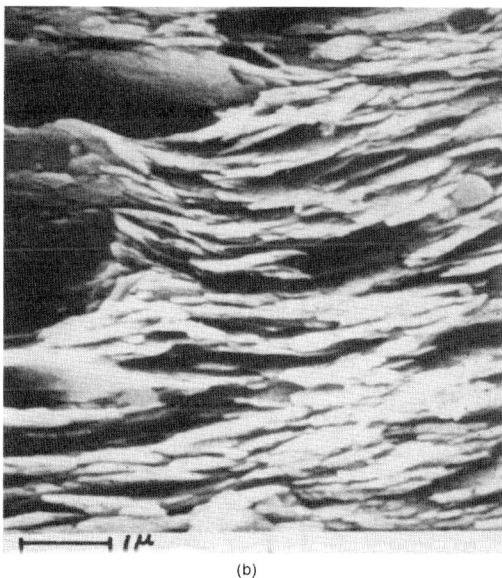
(b)

Fig. 7.36 Microstructure of illite in scanning microscope (according to Sides and Barden). (**a**) Flocculation microstructure. (**b**) Dispersive microstructure

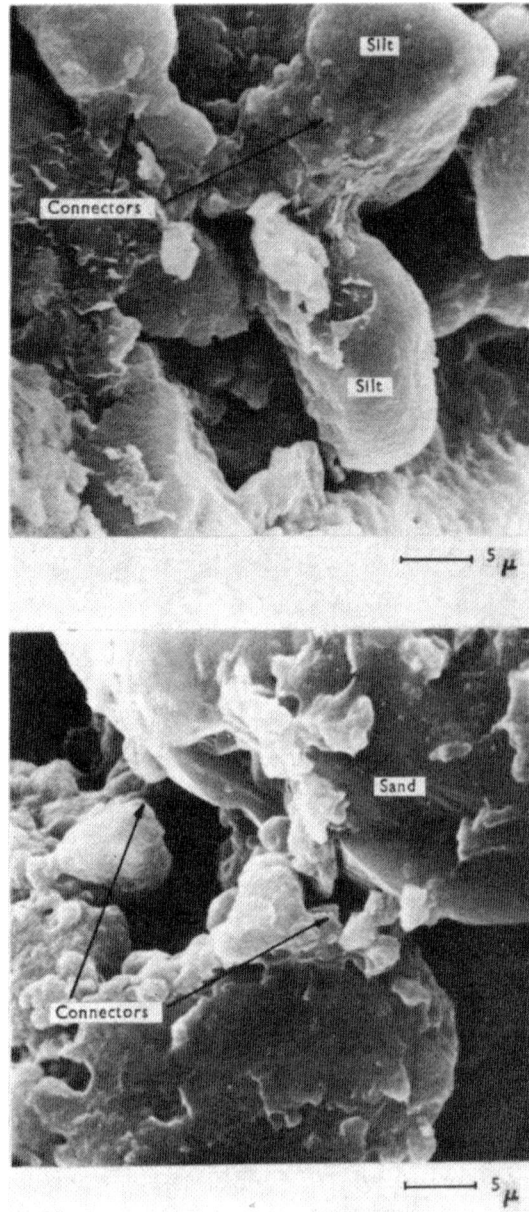

Fig. 7.37 Microstructure of different soils – Connection of coarser grains of sand and silt by bridges from finer particles (according to Collins and McGown)

7.6.4 Dispersive Soils

In spite of the fact that dispersive soils occur very rarely, their localization and identification is very important when used as a construction material in earth structures such as fill dams. With time it was proved that dispersive behaviour is typical for soils with a high content of cation Na. Sodium causes an increase of thickness of diffusive double layer surrounding individual minerals, thereby evokes the decrease of attractive forces between particles and so facilitates a separation of these particles.

Sodium adsorption ratio (SAR) is used as a criterion of proportion of sodium to other cations (with neglecting potassium):

$$SAR = \frac{Na}{\sqrt{0.5\,(Ca + Mg)}} \qquad (7.18)$$

Or with the help of a percentage of sodium – Na%:

$$\%Na = Na/TDS \times 100\% \qquad (7.19)$$

Where TDS is total dissolved salts – total amount of dissolved cations in pore water (TDS = Ca + Mg + Na + K). Metallic cations as Fe are neglected. All in miliequivalents per litre.

Aitchison and Wood (1965) published the flocculation – deflocculation boundary for illitic soils and for montmorillonite – Fig. 7.38a. The boundaries between the flocculated and the deflocculated states are expressed simply in terms of two readily measurable parameters, with total cation concentration of the percolating water and the sodium adsorption ratio (SAR).

7.6.4.1 Problems with Dispersive Soils in Dam Engineering

Aitchison and Wood stated that small earth dams built in Australia from predominantly clay soils show a failure rate of eight percent as a result of piping. This failure rate is higher than is the average rate as discussed earlier. The authors concluded and confirmed that this is due to post-construction deflocculation.

The dams are up to about 6 m high and are constructed from local soil to make storage reservoirs for farm or town water supplies. Homogeneous cross-sections are generally used, without filter zones or drains. In many cases, the dams are built without moisture control (mostly drier then optimum), and with compaction only from construction traffic. A programme of evaluation of dam performance in relation to the characteristics of the soil and of the stored water was established to embrace

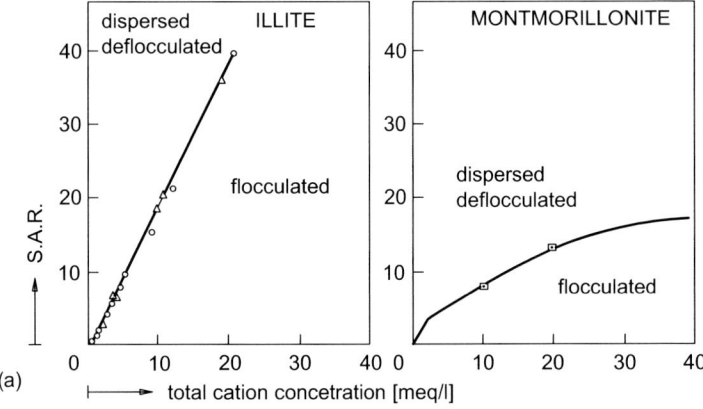

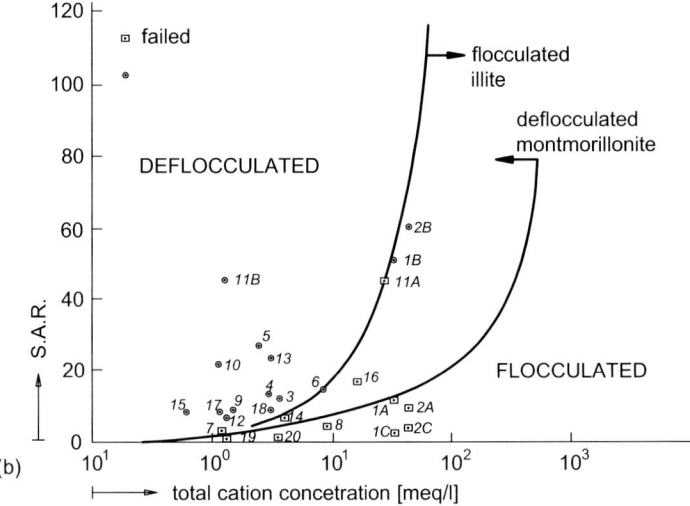

Fig. 7.38 Determination of flocculated and dispersed behaviour of illite and montmorillonite (according to Aitchison and Wood). (**a**) Recommended boundary. (**b**) Results of chemical composition of pore water from materials of dams which failed ● or stayed in order □

a wide variety of climatic and soil conditions. The results of the investigation at 20 dams are presented in Fig. 7.38b. Fourteen of these dams had failed by piping, while six were sound. It is significant that all 14 of the failed dams plot on the "deflocculated" side of the relevant boundary whereas all but one of the six sound dams fall within the flocculated zone. However this dam differs from most of the others reported in terms of its construction history. Compaction in this dam was carried out at a moisture condition wet of optimum. Among other interesting cases belong:

7.6 Soil Behaviour During Seepage Through Crack

- Dam No. 11 was originally filled with groundwater relatively high in dissolved salts – suggesting a condition of flocculation at point 11A. The dam held water with only slight seepage for several years, after which the reservoir water was first replaced by sea water (without an effect on the flocculated condition) and then by river water low in dissolved salts (which tended to create a new equilibrium at point 11B in the deflocculated zone). The dam failed by piping within 3 days of the change to purer water.
- Dams No. 1 and No. 2 were filled with saline bore waters providing in each case an initial flocculated state at point 1A and 2A. However, leaching increased the sodium status of the exchange complex, tending towards a new equilibrium at points 1B, 2B thereby creating a deflocculated condition in the predominantly montmorillonitic clay. Each dam failed by piping after a few months of operation. The breaches were repaired using a high-sodium subsurface soil, but failures occurred repeatedly. Repairs using surface soil richer in calcium (points 1C and 2C) have been successful.

Another case of failure of five small dams in Israel is described by Kassif and Henkin (1967). They have also found the high percentage of exchangeable sodium cations in clayey soil. Sherad et al. (1972a,b), described a failure of small dams in Oklahoma and Mississippi in USA and in Venezuela. They also proved that the failure occurred due to the internal erosion of dispersive soils. However the authors stated that the internal erosion started along the cracks opened by hydraulic fracturing in places of reduced stresses as the result of differential settlement, drying and shrinking. In all cases, increased seepage occurred on the downstream face very quickly after the first reservoir filling. Afterwards Sherard et al. (1976a) recommended a new categorization of dispersive soils, see Fig. 7.39, on the sodium percentage and total amount of dissolved salts in saturated extract. Practically all samples obtained from clayey soils of dams which failed by internal erosion fall into zone A marked as the zone of dispersive soils.

McDaniel and Decker (1979) mentioned the case that was observed for the large dam Los Esteros in the USA. Just before construction started it was proved that the soil in the borrow pit suggested for the central core was dispersive. Two options were proposed:

- To stabilize the clay soil by lime, to neutralize the dispersive behaviour,
- To use a sand filter as a prevention of dispersive erosion along any potential crack, which could be created in the clay core.

The second option was finally recommended with a small exception for the lowest 1.5 m layer of the core, where stabilized soil was applied, Fig. 7.40. This modification was proposed because there were great qualms if the dispersive soils will be in a direct contact with the fractured sandstone bedrock. Similar experiences with dispersive soil modification were described by Melvill (1980) for Elandsfagt dam in South Africa.

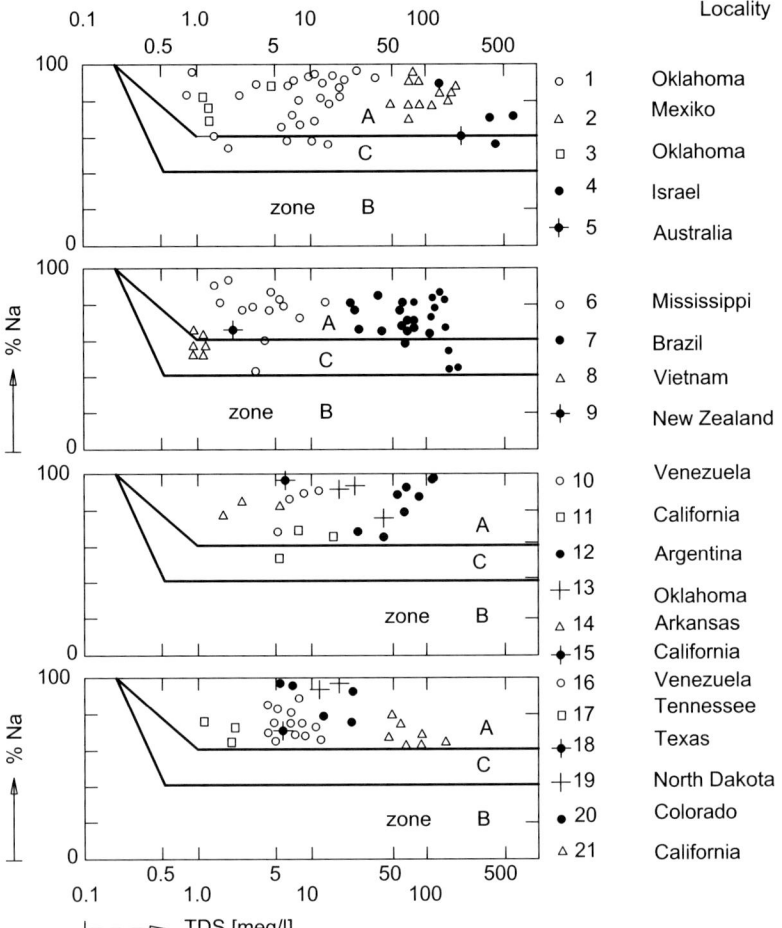

Fig. 7.39 Determination of dispersive soils (according to Sherard et al.): zone A – based on TDS and % Na – with results of chemical composition of pore water for small dams which failed by piping

7.6.4.2 Laboratory Methods of Identifying Dispersive Soils

The above-mentioned experiences showed that a dispersive soils identification test could be very useful not only in dam engineering but also in soil science. The fundamental methods of identification are described by Sherard et al. (1976b), Vaníček (1988).

TDS – Total Dissolved Salts Test

Soil is mixed with distilled water to a consistency near the liquid limit. A pore-water sample ("saturated extract") is sucked out by a vacuum using a filter and finally the

7.6 Soil Behaviour During Seepage Through Crack

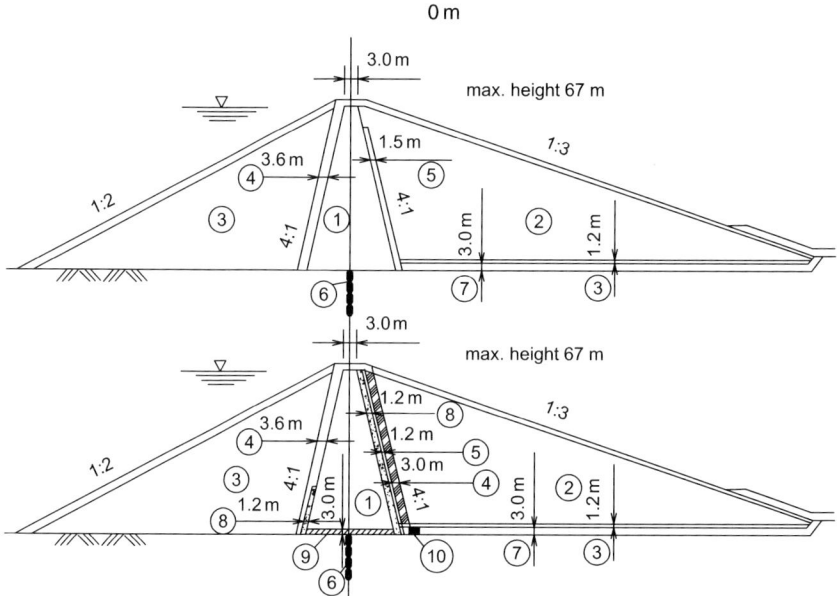

Fig. 7.40 Los Esteros dam with core from dispersive soil – original and new proposed cross sections (according to McDaniel and Decker). 1 – core; 2 – unsorted rockfill; 3 – rockfill; 4 – coarse particles smaller than 100 mm; 5 – sand drain; 6 – injection sealing wall; 7 – sorted rockfill; 8 – coarse grain filter; 9 – chemically stabilized core; 10 – gravel filter

extract is tested to determine the quantities of the four main metallic cations in solution (calcium, magnesium, sodium, and potassium) in miliequivalents per liter. It is worthwhile to mention that soils with a TDS lower than 1 meq per liter are non-dispersive even with a high percentage of Na. It is a positive statement because most natural soils fall into this category.

Double Hydrometer Test

Particle size distribution is first measured using the standard hydrometer test, in which the sample is dispersed in the hydrometer bath with strong mechanical agitation and a chemical dispersant – curve a in Fig. 7.41. A second hydrometer test is then made without strong mechanical agitation and without a chemical dispersant – curve b in Fig. 7.41. Curve b shows less colloidal particles than curve a and the difference is a measure of the tendency of the clay size particles (0.005 mm) to dispersion.

Pinhole Test

Distilled water flows through a 1.0 mm diameter hole in a compacted specimen. For dispersive clays, the water becomes coloured even for small hydraulic gradient

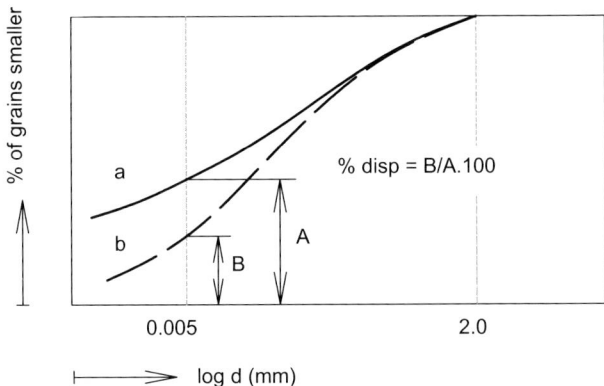

Fig. 7.41 Double hydrometer test for the determination of percentage of dispersivity. a – grain size curve determined with chemical dispersant; b – grain size curve determined without chemical dispersant but with strong mechanical agitation

and the hole rapidly erodes. The layout of the measuring system for a detail of the cross section through a tested sample is shown in Fig. 7.42, Vaníček (1988). For non-dispersive clay, the water is clear and there is no erosion even for a rather high hydraulic gradient. Classification of soils into four groups of non-dispersive a two groups of dispersive soil was recommended by Sherard et al. (1976a). They also performed pinhole tests which continued for a hundred hours, these tests proved that non-dispersives are stable in the long term view. It is in agreement with field observation that most of failures occurred very quickly after the first reservoir filling; on the contrary some fill dams exist where small seepage was observed but this seepage was constant long term.

Crumb Test

A small soil crumb size about 6–9 mm, moisture content of which is natural, is immersed in distilled water and the dispersion is observed directly. One extreme – non-dispersive soil – the water remains clear, for the other – for dispersive soils – the water is turbid and fine particles gather around the crumb.

Index Tests

Although index tests are very often very useful for determining some other soil properties, for identification of dispersive soils this is not valid. Sherard et al. (1976b) stated that there was no significant difference in the clay contents of the dispersive and non-dispersive soils. The data only suggest that soils with less than 10% clay particles may not have enough colloids to support dispersive piping. Also no useful correlation was found between soil plasticity and the incidence of damage to dams.

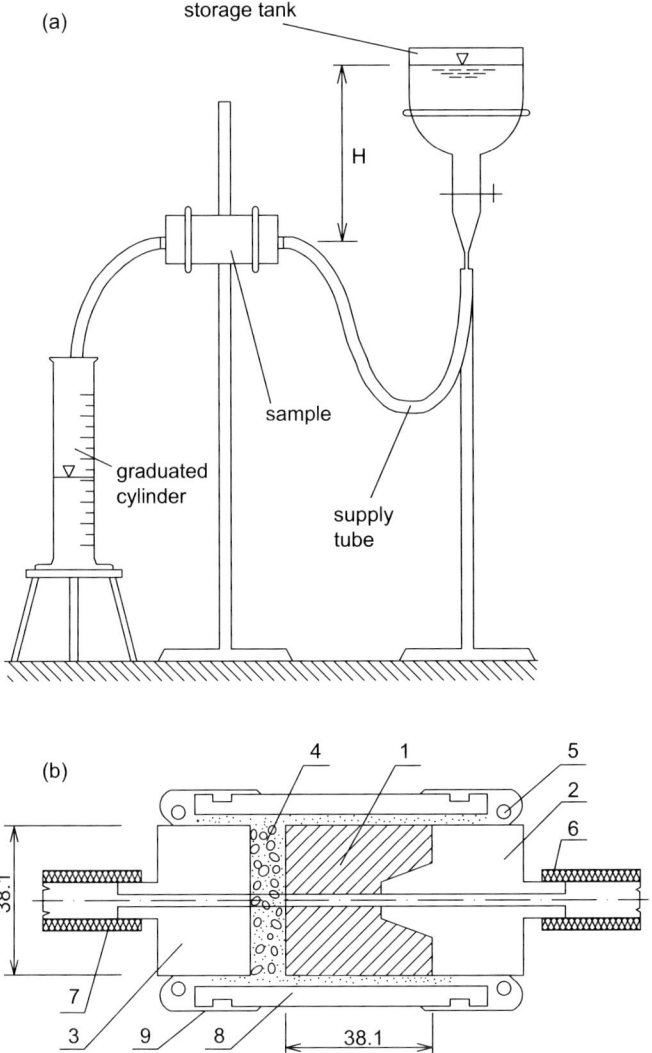

Fig. 7.42 Pinhole test. (**a**) Layout of measuring system. (**b**) Detail of cross section through tested sample (1) with hole in the middle

7.6.5 Flocculated Soils

A very interesting question is connected with the size of individual floccules which can be torn off from the crack walls by seeping water. Great attention to this problem was devoted in the former USSR in VNII VODGEO by a group around Istomina, e.g. Istomina et al. (1975). They used a laboratory experiment – an aperture test, see Fig. 7.43, which is composed of two plates. The lower plate is metal, upper one from

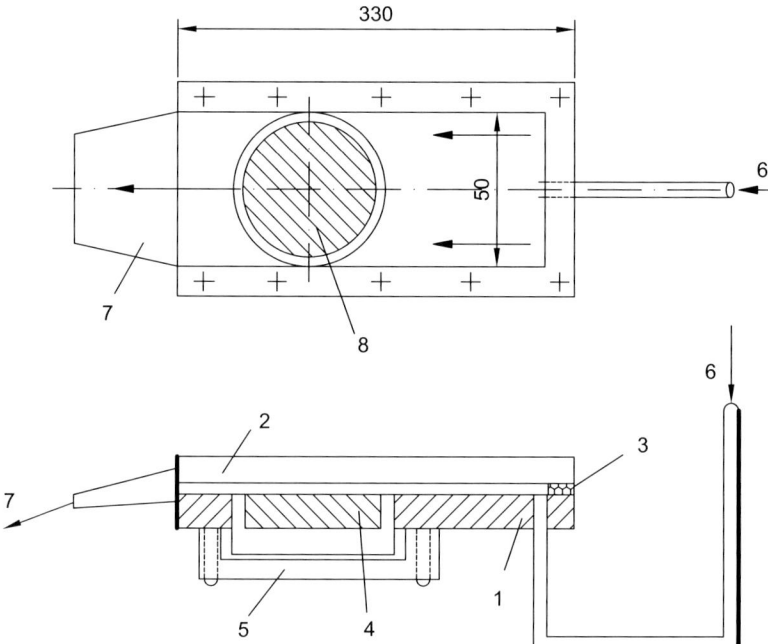

Fig. 7.43 Layout of aperture test observing flocculated particles washed out from crack (according to Istomina et al.). 1 – sleeve; 2 – plexiglass; 3 – sealing; 4 – sleeve with tested soil; 5 – pressure screw; 6 – water inflow; 7 – water outflow; 8 – tested soil

plexiglass, they are connected and laterally sealed. The aperture between plates is changeable between 1.2 and 3 mm. In the lower plate there is circular excision with compacted soil with the height of about 20 mm. Selected water velocity passing through the aperture was in the range of 0.01–$0.6\,\mathrm{m.s^{-1}}$. These velocities were regarded as the most probable velocities in dam core cracks. For a velocity lower than 0.01 no tear off was observed, only swelling. The first washing out started for a velocity higher than this minimum value practically for all samples. Aggregates were collected, dried and their size determined. The size of the aggregates larger than 0.25 mm was determined in sieves, smaller ones with the help of a microscope. After a large set of tests the following conclusion was presented:

- The size of washed up aggregates depends on initial soil parameters; with increasing volume density and moisture content, the size of the aggregates also increases,
- The size of aggregates is significantly higher than the size of original particles determined in laboratory by classical manner,
- The aggregate size distribution curve is inversely proportional to the curve of original soil, see Fig. 7.44.

7.6 Soil Behaviour During Seepage Through Crack

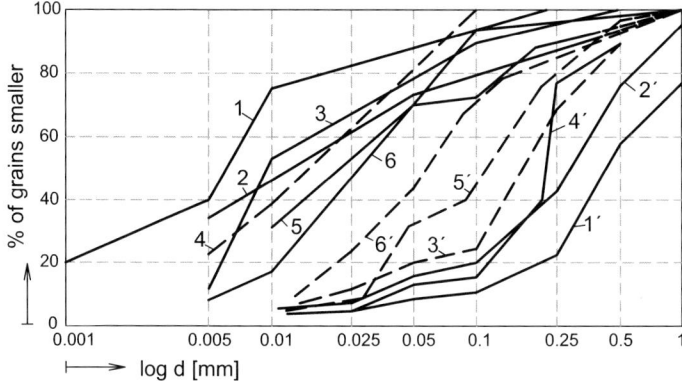

Fig. 7.44 Grain size distribution curves (according to Istomina et al.). 1–6 by standard manner; 1'–6' collected from aperture test

Misurova (1974) published the results of four low plasticity soils, which are presented in Fig. 7.45. The formation of aggregates starts for soils with an index of plasticity around three–four and after that is significantly increasing. Also the influence of degree of saturation is noted for soil with a higher index of plasticity. Finally Misurova states that the new aggregate size distribution curve can be determined with the help of index properties and undrained tensile strength.

Vaughan and Soares (1982) used the test described in previous chapter as double hydrometer tests to determine the aggregate size distribution curve. They did not use a chemical dispersant but they recommended using 15 minutes mechanical agitation and river water from the profile of the proposed dam. For the Cow Green dam, where sealing soil had a similar grain size distribution curve as for the Balderhead dam, differences noted are presented in Fig. 7.46. Aggregate formation started for

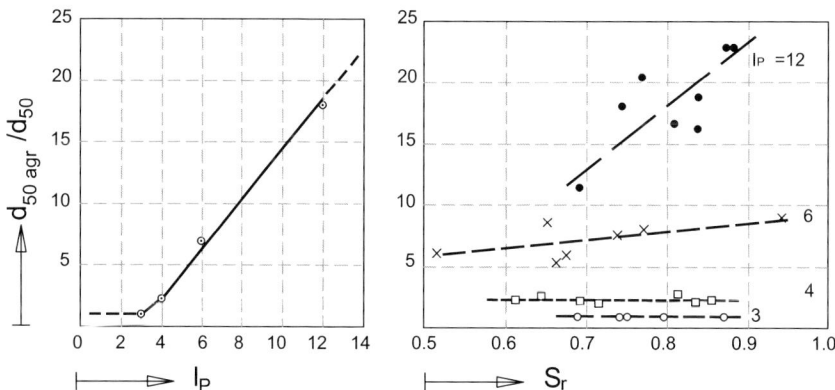

Fig. 7.45 The ability to create aggregates depending on index of plasticity I_P and degree of saturation S_r (according to Misurova)

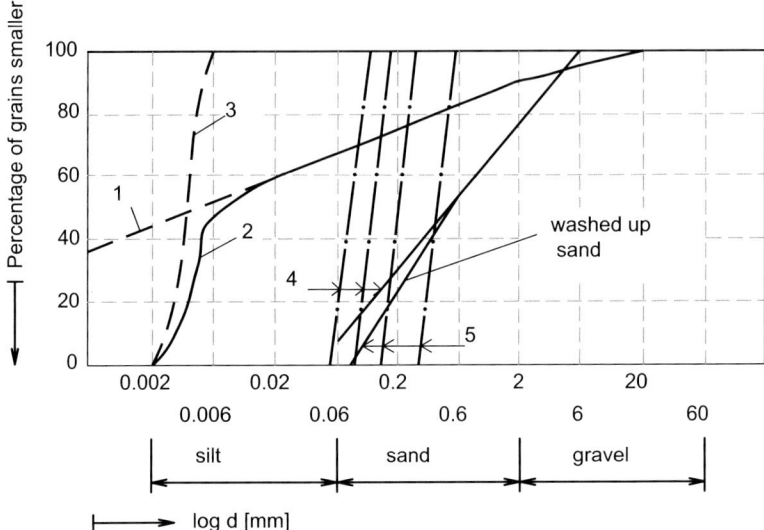

Fig. 7.46 Filter design for Cow Green dam (according to Vaughan and Soares). 1 – clay core – deflocculated grain size curve; 2 – flocculated grain size curve; 3 – clay used during filtration tests; 4 – effective filters; 5 – non-effective filters

particles smaller than 0.02 mm, whereas floccules size is roughly 0.004–0.006 mm. It is obvious the floccules formation is the result of clustering of the finest particles during sedimentation. Coarser particles do not increase in size; fine particles were detached from the surface by mechanical agitation. The approach of Vaughan and Soares is a little bit more conservative compared to the Russian approach. Nevertheless the approach of Vaughan and Soares to design the filters on average size of floccules about 0.004–0.02 mm in them encompasses the possibilities which were not taken into account up to now:

- For the flocculated aggregates transported through a crack to the filter the mechanical perturbation can be higher than that for an aperture test in laboratory conditions,
- The most dangerous situation assumed, known from the Balderhead dam, where water passing through a crack was able to transport to the filter only floccules, while coarser particles settled at the crack bed.

The influence of water salinity, in which a hydrometer test is performed, on the size of floccules is presented in Fig. 7.47. For both tested soils the size of floccules increased up to a salt content 10 meq per liter and after that stayed constant. Both soils contain mostly illitic clay mineral, therefore for different clay minerals different results can be expected.

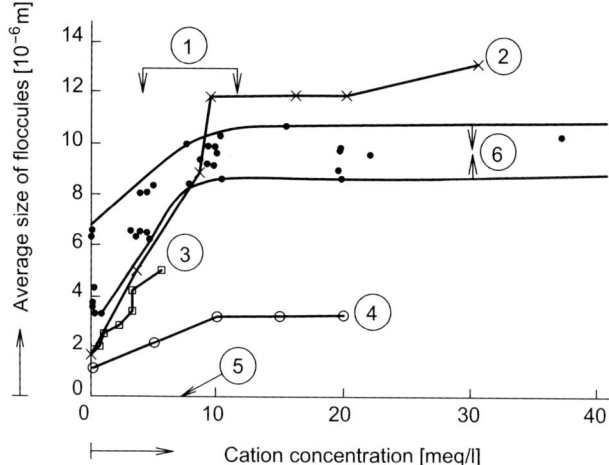

Fig. 7.47 Average size of floccules as function of cation concentration and their nature (according to Vaughan and Soares). 1 – typical range of river water; 2–5 – clay core from Cow Green dam; 2 – +CaCl$_2$ dilution; 3 – + river water – mostly Ca; 4 – + NaCl dilution; 5 – +NaHCO$_3$ dilution, soil deflocculated for all concentration; 6 – clay core from Empingham dam + river water – mostly Ca

7.7 Filtration Contact Stability for Protected Soil with a Crack

The fundamental principles of filter design were presented in Section 5.2.3 – limit state of internal erosion. This chapter will deal with a case where the crack exists in protected soil. Also for this case the filter should be from non-cohesive frictional soil, to protect the crack creation continuously through the core, but also through the filter – directly to the drain.

Two fundamental cases have to be distinguished – what type of soil it is (dispersive, non dispersive – flocculated) and what velocity of passing water will be. For the second case we can count with two assumptions presented in Fig. 7.48:

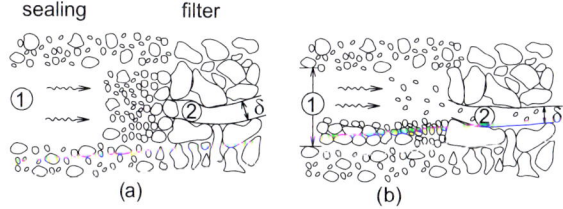

Fig. 7.48 Filtration contact stability for protected soil with crack. (**a**) All washed up particles are transported to the filter – Formation of reverse filter. (**b**) Only finest particles are transported to the filter – Segregation of washed up particles – Coarser particles settled at crack bottom. 1 – open crack in clay core; 2 – continuous seepage channel in filter with diameter δ

- Firstly – passing water is able to transport to the boundary with filter evenly all particles separated from the crack wall,
- Secondly – passing water is able to transport to the boundary with filter only finer particles, coarser particles settles at the crack bottom.

For dispersive soils and low water velocity, the filter is not able to catch the segregated finest particles. Only higher water velocity which is able to transport to the filter contact also coarser particles can stop the process of internal erosion. After that, coarser particles can block off entries to the widest continuous pores. Therefore the application of a filter from the finest particles, but still of non-cohesive character, can be the way how to create stability in the contact zone.

For fine grained flocculated soils the design of a filter is much more promising. The finest particles create the floccules of a higher size than the original smallest grains. Therefore the movement through filter pores is limited.

7.7.1 Protected Cohesive Soil with Crack

As was stated in Section 5.2.3 the design of a filter protecting cohesive soils is not such a great problem. Fundamental conditions were mentioned:

$$D_{15} < 0.4 \text{ mm and } D_{60}/D_{10} > 20 \qquad (7.20)$$

Based on the results of experimental works mentioned in the previous chapter Vaughan (1976) asserted that the size of flocculated aggregates obtained during sedimentation without dispersive agents but with mechanical agitation is in average 0.01 mm for typical cohesive soils tested in the UK. Therefore, the floccules of such size can be detained only by a fine sand filter containing more than 5% of silt particles, which are still non cohesive. This is a stricter requirement than for filters where the crack has not being taken into account. Nevertheless when applying more than 5% of fine particles it is recommended to check the non-cohesive character of such material even by the simplest test, e.g. by observing the shape of a sand sample prepared under natural moisture content after flooding by water.

Vaughan and Soares (1982) recommended checking the design of such filter in laboratory conditions. Flocculated aggregates were separated and their movement through a filter was controlled in the laboratory. Primarily the permeability of filter material was measured under steady state of flow. After that, the flocculated aggregates were added to the water and a change of permeability and turbidity of the effluent was measured. With respect to the chemical composition; water from the proposed dam profile was used. If the filter was not sufficient only a small decrease of permeability was observed. Effluent was turbid. If the filter detained the flocculated aggregates afterwards the permeability decreased and a thin layer of detained material was observed on the surface of the filter. Only in exceptional cases intermediate behaviour was observed – permeability decreased rather slowly and flocculated aggregates continuously clogged pores of the filter.

7.7 Filtration Contact Stability for Protected Soil with a Crack

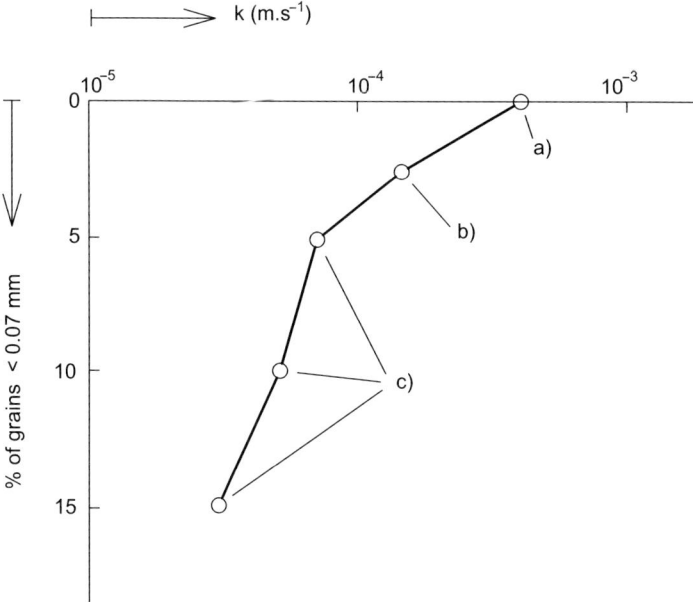

Fig. 7.49 Permeability and filter efficiency as function of fine particles percentage (according to Vaughan and Soares) (**a**) Clay passed through. (**b**) Clay particles contained after 5 hours. (**c**) Clay particles contained immediately

On the basis of series of tests Vaughan and Soares (1982) concluded that filter efficiency is not only affected by the percentage of fine particles, Fig. 7.49 but also by a certain level of filtration coefficient k. For frictional soils the relation between filtration coefficient k and the representative size particles is expressed as:

$$k = A \cdot d_i^2 \qquad (7.21)$$

Where d_i is representative diameter and coefficient A depends on geometrical factors. This equation was used e.g. by Hazen, Terzaghi and many others. Its justification is given by the fact that the size of the particles determines the size of pores and therefore also permeability. Therefore the equation can be rewritten as:

$$k = B \cdot (d_i^0)^2 \qquad (7.22)$$

Where d_i^0 is the representative size of pores. If we are comparing filtration criteria the representative size of pores with the size of a particle δ, which as coarser is passing through the filter, afterwards, analogically, the following expression can be used:

$$k = A_1 \, \delta^2 \qquad (7.23)$$

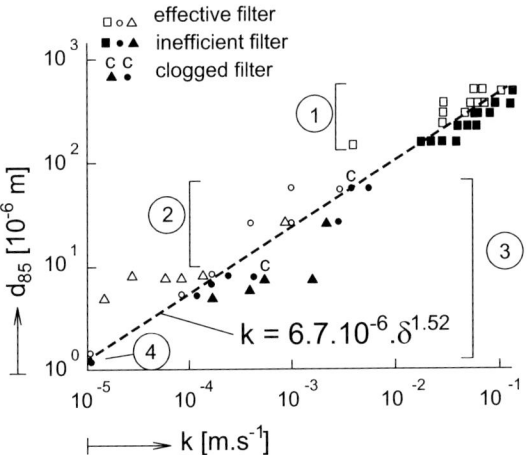

Fig. 7.50 Filter efficiency as function of its permeability and of soil diameter d_{85} trafficability of which was tested (according to Vaughan and Soares). 1 – uniformly graded coarse filters; 2 – crushed quartz; 3 – suspension tests – individual particles or floccules. ○, ● – uniformly graded filter; △, ▲ – non-uniformly graded filter; 4 – deflocculated clay

Vaughan and Soares (1982) recommended the following relation, Fig. 7.50

$$k = 6.7 \cdot 10^{-6} \cdot \delta^{1.52} \tag{7.24}$$

Where δ is in μm and k in m.s^{-1}

Particle size d_{85} from the grain size curve of protected soil, of which movement through the filter was tested, is in Fig. 7.50 on the vertical axes. The presented equation can be used for filter design for the known size of flocculated aggregates. The equation demanded a value of coefficient permeability calculated and then from this value the filter has to be controlled. When using this equation for filter design the demands are stricter than for classical design ($D_{15} < 0.4$ mm). Therefore it is recommended to use a mixture of silt, sand and fine gravel. Such materials can have better retention ability than uniformly graded sands, better workability, on the downstream side can be protected by a coarser second layer of filter. And finally is cheaper when prepared from crushed rock.

In spite of the fact that there are not many conditions for the filters protecting cohesive basic soil, after the publication of Vaughan and Soares equation large discussions appeared in ASCE journal JGED No. 9, 1983 (Kenney, Ripley, Sherard, Truscott, Wilson and Melvill) and new papers have appeared after that describing laboratory tests – Sherard et al. (1984a,b), or field tests – Hillis and Truscott (1983).

From this discussion some opinions are worth mentioning:

– To practicing engineers no such case of dam failure is known where the filter was applied according to the traditional design and containing adequate amount of fine sand (Ripley),
– Examples of failures described by Vaughan in most cases included filters with large maximum grains, where the segregation of the filter is more probable (Sherard),

7.7 Filtration Contact Stability for Protected Soil with a Crack

- The segregation of particles in a leakage crack is a possible case, analogous to the segregation of particles for glacial scours, from which carved clays developed (Kenney),
- It is possible to agree with the proposed recommendation in case of well grained sandy gravels soils with clay particles (e.g. material from glacial moraines) of types GM and GC, where segregation can be the case and also for dispersive soils (Wilson and Melvill),
- Field tests for filters proposed for a group of dams in the Philippines, where crack development by earthquake were discussed, showed a certain conformity with tests of Vaughan. Here the two step filter was proposed, the first one containing from 2 to 10% of fine particles (Truscott),
- Vaughan and Soares in response to this discussion noted that their design of the "perfect" filter is not so much different from the classical approach. For the "perfect" filter; usually the condition $D_{15} < 0.1$–0.4 mm is sufficient.
- Sherard et al. (1984b) published the results of laboratory investigations and proposed $D_{15} < 0.2$ mm and the elimination of a filter for the largest grains.

From this discussion we can feel how sensitive this problem is. However it seems to be useful to apply the filter design as "perfect" in cases where susceptibility to erosion is rather high as well where the probability of crack development is also high. The individual design can be controlled in the laboratory under similar conditions for a real construction site.

7.7.2 Applicability of Geotextile Filters in Dam Engineering

The application of geotextile as a filter element for protection of well-compacted cohesive soils is realistic only for basic cases without cracks. In this case the problem of theoretical sinking of clay particles into pores of geotextile has to be cleared, especially when softening and swelling of clay along the contact with geotextile can be expected. In the case with cracks, the application of geotextile filter is realistic only when neglecting the segregation of eroded particles. When this segregation can occur there the flocculated particles with a size of about $d_{fl} = 0.01$ mm can easily pass through geotextile filter where continuous pores have diameter 0.01 mm or bigger. The condition that $O_I < d_{fl}$ is difficult to guarantee for the present.

Geotextile filters are often used more for small dams and for protective dams. The first review of an application was presented by Giroud et al. (1977). But up to now there is a general opinion, that the application of geotextiles, as only a filtration element for large dams, has to be investigated and tested. The combination of grain filter with geotextile filter is preferred, especially in the case of a shortage of appropriate materials for grain filter or in the case that the filtration criteria are close to the boundary of applicability. A lower thickness of grain filter combined with geotextile is proposed for this case. An interesting example of geotextile application was described by Hollingworth and Druyts (1982).

The discussion in this field can be associated with a combination of geotextile filtration with reinforcement, especially on the upstream face of the filter, which is less critical. Geotextile with significant stiffness can protect clay core against crack creation. Justifiable carefulness is associated with relatively short experience with the application of geosynthetics in civil engineering on one side and long term high risk potential of large dams to the environment on the other.

7.8 Practical Examples from Earth and Rock-Fill Dam Construction

The main aim of this chapter is to show some typical problems of dam construction connected with real or potential risk of tensile cracks. A few fill domestic dams were selected for and are supplemented by the unique example of the Teton dam. At the end of this part there is a summary of quick crack detection methods and possible remediation of defective zones.

7.8.1 Dalešice Dam

Rock-fill dam Dalešice is 100 m high with a central slightly inclined core from loess deposits and with filtration and transition zones, Fig. 7.51. Great attention was given to the possibilities of the development of transverse tensile cracks with respect to the relatively steep slopes of the valley (1:1.3–1:1.55). Theoretically, this problem was treated by Doležalová (1970). Basic properties of core material:

$w_L = 36.3 \pm 4.2\%$,
$w_P = 20.5 \pm 2.1\%$
clay fraction (< 0.002 mm) $= 11$–28%
$w_{opt}(\text{PS}) = 16.9$–$19.0\%$
upstream slope inclination: 1:1.6–1:1.7
downstream slope inclination: 1:1.4–1:1.5

Dalešice dam, as the highest rock-fill dam in the Czech Republic, is very well instrumented. Outside of measurement of pore pressures, earth pressures, deformations, also three electromagnetic sounds type MGS/(VÚIS ČSSR) were installed into expected tensile zones, see Fig. 7.51.

Two monitoring results were discussed very seriously.

Firstly, high pore pressures were observed in the lowest part of the clay core, where the moisture content of the compacted core was higher than optimum according to the Proctor standard test. Pore pressure increase roughly corresponded with increase of vertical stresses. By a series of permeability tests under 3D conditions with an application of back pressure, Vaníček (1978b), it was proved that slow pore pressures dissipation was induced by much lower permeability of the core material than originally expected (up to 10^{-9} m.s^{-1} – about one order lower).

7.8 Practical Examples from Earth and Rock-Fill Dam Construction

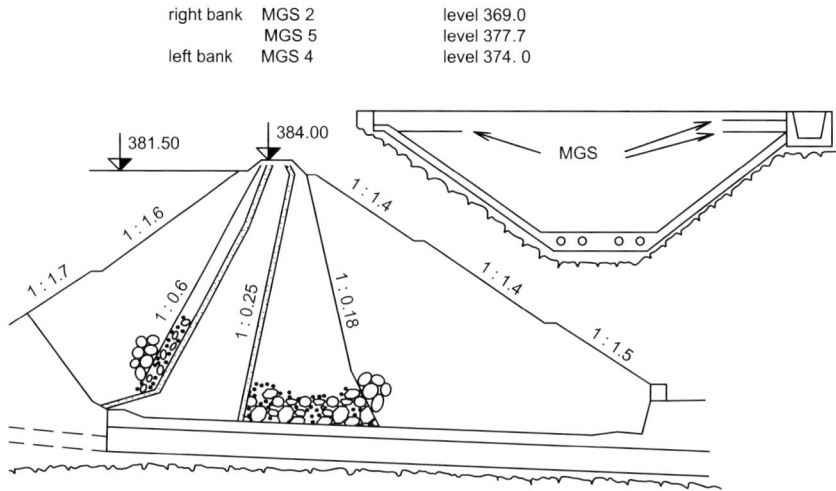

Fig. 7.51 Dalešice dam, maximum height 100 m; cross and longitudinal section; location of electromagnetic probes

At the same time lateral earth pressure was rather high, very close to the vertical pressure. But for some it was not a surprise, because due to high moisture content, high Poisson ratio (close to 0.5) can be expected as well a high coefficient of earth pressure at rest (close to 1.0). Therefore it was concluded that the hydrostatic pressure of reservoir water (even for maximal level) will not be higher than total vertical pressure σ_z in the lower part of the clay core, but at the same time also lower than $(\sigma_z + \sigma_x)/2$ or $(\sigma_z + \sigma_x + \sigma_y)/3$. That is why the risk of hydraulic fracturing for an undrained condition is very low here. Also was proved that high pore pressures have low impact on slope stability.

Secondly the dangerous area was signalled close to the dam crest in the neighbourhood of the spillway. The central core was finished in November 1977 (max level 382.5) when the water in the reservoir was only 10 m lower (372.5). In November 1978 the water in the reservoir was 377.0. In this time the results of the electromagnetic sounds showed the development of tensile zones. Maximum extension measured by MGS 2 and MGS 4 were lower than 0.5% but for MGS 5 – Fig. 7.52 – in part nearest to the spillway the value of extension 1.23% was measured. This value is higher than the general values measured during laboratory tests. But there was no indication of open cracks. Especially, the measured leakage was negligible, practically zero.

In this phase, the fundamental question was connected with sensitivity of further increase of water level in the reservoir, especially from the cracking and internal erosion point of view. To give a final answer on this question the author, Vaníček (1982b), completed supplemented laboratory measurements, which consisted of:

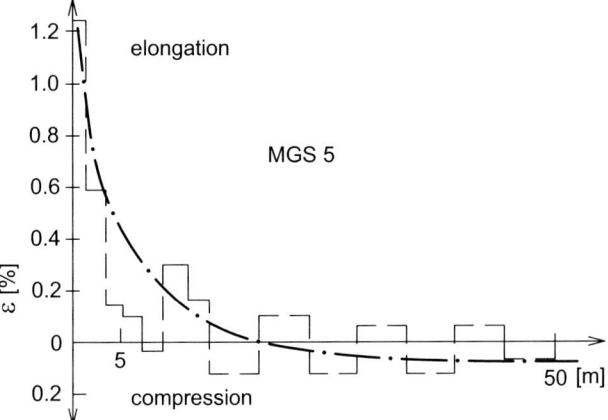

Fig. 7.52 Relative deformation in the individual section of the electromagnetic probe MGS-5

- A series of laboratory bending tests on laboratory prepared and also on in situ sampled beams, which were described in more detail earlier. They really showed that the maximum tensile elongation is roughly in the range of 0.3–0.6% for laboratory prepared beams, or in the range of 0.3–0.4% for in situ sampled beams,
- A pinhole test which proved that the soil is non-dispersive, with a high resistance to internal erosion,
- A series of tests, which were aimed at swelling ability of the core material.

Swelling ability of the clay has two important roles:

- "Dry" crack (which is created before seepage) there is a possibility of self-sealing by swelling,
- If the crack was not opened before seepage, its development as a "wet" crack is strongly influenced by swelling potential, which increases total stresses.

For our clay we expected relatively great influence of swelling potential because the value of clay colloidal activity $I_A = 0.8$ showed that clay particles contained also montmorillonite – Skempton (1953).

Swelling potential was measured in an oedometer. Clay with an initial moisture content $w = 18.5\%$ was compacted by Proctor standard energy and loaded by vertical pressure 100 kPa. After that, the soil specimen was unloaded to the pressure 50 kPa and consequently watered – Fig. 7.53. The results show that swelling was higher than 50 kPa and for open dry crack the increase in the volume by swelling was greater than 0.35%.

Additional information about swelling potential was obtained from a further unsuccessful but useful test. We have tried to model hydraulic fracturing. Small elastic tubes were installed into the central part of compacted clay (total height 210 mm) in a model box. These tubes were connected with a small reservoir and the water head

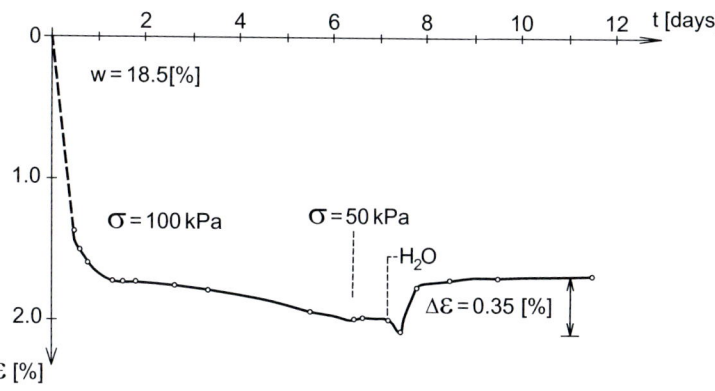

Fig. 7.53 Swelling ability measured in the oedometer test

was slowly increased up to 2 000 mm. No cracking was observed. The level in the reservoir was lowered down to 500 mm and a small loss of water was recorded. After three days a glass (thickness 12 mm) of the model box was disrupted by swelling pressure. There are probably two reasons for this fact:

- Arching effect in the model box (width 160 mm),
- Hydraulic fracturing is difficult to model in unsaturated soils.

On the basis of all these tests it was stated that the risk of cracking and internal erosion is very small for a slow increase in the water level in the reservoir. In 1981, the reservoir was in full operation and no problems of the discussed question were observed. Total measured seepage through the core was very small $-0.0025\,\mathrm{m^3.s^{-1}}$. After another 25 years of operation the dam is in very good state.

7.8.2 Hriňová Dam

Descriptions of problems on rock-fill dam Hriňová are described in detail by Verfel (1979). The dam is 42 m high, with an upstream clay core connected to an injection gallery. Ground elevation of the dam crest is 568.00 m – see Fig. 7.54. The filling of the reservoir started in the summer of 1965. The first leakage was observed at water elevation 563 m at the end of February 1966. The leakage from $0.004\,\mathrm{m^3.s^{-1}}$ grew to $0.10\,\mathrm{m^3.s^{-1}}$. Water elevation was lowered to the 550 m level, when the leakage was around $0.001\,\mathrm{m^3.s^{-1}}$. Because the reason of high leakage was not identified a second round of reservoir filling began. Leakage grew again and in May 1968 the outflow of muddy water of about $0.18\,\mathrm{m^3.s^{-1}}$ was observed below the spillway. After another water decrease below the level 562.5 the outflow was only $0.001-0.0005\,\mathrm{m^3.s^{-1}}$.

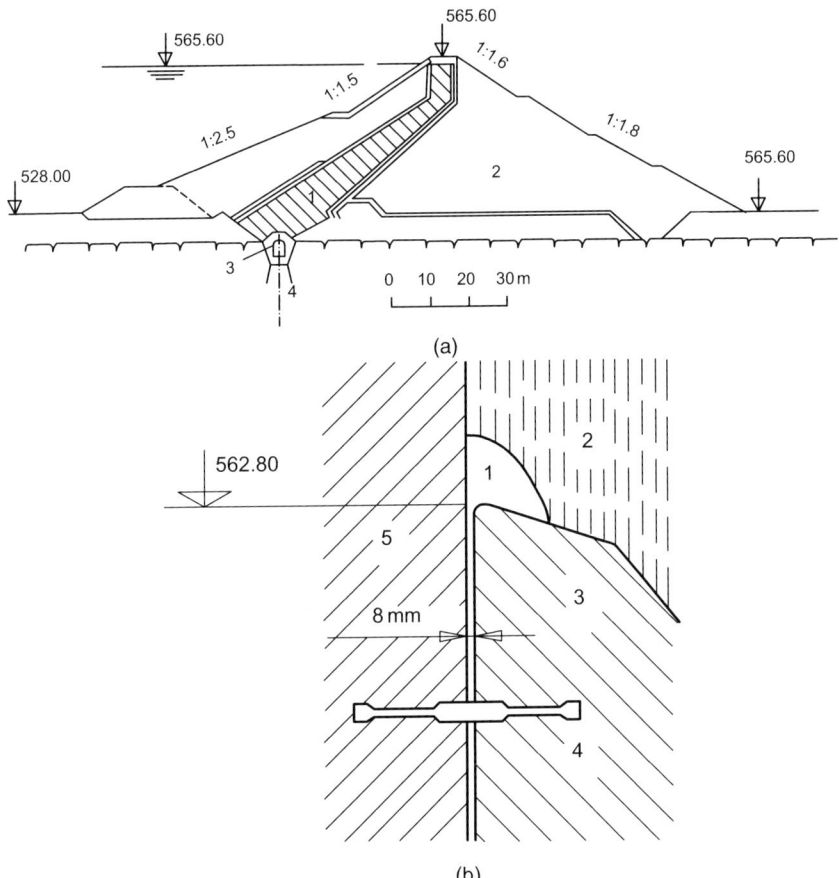

Fig. 7.54 Hriňová dam: height 42 m. (**a**) Cross section. 1 – clay core; 2 – stabilization zone; 3 – injection gallery; 4 – injection wall. (**b**) Detail of leakage place. 1 – cavern; 2 – clay core; 3 – concrete of injection gallery; 4 – sealing rubber; 5 – concrete of spillway

Verfel specified the more probable place of the second leakage at the connection of the injection gallery with the spillway. Sealing rubber in a dilatation joint with a width of about 6–8 mm was not connected to the clay core. Water started to pass through this dilatation joint and as a result of internal erosion eroded caverns above this dilatation joint, see Fig. 7.54b. The existence of a cavern was verified by two core holes and with these holes the cavern was injected a clay-cement mixture. The repair was monitored with the help of a shaft. This shaft helps to clear the way by which water reached the dilatation joint. A system of tensile cracks were observed on the upstream side with a thickness around 2–5 mm, which were very close to each other – about 100 mm. Differential settlement was declared as the reason of these tensile cracks. The reason of high leakage in 1966 was not exactly specified as well as place (if was on the other place), but probably the injection helped to seal this place as well.

7.8.3 Jirkov Dam

The Jirkov dam was selected to show the behaviour of longitudinal outer cracks. Jirkov rock-fill dam is 51 m high and constructed on the river Bílina during the period of 1962–1965, Fig. 7.55. The dam was constructed in a steep valley of trapezoidal shape with a slope inclination of 1:1.5. The crest length is 200 m. Upstream clay core is composed from plastic tertiary clays. The core is protected by a 3 m thick layer from unsorted sandy gravel. Both stabilization zones were filled from biotitic para-gneiss excavated from valley walls. The manner of filling was different; on the downstream side the filling was in very thick layers from 15 to 20 m and compacted by a water gun. In the upstream zone free filling was applied from a height about 6 m. Compaction by a heavy plate (a simple version of dynamic consolidation) was applied only at the upper part, for the downstream zone from level 444 m and for upstream zone from level 449 m – the dam crest is level at 454.30 m. Selected quarrying was performed only for the downstream zone; in the upstream zone all excavated material was used there. The average downstream slope is 1:1.3. The upstream slope was at the beginning close to the inclination of the downstream slope; later on it was reduced, especially in the lower part with help of loading berm. Two periods of excessive deformations were observed, first during the construction period from 1963 to 1965 and the second in the years 1973–1977. Detailed monitoring was described in the periodical reports of technical security supervision. During the first construction period, the first crack was observed at the 448 m level, see Fig. 7.56, Pařízek (1976). The reservoir was emptied and loading berm added. However another longitudinal crack was observed and was visible on the crest at the end of filling in 1965. For roughly 8 years with minimal changes in water level no other cracking on the crest was observed. However, on the downstream slope very close to the crest a deformation was observed in the rock-fill. Average settlement of the crest was about 60 mm per year. Excessive deformations up to 0.2 m were monitored during significant decrease of water level between October 1973 and November 1975. From the monitored deformations and observed faults it was possible to localize the excessive deformation to the surrounding of crest and to the upper part of the upstream slope. The downstream face deformed much less, and dam bedrock practically not at all. After the reservoir was emptied vertical deformation of crest was about 300 mm with the total settlement about 1.5 m.

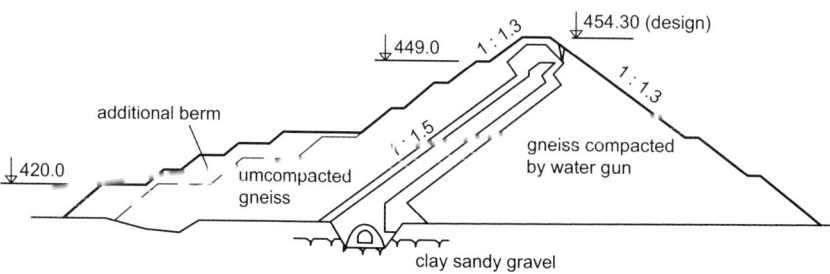

Fig. 7.55 Jirkov dam, height 51 m, cross section

Fig. 7.56 Jirkov dam – Photo of first longitudinal crack on the surface of clay core at elevation 448 m (photo J. Pařízek)

After emptying it was approved that the loading berm settlement was very low and therefore that the observed deformation has no slip character. The bad quality of rock-fill and bad quality of its compaction was the reason of the main problems.

With respect to excessive deformations it was interesting and optimistic that the seepage through the dam was extremely low. No transversal cracks were observed although the crest elongation close to the lateral linking was estimated to be close to 1%. In spite of this positive statement, the dam was reconstructed because the crest was much lower than needed for maximal water level in the reservoir and also because the dam was at this state not fulfilling demands of construction standards for such type of dams.

7.8.4 Low Dams in Southtown in Prague

These dams were selected as proof of the fact that cracks are not only associated with large dams. In connection with the construction of Southtown in Prague a system of small retention dams were constructed. The dams were constructed as homogeneous earth-fill dams. The dam crest was reinforced by asphalt pavement and protected by a railing. The upstream slope is protected by flagstone pavement, downstream is covered in grass.

7.8 Practical Examples from Earth and Rock-Fill Dam Construction

The author investigated the reasons of higher deformations on five different dams on this system. Fundamental problems will be shown for dam R 5 on Chodovec brook, where observed deformations were most visible, Vaníček and Záleský (1981). Maximal height from footing bottom is 11.0 m, length of crest 88.0 m and the volume of stabilization zones about 4 600 m^3. Slopes on both sides are 1:2 and the crest width is 4.0 m.

Geological conditions are rather complicated. In the brook axis there are alluvial deposits of sandy loam character with mud interlayers. The left bank is smooth with fine sand with gravel layers. The right bank is steeper and the bedrock that is on the brook axis roughly to the depth of 6.0 m is now only 1.3 m in depth there. Bedrock is composed from dark grey fine clayey slate, with a laminar structure.

The longitudinal profile is therefore situated on different subsoil, with different properties and thickness. In the outlet axis mud deposits were substituted by compacted fill. Sealing cut-off was completed only in a left-handed wing, where there is cutting permeable fine sand and gravel deposits. A toe drain was applied on the downstream slope. Soil of MS character was used as fill material.

Visible deformations appeared very early after the end of construction. Vertical deformations on the dam crest were in the range of 53.7–84.3 mm. Monitored point situated on outlet settled only 1.9 mm during the same period. The character of visible deformations are obvious from photos, Fig. 7.57a–c. Deformations have a different orientation and width, however longitudinal cracks prevail and two smaller transversal cracks are not continuous through the all dam profile.

It was concluded that the main reason to differential settlements are complicated geological conditions. In the course of a short length of a dam it is first of all different depths of bedrock, significant variability in quaternary deposits. It was stated that too complicated cross-sections were proposed to decrease dam permeability. The secondary reason of higher settlement can be associated with poor compaction, which is a well-known case for a complicated profile.

Individual dams do not represent a critical state, seepage is within allowed limits and with a crest settlement of 100 mm was expected during the design. However it was recommended to correct the cracks and visual irregularities because they provoke an unfriendly impression and provoke distrust in this engineering structure.

7.8.5 Teton Dam

The failure of Teton dam in 1976 is one of the most discussed dam failures simply because no dam of such height had previously failed, was designed by Government agency, failure was very quick that no countermeasure was applied and finally because the high-flood-water wave below the dam profile killed 14 people and material damages only below the dam amounted up to 400 mil. USD.

The Teton dam is located in a steep-walled canyon cut by the Teton river into a volcanic plateau. The walls of the canyon consist of later tertiary rhyolite welded-tuff which is strongly jointed. Alluvium has been deposited in the river channel to

Fig. 7.57 Dam R5 on south town in Prague. (**a**) Longitudinal crack on left-handed wing, which opened again after being filled by asphalt. (**b**) Detail of longitudinal crack on upstream side of left-handed wing under railing. (**c**) Deformed crest and railing (photo J. Záleský)

7.8 Practical Examples from Earth and Rock-Fill Dam Construction   465

(c)

Fig. 7.57 (continued)

a depth of about 30 m and the high lands near the ends of dam are covered with an Aeolian silt deposit of about 9 m thick.

The main cross section through dam is schematically shown in Fig. 7.58. In the central part the height of dam is 90.6 m, when comparing the dam crest with downstream toe, or 123.6 m when comparing the crest with bottom of sealing cut-off trench. A wide core zone of Aeolian silt was supported by upstream and downstream shells consisting mainly of sand, gravel and cobbles. The layout of cross section

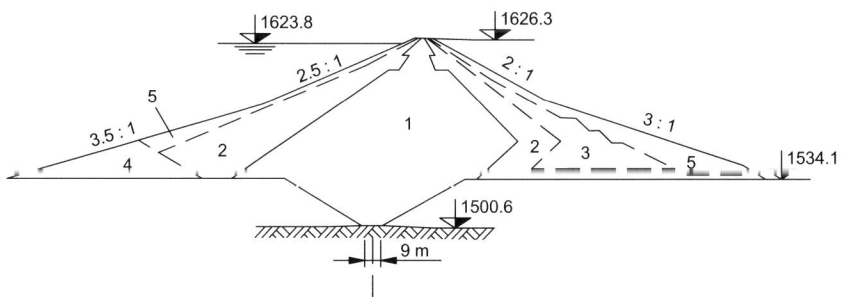

Fig. 7.58 Teton dam – Cross section through center portion of embankment founded on alluvium (according to Seed and Duncan). 1–silt with some clay, sand and gravel; 2–selected sand, gravel and cobbles; 3–miscellaneous fill; 4–selected silt, sand, gravel and cobbles; 5–rockfill

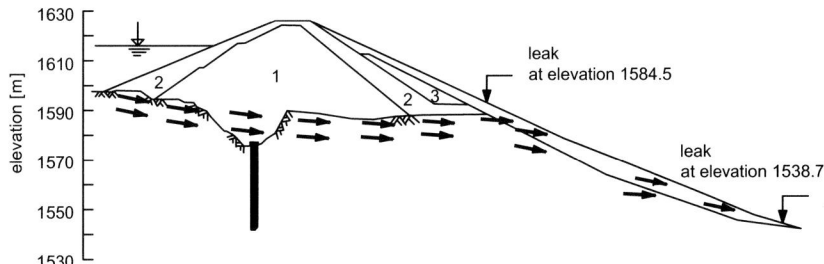

Fig. 7.59 Teton dam – Probable path of water in early stages of leakage (according to Seed and Duncan)

above the level 1 555.5 with 21 m deep key trench is presented on Fig. 7.59. This key trench was proposed because strongly jointed tertiary rhyolite welded-tuff contains joints with width typically between 7 and 70 mm, occasionally up to 300 mm. Due to subsoil high permeability a grout curtain was installed along the full length of the dam, some holes extending to depths of 90 m. Grout holes were along a single line with primary holes 3 m apart, and split spacing where the primary holes did not indicate a tight curtain. However lines of barrier holes, intended to prevent excessive flow of grout from the main grout curtain, were installed on 6 m centres 3 meters upstream and downstream of the main grout curtain. Additionally to help prevent seepage, the key trenches and grout curtain were continued well beyond the ends of the embankment, the curtain extending 300 m into the right abutment and 150 m into the left abutment.

Rate of reservoir filling was little bit quicker than expected, reaching 1.2 m per day during May 1976. By June 5, the day of the failure, the water level stood just 0.9 m below spillway crest and 9 m below the embankment crest.

The first leakage occurred on June 3 roughly 450 m below the dam. The following day other small springs were observed in river valley about 120 m below the dam. During the observation of upstream and downstream slopes at June 4 evening no other abnormalities were found. On June 5 at about 7:00 a.m., when the first workers reached the site, water was observed on the downstream slope about 40 m below the crest at elevation 1 584.5 near the junction of the embankment with right abutment. Outflow increased very quickly, just after midday dam completely failed. Failure gradation was photo documented and was included into "A report of findings" – US Department of Interior (1977).

Two groups of experts investigated the reasons of this failure. Information have spread very quickly, some conclusions were immediately published and made use of, in former Czechoslovakia e.g. by Záruba (1978). Author used the results when investigating an untypical behaviour of Dalešice dam – Vaníček (1979).

A very comprehensive review was published by Seed and Duncan (1981) and this review is still the fundamental information about the Teton dam failure.

Both the official ministerial investigation and independent expert group confirmed that the main reason of failure was insufficient protection of clay core against internal erosion. Connection of sealing clay in the key trench with jointed bedrock was marked as the place where progressive erosion started, due to insufficient preparation of this detail. Additional excavation during investigation disclosed open cracks partly filled by soil. Concentrated leakage followed by progressive erosion arose

– either along crack initiated by hydraulic fracturing in key trench where due to arching effect, the vertical stress was reduced, or
– along grout cap, either owing to higher permeability of injection wall in upper part or just by flow above grout cap where relatively high hydraulic gradient between 7 and 10 had chance to initiate the erosion on contact with jointed rock – this is in agreement with tests of Wolski mentioned in Chapter 5 – Fig. 5.40.

The initiation of failure probably started earlier than observed because water and wash-out material had chance to fill joints in bedrock. The probable path of water in the early stages of leakage is presented in Fig. 7.59. In the case of better instrumented dam the signals about failure initiation would arrive earlier.

However during the investigation other weaknesses were revealed:

– Wet seams in clay core, placement of which corresponds with extreme rain and snow falls in May 1975,
– Filter which was described as sandy gravel was not fulfilling any US filter criteria,
– Assumed good filter permeability was not proved by additional tests, results showed a wide variety of readings from the soil with very good permeability up to nearly impermeable soil.

Publication of all circumstances occurring during the construction followed by verification of properties by field and laboratory tests as well as by laboratory and numerical modelling and finally also by photos of the dam failure, are for dam engineers all around the word, a very beneficial warning, see Fig. 7.60

7.8.6 Failures Remediation

This problem has already been discussed in the previous part when practical examples were described. Effectiveness of remedial actions depends in many areas on an exact localization of place of failure and on the right selection of remediation methods.

Different methods of failure localization and remediation were discussed as question No. 45 during 12th ICOLD congress in Mexico in 1976.

7.8.6.1 Failure Localization

Different monitoring methods are applied which can indicate not only the failure of dam but also the continuous development of such failure:

Fig. 7.60 Teton dam – Present-day view

- Increasing dam seepage,
- Character of seepage water with respect to washed out fine particles,
- Increasing deformations of dam crest and dam slopes,
- Results from the different monitoring devices in the instrumented dam such as piezometers, pressure gauges, pore pressure gauges, extensometers, inclinometers and others.

From the obtained results of monitoring the localization of failure is usually only a rather broad one. For more precise failure localization different methods of sounding are applied. For clay core the core-drilling is preferred, for stabilization zones the exploring shaft.

When searching the seepage path with concentrated leakage three points are important – entry to the dam body, entry to the clay core and finally the place of water discharge, which can be directly on the downstream slope or was collected

7.8 Practical Examples from Earth and Rock-Fill Dam Construction

by drainage system. For generally higher permeability of the stabilization zone, the first two points coincide, at least from the elevation point of view. Therefore usually the first step is connected with reservoir water level change. It is possible to define the level at which the outflow is abruptly increasing or decreasing. The exception is wet crack opened by hydraulic fracturing; its location can be lower than the water level in the reservoir.

Methods observing the properties of leakage can be useful as well, e.g.:

- Thermal analysis,
- Chemical and physical analyses,
- Observation of the natural radioisotopes content.

These methods can help us to distinguish the water from reservoir from underground water as well as water from different levels in reservoir, because water there have different temperature, pH, water hardness, concentration of isotopes etc.

Another problem is connected with zone of failure identification in the longitudinal direction. Different tracers are applied, not only for the identification of water origin and seepage path but also for identification of direction and velocity. Tracers can be soluble such as mineral salts, organic compounds and radioactive isotopes or insoluble, which create water suspension such as clay minerals or artificial fibres. The tracer is applied in a selected place and its content is checked in seepage water; intensity and time delay are very important outputs.

Vestad (1976) describes an interesting case of failure localization for Viddalsvatn dam in Norway. Tracer Rhodamin B was discharged into reservoir water from a boat slowly moving from the one lakeside to the other and back. In the Czech Republic the following tracers are recommended, Poláček (1984) – chloride, iodide, nitrite anions, fluorescein, Rhodamin B and artificial fibres.

Geo electrical or micro seismic geophysical methods can distinguish between different soil layers with different degree of saturation. From other methods the following ones are also often mentioned, e.g. Šimek (1976):

- Ultrasonic detector,
- Inspection by diver,
- Scattering of short plastics fibres with density slightly higher that water density, which are pulled by water to the place of inflow to the dam body,
- Bunches of long artificial fibres embedded to the bottom of reservoir, which are inclined in the direction of water stream.

The utilization of piezometers installed on the upstream face of Balderhead dam (Vaughan et al. 1970), detecting any pore pressure change against hydrostatic one for a given depth, were mentioned earlier.

7.8.6.2 Remediation Methods

During the remediation of developed cracks or eroded caverns this space is refilled by original material which was washed up or by a new material, which will stop the process of erosion and will decrease the permeability there.

In the up to now described cases the injection was applied most often. Verfel (1979) describes the exact composition of injection mixture used for remediation of Hrinova dam: 885 litre of water, 280 kg of tumerit clay and 35 kg of cement per 1 m^3 of injection mixture. He also recommends that the mixture after strengthening should have the stiffness similar to the surrounding soil. Two different steps of injection are generally recommended. In the first step, the mixture has to refill crack and caverns, and in the second one to compact remediate zone. For some dams the three row injection wall are applied. The first row of injection holes are located on the downstream face of clay core, after that on upstream face and finally in the middle. Injection mixture consisted of water, cement, bentonite (roughly 2%) and fine sand. The amount of fine sand is regulated according to the mixture loss in hole. For the central row the sand is usually omitted and content of bentonite increased. Injection pressure is lower than vertical pressure for the given level to prevent hydraulic fracturing.

The reconstruction of Balderhead dam was rather complicated. Instead of injection the construction of new concrete sealing wall was performed, passing along the axis of old disrupted concrete sealing wall. This solution is applicable only for the vertical central core. The reconstruction procedure was very precise, because the new wall with thickness 0.6 m had to be connected to the sealing wall in the dam subsoil.

Failures localized close to the dam crest are reconstructed more easily. Zone of failure is excavated, whereas the direction of cracks is specified, together with their width and depth. When the crack is clean its washing by water is applied and after that the crack is backfilled by slurry of water, sand a clay (bentonite). After drying and shrinkage, the slurry is replenished and in the upper part of trench the classical compaction of stiffer soil is launched.

Last manner of remediation is suitable for small homogeneous earthfill dams with increased leakage through dam subsoil. In the central axis of dam the key trench reaching up to less permeable layer is excavated and backfilled by less permeable soil. Where there are some problems with application of intensive compaction, a soil with plastic up to liquid consistency is used to refill this trench.

A rather old method consists of dissolution of fine grain soils in reservoir. This method counts with successive sealing of seepage path or with creation of fine grain cake on the surface of more permeable material. Very often old ponds have much higher permeability after old settled mud is removed. But with time the decrease of outflow is observed, and in some cases the self-sealing process is supported by phytoplankton. As was mentioned earlier, the dissolution of lime has a positive effect on dams constructed from dispersive soils.

7.9 Summary of Earth and Rockfill Dam Design

At the beginning of the chapter the authors emphasized that our effort to limit the potential risk of dam failure is concentrated on two aspects – firstly to eliminate risk of tensile crack development and secondly to change the approach (logical scheme of the individual steps of the decision making process) to the dam design.

7.9.1 Recommendation for Elimination of Tensile Cracks Development

7.9.1.1 Geometry of the Dam Profile Valley and Dam Alone

The efficiency of fill dams results in a use of dam profiles which are connected with higher risk of crack development. Firstly it is steepness of valley slopes. In the frame of the parametric study the most appropriate dam profile can be recommended and some conditions for selected profile have to be defined. Curvature of longitudinal profile against water is mentioned as one possibility. Nevertheless the problems with Norwegian dams Hyttejuvet or Viddalsvatn, where this solution was proposed, show that not only steepness but, above all, irregularities in the slope are important. Levelling of local irregularities is therefore the first step.

Levelling of bedrock is connected with cleaning or with adaptation of the bedrock, e.g. by applying shotcrete. Clay core is placed on the adapted surface in a thin layer about 0.15–0.5 m with plastic consistency and is carefully compacted, even manually. Most attention is devoted to upper parts, while plastic soil in lower part is obliged to "bypass" irregularities, and in the upper one due to lower pressure, the core can be separated from the surface.

The dangerous influence of subsoil heterogeneity was shown in the example of the Guadalupe dam. In such cases it is more acceptable to leave this profile. Subsoil excavation and replacement by more appropriate material is generally very expensive. In the extreme case pre consolidation has to be applied.

The interaction of fill with stiff structural elements such as injection gallery, outlet, spillway etc., leading to stress redistribution can be partly solved by strict control of geometrical conditions. It is preferred to locate these stiff elements into bedrock or out of dam body. If this solution is not acceptable, concrete elements are proposed with small inclination for example 1:0.1 to 1:0.05 to guarantee that the soil will be pressed to the concrete surface.

Connection of clay core with vertical sealing element (as concrete wall, slurry wall, injection wall, sheet pile wall) of dam subsoil is a special problem. Usually the compressibility of such a vertical wall is lower than the surrounding subsoil, so that at the top of this wall is high concentration of stresses. For large dams the interconnection is realized with the help of control gallery. For elimination of steep changes in subsoil stiffness the subsoil in the upper part can be injected, see Fig. 7.61.

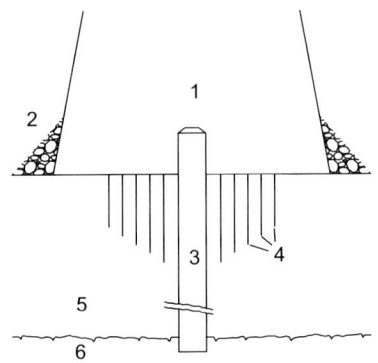

Fig. 7.61 Detail of proposed connection of vertical sealing wall and clay core. 1 – clay core; 2 – filter; 3 – vertical sealing wall; 4 – injection; 5 – permeable and relatively compressible subsoil; 6 – bedrock

From the dam geometry the localization and thickness of clay core is most important. Minimal thickness is limited by maximal hydraulic gradient. Generally recommended thickness is about 25–50% of the water level in the reservoir, in the extreme case only 10%, so that the hydraulic gradient is generally from 2 to 4, extreme 10. Sometimes upstream clay core is preferred largely from technology of filling point of view – construction area is not divided into two parts and stabilization zone can be constructed in advance. However the central core is preferred from the stability point of view and usually total volume is lower. Also length of vertical sealing wall is shorter. To eliminate arching effect in clay core, and thus to reduce the crack development the central core should be wider and slightly inclined against water.

7.9.1.2 Selection and Adaptation of Fill Material

Clay core, filtration and stabilization zones are the main components of fill dams. The approach to individual component cannot be separated from others, because the final state of stresses is the result of interactions between these main elements. Therefore, before the design of final cross section it is necessary to know fundamental properties of the individual materials and possibilities for their modification. However, the selection of suitable material is always limited by need to find this material in close vicinity of construction.

Clay Core

Clay core represents – concerning the safeness of all dam – the most important element. During the selection of clay material not only index properties, shear strength, compressibility and permeability have to be measured, but also deformation characteristic in tension and as well as susceptibility to erosion namely as a function of moisture content, compaction effort and time. Up to now soil, with continuously variable grain size distribution, dispersive soils and silts with low index of plasticity were mentioned as highly susceptible to erosion.

Soils with higher moisture content are more flexible and can withstand without cracking higher tensile strains as a result of differential settlement.

7.9 Summary of Earth and Rockfill Dam Design

From the view of inner cracking above all by hydraulic fracturing two different approaches exist but with the same aim to reach higher total stresses in the clay core. The first one prefers stiff clay core which will settle less than surrounding filter and stabilization zones. The second one prefers very soft clay, which behaves like a liquid and reduces the effect of arching. The second one is supported by Penman (1977) and was discussed by authors more closely for the Dalešice dam. The first approach, especially for sealing material filling trenched or any other irregularities in the dam bottom, is supported by Wilson (1977). For trenched and lowest irregularities he recommends sealing materials with nearly the same compressibility as for surrounding material. However, for upper parts, he recommends also more flexible sealing material.

From the compaction point of view, the moisture content slightly higher than optimum is recommended, but only with application of compaction energy slightly higher than energy of Proctor standard test, to obtain higher degree of saturation and dry density. The compaction should guarantee good cross connection of individual layers, because higher permeability in the horizontal direction increases the risk of hydraulic fracturing.

Filter

Rather high demands for filters from natural materials were discussed in previous chapters. Discussed demands on higher percentage of fine nonplastic particles than generally accepted 5% should be included into national codes of practice, especially for cases, where the probability of cracking is higher.

Filter is a significant component of mutual interaction. Their compressibility compared to sealing and rockfill is generally lower. Since the filters are mostly composed by sands and sandy gravels they are well compacted. However over compaction can leads to the worsening of the state. Therefore the filters, or transition zones generally, should be wider, symmetric and their compressibility not so different from other components.

During filling and compaction it is necessary to restrict the segregation of particles into finer and coarser layers, because after that this aspect should be included into filter design.

Stabilization Zone

For rockfill dams the quarrystone is used in the majority of cases for stabilization zones. The compressibility is generally higher than for filters. To limit negative effect of interaction on the boundary of these two zones, the following counter measures can be applied:

– To limit maximal size of large grains,
– To improve grain size distribution curve – for stepless curve more intergranular contacts are reached so leading to lower compressibility,
– To decrease the thickness of individual layers,

- To increase compaction effort,
- To add water in-process of compaction.

Adding water in-progress of compaction is meaningful, decreasing future small deformational collapse after wetting by first reservoir filling. This aspect relates to possible reservoir filling during dam construction, because not only does this collapse on grain contacts, but also continuous core loading by hydrostatic pressure can go through in advance, during reservoir filling.

Estimation of deformability of rockfill is usually a difficult task – Penman and Charles (1976), requiring large laboratory devices, large diameter oedometer, Vaníček and Vacín (1968), see Fig. 7.62 – or even triaxial apparatus, e.g. Marsal (1973).

Because deformation characteristics of rockfill are strongly influenced by crushing process on the contact of individual grains, one proposed way, proposed by author, Vaníček and Vacín (1968) is to measure so called intergranular equilibrium pressure. This method was applied in the phase of selection of most appropriate material for construction of the Přísečnice dam, where three different quarries were employed. Fixed sample of rock with size about 100 mm was loaded in press with increasing pressure and, for selected load level, the contact area of sample with loading plate of press was measured. Investigated relation was practically linear, defining intergranular equilibrium pressure. Different sample orientation and contact wetting were observed.

For rockfill material of the Jirkov dam it was approved that weathering processes proceeded in time, lowering strength of rock with time, especially in area with fluctuating water level in reservoir and with temperature changes.

Fig. 7.62 Large scale oedometer for compressibility of rockfill testing

7.9.1.3 Other Aspects

Two other points will be mentioned. First is the speed of reservoir filling. For the El Infiernillo dam it was shown that after rather quick filling first crack appeared for relatively low elongation, about 0.2%. However, after the reparation, new crack was not observed even though the continued elongation reached 1%. Other experiences show that quick filling is connected with higher potential risk of crack development. Swelling effect is certainly playing an important role. Therefore it is recommended to slow down speed of filling at least for last 10–20 m. Unfortunately this condition is difficult to fulfil for dams with main anti flood function.

Secondly it is recommended to limit any soundings in realized clay core by virtue of the hydraulic fracturing potential risk limitation.

There is also discussion about possibility of reinforcing potential tensile zones by geosynthetic reinforcement, e.g. Chabal et al. (1983). However, a non answered question is connected with preferential seepage path along the contact between clay and reinforcing element. Maybe that scattered reinforcement can solve this question.

7.9.2 Recommended Logical Way for Fill Dams Design

Recommended logical way for the fill dams design starts:

From the necessity to conform the design to the potential risk, to the risk which the potential failure of structure can bring about – this approach is at the present time supported by Eurocode 7 – Geotechnical design: "Geotechnical design shall be identified together with the associated risks". It is the way strongly supported by Whitman (1984) in his Terzaghi Lecture.

From the recognition mentioned at the beginning of this chapter, the design of fill dams is not governed by slope stability with subsequent verification of others modes of failure, but to a greater extent by the danger of tensile crack creation, followed by the internal erosion or by other limit states.

Therefore the authors recommend the logical way, which is taking into account all findings discussed in previous chapters and which each designer of fill dams should implement. For visual illustration of this way we recommend the logical scheme presented by Whitman (1984) and which is in slightly modified version shown in Fig. 7.63, Vaníček (1985). This logical scheme (event tree) deliberates fundamental geotechnical aspects affecting the safety of fill dams with clay core. This scheme is not taking into account the hydraulic aspects, for example the risk of crest overflow or failure of spillway during floods.

Primordial question relating to the crack development includes all reasons which can have some relation to the crack development. In the broader sense it is possible to extend this question to the problems of existence of opened fissures in valley slopes, insufficient sealing of subsoil, seepage canal along outlet etc.

The approach to the logical scheme presented in Fig. 7.63, especially to the primordial question enables the use of numerical methods for the calculation of

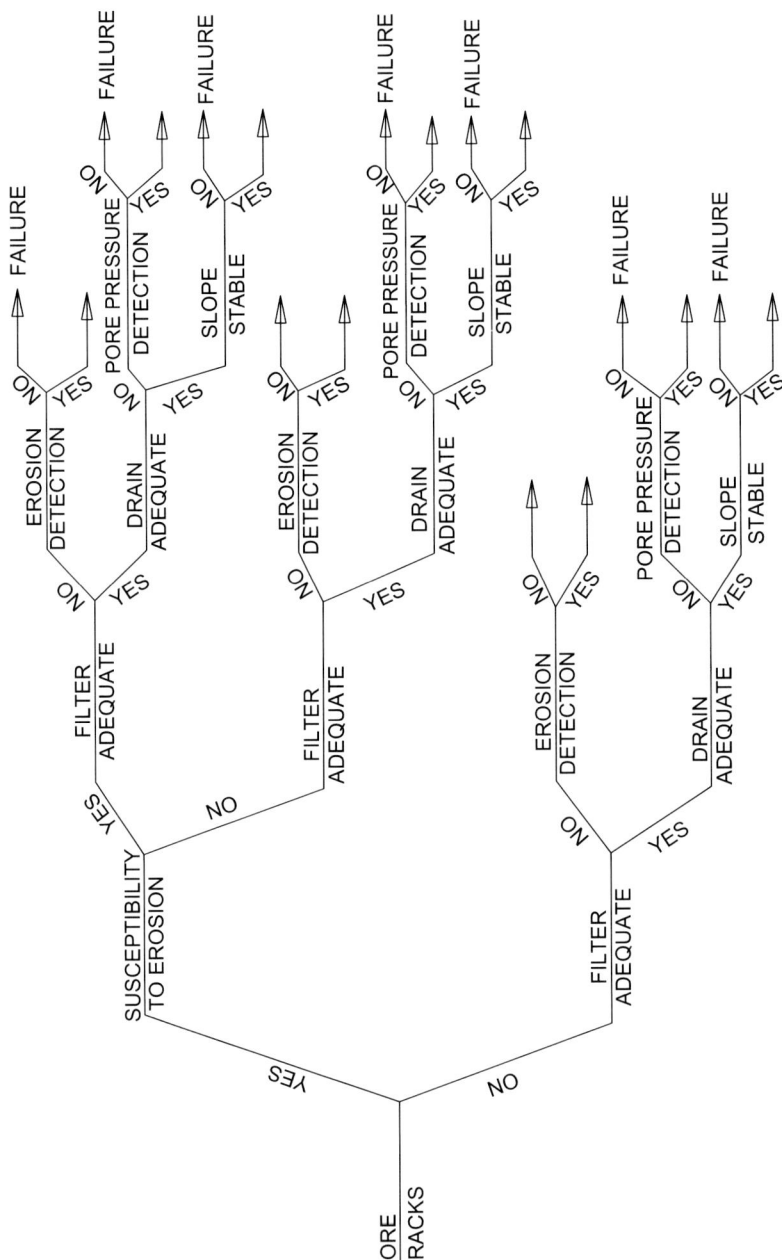

Fig. 7.63 Logical scheme of fundamental geotechnical conditions affecting safety of fill dams with clay core

7.9 Summary of Earth and Rockfill Dam Design

stress-strain state or to apply second logical scheme to lower degree. Some specification of the logical scheme of second degree is presented below.

The crack development is influenced by:
– *geometrical conditions*:
- dam valley < regular / irregular
- valley slopes < flat / steep
- longitudinal axis < incurvate against water / straight line
- clay core < wide, slightly inclined / thin, vertical

– *material conditions*:
- clay core < flexible / brittle
- interaction of individual elements < low / high
- compressibility of stabilization zone < small / high
- subsoil compressibility (incl. differences) < small / high
- swelling potential < high / low

– *other conditions*:
- speed of filling < slow / quick
- soundings in core < no / yes

The answer to the primordial question (though with a certain probability) strongly influences the approach to other questions.

The answer the question concerning *susceptibility to internal erosion* depends on protected soil:

– protected soil (core) < flocculated < resistant to erosion / susceptible to erosion

dispersive

swelling potential < high / low

The answer to the question concerning the *filter adequacy* depends on:

– protected soil (core) – susceptibility to segregation < small / high

– filter properties – filter < granular / geotextile

– filtration criteria considering < perfect filter / classical filter formed of aggregates

– other conditions – hydraulic gradient < small / high

– supervision < careful / practically none

As to the question about filter adequacy, the case of cracked or intact cores have to be differentiated, because in the first case, the hydraulic gradient will be more important while, in the second case, the susceptibility to segregation.

Another question is connected with *adequacy of drain* that is with all measures leading to the limitation of pore pressures at the downstream part of the dam. Recommended factor of safety defined as ratio of designed capacity to calculated seepage, will depend on the susceptibility to cracking, quality of workmanship and on potential risk of choking (including a chemical one).

Questions about the systems of *detection of internal erosion initiation* or of *detection of pore pressures*, relate to the dam monitoring and visual inspection. The answer to these questions can be yes only if it is guaranteed, that the countermeasures, applied after the detection of these dangerous events, will be in such a degree operative that will be able to prevent the dam failure. The answer depends also on monitoring and inspection frequency, whether the countermeasures are prepared in advance or on the sensitivity of clay core to the internal erosion.

In this case the question connected with *slope stability failure* is on the last place. The calculation of this limit state has to take into account all previous answers, or their probability.

7.10 Specificity of Small Fish Pond Dams

Roughly between 400 and 600 years ago many small earth dams were constructed mainly in the south part of the Czech Republic. They were used mostly for fish production and flood protection. To present day roughly one third survived, leaving the number close to 25 000. Dam crest height is between 3 and 15 meters and these earth-fill dams were constructed as nearly homogeneous dams from the local soils. Their specificity will be described on the ground of heavy floods in 2002.

7.10.1 Rehabilitation of Old Earthfill Dams Failed During Heavy Floods

During catastrophic floods in 2002 many of these 25 000 dams had some problems but less than 0.3% failed. The highest concentration of failures were connected with small river Lomnice, Fig. 7.64, situated roughly 100 km south of the capital Prague, where five dams failed and two others were damaged, mostly by the so called domino effect, I. Vaníček and J. Vaníček (2004). The reasons and manners for dam failure are discussed together with a new approach for the overall stability. Also some practical results from reconstruction are added.

Heavy rainfalls during 7 and 9th August 2002 did not cause any problems; pond reservoirs were nearly full and fulfilled their flood protection role. The situation was improved during Saturday and Sunday (10–11th August) and with the help of directed outlets a certain reservoir volume was prepared, for example in three cases water level was 0.3–0.6 m under normal level.

New heavy rainfalls started on Monday, which had very quick effect due to the full saturation of surrounding land. On Monday evening the flow volume reached 100 years flow rate. But still all dams were able to catch all these volumes.

The critical situation started at around 4 am of the next day when – due to overflowing – the first dam (Melín) on this river-basin failed. Additional wave reached the crest of the next dam "Metelský" in few minutes and very quickly this dam also failed in two places, Fig. 7.65. Relatively large volume from this reservoir (about 1 mil.m^3), with the estimated (and recalculated) outlet more than 500 m^3.s^{-1} (which was much higher than 100 years flow rate – ca 20 m^3.s^{-1}), affected the villages Metly and Předmíř, where some houses failed and one man died. The embankment of the next pond "Veský" survived due to the very wide crest with asphalt pavement but was strongly eroded on the downstream side. The same situation was observed on the next pond "Zámecký", the crest of which is also protected by asphalt pavement – European road E 49 is passing through it, Fig. 7.66. The embankment of the fifth pond (Podhajský) resisted for a few minutes, even when the water went over the crest (the calculated 100 years capacity of spillway was not sufficient), but finally also failed in the place where the reconstruction was realized roughly 13 years ago.

Flood wave after that reached the railway bridge which survived but the continuing embankment on both sides of the bridge was eroded, leaving railway track in the

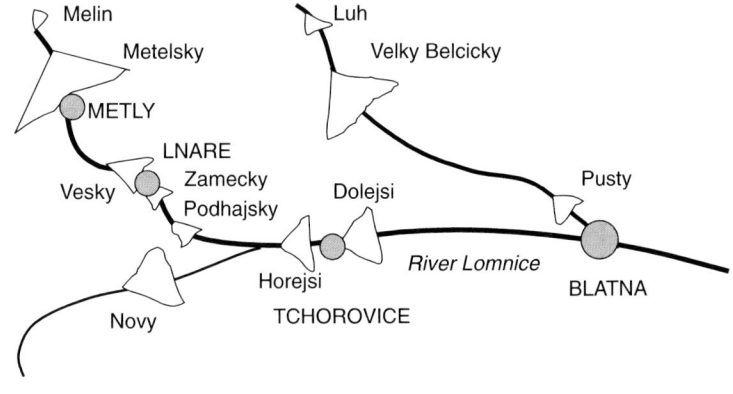

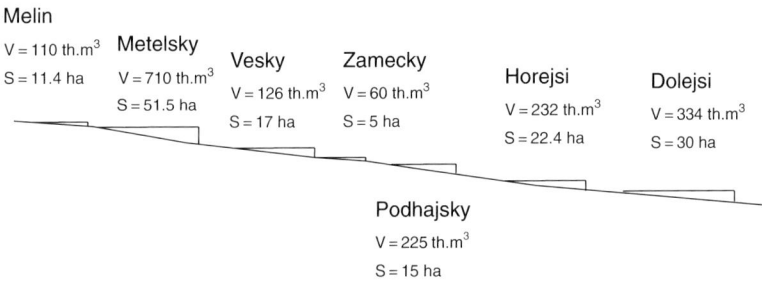

Fig. 7.64 Little river Lomnice – Situation and scheme of dams

air. The last two dams also failed, when the leading one (Hořejší) strongly affected the village Tchořovice. After the failure of the last one (Dolejší) the flood wave was getting flatter in much wider and flatter valley so that the impact on the town Blatná was diminished. But even there the damages were very high, because on the brook, joining Lomnice in Blatná, three dams (Luh, Velký Bělčický and Pustý) failed as well, Fig. 7.67.

7.10.2 Limit States of Failure

Usually the limit state of stability receives the priority during the cross section design. Such sort of the failure was not observed in any case. That is in agreement with a general opinion, that the most of the earth and rock fill dams have problems with other limit states as is the limit state of deformation (leading to the development of cracks – see previous chapters) and limit states of surface and internal erosion.

In all observed cases the surface erosion played the most important role, but in two cases the internal erosion also probably participated in the final collapse. For

7.10 Specificity of Small Fish Pond Dams 481

Fig. 7.65 Dam Metelský. (**a**) Gash with old wooden outlet. (**b**) Second gash

example in one place of failure – gash – of the embankment of the pond Metly, the part of the old wooden outlet was found, see Fig. 7.65a. From the historical records this old outlet was left there roughly in seventeenth century, because from these days a new place for new outlet was selected. Approximately, in this profile, on the downstream side the place with higher humectation (moisture content) was observed for last decades, probably as the result of higher permeability in this

Fig. 7.66 Dam Zámecký in village Lnáře – Road E 49

Fig. 7.67 Town Blatná – On 13 August 2002 morning

7.10 Specificity of Small Fish Pond Dams

profile. Due to higher hydraulic gradient, the clogging of this old outlet failed and internal erosion started in this place. It is interesting to note that the erosion started with a short delay when the water level in the reservoir went down due to the first failure. The second case, where probably the internal erosion also played the significant role was the pond Podhajský, Fig. 7.68, the embankment of which was reconstructed 13 years ago and sealing wall formed by jet grouting was applied in one part of the embankment. All of the downstream side was eroded; nevertheless

Fig. 7.68 Dam Podhajský. (**a**) Total gash with eroded downstream side. (**b**) Old clay-cement jet grouted sealing wall

the total failure occurred in the same place as sealing wall was installed. Probably the internal erosion started along the periphery of this sealing wall as the result of higher hydraulic gradients there and as the uncertainty of the connection of this wall with the rest of the embankment and also with the dam bedrock where large granite boulders were disclosed.

7.10.3 Principles of the Reconstruction Design

The reconstruction started very quickly after the failure. Two dams, which did not fail completely and the crest of which is used as a main roadway, were reconstructed during 3 months, in the autumn of 2002, with financial support of the Roadway directory. All others were reconstructed during the following 2 years with the financial support from the state budget, covering flood damages via Ministry of agriculture. For two of them situated just above villages the first step was provisional cofferdam to catch increased outlets. The detailed attention was directed to the following items, I. Vaníček (2004a):

- Verification of hydrological conditions, recalculation of spillway capacity for 100 years outlets and supplemented increase of this capacity,
- Selection and recommendation for most appropriate type of soil used for reconstruction,
- Unification of the crest vertical alignment in the all length, because some differences up to 0.3 m were observed.

7.10.3.1 Hydrological Data and New Design of Water Diversion Systems

After the floods the Czech Hydro Meteorological Institute recalculated all new data but the new recommended 100 years outlet was very close to the previous ones. It was proved that the capacity of water diversion systems for all ponds were good enough even for new data. Nevertheless the new proposal recommended to increase this capacity with the following points J. Vaníček (2004), Fig. 7.69:

- Where two defrayable spillways existed for the first one the spillway frontal crest was decreased about 0.2 m,
- Capacity of both spillways were increased by pulling down the intermediate column for the increase of flow profile and for the decrease of the possibility of the blockage by timber, small trees etc.
- New arrangement of the lifting facilities allowing to rise of flood gates above the flow profile,
- Construction of an additional unprotected spillway with the crest roughly 0.2–0.4 m below the crest of the embankment – so that floods higher than 100 years flood will go firstly by this profile.

7.10.3.2 Geotechnical Aspects – Selection of Fill Material

The preliminary estimation of the volume fill material for the reconstruction of all three dams was roughly 63 000 m^3. Firstly we have started with the tests on the existing dams. From the exposed profile of the gash, which was practically vertical, the tests proved that the material is composed from granite eluvium, soils which are typical for this region. Although the vertical walls stayed stable for nearly 2

Fig. 7.69 Dam Dolejší. (**a**) Blocked protected old spillway (8 pcs). (**b**) New spillway (4 pcs) – without middle pillar. (**c**) New additional unprotected spillway

(c)

Fig. 7.69 (continued)

years, grain size distribution curve had typical sandy character. The main reason for some cohesion and relatively lower permeability, than should be judged from the grain size distribution curve, is probably high content of mica. The surface of such compacted soil is very glossy, similar to the clayey soil. Some results on the material from the existing dams:

- Percentage of fine particles (lower than 0.06 mm) – between 10 and 35%,
- Liquid limit w_L relatively high – 30–37%,
- Plasticity limit w_P – 23–28% (but in some cases not detectable due to high sandy content),
- Optimum moisture content according to the Proctor standard test – $w_{opt} = 12.5$–17.0%,
- Coefficient of permeability after the compaction – $k = 10^{-8} - 10^{-10}$ m.s^{-1}.

Borrow pit was selected at the edge of field – after the removal and storage of humus layer the samples were tested from the different depth. Degree of weathering was in very different stages, but usually material closer to the surface had higher plasticity. Finally it was decided to construct the dams as zoned ones; more plastic material was situated on the upstream side, sandier one on the downstream side. Slightly higher moisture content was recommended than the optimum one to decrease the coefficient of permeability. During the summer time it was necessary to increase the moisture content by sprinkling, storing the material in stockpile before using in the dam body. Heavy sheep foot roller was used and good experience was obtained, due to high disintegration of weathered material – also in the winter time, work continued up to $-4°C$ in the night time if daily temperature was slightly above zero.

Great attention was paid not only to the compaction control – samples were taken each $500\,m^3$ – but also to the connection of new part of the fill to the old one. Eroded slopes were arranged to the slope inclination at the least 1:7 to get good connection between old and new parts and to avoid cracks development. In longitudinal direction in the central part of the cross section the contact was done with the help of cut-off, in the bottom as well. The depth of this vertical sealing wall was selected according to the permeability of the excavated material. For the embankment of the pond Podhajský, a special arrangement was recommended and applied in the part where in slope and in the bottom of the gash great granite boulders were disclosed – these boulders were coated with bentonite slurry and most plastic materials was compacted there in small layers by smaller compaction tamp.

7.10.3.3 Application of Geosynthetic Materials

In spite of tendency to restore the dams close to the old standard some new materials were also applied – for example the geosynthetic materials. Not only as the separation layer on the contact of the fills with upstream side protection layer composed from gravel and boulders, but also for slope reinforcement. Downstream slope on the pond Hořejší was composed from the concrete gravity retaining wall on part where fish-pond for nursery fish was very close to the embankment. After the failure of this retaining wall a new one from the reinforced soil was constructed there. Small prefabricated blocks were anchored to the slope with the help of reinforcing geogrid from polyester, see Fig. 7.70. Not only price but also aesthetical incorporation was better in this case than the old solution.

Fig. 7.70 Dam Hořejší – Retaining wall from the reinforced soils

7.10.4 Some Recommendation Obtained from the Reconstruction

To minimize the negative impact of the heavy floods in the future, greater attention should be devoted to the following points, Vaníček (2004b).

7.10.4.1 Domino Effect

Our engineering practice is not taking this effect into account, even as would be shown that this effect is playing the very important role. Our codes of practice are rather unambiguous – to ensure safe diversion of 100 year flood. With this the following notes can be added – possible demand for safe transfer of the higher flood can have following impacts:

- Disproportionate increase in price, which from the long term look can significantly exceed pertinent price for the reconstruction and loss – this approach deserves separate attention – definition of acceptable risk,
- Securing the safe transfer of the higher flood can cause worsening of the situation below the dam – especially when this area is highly populated,
- Therefore we recommend the greater attention to be paid to the increase of the storage capacity – first of all with the help of mud clearing in places where the fluvial deposit step down reservoir area above the level of dead storage.

However even increased capacity of spillways cannot be sufficient in the case which happened on the river Lomnice. In this case the second pond on this river-basin – pond Metelský which failed – had higher storage capacity than the rest of following ponds below it. Therefore the flood wave from this pond had such a great negative impact on the safety of the ponds below it, see also Fig. 7.64. Practically all failures started with time delay when this wave reached the individual pond. Generally we have a "classical case" when the reservoir volume is increasing along river-basin. Therefore we recommend for the pond system in the Czech Republic to define these untypical cases and preferably for the most important pond with critical reservoir volume, to define stricter conditions from the stability point of view.

7.10.4.2 Interaction of the Dam with Road on Its Surface

Positive effect of the road pavement (and usually with wider crest) was clearly observed for ponds Veský and Zámecký, where flood wave only destroyed part of the downstream side. But in some other cases the negative impact of roads was observed. For the historical dams, the width of crest was relatively a small one. During the last decades the asphalt pavement was applied with two negative impacts. The first one is connected with increasing transport especially with heavy tracks which causes additional settlement, very often a differential one – because the compaction in the past was not so efficient. Very often there is additional dynamic effect (often caused by trailer), which not only influences the settlement but also the connection between old (very often wooden) outlet and fill or between this old wooden outlet and the new concrete feed-pipe – these contacts can be a preferential path for seepage water, which can be reason for an initiation of the internal erosion.

The second negative impact is connected with the effort to widen the crest width, when very often the asphalt pavement spills over the original profile and when the longitudinal cracks are the result of heavy loading at this part. Potential risk of accident is therefore very high. The discussion between pond and road owners is therefore a significant step which has to start.

7.10.4.3 Utilization of Local Soils – Granite Eluvium

These soils were used in spite of the fact that their properties, especially from the point of view of grain size distribution and plasticity are on the boundary of materials which are generally accepted for earth-fill dams. This approach was applied due to the negative experience with application of additional sealing element (jet grouting sealing wall) along which perimeter the internal erosion probably started due to the high hydraulic gradient there. According to our opinion, more coarse materials can be used when they are able to fulfil sealing function and the total seepage is small enough. Because for all historical dams no special problems in this field were observed, and the permeability tests proved the sealing function, the local material was used which also guaranteed the same cross section as for the original dam. High content of the mica in granite eluvium is with high probability the reason for low permeability.

7.10.5 Conclusions

Limit states of surface and internal erosion were critical during small historical earth-fill dam failures during heavy floods in 2002. The high concentration of failures in little river Lomnice was probably caused by the domino effect, due to the fact that the dam with the highest reservoir volume is at the beginning of river-basin. Reconstruction approach had as the main aim to decrease the probability of failure in the future. From the geotechnical engineering point of view the utilization of the original local soils, even when they are on the boundary of soils generally accepted for earth-fill dams, were recommended. From the hydro engineering point of view, the reconstruction of outlet and spillways were performed in such a way that nowadays they are capable of transferring safely not only 100 year flood but even higher. Additional, unprotected, spillways were constructed only a few decimetres below the dam crest on places where the surface erosion is expected to play a minimum role.

The final note to this chapter is connected with the present approach to the construction of new dams. Not only some environmentalists but probably the majority of the population do not like the idea of new dam construction even when the main aim of this dam will be anti flood protection. On the contrary most of them have seen the failure of small dams as a deep scar on our environment and were very satisfied when all these dams were reconstructed and water level and surroundings were recovered to the "original" state, Fig. 7.71. Most of them are unaware of the fact that this nice countryside with many ponds and typical flora and fauna was created by human activity roughly four or five hundred years ago.

Fig. 7.71 Dam Dolejší. (**a**) View after floods – Downstream slope eroded along the entire crest before the dam failed in two places. (**b**) View after the reconstruction

7.11 Specificity of Flood Protection Dams (Dikes)

Floods belong among the most important and most dangerous natural hazards. Roughly one third of all damage caused by natural hazards are floods. There is little difference in material damage, number of victims or number of civilians affected; all are greatly affected. In such countries where the number of other natural hazards as earthquakes, extreme draughts, hurricanes, landslides is substantially less, for example in the Czech Republic or The Netherlands, this ratio is much higher.

Natural hazards generally represent a significant impact on human life, inductive of serious social consequences and large economical damage. Natural hazards constitute the limitation of sustainable development, affecting its three main components: economical, social and environmental.

Despite the steps taken during the last decade (The International Decade for Natural Disaster Reduction), the number of natural hazards are in fact increasing; causing at present-day, in average yearly damage, about 50 billion US dollars or 40 000 human lives.

Fill dams are the most often-used protection element during extreme hydrological situations. All classical earth and rock-fill dams also provide anti-flood protection, but for dikes, levees, and dry dams these are their main role. Floods are wide spread, practically all over the planet, with exception for extremely dry regions, but even there during short period of rains some problems can occur. Each country has a different approach to this problem, which is sometimes very specific. It is worth mentioning that in ancient Egypt, regular spring floods were welcomed, bringing fertile soils and taxes depending on the maximum water level. Today the situation is inverted; the main attention is devoted to the protection of inhabitants and their property.

7.11.1 Details for Individual Countries

7.11.1.1 The Netherlands

As a low-lying and highly developed country The Netherlands continuously faces a potential flood disaster. Large parts of the country lie below levels that may occur on the North Sea, on the rivers Rhine and Meuse and on the Ijsselmeer. Flood protection measures have to provide sufficient safety to a large number of inhabitants and ever-increasing investments. Construction, management and maintenance of flood protection structures are essential conditions for the population and further development of the country, Pilarczyk (1998).

Most of The Netherlands is protected by flood protection structures. Along the coast protection against flooding is principally provided by dunes. Where the dunes are absent or too narrow or where sea arms have been closed off, flood protection structures in the form of sea dikes or storm surge barriers have been constructed. Along the full length of the Rhine and along the lower parts of the Meuse river, protection against flooding is provided by dikes. This adds up to a total of about

2 500 km of flood protection structures, which are vital for the existence of the Dutch nation.

The coast of the Netherlands consists of about 300 km of dunes and about 100 km of dikes and dams. Already under present climate conditions (over the last 100 years) the relative sea level rise (due to sea-level rise and subsidence of mainland) is about 200 mm per century.

Sea level rise increases the risks of overrunning the dikes and the ultimate collapse of these structures during storms.

In February 1953, a storm surge level reached 3–3.5 m above normal and exceeded design storm surge levels by about 0.5 m at some places.

At several hundred places the dikes were damaged and/or broken, over a total length of 190 km. This resulted in catastrophic flooding, through nearly 90 breaches 150 000 ha of polder land inundated. It also caused the deaths of 1 850 people and 100 000 people had to be evacuated.

The Netherlands also has a unique feature that it must try to protect from and that is the combination increased sea level and increased level in rivers. During western winds the level of sea and waves increase to such a degree that it protects to the outflow from the rivers. However when both levels rise they increase the potential of overflowing the crest of protection dams as along the seashore so along the rivers and polders. After the 1953 heavy floods, the Dutch Parliament declared the so-called "Delta Act". Under this program the situation has improved.

Another unique detail of The Netherlands is the fact that the fundamental material of protection dams is fine sand, the protection of which creates only a relatively thin clay layer. Taken together with topsoil it is possible to grow vegetation, which has better erosion resistance.

An extensive discussion in The Netherlands is associated with dikes maintenance, van Staveren (2006). He states that because there will be limits to maintenance budgets, a risk-driven approach may help to establish cost-effective dike maintenance programmes. Ideally, dike maintenance and reinforcement are restricted to those parts of the dike demanding it. The conventional confirmatory approach of visual dike inspection, appraising the main failure mechanisms, providing strength and stability calculations and executing the resulting maintenance and reinforcement programmes may need support. Therefore a new approach, to consider the failure risk of existing dikes throughout their entire life time, has been developed. This method, "Rational risk management for dikes" calculates the failure risk of a particular dike section for a number of scenarios – Beetstra and Stoutjesdijk (2005).

7.11.1.2 USA – River Mississippi

Although 3 200 km of protection dams were constructed along the Mississippi River, still some problems exist. In 1953, the heaviest flood during last 150 years affected the Mississippi river basin. High water level broke through or overflowed two-thirds of dams, killed 50 people, 70 000 of habitants lost their houses and water flooded the area (approx. 44 000 km^2).

7.11 Specificity of Flood Protection Dams (Dikes)

New Orleans, constructed in the Mississippi delta, is mostly surrounded by the protection dams of Pontchartain Lake and by the dikes of the Mississippi. Many water canals pass through the city, connected to the lake, the Mississippi and the Gulf of Mexico.

The largest part of the town is roughly between 1 and 6 meters under sea level. Normal water level in Pontchartain Lake is less than half meter above sea level.

In September 2005 water from hurricane Katrina sharply increased both levels in Pontchartain Lake and in the Mississippi, in the lake 5.5 m above normal level. Water level in canals increased as well. Dam failures started; nearly 80% of the town area was flooded, leaving 1 037 deaths and over 200 billion US dollars in damage.

7.11.1.3 Germany, Austria, Slovakia – Danube River

Big problems are connected with the Danube river in the central part of Europe, where heavy floods occurred in 1954, 1965, 1997, 2002 and 2006, e.g. Peter (1975), Brandl and Blovsky (2003), Hulla and Kadubcová (2000), Milerski et al. (2003).

The construction of protection dams along the Danube River is documented from the thirteenth century in Slovakia. The flood in 1965 was lower than that in 1954 but lasted longer (from March up to June) and created a larger catastrophe. Around 2 300 outflows were observed close to the downstream abutment, but some of them even 350 m away from the dam.

The left side of the dam failed at two places. The failures had a common character – they were triggered by internal instability of fine sand induced by a high hydraulic gradient. Fine sand particles were washed out from the sandy gravel layer, producing higher water velocity and also a higher hydraulic gradient. This higher hydraulic gradient washed away larger particles and so progressively created a seepage canal. After these canal walls collapsed the protection dam eroded. It was a classical example of piping. One thousand square kilometer of high quality agricultural soil was flooded, 3 910 houses were destroyed and another 6 180 houses were damaged.

Heavy floods in July 1997 in the Czech Republic, mostly in Moravia, affected 538 towns and villages; about 136 km of protection dams were damaged, with 55 failures, mostly due to overflow, leaving 50 deaths, 60 billion CZK in damage. Nine hundred and forty six kilometer of the railway track were destroyed as well as 1 850 km of roads and 2 151 houses.

The heavy floods in 2002 were even worse. This flood affected about 1,333 million inhabitants, which is 15.7% of the country's population.

7.11.2 Fundamental Specificity of the Protection Dams

Fundamental specifications of protection dams in which they differ from classical fill dams are:

- Loading by water is strongly time limited – therefore during the flood the soil is partly saturated, seepage is limited by air bubbles. The demands on permeability are not as high as for classical small dams, allowing construction of protection dams from more permeable, local soils.
- Dam localisation – roughly the line of protection dam is predestined (whereas for classical dams the profile is carefully selected according to the quality of subsoil) therefore the subsoil conditions are often very complicated because the quality of river sediments change from gravels up to the sullage.
- Construction materials – local soils are preferred which can cause some problems with high heterogeneity and possibly influence the seepage path in the future.
- Dimensions – the third dimension – dam length is extreme – one weak place can cause a large catastrophe.
- Protection dams along the seashore – extreme demands caused by surface erosion from high sea waves.

All relevant limit states have to be checked during the design and reconstruction of protection dams. From the limit states mentioned in Chapter 5 the limit states type UPL and HYD are more important. The fundamental scheme of a protection dam along the river is shown in Fig. 7.71. Besides of overflowing causing surface erosion, the floods in 1997 and 2002 showed that the area of the Czech Republic has two basic cases that need to be discussed separately:

- The failure of the protection dam body – the highest erosion hazard is connected with the internal erosion starting at the downstream toe of the dam – a place at which the depression curve reaches the downstream face – this is valid for old dams, for new ones the toe drains are not missing. A certain potential danger is connected with animals; otters, beavers, can cause preferential path for seepage.
- The failure of dam subsoil – failure starts behind the dam on the downstream side – either by uplift, when the water pressure on the impermeable layer is higher than geostatic pressure there or at even a higher distance from the dam where the impermeable layer is missing and where upward seepage pressure is able to erode fine particles. The process of piping begins there leading to the final phase of dam collapse (the case described also applies for protection dams below Bratislava on the Danube River).

Schematic cross-sections in Fig. 7.72 show different ways of increasing safety:

- Extension of seepage paths to decrease the hydraulic gradient – with the help of extending sealing carpets on both sides of the dam,
- Restoration of the top sealing layer which was depleted, for example, by excavation of soils for the dam body,
- Sealing of horizontal seepage paths in dam subsoil either by the vertical sealing wall going through the whole the dam profile and which in the upper part can

7.11 Specificity of Flood Protection Dams (Dikes)

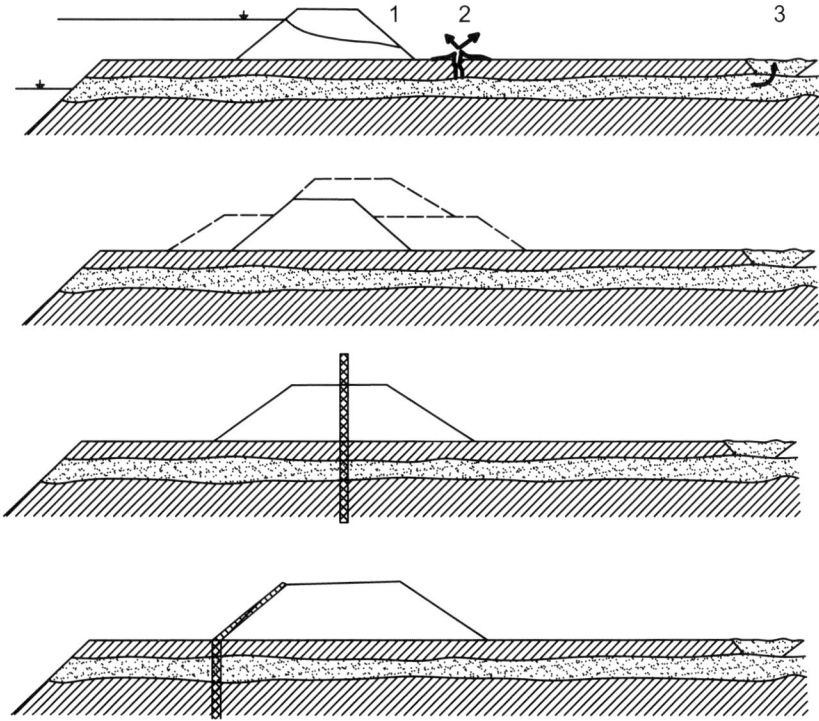

Fig. 7.72 Protection dam – Schematic cross sections showing the different ways of safety increase

increase the total height of the protection dam or by the vertical sealing wall on the upstream side which is connected at the dam toe with the upstream sealing layer, which can also be constructed from geomembranes.

At the end of this subchapter a few words about the philosophy of flood protection can be added. Protection dams constructed very close to the river basin decrease flow profile; therefore protection dams have to be very high. Therefore, there is a legitimate question on what flood type the protection dam should to be constructed for, whether for 100 years type, 200 years or even higher (with probability of occurrence once during 100 or 200 years). Where the morphology and character of the area utilization allows it, smaller protection dams along the river basin are constructed lets say for 20-year flood type, securing the protection for some years (theoretically for 19). Only during floods higher than a 20 year-flood is this dam "opened" – water is passing behind by opened outlets and this outer space is protected by a second row of protection dams. After that, the river-basin profile is larger so that the new level cannot be significantly higher for higher floods than the previous one (in some case even lower). In this case, most of the population lives behind the second row of protection dams and the area between dams is used mostly for agricultural purposes.

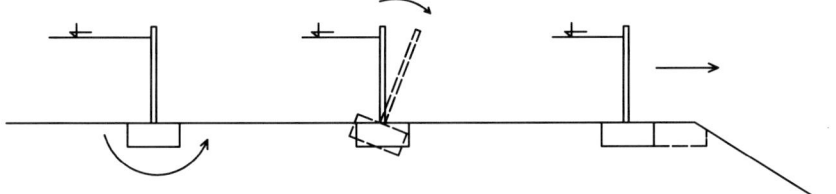

Fig. 7.73 Additional protection barrier – Different types of possible failure

Fig. 7.74 Additional protection barrier applied during heavy floods in Prague
(photo courtesy of Hydroprojekt Praha)

However this system cannot be used in cities, therefore in many cases the additional mobile protection barriers are proposed for use in cities. Great attention has to be devoted to the potential risk of failure due to internal erosion caused by high hydraulic gradient at the place of the barrier connection to the subsoil. Figure 7.73 shows such examples and overturning and sliding is also mentioned there. How these additional mobile barriers can be loaded is shown in the following photo – Fig. 7.74 – taken during 2002 heavy floods in Prague.

Chapter 8
Earth Structures in Environmental Engineering

Geotechnical engineering always had a very close relationship to nature. However gradually the interest in and problems associated with environment protection is acquiring a special position in the wider branch of classical geotechnics.

A very significant step in this process was State of the Art Report "Environmental Geotechnics" presented by Sembenelli and Ueshita during the Xth International Conference SMFE in Stockholm 1981. From 1994 international congresses of Environmental Geotechnics are organized by ISSMFE (ISSMGE) – Edmonton 1994, Osaka 1996, Lisbon 1998, Rio de Janeiro 2002, and Cardiff 2006.

With respect to the regard of the earth structures and environmental problems Sembenelli and Ueshita defined four main spheres:

– Solids removal from surface,
– Solids extraction from underground,
– Solids accumulated on surface,
– Underground deposits.

In the last part of this publication only the problems associated with "Solids accumulated on surface" will be specified in more details, whereas the emphasis will be given to the large-volume waste materials and to new artificial earth structures which they create as are spoil heaps (mine waste piles, dumps), tailing dams and landfills. To the other spheres only short notes will be given.

Solids removal from surface – in Chapter 6 this problem was partly covered during discussions about cuts of transport infrastructure. In principle the problems are similar as for cuts during open pit mining or during quarrying of soil and rock for earth structures and other purposes.

Differences are given by higher height and also by greater influence on underground water for open pit mines and borrow pits. An interesting example of the influence of mining activity on the stability of hillsides of the Krušné hory mountains are presented by Marek et al. (2003). The time for which the slope is exposed is very often strictly limited. For short term slope stability problems the observational method of the slope inclination can be applied. The experiences from previous activities are usually applied and verified by field monitoring. With the respect to this feedback the design can be slightly corrected and modified.

Solids extraction from underground – problem connected with mining, tunnelling and other underground facilities – with a certain impact on underground water level, subsidence of surrounding surface.

Underground deposits – above all underground waste disposal. This problem is very challenging from the point of view of Environmental Geotechnics, but for the general population it is very sensitive. Filling of underground spaces by large-volume waste as for example flying ash from electric power stations can on one side limit the surface deformations but on the other side this waste material is in direct contact with underground water and therefore it is necessary to prove that this interaction is not dangerous for environments, Vaníček (1995b). A very specific problem is associated with high level radioactive waste disposal in underground spaces, e.g. Gens (2003), Hoteit and Faucher (2003). It is an up to date fundamental concept of disposal of spent fuel from nuclear power plants. Principle of multi-barrier protection is applied there and a natural geological barrier is a fundamental element of this protection. Spent fuel is fixed in special containers, which are in underground spaces separated from surrounding rock massive by special sealing zone. Up to date conceptions proceed from utilization of excellent sealing properties of bentonite. Prefabricated blocks of bentonite have to fulfil the space between containers and rock massive, and to the excellent sealing not only extremely low permeability is used, but also swelling properties. It is not our intention to go into any details, but just to show on this example the fundamental logistic approach to the design of such difficult geotechnical structure. The design with the help of numerical models have to prove that even after tens and hundreds of thousands of years when the radio nuclides with long term half-life from the deposited spent fuel can reach the surface of terrain, their activity will be so low that it cannot negatively affect our environments. In fact we are designing the structure with lifetime extremely exceeding our up to date experiences. Numerical model, describing not only radio-nuclides spreading in sealing layer and rock massive but also taking into consideration mechanical and thermal properties of the all multi-barrier system, has to be tested and verified, so to prove its credibility. In underground rock laboratories, all the above mentioned processes are simulated and compared with proposed numerical models. The credibility of the models can be performed there even if just for a much shorter period, from today's view roughly 20 years. Therefore the significance of the underground rock laboratory is so important, e.g. Vaníček (1999).

Hereafter our attention will be devoted to the deposition of large-waste material on the surface. The volume of such deposited material is extremely high, only in the Czech Republic it is hundreds of million tons per year, mostly deposited in the form of spoil heaps. Present-day generations are thus in such a significant way reshaping our landscape, that we can speak even about new artificial geological period, about period of Quinternary. A fundamental overview of the problems associated with deposition of waste material on the surface was summarized e.g. by Vaníček (2002b), respectively by Clark and Weltman (2003).

The basic difference of earth structures of Environmental Engineering from the other ones is that except of basic limit states for classical earth structures, also all environmental aspects (as e.g. contaminant spreading, methane development potential danger etc.) will be limited below acceptable limits.

8.1 Spoil Heaps

This chapter deals with principles of earth structures which originate as by-products during mining activity. Into this group belong materials which overlay measures of raw materials, e.g. during open pit coal mining and also materials, very often called waste rock (tails) which are result of separation of excavated material from basic raw material irrespective of ores raw materials or other raw materials. In the Czech Republic typical spoil heaps are composed from clayey soils which overlay brown coal in the northern part of Bohemia. Much lower extend reach waste rock dumps in Ostrava region as the result of deep black coal mining or waste rock dumps in Příbram region where uranium ore was excavated. For all new earth structures composed from these different materials some common properties can be specified:

– Free fall filling is more often a manner of spoil heap construction,
– Deposited material has high porosity (macro-porosity) between individual clods, grains,
– Individual clods are sensitive to weathering,
– Properties are time dependent, not only as function of weathering but also as the result of water saturation, depth of deposition etc.

It is obvious that there is a general tendency to store on 1 m^2 utmost of materials, therefore to construct spoil heaps as steep and as high as possible. This tendency brings high risk of slope instability, creating landslides with high potential risk to the surrounding area. Very often in this connection the landslide in Aberfan in Wales is mentioned. The landslide which occurred there in 1966 claimed the lives of 144 people, of which 116 were children – e.g. Bishop (1973).

Because deposited materials were excavated from natural geological environment the problems with possible contamination are generally limited, nevertheless this problem always has to be more closely specified as for cases of waste rock coming from uranium ore mines, or for clays which are in the vicinity of coal with a high percentage of sulphur. Attention is after that concentrated on quality of water discharge from spoil heap body, because water can be generally acid, can show slow radioactivity or can contain higher percentages of sulphur. If this potential risk is high, one of the possibilities how to protect our environment is the capping sealing system limiting the inflow of water into the spoil heap body.

To describe properties for all different materials and also for different time segment or climatic conditions is very difficult, however some specificity is prevailing. Their demonstration will be shown for spoil heaps from clayey materials which overlay brown coal deposits in north part of Bohemia, Vaníček (1986, 1987). This example is not only typical for this region but for more places; from surrounding states it is possible to mention Germany and Hungary. Some experiences with clayey spoil heaps describe also D'Elia et al. (1979, 1983), or Charles et al. (1979). The orientation on this type of spoil heap is not only as the result of huge volumes of such new earth structures but also due to the fact that the change of properties is extremely high – they can change properties from one side of soil mechanics spectrum – from properties of rockfill – up to the other side of this spectrum – up to the soft clay.

Roughly 200 mil.m^3 per year of clayey material overlaying brown coal is deposited into spoil heaps in the Czech Republic even when volume of brown coal is decreasing. Stripping ratio – volume of clay to the volume of coal is steadily increasing with time, now reaching in average a value of 6:1. If in average 40 m of material is stored on 1 m^2, after that each year spoil heaps are covering the area of 5 km^2. It means that in this region large part of territory is affected not only by mining activity but also by spoil heaps construction. Therefore new constructions on the top of these new earth structures is practically a necessity.

Spoil heaps are principally divided on inner and outer spoil heaps, see Fig. 8.1. Inner spoil heaps are backfilling excavated space, therefore in the first phase of open pit mining outer spoil heaps prevail. Because the range of outer spoil heaps is limited due to strong pressure to limit other annexation of agricultural land a percentage of inner spoil heaps is increasing. In the last phase of open pit mining in this region the inner spoil heaps will prevail. Elevation of spoil heaps is more often in the range of 40–150 m.

8.1.1 Behaviour of Stored Clayey Material with Large Voids – Macropores

8.1.1.1 Specific Properties of Clay Fills

The thickness of the tertiary sediments reaches 150 m in average in the North-Bohemian Brown-Coal Basin. Besides brown coal this layer is composed mainly of clays and of claystones. But sand, underclay, slate coal, sandy clay and sandy claystone are also present.

Clays and claystones have the following characteristics:

– Plasticity limit $w_p = 30$–35%
– Liquidity limit $w_L = 70$–80%
– Plasticity index $I_p = 35$–50

From clay minerals mainly the kaolinite and illite are in abundance, montmorillonite is also present but with variable low percentage. The proportion of clay particles

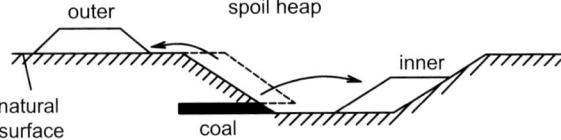

Fig. 8.1 Scheme of the construction of outer and inner spoil heaps

ranges from 10 to 40% and silt particles from 20 to 60%. The other parameters are strongly influenced by the depth of deposition:

- Bulk density ρ increases from the value of $1\,800\,\text{kg.m}^{-3}$ near the surface to $2\,100\,\text{kg.m}^{-3}$ for the depths of 70–100 m,
- Moisture content decreases approximately from 20 to 45% for the depth up to 30 m to the value 13–32% for the depths 30–60 m,
- Porosity decreases from roughly 45% near the surface to 30–37% for the depth up to 100 m,
- Degree of saturation S_r is close to 1,
- Unconfined strength increases with depth and for the depths around 40–80 m is in the range of 1–3 MPa.

Generally we can say that the overlying clays have a character of overconsolidated fissured soils with stiff to hard consistency.

8.1.1.2 Change of Properties During Transport and After Filling

From the place of excavation to the place of deposition into spoil heap body the clayey material is transported by train or by belt, during the last period preferably by belt. Before the belt transport (width of belt roughly 2–3 m) the excavated great clods are in some cases crushed down. During this transport the individual clods are round off and during wet weather their moisture content is increased. At the end of the transport the individual clods are partly compacted by free fall (for overburden conveyor bridge it is up to 20 m), see Fig. 8.2. By free fall individual clods are partly crushed and compacted but their bulk density is approximately 1500–$1600\,\text{kg.m}^{-3}$ and so the macro porosity is around 30%. The individual macro pores between the individual clods are interconnected and air is in continuous form and so the permeability of soil for air is relatively high. Character of the fill is close to rock fill – Fig. 8.3. Under this condition an air pore pressure is relatively quickly equalized

Fig. 8.2 General view on spoil heap during its construction – filling

Fig. 8.3 Character of clay clods after deposition

to the atmospheric pressure. On the contrary the pore water pressure inside of individual clods is negative due to great unloading. Under this condition such soil will easily absorb water – free water or water from saturated air – even from air inside of the spoil heap's body. This last statement deserves a short note. The temperature inside of a 15 year old spoil heap only 1 m below surface was measured for all year in the range of 6–12°C. Therefore it is possible that air with high humidity entering the spoil heap body can lose part of its humidity by precipitation. Cycle of continuous increase of humidity is enabled by variation of the barometric pressure. During high barometric pressure air with high humidity is entering spoil heap and during low pressure is outgoing from the spoil heap body with lower humidity. Both processes so enabled gradual increase of moisture content of clay particles, preferably along the periphery. But this is valid only in surface envelope of spoil heap body where macropores are interconnected. Herewith shear strength is dropping along the contact of individual clods allowing small shifting in relative position. At the same time surface of clods is kneaded, process leading to softening and yielding of clay fill and therefore to other decrease of strength.

The properties of the deposited clayey soils are therefore changing with time due to two basic contradiction aspects, Vaníček (1995a):

– Process of softening as a result of weathering, moisture content increase and kneading,
– Process of hardening as a result of surcharge by new deposited layers.

Result of these contradictory processes is a significant heterogeneity of the deposited material.

These two processes are schematically described in Fig. 8.4. In the first case it is necessary to take into account that by unloading of soil from the great depth the total stresses drop to zero. As a result of this the closed cracks in fissured overcon-

8.1 Spoil Heaps

Fig. 8.4 Two basic changes of clay fill – process of softening and process of hardening – typical problem of double porosity

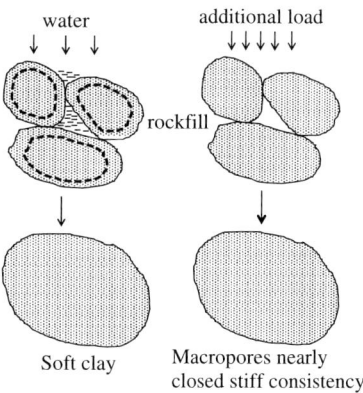

solidated soils are partly opened and process of weathering can start. After a certain time the size of the individual clods, especially if excavated from great depth, can look like those on Fig. 8.5. Small note – in laboratory behaviour of clods with dia 30–40 mm and initial moisture content 12.5% were observed. Clay clods submerged in water for 3 months increased moisture content up to 43.2% (more than plasticity limit) but the shape of the individual clods remained nearly unchanged. Only at the beginning small parallel cracks were seen on the clod surface. This means that due to suction and swelling potential additional water was absorbed without significant disturbance of individual particles. On the contrary, clay clods which were frozen down to $-20°C$ and after defrosting sprinkled by some water, degraded very quickly, original size was changed to the size of 1–5 mm and even less. Moisture content after 5 cycles of freezing and refreezing increased up to 16.3% but the consistency remained stiff to hard. Two examples of the view on spoil heap areas are shown in Fig. 8.6 and Fig. 8.7, Štýs (1998).

The process of softening is not only connected with spoil heap surface. As mentioned before there is a great suction potential due to negative pore pressure and so

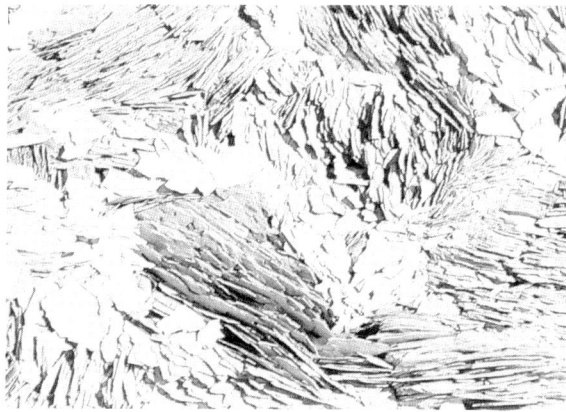

Fig. 8.5 Character of clay clods excavated from great depth after weathering, (photo Štys)

Fig. 8.6 Typical view on the spoil heap surface after a certain time, (photo Štys)

the individual clods are absorbing either free water flowing into the spoil heap body (e.g. rain water) or condensate water from air in soil massive.

For the calculation of the amount of absorbed free water it is possible to start from atmospheric precipitation. If maximum infiltration for partly levelled surface is roughly 20% of annual precipitation 500 mm then the increase in moisture content of about 1% is realized for upper 7 m of soil. Then far greater influence will have an infiltration in places of depression where surface waters are collected and from where they are soaking into spoil heap through the permeable surface. In this case a certain region inside of the spoil heap body can be fully saturated.

Fig. 8.7 Dumpsites from bird's-eye perspective, (photo Štys)

8.1 Spoil Heaps

The least favourable situation exists if individual macropores are from the beginning fulfilled by water. This situation can arrive for the bottom part of inner spoil heap which is not well drained or for the outer spoil heap along its periphery, because spoil heap is changing streams of surface waters. We are therefore speaking further about undrained (wet) basement or about drained basement when this situation is protected. For the case of undrained basement the consistency of all volumes of the individual clods can be changed with the help of kneading to soft or even liquid consistency. Both cases were really registered by site investigation on realized spoil heaps. Barvínek and Havlíček (1977) explained this process by numerical calculation. As a basic case they used soil with the following parameters:

$$w = 31\%, \ S_r = 1, \ \rho_d = 1475 \ \text{kg.m}^{-3}, \ I_c = 1, \ c_u = 80 \ \text{kPa} \tag{8.1}$$

If macroporosity $m = 20\%$ and all macropores are saturated by water, then total moisture content after process of homogenization will grow up by about 13.6%, consistency index will be $I_c = 0.59$ and c_u will drop to the value of 40 kPa – they used an empirical equation

$$c_u = (I_c - 0.2) \cdot 100 \ [\text{kPa}] \tag{8.2}$$

In their work they registered local regions in the spoil heap where measured value of total cohesion c_u was less than 20 kPa. This case really can be true, because of measured macroporosity $m = 37.3\%$ on the surface of the spoil heap body.

What is the main conclusion from this described process? Clay fill from the character of rockfill can be very easily changed to the character of soft clay. It is probably the only example where such a great change from the one extreme spectrum of soil mechanics (rockfill) to the other one (soft clay) can be observed. Our main aim is to slow down this process to reduce the negative effect of the process of softening on stability and deformation of spoil heap body.

In reality it means that we have to use all *possibilities to slow down the degradation process*, generally with the help of the following measures:

- To protect filling into free water, therefore such great attention is devoted to the drainage of the spoil heap basement,
- During filling to decrease the number of depressions where rain water can be collected,
- At the end of filling to smooth the surface and if there is the possibility to compact it to decrease the water infiltration into spoil heap body.

Great attention was devoted to the depth at which macropores are closed by surcharge of new deposited layers, because in this case the permeability of such soil even for the air is a very low one. The different possibilities used to investigate this phenomenon were:

- By laboratory testing – in large oedometer or in triaxial cell clay clods were loaded by increased pressure and speed of water infiltration and deformation after wetting were observed for different densities and moisture contents – e.g. Vaníček (1986),
- By in situ tests – by measuring permeability for air,
- By in situ tests – by measuring pore water pressure.

Now we can conclude that this depth is between 10 and 40 m mainly as a function of initial water content, that's the depth from which material was excavated. Therefore clay clods which are covered very quickly during filling by such height of additional material will improve the stiffness as a function of depth–volume of macro pores will steadily decrease.

8.1.2 Stability of Spoil Heaps

Slope stability of clayey fill calculation for spoil heaps is more complicated than for classically compacted fill due to the following aspects:

- Extreme height of spoil heaps,
- Decrease of shear strength as a result of moisture content increase,
- Possibility of development of local zones with extra decreased strength.

Therefore existing slopes of spoil heaps have relatively flat dip, roughly between 1:6 up and 1:12, with lower value for the case that filling was applied on undrained basement with surface water. Only low spoil heaps with height up to 20–25 m can be stable for some years in inclination of 1:2, due to rockfill character.

The progressive decrease of slope inclination is connected with landslides which can have deep slip surfaces or shallow ones. To decrease the possibility of slip failures occurrence it is now recommended to construct spoil heaps with general inclination roughly 1:6, where steeper lower slopes are combined with benches.

The position of deep slip surface is directed by weakest zones; see Fig. 8.8, where slip surfaces are passing through:

a) Basement,

b) Soft subsoil,

c) Weakest zone which is inside of spoil heap body.

The case ad a) is typical for undrained (wet) basement, where the consistency and thus strength are lowest. For inclined basement this potential risk is particularly unfavourable.

The case ad b) is realistic for subsoil composed from saturated clay soil. Increase of pore water pressure is much higher there than in the drained basement. The case

8.1 Spoil Heaps

Fig. 8.8 Deep slip surfaces for spoil heap

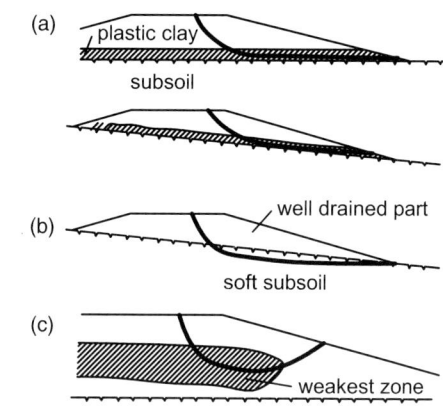

ad c) is less typical. It can develop only for drained basement if inflow of outer water is allowed in.

In the eighties of the last century four large landslides of spoil heaps occurred in the north part of Bohemia each with a volume of moving mass exceeding 50 mil.m^3. Landslide of inner spoil heap Vršany affected mining activity there. Three others which occurred on outer spoil heaps affected buildings (Vřesová) or transport infrastructure (Loket spoil heap). Velocity of mass movement was maximally a few meters per day. Therefore it was possible to stop the movement in Vřesová by a huge loading berm with a volume of 0.75 mil.m^3 composed from coarse material at the moment when the central part of landslide was in average inclination 1:18. This last case proved the importance of spoil heap monitoring. With high probability this landslide was initiated by the quick construction of the last etage of filling. If the pore water pressure would have been monitored and a speed of dissipation controlled there, the landslide would probably not developed under slower speed of filling.

Potential risk of *shallow slip surface* is highest shortly after the phase of filling. By free fall the slope inclination is close to limit equilibrium. Therefore even small changes of properties on the clods contacts can initiate the movement.

Before speaking about slope stability, a few notes on the problem of shear strength have to be added. At the beginning, when dry clay clods behave as a loose rockfill, the angle of internal friction is a high one, there is no distinction between drained and undrained tests. For higher normal stresses the angle of internal friction can even increase – material behave as a dense coarse material. But this is valid only up to the stage at which macro-pores are closed. $\varphi_u = 0$ approach can be applied for undrained test when practically all macro – pores are closed. Větrovský (2006) states that this situation occurs for pressures in the range of 0.6–1.1 MPa. These pressures correspond to the depth in spoil heaps between 40 and 70 m. For this situation effective angle of internal friction is between 13 and 16°. Feda et al. (1994) described this behaviour in more detail on the laboratory prepared samples of coarse clayey material from spoil heaps on the base of the series of CID and CIUP triaxial tests.

The range of effective angle of internal friction for peak strength was approved not only by laboratory tests (e.g. by large diameter shear box) but also during field shear tests. But even for these tests a rather great scatter of results were obtained, due to different conditions under which the individual tests were performed. More often mentioned value of residual angle of internal friction is close to 7°.

Design of slopes inclination is based on:

– Previous experiences,
– Numerical calculation.

For North-Bohemian Brown-Coal Basin large amount of information about stable and unstable slopes were obtained and were back analysed according to the:

– Character of spoil heap basement (deformation characteristic, shear strength and first of all extend of drainage),
– Type of the deposited material,
– Degree of saturation of spoil heap body,
– Methods of filling.

Extensive set of these back analyses with high correlation index became a feedback for new spoil heap construction. One of these back analyses is shown in Fig. 8.9, where results are summarized of the stable slopes for different composition of deposited soils, Dykast et al. (2003).

Numerical calculation preferred limit equilibrium methods using either analysis in total or effective shear parameters.

Calculation based on the utilization of undrained strength seems to be easier. Heavy penetration is used for investigation of spoil heap and undrained strength is recalculated using some correlation factors. On one side many results are obtained, but usually the scatter is very high. Usually more representative value $s_u = 90$ kPa is referred as reaching highest percent occurrence. Due to high scatter of results the probabilistic approach was also applied there. Nevertheless the discussion is about the correlation between results of penetration tests and undrained shear strength especially for unsaturated zones.

Fig. 8.9 Back analysis of slopes for different composition of deposited soils. (**a**) Spoil heaps containing soils excavated from low depth. (**b**) Spoil heaps containing only limited amount of soils excavated from low depth

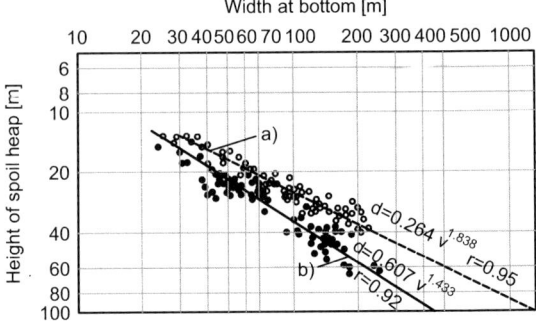

8.1 Spoil Heaps

Calculation based on drained shear strength have main problems with determination of pore water pressures along slip surface. On the other side the drained peak angle of internal friction is much more precisely defined, as mentioned before with lowest value around 13°, more often in the range of 13–16°. During the last period pore pressure measurement is not the exception, using not only piezometers described in Chapter 5, but also special measuring devices. Herštus and Šťastný (1992) or Feda et al. (1994) describe polyhedral device for measurement of stress tensor components and pore pressures in situ, see Fig. 8.10. The main reason for construction and application of such a unit was to investigate the actual stress-strain behaviour of clayey spoil heap in the field. The obtained results should define the future direction of the laboratory investigation and numerical modelling. First field measurement proved that the deformation of spoil heaps approximates plane-strain conditions.

Back to pore pressure measurement. The observation proved that first signals about positive pore pressure starts at depth h_i at which macro-pores are closed.

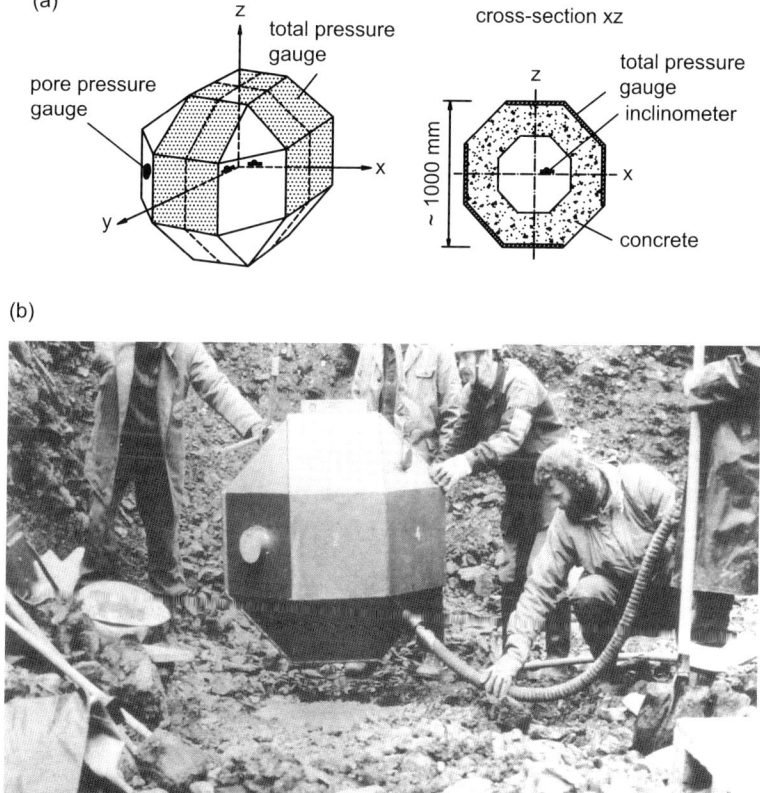

Fig. 8.10 Polyhedral device for measurement of stress tensor components and pore pressure in spoil heap body, (according to Herštus and Šťastný). (**a**) Cross section. (**b**) installation

Fig. 8.11 Results of pore pressure measurement for spoil heaps Libuš and Merkur, (according to Smolík)

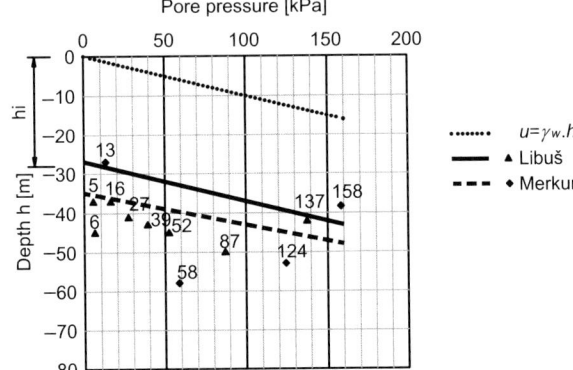

Further increase is nearly linear, see e.g. Fig. 8.11 – Smolík (1996), so that the pore water pressure can be expressed by equation, Herštus et al. (1992):

$$u = C \cdot \gamma_{fill} (h - h_i) \qquad (8.3)$$

where

> h – depth below surface,
> C – coefficient of pore pressure

When compared with results presented in Fig. 8.11, after that for spoil heap Libuš: $h_i = 27$ m, $C = 0.5$, for soil heap Merkur $h_i = 35$ m, $C = 0.35$. For both of these spoil heaps, relatively high value h_i is attributed to high quality of claystone clods.

For known materials deposited into spoil heap body, both basic values affecting pore pressure can be estimated with sufficient preciseness and calculated pore pressure can be used for slope stability calculation.

As was said before, spoil heaps are very sensitive to instability. Therefore monitoring of their behaviour is very important for future prediction. For example a great attention is devoted to Merkur spoil heap which with total volume 650 mil. m³ occupies an area 16.7 km². Spoil heap Merkur belongs also to spoil heaps which were affected by slope instability in eighties. Now a great part of its surface is reclaimed and only part of the inner spoil heap is active. First signals of slope instability of spoil heap Merkur were observed in 1979. A critical situation was in the autumn of 1985 when movement at the toe reached 2–3 m per day in all front of slope in length 2.5 km. Total volume of moving mass was judged on 140 mil.m³. Applied counter-measures included – drainage, trench drains and loading bench. From 1986 monitoring system is in function and includes:

- Geodetic monitoring of selected points,
- Inclinometric bore holes,
- Pore pressure measurement.

In 1992 another fill was deposited on the surface of this spoil heap. All monitoring points registered this fact. For example piezometers signaled pore pressure increase. From measured data it was back analyzed that $h_i = 15 - 20$ m, $C \sim 0.7$. After the end of filling piezometers showed pore pressure dissipation and shear movements indicated by inclinometer installed at the depth of around 45 m signaled decreasing value as well. Therefore it was possible to conclude that the situation is becoming stabilized.

A monitoring system is also a useful tool for the final statement that the spoil heap is stable and therefore foundations of new buildings or transport infrastructure on its surface can start.

As mining activity is slowing down now it is clear that not all pits will be backfilled by excavated material. Some of these pits will be flooded by water to create new artificial water reservoirs in the North-Bohemian Brown-Coal Basin. Because surrounding natural surface with inclined terrain is partly covered by outer spoil heaps, monitoring in this area is fixed on observation of water level increase in pits and on the influence of this water level as well as of rain water infiltrated into outer spoil heap body and natural subsoil on pore pressures in the whole area. Pore pressures are changing there with time initiating only negligible creep movement in best case or landslide in worst case. Some results of this measurement are described by Záleský in Vaníček et al. (2005).

8.1.3 Deformation of Spoil Heaps

It is very difficult to define deformation characteristics of spoil heap clayey materials because they are very variable in dependence on many aspects as are initial moisture content, depth from which material was excavated, what consistency they have, what is size of individual clods, whether water have chance to flow in and so on. Therefore it is very difficult to unambiguously define the deformation characteristics as well as numerical models of such earth structure.

Determination of these deformation characteristics with the help of investigation or via back analysis can be performed on the basis of:

- Laboratory measurement on samples from clay clods before they are deposited into spoil heap body – with respect to the size of measuring devices it is necessary to decrease their size – but with this the number of intergranular contacts per square unit is changed even in the case when devices have non-standard proportions – e.g. oedometers with dia between 0.5 and 0.7 m.
- Laboratory measurement on samples obtained from boreholes – realization of borings is very complicated – better results with respect to the core revenue are obtained from greater depths – but even there a great heterogeneity is observed – Along depth an alternation of relatively stiff and hard layers with soft ones are observed or between greater part of hard clay (claystone) there is much softer zone from smaller shreds – see Fig. 8.12.

(a)

(b)

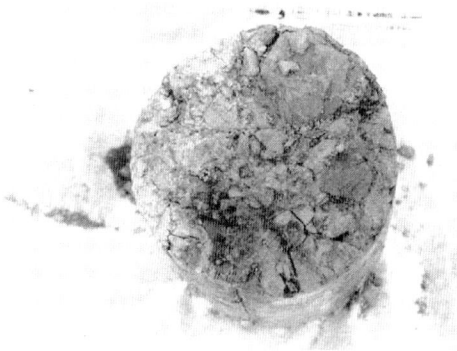

Fig. 8.12 Heterogeneity of spoil heap, (according to Větrovský). (**a**) boring core from the depth 34 m. (**b**) core from smaller depth

- Back analysis from the monitoring of spoil heap body,
- Indirect tests as penetration tests, pressiometric tests etc.,
- Determination of deformation characteristics of the spoil heap surface with the help of loading plate test, spiral loading test or with the help of loading bench.

With the exception of the first case deformation characteristics are obtained for a specific situation. On the contrary during laboratory tests the concentration is aimed on tests where boundary conditions are well defined:

- Clay clods are tested with natural moisture content without possibility of additional water increase – fundamental case for which tested sample has significant macro-porosity. Feda (2003) is using the term double porosity and states that

8.1 Spoil Heaps

- The strange behaviour for this case is preferably produced by grain crushing which is caused by the porous nature of their grains. The effect of crushing makes the compression curve of a garlandlike nature see Fig. 8.13. Strain softening reflects the phase of crushing, strain hardening the newly acquired structural stability (increased number of contacts) and subsequently their reconfiguration into a stable position. For constant pressure, which is valid for the end of filling of spoil heap, it is important to observe deformation as a function of time. Even for this state some local collapse of structure can be observed. These results are important for estimation of the spoil heap surface settlement.
- Clods which are saturated e.g. by sample inundation when macro-pores are under loading practically closed – for this case consolidation characteristics (Terzaghi theory of consolidation) are well defined – from the 1D deformation point of view (with lateral confinement which can be valid for inner spoil heap) deformation characteristics are very high (theoretically infinitely), whereas for 3D conditions lateral extension is possible and the situation can lead to the failure.
- Clods are tested for natural initial state and the influence of inundation is observed – with stress level increase the value of additional settlement due to wetting is decreasing because water inflow is strictly influenced by macro-pores connections enclosure. These results can help to define appropriate height of inner spoil heap at which pumping of underground water can be stopped. Charles and Watts (1996) emphasize the significance of this phenomenon because about 20% of low-rise constructions in Britain take place on filled ground.

Briefly summarize – even only for well defined boundary conditions the following three different collapses phenomena can be observed:

- Collapse due to load increase,
- Collapse for constant loading which can occur during a certain time,
- Collapse due to sudden moisture content increase.

Many different oedometer tests were performed with time. From a classical diameter of 120 mm to new ones with diameters of 250, 500 and 667 mm were developed. For an elimination of skin friction the independent force measurement above and below

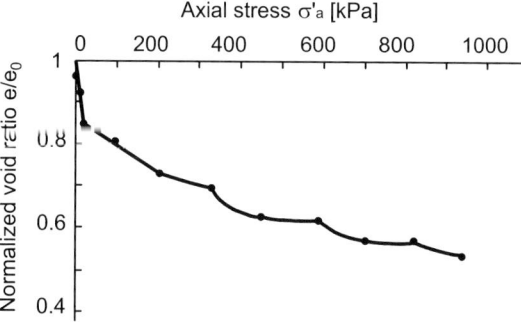

Fig. 8.13 Oedometric compression curve of a granular clay (2–4 mm) – compression curve of a garlandlike nature, (according to Feda)

of the sample was used. Dead load up to the pressure of 1.0 MPa was applied and for higher pressures – up to 6 MPa – a compression machine was used with the ability to hold pressure constant for a long period. And finally the measurement of pore pressure at the bottom of the oedometer was used. For this last case the attention should be given to the contact of fill with pore water measurement membrane – see Fig. 8.14.

Some main conclusions can be summarised. For relatively dry clay clods (with low natural moisture content due to excavation from a great depth) with a high macroporosity the oedometric modulus is very small for low pressures (0.1–0.4 MPa) and significantly increases for higher stress. The measured settlement for a new loading step is very quick and after that is nearly linear, with constant increase for the time unit. Some results for a high initial macroporosity are summarized in Fig. 8.15. Clay clods with dia of 30–40 mm and with stiff to hard consistency (w = 12.3%) were used there as a basic sample, Vaníček (1986). The sample from these clods was compressed in the oedometer (dia 250 mm, height 113 mm) using a compression hydraulic machine with an automatic pressure regulator. Clay clods have had a very low density after being put into oedometer (natural density 861 kg.m^{-3}, dry density 768 kg.m^{-3}). Loading was increased in steps – each step lasted 3–4 h. Mostly overnight, the sample was unloaded to zero. The following day the last pressure was restored and higher pressure applied after 15–30 minutes of reconsolidation.

8.1.3.1 Spoil Heap Monitoring – Ervěnice Corridor

Ervěnice corridor is a part of the future inner spoil heap, situated close to the town Most, roughly 100 km north-west direction of the capital Prague. Construction started in 1962 and was caused by mining activity. The road connecting towns Most and Jirkov went together with the river Bílina between two mines ČSA and JŠ. The river and road were diverted and coal was excavated in this central part together with a destruction of the village Ervěnice. A filling of the overlaying material started on the excavated part. After the filling of all excavated area a new motorway, railway and the river Bílina in 4 tubes were situated on the top of this corridor. Three basic steps of this process are demonstrated in Fig. 8.16. Cross section is

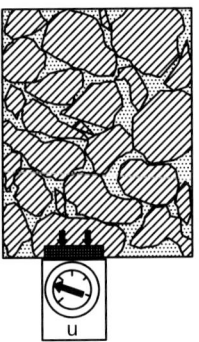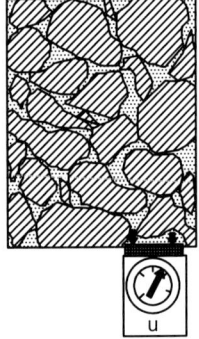

Fig. 8.14 Significance of the pore pressure gauge contact with clay clods

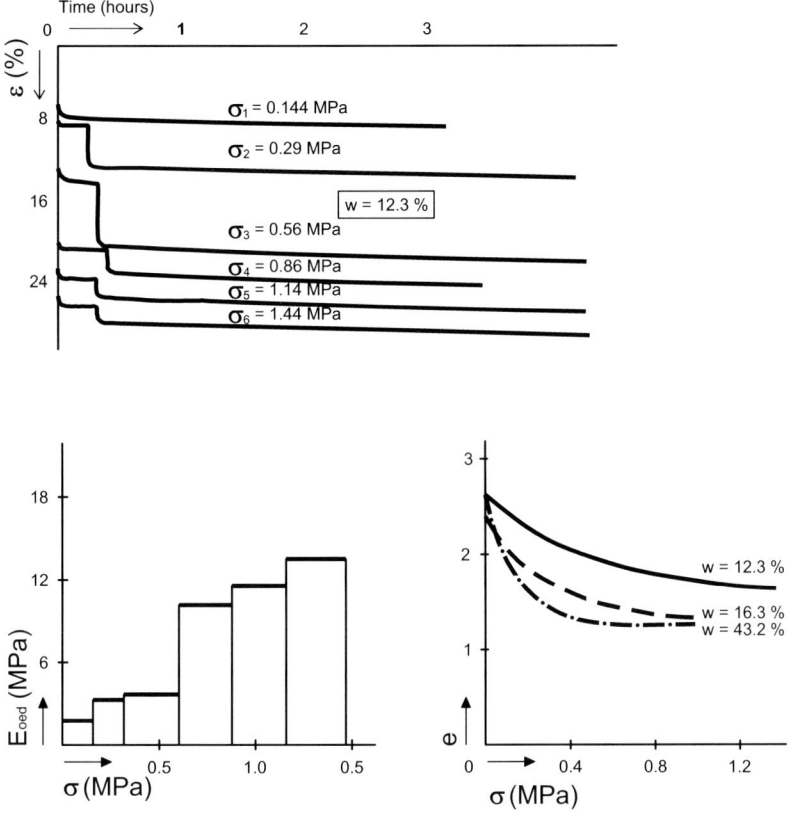

Fig. 8.15 Results of 1D compression tests for spoil heap material with high initial macroporosity

shown in Fig. 8.17. The maximum elevation is 171 m, slope inclination 1:7 on one side and 1:6.2 on the other one. Total volume of the material situated into corridor was at the end of filling 540 mil.m^3, what is probably the biggest earth structure in the world. The width at the bottom was roughly 2 600 m and at the top 260 m, the length at the top 3600 m. Deformation of Ervěnice corridor was monitored, namely settlement of surface as a function of time and also settlement of deep fixed points in the predefined levels. Some results will be described in the next chapter. After 20 years of failure-free operation of the structures situated on the top of this corridor it can be concluded that all technical issues during the construction and also in the operating phase were mastered very well – Dykast et al. (2003).

8.1.4 Experiences with New Construction on Spoil Heap Surface

Construction on the spoil heap surface represents one of the examples of new construction on building plots previously affected by human activity, Vaníček (1990, 1991). In previous chapters some examples of such constructions were mentioned

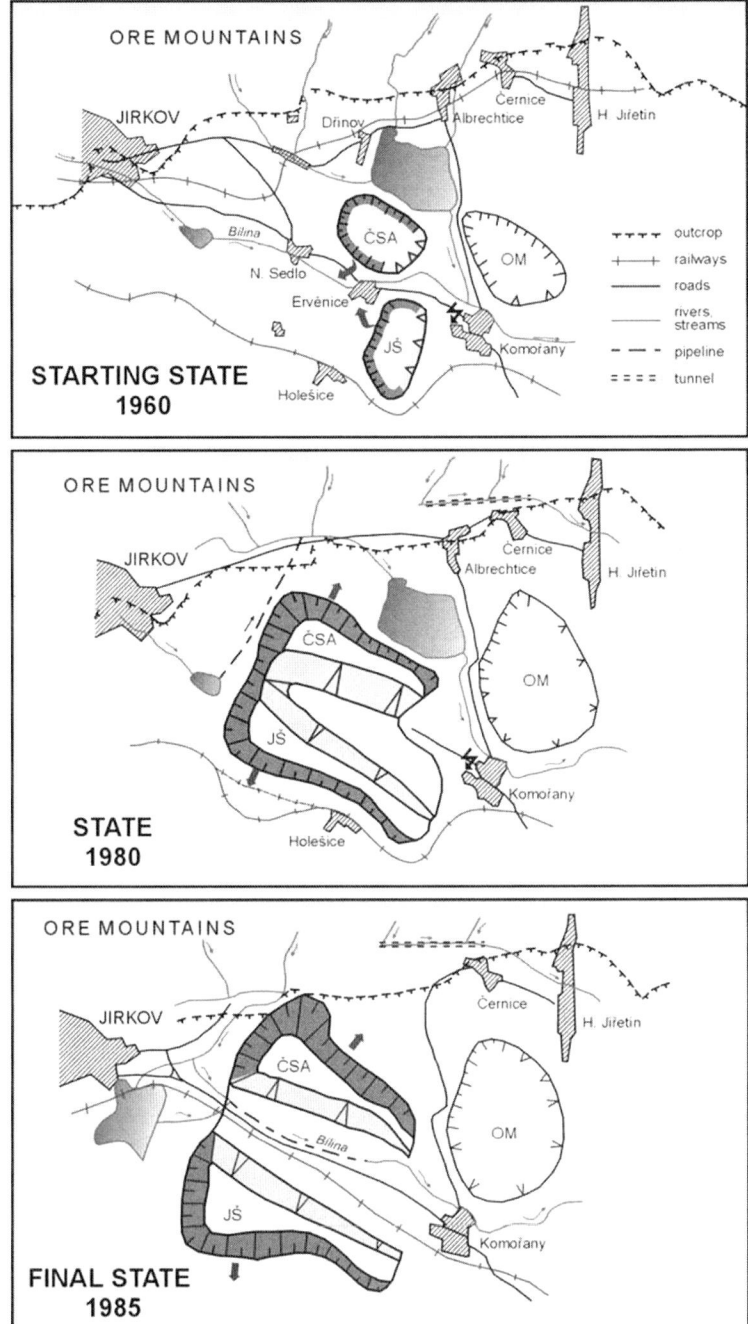

Fig. 8.16 Ervěnice corridor – three basic steps of corridor construction

8.1 Spoil Heaps

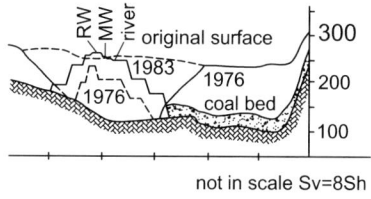 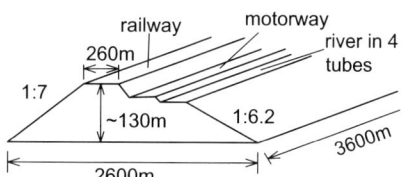

Fig. 8.17 Ervěnice corridor – cross section

such as construction of high speed trains on land affected by old mining activity, or as construction of airfields on the sea bed. In these cases the settlement of foundations of new constructions are composed not only by settlement in the active zone below foundations, but also by settlement of fill and settlement of its basement, or by collapse of underground spaces etc. For the new construction on spoil heap surfaces it is possible to divide the total settlement into three individual parts. Besides settlement in the active zone there is settlement of spoil heap body and settlement of spoil heap basement. The last two components can be put together, because the differentiation is not simple, and their sum corresponds to the spoil heap surface settlement. Components of settlement as a function of time are shown in Fig. 8.18 and expressed by equation, Vaníček, (1990, 1995a):

$$s_t = s_{a,t} + s_{s,t} = s_{a,t} + s_{v,t} + s_{b,t} \tag{8.4}$$

when $s_{s,t} = s_{v,t} + s_{b,t}$
Where

- $s_{a,t}$ – settlement of foundations in active zone in time t
- $s_{s,t}$ – settlement of spoil heap surface in time t
- $s_{v,t}$ – settlement of spoil heap body in time t
- $s_{b,t}$ – settlement of spoil heap basement in time t after the end of spoil heap filling.

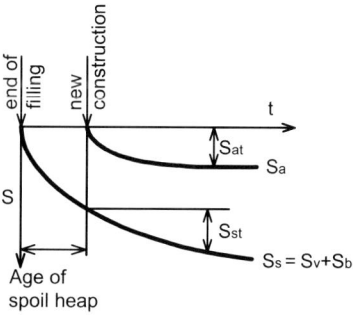

Fig. 8.18 Construction on spoil heap surface – components of settlement as a function of time

Settlement of new spread foundations on unaffected building lot has only single component, settlement in the active zone. Computed settlement is compared with allowable limits which are:

– For total settlement $s \leq 40\text{–}100$ mm
– For differential (distortional) settlement $\Delta s / \Delta L \leq 1/150\text{–}1/600$

Fulfilment of these limits for new construction on spoil heap surface is very problematic, and very often unrealizable. Therefore reliable estimation of the settlement of spoil heap body and its basement is so important.

8.1.4.1 Settlement of the Spoil Heap Basement

There are great differences between inner and outer spoil heaps. For inner spoil heap the situation is much better because the basement is composed from stiff to hard materials, and loading by spoil heap body is in general lower than the original geostatic pressure which was there before coal excavation. In principle the settlement is caused by reconsolidation on nearly original pressure only.

Substantially a more unfavourable situation is valid for outer spoil heap. Very often the subsoil is created by clay materials, which were described as clays overlaying coal seam. To be able to estimate this settlement two relatively exceptional analogous cases were used, for which the subsoil settlement was measured, Dykast (1993). In the first case subsoil of earth-fill dam Nechranice was monitored. This dam is situated on similar subsoil as spoil heaps are. The second case is describing the spoil heap Sophienhohe of the mine Hambach in Ruhr area (Kuntsche 1992).

Nechranice earth-fill dam (3.2 km long) is founded on tertiary clays and silts. Settlement of subsoil was measured with the help of marks embedded in different depth under the footing bottom, Seyček (1975). The deepest situated mark showed settlement of 80 mm after 4 years. Measured settlement of the footing bottom was compared with the calculated one. Comparison showed that total expected settlement 1630 mm (after 100 years) is comparable with prognosed settlement 1370 mm (after the extrapolation of results obtained during first 14 years of measurement). However the differences between calculated and measured values for the first 10 years were up to 100%. Subsoil soils consolidated faster than calculated. Oedometric moduli of deformation were back analysed from the subsoil deformation:

– For clayey soils for range of stresses 0–0.6 MPa $\quad E_{oed} = 14$ MPa
$\quad\quad\quad\quad\quad\quad\quad\quad$ for range of stresses 0.6–1.0 MPa $\quad E_{oed} = 22$ MPa
– For silts for range of stresses 0–0.6 MPa $\quad E_{oed} = 24$ MPa
$\quad\quad\quad\quad\quad\quad\quad\quad$ for range of stresses 0.6–1.0 MPa $\quad E_{oed} = 33$ MPa

Settlement of spoil heap Sophienhohe basement was calculated as a circular flexible foundation with diameter $r = 1\,500$ m and with uniform contact loading in footing bottom 2.5 MPa. The thickness of the deformable zone under contact loading was assumed to be in the range of 500–600 m. Subsoil is composed mostly from sandy soils ($E_{oed} = 200$ MPa), partly from clayey soils ($E_{oed} = 40$–200 MPa according to depth) and from coal ($E_{oed} = 25$ MPa). Before the spoil heap filling started the steel plates with size 3 ×3 m were situated on the spoil heap basement. After the end of filling (2.5–4 years) these steel plates were reached by monitoring wells and from this time monitoring of their settlement started. It was recorded that roughly 50% of the total calculated settlement was realized during spoil heap filling. This portion of settlement roughly corresponds with calculated total settlement of sandy soils in the deformable zone. For the observation point for which the height of spoil heap was about 180 km the total settlement of the steel plate was nearly 11 m and continued with a rate of 250 mm per year. In a given point the calculated settlement was 10.89 m so it means that the real total settlement will be higher, roughly about 20–30%. For the spoil heap composed mostly from sand the settlement of the spoil heap basement is much higher than the settlement of spoil heap body. For spoil heaps composed from clay clods the settlement of spoil heap body will be more important, for inner spoil heap dominant.

8.1.4.2 Settlement of the Spoil Heap Surface

Reference values of spoil heap surface settlement based on experience gained in various sites of North Bohemia are:

– Final value of settlement $\quad s = (0.02$–$0.03)H$

Where H is total height of spoil heap

Lower values are valid for lower soil heaps roughly $H = 30$ m and higher values are valid for higher spoil heaps $H = 100$ m.

From the time path the settlement is estimated as:

– 25–50% of the final value during 1st year,
– 70–75% within 5 years,
– 85–90% within 10 years.

Settlement prediction for surface of spoil heap for Ervěnice corridor was based on the results of compression tests realized in the large oedometer device. In 1981 2 years before the end of filling of the Ervěnice corridor, a prediction of the settlement rate in the places of proposed new railway track was realized. Field geodetic measurement of the actual settlement carried out in subsequent years confirmed that it was quite an accurate settlement prediction, Fig. 8.19.

It is obvious that the prediction of spoil heap surface settlement based on the results of oedometric tests brings some simplification. For example oedometric tests are realized under certain conditions which do not have to be fulfilled in reality. First of all it is due to water saturation not only during construction but also after the end of filling. For example certain correlations were observed between rate of settlement and annual precipitation, see Fig. 8.20.

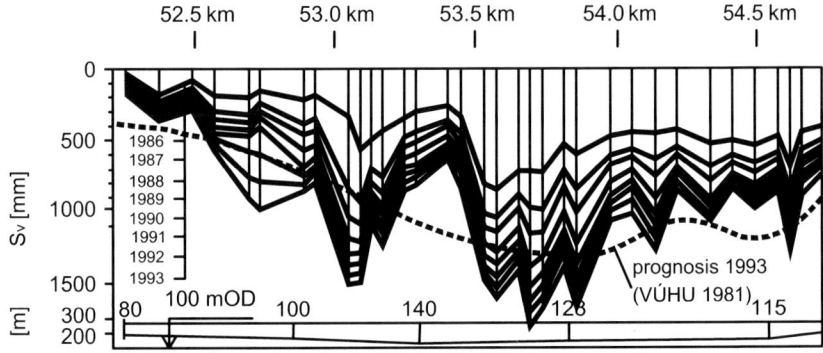

Fig. 8.19 Ervěnice corridor – predicted and measured settlements along longitudinal axis, (according to Dykast)

When the construction of spoil heap is completed then the most specific forecast of further course of the surface settlement could be based on geodetic monitoring. Theoretically this monitoring can start immediately after the end of filling but in fact the initial (zero) measurement is usually done after certain time delay. The accuracy to forecast the further settlement development depends on the duration of monitoring. Five years seems to be the minimum time.

Different functions $s_t = f(t)$ was checked, especially these in which the accession of settlement drops down logarithmically of exponentially. Now the equation

$$s_t = a + b \cdot \ln t \qquad (8.5)$$

is recommended in which a and b are parameters determined from the first measurements, Dykast (1993).

8.1.4.3 Possibilities of Settlement Reduction of the Spoil Heap Surface

Possibilities of reduction of spoil heap basement settlement are relatively small ones. For outer spoil heap the subsoil layer can be removed or the consolidation

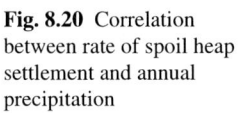

Fig. 8.20 Correlation between rate of spoil heap settlement and annual precipitation

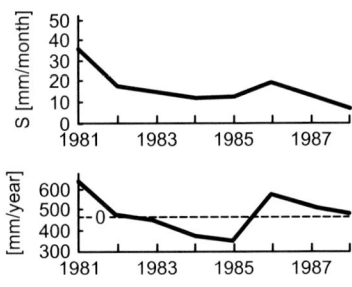

rate can be speeded up by vertical drains. With respect to an enormous area covered by the spoil heap these countermeasures are financially very demanding.

There are better possibilities for reduction of spoil heap body settlement. Primarily the complexity of countermeasures which was previously mentioned can be applied as a prevention of filling into free water, good drainage of the contact of spoil heap body with basement, and limitation of infiltration to the spoil heap body by smoothing of surface of individual decks. Also it is recommended to store into the lower part of spoil heap material excavated from a greater depth.

For Ervěnice corridor the method of preloading was also investigated in the laboratory. The aim of oedometric compression test was to find how many meters to add to the final level, how long to leave this additional loading there with the view to minimizing the deformation of the final level after removal of the additional loading. But with respect to demands to speed up the utilization of this corridor as soon as possible this method was finally not applied.

In line with Fig. 8.18 the basic possibility to reduce the influence of higher settlement on new construction is to start with this construction with some delay. We can speak about "age" of spoil heap in which a new construction can start. On the base of experience gained in various sites a simple relation is recommended – new construction can start when the age in years corresponds with H in meters. This recommendation with respect to the present-day heap elevation is hardly compliant. Nevertheless each year for which the new construction can be delayed is important. For example for the spoil heap of 100 m high, where new construction starts 10 years after the end of filling, the further expected settlement of the surface can be roughly 300 mm.

8.1.4.4 Possibilities of New Foundation Settlement Reduction in its Active Zone

Herein it is necessary to state that for the foundation of new structures the spread foundations will prevail. It is given by two aspects. Firstly it is elevation of spoil heaps, because the practical length of piles is usually lower than present-day height of spoil heaps. Secondly, it is the fact that the bearing capacity of piles is lowered by negative skin friction. Deformations of surrounding soil are greater than the deformation of piles. Nevertheless the piles cannot be excluded, because mainly displacement piles are improving properties of surrounding soil by lowering the macroporosity.

In the case of spread foundations the main attention is devoted to the lowering of macroporosity of spoil heap body in the active zone. With this lowering deformation modulus are increased. In the first place the following methods and improvements can be applied from the surface of the existing level of spoil heap:

- Compaction of the surface by dynamic consolidation method with efficiency reaching few meters,

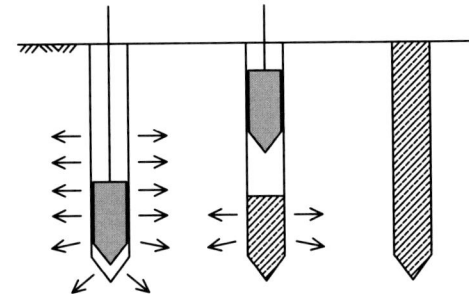

Fig. 8.21 Lowering of the macroporosity of the spoil heap body in the active zone – compaction with the help of "clay" piles

- Compaction of the active zone with the help of "clay" piles – when the pre-driven profile is backfilled by clay of similar properties as is surrounding material and subsequently compacted here, see Fig. 8.21.

If it is possible to construct the last few meters of spoil heap by different methods than by free fall, then it is possible to think upon:

- Classical compaction – when the last meters in the active zone are compacted by rollers,
- Classical compaction supplemented by other improvement as adding the ash which can fulfil macropores between individual clods or as adding lime or cement for strengthening of the top layers,
- Reinforcement of the last layers by geosynthetic reinforcing elements.

Method of preloading can have some advantages in this case when for active zone the ratio of new loading to the previous one (overconsolidation ratio) is the highest.

A limiting of the settlement in the active zone can be also positively influenced by the selected method of foundation. Foundation slabs situated in such a depth that the overloading there will be practically zero, is strongly reducing settlement in the active zone – new loading will be roughly the same as previous geostatic loading – principle of "floating" foundations.

8.1.4.5 Limitation of the Impact of Higher Settlement

Nevertheless even under the application of the mentioned techniques it is not possible to guarantee that the settlement will be in allowable limits. Among the active measures the different approach is usually chosen for total and for differential settlements, Vaníček (2003). A higher value of total settlement can be accepted if:

- Special technical solution is applied for engineering services as electricity, gas, sewage...,
- Rectification can be applied e.g. for railway track, pipelines etc.

But most sensitive questions are connected with differential settlements with direct impact on damages to the structural elements and to the manner of the practical use of the structures.

For a classical case Bjerrum (1963) summarized the following factors affecting differential settlements:

1. Variation of thickness of compressible strata.
2. Variation of loading of different parts of buildings.
3. Distribution of stresses in the soil beneath of foundations.
4. Inhomogeneity of the compressible strata.

For structures on man-made deposits as is soil heap, tailings dam, and sanitary landfill the following two factors should be added:

5. Inhomogeneity of the fill.
6. Inhomogeneity of the fill basement.

All these factors have to be taken into account when designing a new foundation on spoil heap surface. However on the contrary to the total settlements, where we have a certain possibility to accept higher values, the differential settlements have direct impact on damages of structures and on manners of their utilization. Therefore if the probability that the expected differential settlements will be higher than the accepted ones this situation will have to be solved with the help of the following steps:

– To select such construction system which is not so sensitive to the differential settlements or
– To improve the subsoil beneath the foundations.

The second case was mentioned before, only it is possible to add, that any improvement in the active zone is not only reducing the total settlement but also reducing the impact of differential settlements of layers below active zone on new foundations.

For the first case one extreme is the wooden skeleton (where closing of doors and windows can be the limiting factor) and on the other side of this spectrum is the stiff "box type" of structure (where a limiting factor can be connected with inclination of floor or where tilting of high rigid buildings might become visible). For motorways the asphalt surface is usually preferred.

8.1.4.6 Practical Applications

Very good experiences were obtained from the Ervěnice corridor. Transport infrastructure on its surface started operation in 1983–1984, 2–3 years after the end of fillings, Dykast et al. (2003). The railway track is situated on the top of the cross section of this corridor. To counteract the differential settlement for the last 8 m the coarse material (debris) with classical compaction was applied underneath the rail track. The construction of four rail tracks enabled uninterrupted two-rail traffic in both directions and the rectification of settlement in free railroad lines. Feeding ducts have been installed to be able to adjust the height of the trolley line above the rail crown. A speed limit of $40 \text{ km} \cdot \text{h}^{-1}$ has been imposed on the western part (where the situation from the differential settlements point of view is the worst).

With respect to the 4 lane motorway diversion over the Ervěnice corridor the following notes can be added:

- Motorway crown was situated on 7 m thick compacted layer from sandy-gravel,
- Crown was drained with linear stone ribs of trapezoid cross-section using geotextile,
- Transverse gravel drains were built in the same way every 25 m to divert water off the road,
- The upper part of the motorway is built of asphalt-concrete coat.

No special speed limit was applied there (allowed is $90\,\text{km.h}^{-1}$), and differential settlements can be seen on the surface, but it is rather very smooth, Fig. 8.22. In 1990 a general repair of all lines had to be carried out, equalizing the levels with compacted gravel layer and laying a new AB coat.

River Bílina was situated on the surface of this corridor in four tubes. Only two pipes have been in operation with respect to water flow. The remaining two pipes are on standby and have been rectified in the horizontal and vertical directions according to the regular height measurements checks. The steel pipeline is embedded in trapeze channels on wooden sleepers, which stabilize their lateral location. The sleepers are located on the gravel subbase. A report about the water transfer from the pipeline into the natural flee-flow channel is now in preparation.

Relatively very good experiences were gained with construction on soil heaps in places where reasonable care was devoted to the improvement of the material in the active zone. Subsequently by that way the constructions even very sensitive to the differential settlement were realized there. As such examples the motor-racing circuit, race-course and small airfield in the vicinity of the town Most can be mentioned.

On the other side negative experiences were obtained even for construction of small buildings but situated on subsoil the improvement of which was performed with much lower care. For example single-storey garage row showed a large amount

Fig. 8.22 Ervěnice corridor – photo of motorway with visible signs of differential settlement

of cracks affecting even total stability; as well temporary buildings realized by mining companies directly on the surface of spoil heaps. Such an example is the crossing of belt transport with rail transport, where foundation of separate footings suffered by extreme deformations. Accordingly other structures such as transformer and storerooms have been damaged.

One example of the failure of industrial hall will be described in more detail. This industrial hall was founded on spoil heap 12 years old and of about 20 m height, Fiedler (1990). From the beginning of the design stage some points for settlement measurements were installed on a surface of the fill. Their settlement was about 10–20 mm per year. The dimensions of the object were about 60 × 60 m, Fig. 8.23a. The schematic section of the foundation is in Fig. 8.23b. The pressure on the top of the sand layer due to the construction was 117 kPa. Deep footing bottom significantly decreased load increase there, foundation strip was heavily reinforced and sand cushion was applied to guarantee high deformation modulus just below footing bottom. But with high probability this permeable sand cushion

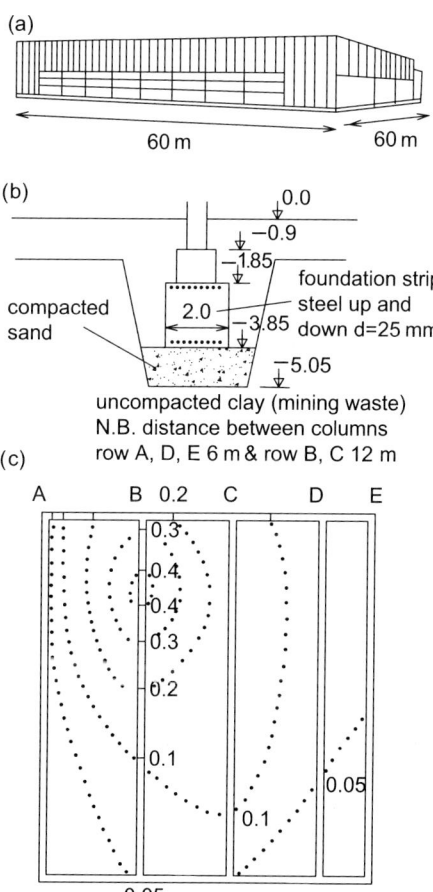

Fig. 8.23 Industrial hall constructed on soil heap, (according to Fiedler). (a) hall sketch. (b) schematic section of the foundation. (c) results of the deformation measurement

was the reason of problems which started there. Permeable sand allowed water access and bottom of excavation was badly drained. Water was retained in the lowest place and caused there the highest settlement of the fill. The shell of the hall was realized in 1981, and measurement of settlement started immediately after some differential settlement was observed. Results of the deformation between 1.4.1981 and 13.8.1981 are shown in Fig. 8.23c. Maximum settlement of column B IX was 0.46 m, for neighbouring one B VII was 0.26 m. Distance between columns is 12 m. Cracks in cross-beam were observed as well as in the columns. Some protection measures were applied and the operation of this industrial hall was only limited. Deformation was measured up to 1989 and the additional settlement was up to 240 mm. Nevertheless the results also showed that our up to date allowable limits for differential settlements are rather strict. For row of columns A where distance between individual columns was 6 m the measured differential settlement was up to 40 mm. But no failures and cracks were observed there. This differential settlement 0.007 is much higher than the allowable 0.002.

8.2 Tailings Dams

Tailings dams are defined as a natural or artificial space on the earth surface, serving for permanent or temporal deposition of hydraulically transported tails; component part of tailing dams is also an embankment system.

According to the types of deposited sediments the tailing dams can be divided as:

– Tailings dams with sediments resulting from burning of solid fuel – e.g. flying ash from electric power stations,
– Tailings dams with sediments from concentrator factories,
– Tailings dams with sediments from chemical factories,
– Tailings dams with other sediments.

Herein it is necessary to state the basic principle of sustainable development (construction). Before deposition we have to try to use "waste" as much as possible to limit the volume of deposited material. But up to now not all "waste" material can be used and therefore the deposition is a necessity. Nevertheless in some cases the deposition is temporary because with development of new technologies in future the deposited material can be used as a raw stock.

With respect to the great variability and specificity of deposited tails on one side, respectively with respect to the fact that frequently the construction and operation of tailing dams are realized with less qualified workers as is common for dam construction, there are many incidents and failures registered. For example during the earthquake with the intensity of 7–7.5° of the Richter scale 22 different tailing dams with sediments from concentrator factories, mostly from copper, failed in Chile in 1965. Large failures occurred in Val di Fiemme Valley on the foothill of the Dolomites in north Italy in July of 1985. Two tailings dams with sediments after extraction of fluoride subsequently failed there. Mixture of sediments with water created a wave with the elevation up to 30–40 m which run through the valley during 20 s. The village of Stava with tourist hotels was completely destroyed and the valley with a width of 150 m along the length of 6 km was covered by a thick layer of mud. The cost was dreadful, 300 people died.

The failures of tailing dams cannot only be connected with the just described negative impact, but can also bring a great potential negative impact on our environment. This negative impact depends on chemicals which are used in concentrator or chemical factories. In this direction the failures which occurred on the tailing dams in Aznacollar in Spain are very often mentioned, where sediments from zinc and lead concentrator factory were stored or the failure in Romania where sediments from a gold concentrator factory were stored and which after failure strongly contaminated waters of the river Tisza in Hungary. Hereafter failures which also occurred in the Czech Republic will be described.

Morgenstern (1999) expressed his view after summarizing some recent failures and incidents coming from tailings dams construction for storage mine waste as follows. "Experience in recent years within the mining industry, particularly with tailings dams, indicates that the technical and managerial challenges of responsible mine waste management is not adequately recognized. Geotechnical engineers have many contributions to make to meet this challenge and this is often underappreciated. However there are too many errors arising from recent geotechnical practice for mine management to feel totally secure in the hands of their geotechnical advisors. Resolving this issue is a serious and immediate issue for Geotechnics in the New Millennium".

8.2.1 Types and Construction of Tailings Dams

Selection of the best location of the tailing dams is always a very difficult problem. It is necessary to take into account not only annexation of land, possible contamination of surface and underground water but also the direct impact on inhabitants, e.g. with respect to possible dusting.

With respect to the interaction with surface water the tailing dams are at the present time designed as not allowing surface water to pass through. Entire water from the catchments' basin has to be diverted. Therefore the construction of tailings dams in a valley, as is a typical case for classical dams, brings some problems with this diversion although on the other side the volume of the embankment would be the smallest one, see Fig. 8.24. Tailings dams situated on inclined planes require the construction of the main dam, side dams and dividing dams with help of which the total area of tailings dams is divided into at least two fields. These fields are alternately hydraulically filed or set away. During the second phase additional dams are constructed. Planar tailings dams require the construction of dams around all peripheries. Compared to the classical dams which are constructed on full height and after the construction are operated the construction of the tailings dams on full height is rather an exception, for typical cross section see Fig. 8.25.

Fig. 8.24 Different types of tailing dams according to location. (**a**) valley type. (**b**) type with lateral dams. (**c**) planar (horizontal) type – with dams along whole circumference

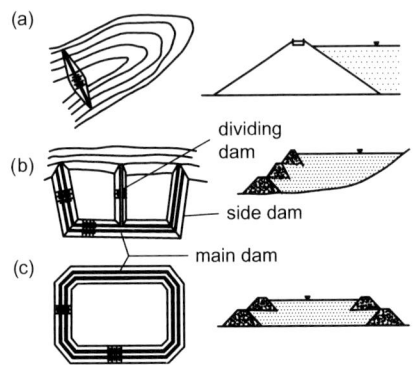

Nevertheless to help to define the basic principles of construction, the main dam on full high is assumed, see Fig. 8.24a. The storage of as much waste as possible is the basic aim of the tailings dams design. It is given by:

- Price of land on which the tailings dams are constructed,
- Fact that the tighter arrangement of grains is from the point of view of safety much safer.

Therefore the tails transported by hydraulic way should sediment on part of the surface of deposited tailings which is closer to the main (or additional) dam. This part is often called beach. In this case the vertical force acting on individual grains during sedimentation is not given only by gravitational forces but also by seepage pressure. Only the smallest particles reach the part of tailings dams with free water level. During sedimentation in free water the gravitational force is decreased by uplift and therefore settling particles are deposited under very loose state – with high porosity. But this loose state has very low shear strength, high susceptibility to structural collapse especially during dynamic loading. This state can easily lead to the liquefaction.

During nearly uniform hydraulic filling with the help of enough outlets, the biggest and heaviest particles sediment close to the main (additional) dam and create there the zone with highest permeability. It is not our aim to construct the main dam – as it is the general case for classical hydro structures – with very low permeability, but on the contrary it is necessary by proper selection of material of the

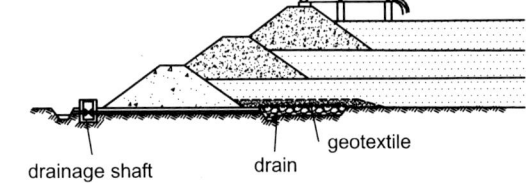

Fig. 8.25 Cross section through the upstream constructed tailings dams

8.2 Tailings Dams

main dam and adjacent settled material to guarantee (together with good drainage system) that:

- Phreatic line is sufficiently far off the downstream face of the tailings dams,
- Free water level is covering roughly two thirds of tailings dams surface – this condition together with low inclination of the settled waste – about 1:100 – can guarantee that the area of the beach will not be dried-up. Therefore this condition can protect the windy erosion.

Fulfilling of these conditions is ensured by drainage systems and according to the demands also by spillway on free water level. This spillway is gradually raised up as sedimentation continues. Floating pumps can also be used. Water collected by both these systems is returned back to the starting point of the closed circuit.

Settled waste always has certain anisotropy of permeability because the formation of small finer and coarse layers is difficult to protect. Therefore during modelling of the seepage inside of tailings dams this aspect should be taken into account, see Fig. 5.53. Anisotropy of permeability can be partly reduced by the above mentioned uniform hydraulic filling with sufficient number of outlet respectively by application of hydro cyclones. In hydro cyclones the coming tails is separated on finer and coarser components. Coarser material is after that deposited to the main (additional) dam, finer particles into the inner space of tailings dams body.

Additional dams are constructed either from imported soil or from the already deposited waste. Second variant is in agreement with up to date tendency of sustainable construction. As will be shown later the deposited materials have relatively high shear strength, very close to the properties of sand. However for the design of the general slope inclination the empirical findings for earth structures constructed from sand are hardly to be applied in this case. The main reason for this distinction is the difference in volume density.

Two main problems will be therefore discussed in more detail – possibility to construct main (additional) dams from deposited waste material and the significance of the drainage system for overall stability.

During the application of tails (ash or slime from concentrator factories) the following main conceptions of constructions are used:

- Dams are constructed from widespread unsorted waste compacted by classical manner in individual layers. In exceptional cases the conception is applied also for the main dam. A highly discussed question deals with determination of the optimal moisture content and with subsequent compliance of the admissible range of it. This problem in most cases deserves the individual approach.
- Dams are constructed from partly sorted and also partly compacted waste (e.g. by bulldozer). Waste material is pushed up on the dam from the beach where coarser material is deposited.
- Dams are constructed by direct floating of the separated coarser material from hydro cyclones. Upon application of the two steps hydro cyclones the coarsest fraction is deposited directly into profile of additional dam with the help of

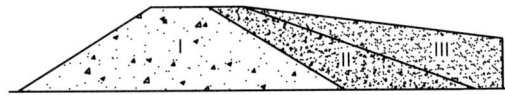

Fig. 8.26 Construction of tailings dam by direct floating from separated ash in two-step hydro-cyclone; I – coarsest; II – middle; III – finest

secondary mobile hydro cyclone. Floating of thickened tails makes it possible to create the designed profile of the additional dam, see Fig. 8.26. Authors also proposed the construction of main and additional dams with the help of geotextile sleeve, where this sleeve is filled by waste material and geotextile is satisfying the drainage and filtration role, see Fig. 8.27. During the design and realization of the drainage system it is necessary to respect the following demands:

- Phreatic line (depression curve) should be at least 2 m far from the downstream face of the dam,
- Capacity of the drainage system should guarantee the safety factor at least 3,
- Drainage system must be protected by well designed and performing filter, to protect drainage against clogging and so lost of serviceability.

Few types of drainage systems are shown on Fig. 8.28. Here is seeing that the most sensitive place is the connection of drainage systems of main and additional dams. This connection can be either on the surface of tailings dams or inside of its body. Great attention to the design of drainage is devoted in Bulletin No. 97 – ICOLD: Tailings dams, design of drainage, ICOLD Paris 1994.

8.2.2 Tailings Dams with Sediments Resulting from Burning of Solid Fuel

Burning of solid fuel for production of electricity and heat is still the prevailing manner of the production of energy in the world. Also in the Czech Republic this rate is very high, reaching up to 75% of energy consumption. Therefore the production of sediments resulting from solid fuel burning is high per habitants, roughly 1.1 t per year. The reason is either high ash content coming from brown coal or high energetic demand of the industry in the ČR.

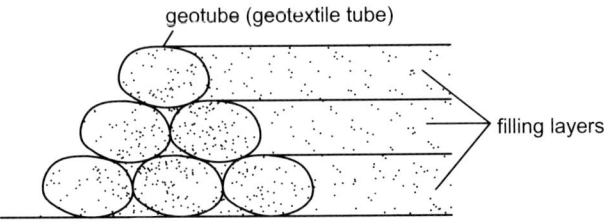

Fig. 8.27 Dams constructed with the help of geotextile sleeve filled also by waste material

8.2 Tailings Dams

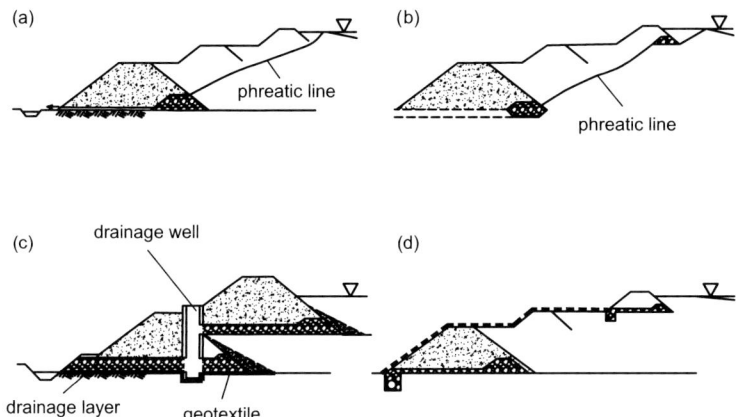

Fig. 8.28 Examples of drainage systems. (**a**) toe drain in main dam. (**b**) stepped toe drains. (**c**) planar drains with drain wells. (**d**) planar drains with surface pipe collectors

All of these plants (mostly slag – type boiler) produced yearly between 12 and 16 mil.tons of ash and slag with decreasing percentage of slag – for year 1985 it was 25%, Vaníček (1998).

Roughly between 10 and 20% of the remaining parts after coals burning were used and the rest were stored by wet way in tailing dams. The biggest utilisation of the remaining parts after coal burning consisted of the following areas:

- Construction materials (partly as a substitution for cement, artificial light gravel – aggloporite, light prefabricates,...),
- In earth structures (motorway embankment,...),
- In deep mines, especially in Ostrava black-pit coal region. Nearly 40% of the produced ash in power and boiler plants in this region was stored inside of the closed down deep mines. Horizontal galleries after filling are more stable with positive impact on surface, limiting the subsidence of the terrain. After that only 24% of the ash remained unused there.

To the last point the short note can be added. The utilisation of the remaining parts after burning from plants using brown coal is lower than from using black coal. The problems there were always connected with low content of basic compounds CaO and with higher content of acid compounds – Al_2O_3 and SiO_2.

The overview of coal ash utilisation is also given by Tyson (1994) or Cabrera and Woolley (1994).

Up to date utilization is higher and relates to change of the desulphurization processes used now for electric power and heating stations. Nevertheless tailings dams were and still are part of each power and bigger heating stations. The largest area for individual tailings dam is around 200 ha, when total area for all tailings dams in CR is roughly 13 km^2. Greatest concentration of tailings dams is in the north part of Bohemia, where there is a main brown coal basin.

8.2.2.1 Ash Properties

Properties of ash are very variable; they are changed not only for different electric power stations but also for individual tailing dams. In this case there is a great heterogeneity due to the sedimentation of the finer and coarser particles in horizontal layers. Particularly it is typical for the old electric power plants with higher percentage of slag. For the tailings dam of the electric power station Prunéřov three different samples were tested – Vaníček et al. (1989). These samples represent the fine, coarse and the middle ones. Grain size curves are presented in Fig. 8.29. It is obvious that particles above 2 mm are presented only in the coarse sample but with low percentage. Significant ratio of the fine particles and also the particles lower than 0.002 mm cannot be considered to be as clay particles. Due to the absence of active clay particles (clay minerals) the plasticity is practically zero. Sometimes the reddish colour indicates the presence of clay material in coal, but after the grinding and burning these clay minerals are not active. Specific gravity ρ_s is lower than is for typical soils, roughly around 2 350 kg.m^{-3}. For one sample the particles were crushed down and after that we obtained value 2 363 kg.m^{-3} instead of previous 2 293 kg.m^{-3}. This result proves the fact that ash particles contain also a certain amount of closed pores. The results of specific gravity together with maximum and minimum porosity are presented in Table 8.1.

From the results of porosity we can conclude:

– High values for minimum porosity,
– Small differences between maximum and minimum porosity,
– Practically the same results for all samples.

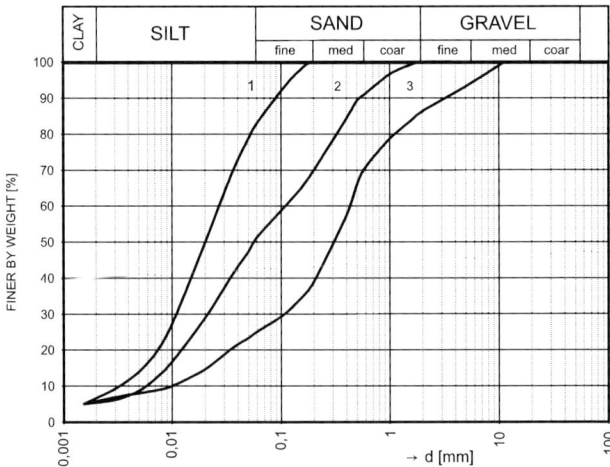

Fig. 8.29 Typical grain size curves of ash

8.2 Tailings Dams

Table 8.1 Specific gravity and porosity for 3 grain size curves of ash

	ρ_s [kg.m^{-3}]	n_{max} [%]	n_{min} [%]
coarse	2391	63	51
middle	2293	64	52
fine	2354	62	50

From the other results:

- Natural volume density $-\rho = 1\,500 - 1\,600$ kg.m^{-3},
- Dry volume density $-\rho_d = 950 - 1050$ kg.m^{-3},
- Deformation parameters $-E_{def} = 4 - 30$ MPa,
- Poisson's ratio $-\nu = 0.23 - 0.36$
- Coefficient of consolidation $-c_v = 130 - 1510$ mm^2.s^{-1}
- Coefficient of filtration $-k = 1.10^{-6} - 1.10^{-7}$ m.s^{-1} – Note: This measured range was rather small, for other tailings dams the range was from 1.10^{-4} to 1.10^{-7}.
- Shear parameters $-\varphi_{ef} = 38-47°$, c_{ef} usually zero.

Shear strength measurement in triaxial apparatus under different conditions (change of porosity, CID or CIUP tests) showed that:

- Friction angle is decreasing with level of stresses – for coarse sample it was more significant than for fine samples – during shearing larger grains are crushed down,
- For expected level of stresses corresponding to the deposited material around 25–30 m the ash showed dilatation behaviour,
- During the phase of expansion the coefficient of pore pressure A was in the range of –0.1 to –0.3.

In literature more similar results can be find, e.g. for tailings from the electric power station Nováky utilising also brown coal, see e.g. Slávik et al. (1994). Tailings dam for this power plant failed in 1965, strongly affecting river Nitra, transport infrastructure and the village Zemianske Kostolany. Some papers are concentrated on the structural collapse during shearing and loading, studying the influence of microstructure and shape of individual grains, e.g. Skarzynska et al. (1989).

8.2.2.2 Limit States

During the design of tailings dams the all potential limit states should be checked as showed in Chapter 5, but for tailings dams the most sensitive are, Vaníček (1998):

- Slope stability,
- Internal erosion,
- Surface water erosion,
- Surface wind erosion,
- Deformation.

Slope Stability

From the beginning the angle of the internal friction was the main design parameter. According to the similarity with typical results for sand and gravel, the rather steep slopes were designed. Later it was recognised that the unit weight γ has priority. For the simplified case where water is flowing parallel to slope with angle β – see Fig. 8.30, factor of safety can be expressed from the equation:

$$F = (1 - \gamma_w/\gamma_{sat}) \, \text{tg} \, \varphi_{ef}/\text{tg} \, \beta \qquad (8.6)$$

where

γ_{sat} – unit weight for saturated sample,
γ_w – unit weight for water.

For typical values of $\gamma_{sat} = 14 \, \text{kN.m}^{-3}$ and $\varphi_{ef} = 35°$ and factor safety F = 1.5 the slope inclination should be:

$$\text{tg} \, \beta = (1 - 10/14) \, \text{tg} \, 35°/1.5 = 1/7.5 \qquad (8.7)$$

The influence of γ_{sat} is visible from its small change. For $\gamma_{sat} = 15 \, \text{kN.m}^{-3}$ and $\varphi_{ef} = 37°$ and for the same demanded factor of safety the slope inclination can be:

$$\text{tg} \, \beta = (1 - 10/15) \, \text{tg} \, 37°/1.5 = 1/6.0 \qquad (8.8)$$

Now the general inclination for tailings dams with sediments resulting from burning of solid fuel is roughly between 1:6 and 1:8. Great effort is therefore devoted to the increase of unit weight. A much better situation is for tailings dams with sediments from concentrator factories, where unit weight can be even higher than for sand and gravel, very often around $21 \, \text{kN.m}^{-3}$. For such tailings the general inclination can be about two times steeper.

When solving slope stability the real estimation of pore pressures are very important. Therefore in the phase of design the flow net is constructed with the help of FEM where the heterogeneity for permeability is taken into account – see e.g. Fig. 5.53.

However flow net is valid for steady flow, and therefore the estimation of pore pressure changes in different phases of construction are important as well.

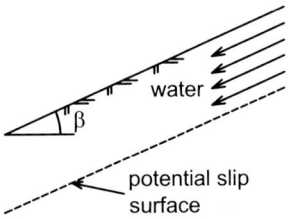

Fig. 8.30 Sketch for defining general slope of tailings dams

Slip surface can pass through subsoil usually composed in north Bohemia from tertiary clays to limit infiltration of water from tailings dam to the underground. Therefore it is necessary to count with pore pressure increase in these clays induced by load increase. With respect to fine particles of ash the question of pore water increase and its dissipation was checked for bottom of tailings dam with stiff and impermeable subsoil. For fine ash it was concluded that pore pressure increase in contact with impermeable subsoil induced by additional deposits can dissipate when load increase correspond up to 3 m height of deposited material per year up to the total height of 50 m. For higher tailings dams this possibility has to be proved based on the estimation of the coefficient of consolidation and the theory of consolidation counting with triangular (geostatic) distribution of loading.

In some localities a special case was observed. At the end of hydraulic filling, when additional dams were constructed, water in tailings dam reservoir went down creating upper part of tailings unsaturated. After the renewal of hydraulic filing, a significant part of unsaturated tailings is surrounded by water, see Fig. 8.31. Air pore pressure is higher than pressure in water and therefore air bubbles have a tendency to go up, creating a certain boiling of ash accompanied with shear strength decrease. Havlíček and Myslivec (1965) explained the behaviour of air bubbles when surrounded by water. Havlíček (1972) describes essential cases of water and air arrangement between individual grains for unsaturated soils, see Fig. 8.32. Here he explained how air bubbles can cause the fine sand liquefaction with the help of laboratory experiments. He stated that "when vertical effective stress in sand is dropping to zero a process of 'swelling' is starting. At this moment star shape of air bubbles is changing to spherical one, magnitude of suction is dropping down. Pushing of air bubbles on individual grains is causing sand volume increase, which for 1D condition is observed as surface lifting. Sand volume increase means a decrease of contacts between individual sand grains and consequently is leading to the decrease of friction between them. So this process can finally cause the sand liquefaction". Such special behaviour was for unsaturated flying ash proved by series of the triaxial tests, where attention was devoted not only to the phase of shearing but also to the phase of deformation where pore pressures of water and air were measured for nearly isotropic loading and loading under K_0 conditions, Vaníček et al. (1990). To protect the case of ash liquefaction in the tailings dams it is better not to allow such water decrease to create such a significant part of unsaturated ash inside of settled material. Probably this observation can help to clear up the examples of slope instability during heavy rains coming after dry period as mentioned also in

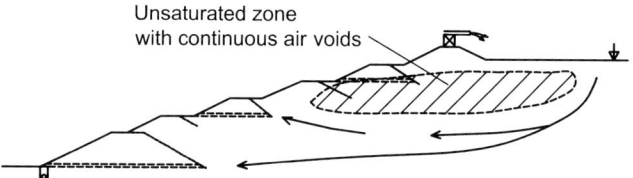

Fig. 8.31 Development of closed unsaturated zone after renewal of hydraulic filling

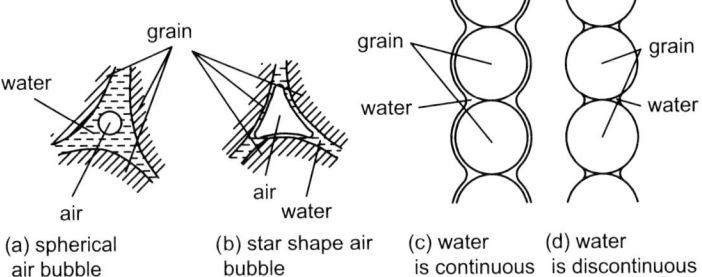

Fig. 8.32 Essential cases of water and air between individual grains for unsaturated soils, (according to Havlíček). (**a**) spherical air bubble. (**b**) star shape air bubble. (**c**) water is continuous. (**d**) water is discontinuous

Chapter 7 for flood protection dams. Zone of unsaturated soil with air bubbles are closed not only from the surface but also from the bottom, where underground water is going up. Therefore some slip surfaces pass through this unsaturated zone where high air pore pressures are the result of the underground water increase.

Filtration Stability

Main attention is always devoted to the protection of drainage system. Filtration criteria are used for both grain and geotextile filters as mentioned in Chapter 5. Nevertheless the laboratory experiments for checking the design especially for geotextile filters are recommended in the Czech standards. In all cases it is forbidden to pour slurry directly on filter. Filter must be covered before flooding by so called average sample of ash.

Surface Water Erosion

Usually two principal situations are leading to this sort of failure:

– Rain erosion of the downstream slope,
– Overflowing of the dam crest.

For main and additional dams composed from fine ash this limit state is one of the more dangerous, because particles are very susceptible to erosion, not only as the result of small particles with zero interparticle forces but also as the result of low dry density. Therefore the protection of surface with natural soil with thickness between 0.3 and 0.5 m is a necessity. Protection or separation geotextile is also useful.

In Neratovice area the main dam from compacted ash failed due to overflowing. Height was 5 m. Width of the gash was relatively narrow at roughly 12 m. Walls of this gash were nearly vertical at all heights, see Fig. 8.33. A significant part of the ash deposited in tailings pond flowed out. The area of the tailings pond was about $0.065 \, km^2$.

Fig. 8.33 The gash in the main dam constructed from ash

The main reason for this failure was human factor. The accident happened during a long weekend. The attitude of the water outlet (spillway regulating the water level) was arranged roughly 1.3 m above the permitted level. The dam crest was in critical place around 0.4 m lower than the surrounding one. Probably there were higher settlements due to very soft organic subsoil (old brook). Soil cover of the crest was only 0.1 m, and slopes were covered a little bit more, with maximum 0.3 m.

In the course of subsequent examination two facts bearing on the issue were observed:

- Very high heterogeneity of the deposited material coming from old slag type boiler, see Fig. 8.34,
- Relatively deep holes on downstream slope generated by animals, mostly by rabbits.

Fig. 8.34 Typical heterogeneity of the deposited material

Surface Wind Erosion

As mentioned before larger tailings dams are usually divided into two parts. First part is filled by ash and at the same time the additional dam is constructed for the second one. During this time the water table is going down and particles on the surface are dried. Dried particles are very sensitive to wind erosion and therefore have to be protected. The natural way is to use water to spread over the dry area to increase moisture content because the higher resistance due to capillary forces is increased again, see Fig. 8.35.

More than 20 years ago the anti-dust protection geotextile was used for the first time. A fundamental problem was connected with the question if this geotextile has to be removed before hydraulic filling will start again or not. What will be its influence on filtration conditions. The experimental place was selected and the collection of test samples started before renewal of hydraulic filling and after it. Nonwoven geotextile from natural materials was used there. Before renewal of hydraulic filling, but after a few months of exposition on the tailing surface, the tested geotextile showed higher permeability than had surrounding ash even when the pores were partly filled by fine ash. With time the permeability was increased nearly to the values for clean geotextile but only when water had the chance to wash out the particles connected to the one side. For the opposite direction when water had the chance to plug down more particles into geotextile pores the permeability decreased with time but still was the same or higher order than the surrounding ash – Vaníček and Kudrnáčová (1987). After that it was recommended to leave the geotextile on the place. After 2 years a 5 m deep pit was excavated (underground water table was lowered) and some samples collected. Again it was possible to see a great heterogeneity in the pit walls. Permeability of the undisturbed sample with geotextile inside had the similar permeability as a surrounding ash – above or below

Fig. 8.35 Spray irrigation of tailings surface to prevent wind erosion

the geotextile layer. More differences were obtained for the samples collected from the fine or coarse layers. Tests proved that this anti-dust geotextile has no special negative effect on hydraulic conditions inside the tailings dam and can be left on the place.

Deformation

Generally the problem of deformation is not as sensitive as the above mentioned examples. Nevertheless the settlement calculation should be performed in the cases where justifiable suspects are from a negative impact of deformation on the drainage system or on the pipelines going on tailings dam surface, because after the failure they can cause internal and surface erosion. A more cautious approach to this limit state is in the areas with natural seismicity, as deposited materials can be liquefied.

8.2.2.3 Ageing: Problems with Old Tailings Dams

First tailings dams storing flying ash started to operate in the Czech Republic at the end of the fifties so most of them are 40–50 years old. Main electric units started to use new methods of the desulphurisation process during the last decade so that the hydraulic way of storing was significantly reduced. Nevertheless most of these tailings dams are still in partial operation and are potential threat to their environment, Vaníček (2002a).

One of the most important construction elements of the tailing dams is the drainage system very often in the connection with the filters. Both have a strong influence on pore pressure control and therefore on slope stability as well as on filtration stability. The operating life of these elements is limited independently of type of material – soils, geotextiles, concrete. It is mostly due to clogging of filters and drains, degradation of concrete etc.

Although after ending with hydraulic sluicing into the tailings pond space we can expect an improvement from a filtration point of view, it is necessary to give careful attention to the monitoring system, which can in time (but not in all cases) inform us about worsening of the situation and about the urgency to initiate some protection measures. A selected practical example describes the problems, which started after the failure of the concrete drainage gallery in a 40 years old tailings dam.

Tailings dam Debrné was and still is used for deposition of combustion residues from the burning of coal in electric power and heat stations. Ash was transported by hydraulic method to the distance of 2 km. The construction of the tailings dam Debrné started at the end of fifties (Vaníček 1998). The main earth-fill dam with a height of 25 m was constructed across the valley. Small brook Debrnka was situated into concrete gallery at the bottom of the valley. But very early this brook was diverted out of the area of tailings dam and this gallery was connected with sideway drainage channels, which collected water from the tailings dam surface, Fig. 8.36. Each sideway drainage channel was a double conduit from reinforced concrete, which was gradually, according to the water table in tailing pond, covered by prefabricated concrete slabs. But with time even this dewatering system was closed

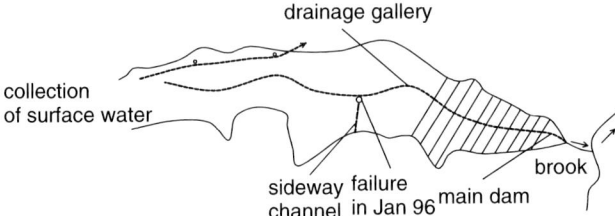

Fig. 8.36 General plan view of tailings dam Debrné

and the other system – on the surface – was used. So finally this concrete gallery was used only as a drainage system. Infiltration used first of all the existing seepage pathways from all sideway drainage channels and additionally from new drainage holes, which were drilled in the main drainage gallery. These small drainage holes were described during the quality control of concrete of this main gallery in 1965. Later, roughly in 1980, two concrete plugs were constructed in the lower part of the concrete drainage gallery with steel tubes passing through with the diameter of 200 mm. After filling the space behind the main dam I/1, the additional earth-fill dam with a height of 3 m was constructed. Up to now 13 additional dams I/2–I/14 have been constructed so that the total height of the tailings dam is now 64 m. From the cross section – Fig. 8.37 – can be seen that the general inclination was steeper for the first steps. General inclination was lowered after that and some additional material was used at the bottom of the main dam. At the same time the concrete drainage gallery was prolonged.

All the changes and improvements after 1960 were made with the main aim – to improve the dam stability and the impact on the environment according to the new findings from our own domestic experiences and also from the international ones e.g. Ist Int. Tailing Symposium, Tuscon, Arizona, USA 1972, XIIth Int. Congress on Large Dams, Mexico City, Mexico, 1976. Before the first problems in 1996 all drainage systems collected roughly $30 \, l.s^{-1}$ from which one third was collected by the concrete drainage gallery, roughly $6 \, l.s^{-1}$ by the system installed close to dam I/5 based on a siphon tube and the rest by the drainage system at the bottom of the main dam.

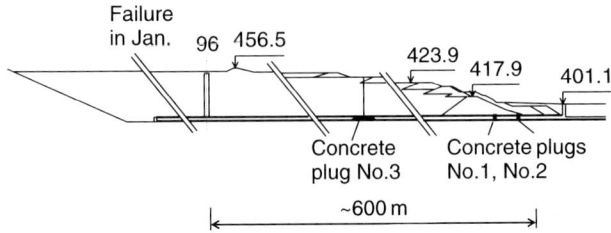

Fig. 8.37 Tailings dam Debrné – cross section

8.2 Tailings Dams

On the day 18.1.1996 suddenly not only an increased outlet of water from drainage system was observed (roughly $42\,l.s^{-1}$), but primarily a significant increase of flying ash in the water. The amount of insoluble particles went up from nearly zero to the orders of $10^2-10^3\,mg.l^{-1}$. Immediately some provisional countermeasures were taken to protect nearby river from pollution. After 2 days a big crater developed on the surface of the tailings dam, roughly 600 m from the downstream end of the concrete drainage gallery. The diameter of the crater was 20 m and the depth ca 12 m.

Water outflow went up to the values $\times \cdot 10^2\,l.s^{-1}(\times \cdot 10^{-1}\,m^3.s^{-1})$ and the amount of ash particles reached extreme values. After a detailed survey it was concluded that the crater was above the place where the sideway drainage channel is attached to the concrete drainage gallery. The positioning of this attachment or the failure of the upper prefabricated slab of sideway channel was marked as the main reason for ash outflow.

Remediation Measures

To prevent further pollution of the nearby river it was decided to close the tube in the first plug (No. 1) and used there a controlled stopper (piston valve) with dia 150 mm. A water pressure transducer was installed in front of this plug at the same time. After closing the piston valve water pressure went up very quickly. Reaching maximum value 110 kPa, this pressure was stabilized around the value 56 kPa. Total outlet of drainage water from the other drainage systems stabilized around $28\,l.s^{-1}$ ($2.8.10^{-2}\,m^3.s^{-1}$) and concentration of insoluble particles drop below $300\,mg.l^{-1}$ ($300\,g.m^{-3}$). But with time the colour of this water changed to a reddish one indicating that also red soil from the main dam was being eroded.

At the same time drilling works started to reach the concrete drainage gallery below the point of failure to stop more outflow of fine particles from this place. Drilling and also other remedies were complicated by very bad climatic conditions, with heavy snowfall and temperature minus 15°C. The first well was successful; the roof of the drainage gallery was reached. Before perforation the contact area was injected. After perforation the well was observed by a TV camera. The lower part of the drainage gallery was filled by fine ash, the upper part by water. The water was under pressure, no significant flow was observed. The second well did not reach the gallery and was used as a monitoring well. Therefore the first well was used for plugging – plug No. 3. Concrete was transported to the drainage gallery and subsequently this part was injected.

The water table was observed in monitoring wells for all this period. Most of them were drilled during the tailings dam construction. This monitoring system was only partially supplemented by additional wells. Already a long time before the failure the limit values of the water table in these monitoring wells were defined according to the slope stability calculation. During remedial work limit values were reached in none of these monitoring wells. Nevertheless a small increase in the water table was observed. With time the amount of insoluble particles was decreased and was stabilized around value ca $100\,mg.l^{-1}$ ($100\,g.m^{-3}$).

After finishing the plug No. 3 it was possible to open piston valve on the plug No. 1. Very quickly outflow was stabilized on the value $8 \,l.s^{-1}$ ($0.008 \,m^3.s^{-1}$) and concentration of insoluble particles reached $50 \,mg.l^{-1}$ l ($50 \,g.m^{-3}$). But later on also here water started to become a little bit reddish, indicating on one side that the drainage function was operating again but on the other side indicating continuation of internal erosion around the drainage gallery. With the help of some archive material and after data evaluation from a video camera, which was inserted into the drainage system below the main dam, the right lateral wall of the drainage gallery was perforated in two places and the cavern with flowing muddy reddish water was confirmed there. Also geophysical investigation methods were used for cavern verification. Grain size distribution of insoluble particles was obtained on the samples collected from this muddy reddish water. After that a grain size distribution for grain filters were proposed to achieve filtration stability in this system. But with respect to the relative steady state (concentration of insoluble particles less than $50 \,mg.l^{-1}$–$50 \,g.m^{-3}$) remedial work did not continue quickly enough in this direction.

The whole situation deteriorated in autumn 1996, when one after the other five craters were registered on the surface of the tailing dam – all of them were situated above the concrete drainage gallery and also above the plan of the main dam, (see Fig. 8.38.) The biggest crater had dia of 6 m and depth 2.5 m. These craters proved that the roofs of the caverns have fallen down and their space was filled by softened soil. The situation was evaluated as critical primarily from filtration stability point of view and therefore the following measures were recommended:

- Implementation of four pumping wells from the level of additional dam I/4 to decrease water table in the tailing dam body – to reach lower hydraulic gradients in area where part of water from tailing dam is flowing into drainage system and so decrease the amount of wash out particles.
- Implementations of drillings above the plug No. 2, to check the conditions in the drainage gallery and after that to start with new additional plug No. 4 – which can permit the removal of old plugs No. 1 and No. 2 and after that to restore the drainage function of the gallery in this part. Place of this new plug No. 4 was selected below the place where the main drainage concrete gallery is connected to the other sideway drainage channel, where the probability of failure is significantly higher.

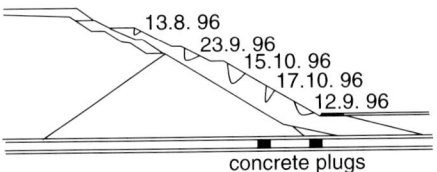

Fig. 8.38 Five craters on the surface of the main dam

- Partially to implement additional loading of the main dam surface at the toe of the slope by coarse material for slope stability improvement.
- Extension of the monitoring system – especially for monitoring of the character of insoluble particles in the drainage system below main dam – the critical case was defined when ash particles could be observed, indicating the spread of caverns up to the area of deposited ash.
- Greatest attention was devoted to the implementation of the plug No. 4. But before its implementation a higher concentration of ash particles was observed on the outlet from the plug. No. 1. Shortly before that the outlet dropped down to nearly zero value indicating that the tube with dia of 200 mm in the plug No. 2 was closed by ash particles. But due to the water pressure increase above this filling a function of tube was restored again with water outflow increase together with increase of insoluble particles in this water. Based on calculations it was concluded that the total amount of ash particles is higher than should be in the drainage gallery from the failure from January 1996. Probably the plug No. 3 was not perfect or there existed another place of smaller inflow. Implementation of the plug No. 4 was similar to the plug No. 3 only greater care was devoted to the cleaning of the bottom of the drainage gallery. After the control that there was no pressure water above the plug No. 2 the destruction of plugs No. 1 and No. 2 started. From that practical finding it is useful to mention that the ash, which reached plugs No. 1 and No. 2 was very fine due to sedimentation of coarser particles in the upper part of the drainage gallery. This very fine ash embodied some cohesion and also high resistance to water erosion. Plug No. 4 was finished according to the Fig. 8.39. In the roof of the gallery, especially in one place, there were cracks, concrete was partly degraded and in some places steel reinforcement was also visible. Therefore the state of this gallery was compared with the state from the year 1965 and it was concluded that cracks in the roof were wider; their width was 1–4 mm in 1965 and 10–15 mm in 1997. During controlled measurements no additional cracks or openings were observed for a period of 2 months. After that remedial work on this drainage gallery started to improve the overall stability and also to restore its drainage function. For the drainage hole it was recommended to apply a filter, which can be replaced. In this final stage main attention was concentrated on the monitoring system:
- Monitoring of water table in the tailings dam body,

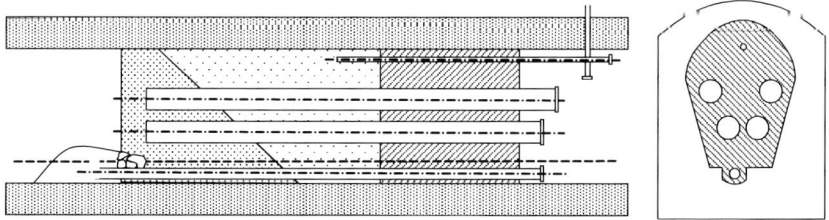

Fig. 8.39 Plug No. 4 – outlet through filter

- Monitoring of the amount of water flowing out of tailings dam body together with concentration and character of insoluble particles,
- Monitoring of deformation on the surface of the tailing dam, where five craters in the autumn of 1996 were observed.

Conclusion

The most sensitive element of tailings dams is the drainage system because its operating life can be a limiting factor in the lifetime of the tailings dams. The experiences from the tailings dam Debrné indicate that the collapse of the drainage system can be relatively quick and following remedial work very complicated, expensive and time consuming. In spite of this after 2 years the overall situation was under control and the current monitoring indicates that the water table is going slowly but steadily down with a positive impact on the filtration stability. The amount of insoluble particles in the drainage water is lower than the limit value is.

8.2.2.4 Sediments Resulting from Burning of Solid Fuel After the Application of New Methods of Desulphurization

During the last decade most of the electric power stations in the Czech Republic installed new desulphurization systems. The most often used systems are:

- Wet lime washing method – which is preferred for bigger electric units. There combustion gas is mixed with water and a mixture of fine milled limestone,
- Fluid burning – which is preferred for smaller units. There fine milled coal is mixed with fine milled limestone and this mixture is burned together.

Remaining parts after fluid burning contain relatively high percentage of free CaO. The resulting product has a greater potential to hardening and can be used directly in some civil engineering application, and is partly stored on the surface of existing tailings dams, creating a capping system.

Remaining parts coming from wet lime washing method contain the slag, ash, energygypsum and waste water. The amount of particular remaining components and the ratio of them are determined by the size and the type of boiler. These amounts and the ratios of particular components may, however, be substantially changed if some particular component is either fully or partially utilised as a secondary raw material. First of all the energygypsum may be used for manufacturing of plaster-board construction units. The mixture of the remaining parts is often called deponate. The main aim is to solidify waste water and create material the leachate of which is friendlier to the environment than the leachate from individual parts. Deponate from electric power stations situated in the brown coal basin is used as material with which the new regular morphology of terrain after coal mining activities is created. The mixture of the remaining parts with additional additives as cement or lime can be used in civil engineering applications as for earth structures of transport engineering, as a part of capping systems of old landfills and so on. The amount of stored materials is after that strongly reduced and dry way of deposition is preferred.

8.2.3 Tailing Dams with Sediments from Concentrator Factories, Uranium Tailings

From the view of limit states of earth structures there are very close similarities as for ash deposits. Only as was mentioned before, the general slope inclination can be steeper due to higher density of deposited material, which is therefore also a little bit less susceptible to erosion. But on the other side, due to some chemicals used in the concentration factories, the impact on the environment is more dangerous.

Because during the construction of the old tailings dams the care about environment was on much lower level than today, most of the old tailings dams with sediments from concentrator factories are not fulfilling up to day demands. The potential risk of failure is therefore very sensitive, amplified by the fact that these tailings dams are mostly situated in mountain areas, so it means that failure is affecting water streams and settlements in valleys below these dams.

Troncoso (1998) gives an overview on how these problems can be solved. In the first place he recommends reprocessing of tailings deposits. He states that the reason to reprocess old deposits may be: safety, to prevent failures, breaching and flooding, as consequences of flow failures or overtopping; economy: to obtain commercial benefits from old residues rich in valuable metals; environment protection: to eliminate sources of acid generation, groundwater contamination or aerial pollution.

Nevertheless many new tailings dams are constructed all over the world due to increasing demand for raw materials. Variation in the types of tailings by different industries and variety of site conditions makes every tailing dam an individual. Tailings dams so represent significant earth structures, some are higher than 150 m, a storage volume of some impoundments exceed 100×10^6 m^3. Therefore ICOLD is giving a great attention to the tailings dams, publishing special Bulletins – ICOLD (1994), (1996), (2001) as well as ISSMGE on different conferences, e.g. Wolski (2003), Rodriguez-Ortiz (2003), Masarovičová et al. (2003), Troncoso (1996), Blight (1994) and Blight et al. (2000).

But in some countries with small mines, as is also the case of the Czech Republic, the mining is less effective and therefore most of them are closed now. Therefore only two additional examples will be presented.

First one represents a special case of surface erosion. During extremely low temperatures the hydraulically transported tailings in Zlaté Hory (Golden Mountains) created an ice barrier on beachside. When high of this ice barrier exceeded the attitude of additional dam, transported fill started to overflow this additional dam and created there an erosion gash, Fig. 8.40. Very quickly this problem was registered, and hydraulic filling stopped. The failure had therefore only confined extent.

Fig. 8.40 Ditch created by surface water erosion at Zlaté Hory tailings dam. (**a**) general view. (**b**) detail of gash

The second example is typical for countries where uranium ore are produced. This example will be described in more details because it is bringing some specific demands, Kolářová and Vaníček (2006).

8.2.3.1 Uranium Ore Tailings

Extensive investigation and mining activities for the uranium ore started in the Czech Republic only at the end of the Second World War, firstly for military and later for energy purposes. In the new political and economic situation of the eighties and nineties the curtailment of these activities began and most of the mines and concentration plants (mills) were closed. The paper concentrates on the uranium mine tailings from the uranium mill MAPE Mydlovary, (the other ones are Stráž pod Ralskem and Dolní Rožínka, which is still active), especially on the capping system, which have to fulfill such different demands as low permeability, surface drainage, shielding gamma radiation and aerial emission of radon.

The construction of the uranium mill in Mydlovary, situated in the South of Bohemia, Fig. 8.41 started in 1959 and this uranium mill operated from 1962 up to 1991. The capacity was increased step by step from the 300 000 t per year up to 600 000 t per year. Between the years 1962 and 1991 nearly 17 mil.tons of the uranium ore with average uranium content 0.184% was milled and processed there and the tailings after that were deposited by the hydraulic method to the tailing ponds. Roughly 36 mil.tons of the tailings are stored there still with some content of uranium (2 320 tons) with radioactivity E14 Bq Ra 226. Two different methods of processing were used there – by acid leaching (12.8 Mt), or by alkaline method

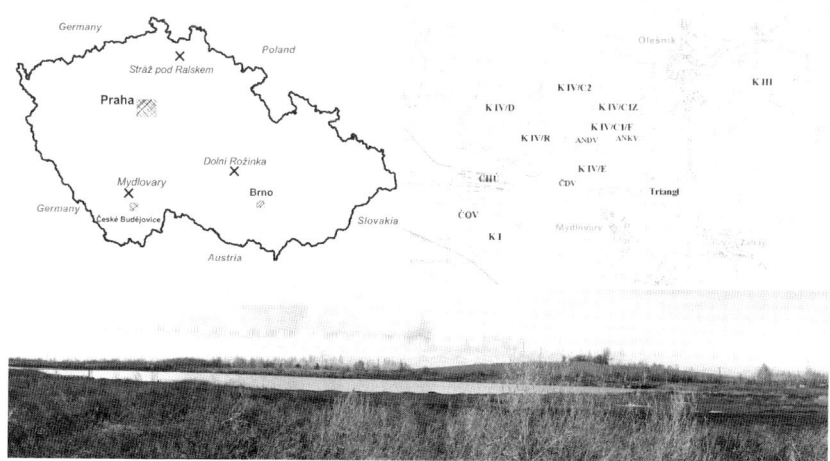

Fig. 8.41 Location of the concentration plant MAPE Mydlovary on the map of the Czech Republic and detailed placing of uranium tailing dams (KI, KIII and different KIV), concentration plant (CHÚ), sewage plant (ČOV), mine water plant (ČDV), and tailing dam for ash from the power plant Mydlovary (Triangl)

(3.9 Mt). Tailings were stored into eight different tailing ponds, the first one is the classical tailing dam (KI), constructed on the surface, the rest were deposited into depressions after lignite surface mining. Total inner area is roughly 23 km², with the infiltration area about 28 km².

Remediation activities started at the end of the eighties. From the beginning some theoretical demands from the environment protection point of view were defined and after that the practical application started, which had mostly an experimental character. The first tailing pond is partly remediated, but during the application of the protection methods some problems appeared such as:

- Slope instability, especially on places, where the periphery of the tailing dam is very close to another pit after surface mining,
- Leachate spreading was recorded as a result of not so good a sealing system based mostly on the natural geological barrier,
- High settlement of the capping system.

For other tailing ponds the two initial points are not so important, due to the fact that the deposited material is situated in old excavated pits of lignite mines with relatively good clayey subsoil. Therefore the main focus is now on the capping system. The design should eliminate as much as possible the differential settlement, to prevent tensile cracks of the sealing clay liner and to guarantee surface drainage. The remedial works are moving so far relatively slowly. In principle there are two reasons. Firstly, water, which now covers the surface of tailing ponds, creates a sufficient protection barrier. Secondly, the total remedy effort requires a huge amount of materials. If the capping system is composed on average from 1.5 m of material, then the total volume of the material for the remediation is close to 40 mil.m³. Technically feasible is the transfer of the deposited tailings from the smaller, and initially, from shallower tailing ponds (as e.g. K IV/D) to the bigger ones. In this way it is possible to reduce the amount of the material for the capping system and to remedy excavated places very promptly. Against this solution there are two concerns. Firstly, the environmental one, because re-excavation and transport can bring new problems with the evolution of radio nuclides into the environment and can also lead to the contamination of the traffic way. The second concern is a financial one because the price of this solution is higher.

8.2.3.2 Basic Requirements for the Capping System

The main aim of this area remedy effort is to find an optimal solution for the capping system which is able to fulfil all demanded technical criteria with minimum claims on the protection material. The demands on the capping system can be as follows.

- Shielding of the gamma radiation.
- Restriction of the aerial emission of radon – Rn.
- Prevention of dust spreading with different radio nuclides as the result of wind surface erosion.

- Limitation of rain water infiltration.
- Securing of the surface drainage of rain water.

Shielding of Gamma Radiation

One of the basic conditions is the reduction of gamma radiation on the surface of the remediated area. According to the Czech codes the recommended maximum value is 0.1 µSv per hour plus natural background, altogether maximally 0.4 µSv per hour. Another possibility is to restrict access to this area. The estimation of the thickness of the shielding layer for deposited material after uranium ore mining or milling and processing is very often close to 1 m, with a maximum up to 2.5 m. Proposed thickness has to be checked by equation for the radiation decay during passage through the capping system.

Restriction of the Aerial Emission of Radon – Rn

The aerial emission of radon from remediated areas is directed by the content of Ra^{226} in deposited material. Only a limited part of radon can reach the atmosphere. How big it will be depends on the values of the emanation coefficient and permeability for gases. Therefore the capping system has to be composed from material with low permeability for gases. According to international standards the accepted level of aerial emission should be lower than $0.8\,Bq.m^{-2}.s^{-1}$. From this point of view the most problematic tailing pond is K IV/R where the aerial emission of radon directly on the surface of deposited tailings is between 20 and $24\,Bq.m^{-2}.s^{-1}$ above the natural background. For the rest of the tailing ponds it is much lower, in the range of $5-9\,Bq.m^{-2}.s^{-1}$. The basic equation for the radon emission on the surface of the deposited material is:

$$F_t = A \cdot E \cdot \rho \cdot (\lambda \cdot D_t)^{1/2} \quad [Bq.\,m^{-2}.s^{-1}] \quad (8.9)$$

where

A – activity of Ra^{226} $(Bq.kg^{-1})$
E – emanation coefficient
ρ – volume density of the deposited material $(kg.m^{-3})$
λ – decay constant Rn $(2.1 \times 10^{-6}.s^{-1})$
D_t – diffusive coefficient of the deposited material Rn $(m^2.s^{-1})$

For a multilayer-capping system the equation is more complicated, as it is necessary to know the diffusive coefficient of each layer.

Prevention of Dust Spreading with Different Radio Nuclides as the Result of Wind Surface Erosion

This condition can be fulfilled relatively easily by creation of a re-vegetation surface, composed for example from 0.1 to 0.2 m thick biologically active layer.

Limitation of Rain Water Infiltration: Securing of Surface Drainage for Rain Water

Both these conditions are very similar to the conditions for classical municipal waste landfills. On the one hand it is necessary to construct a special sealing layer, usually under the condition that the coefficient of permeability is equal or lower than 1.10^{-9} m.s^{-1} and on the other one to guarantee that the dominant amount of rain water will be drained from the surface or via a drainage layer or will be evaporated from the remediated surface. To guarantee this condition the inclination of the remediated surface should be at least 1%. So it means that for the tailing ponds with relatively large areas the thickness will be strongly affected by this condition especially in the middle of the tailing pond. A minimum condition for the thickness along the periphery of the tailing pond will be in most cases directed by aerial emission of Rn.

Basic Composition of the Capping System

For fulfilment of the above mentioned conditions the basic composition of the capping system should consist of, Vaníček (2002d), Gens (2003), Fig. 8.42:

- Deposited tailings,
- Sub base layer, minimum thickness of which around the tailing pond should fulfill the requirements for protection against aerial emission of Rn, and increasing in such a way that even after the settlement (with respect to the different thickness of deposited tailings and with the different thickness of the capping system eliciting surcharge loading above layer of tailings) will guarantee the required inclination of the surface of the capping system.
- Sealing layer,

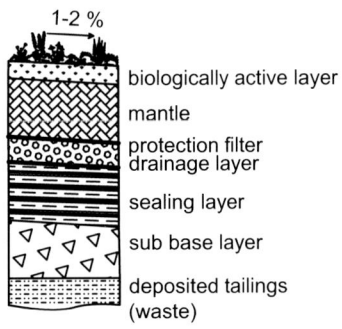

Fig. 8.42 Basic composition of the capping system

- Drainage layer, Collected water will not be affected by deposited tailings and can be drained into the water course to restore the original hydro geological system,
- Protection filter,
- Mantle, partly biologically active, appropriate for plant development,
- Final biologically active layer – humus, top soil.

The sub base layer, which can consist of unsorted soil, probably will reach the largest volume. Total volume will depend on the recommended arrangement of the surface. Here we can speak about roof-shaped dewatering or about valley dewatering, Fig. 8.43. Preferably this layer should be permeable to fulfil the condition that the deposited and saturated tailings should be drained during consolidation even if from the upper part. Around the tailing pond the collection of water from the process of consolidation has to be arranged and collected water treated in the sewage disposal plant. In our case we suppose that the sub base layer will be deposited directly on the surface of the uranium tailings. In Germany, the strengthening of this contact by geosynthetics is proposed to eliminate low bearing capacity of the deposited uranium tailings, Leupold et al. (2003).

Under the effort to optimize the total thickness of the capping system not only from the point of view of protection against aerial emission of radon, but also to minimize the surcharge loading to decrease the total settlement the calculation of radon emission was done for two cross sections. The first cross section uses the compacted clay liner and the second one uses the geosynthetic clay liner as a main component to fulfil the condition of the permeability of the sealing layer.

Cross Section Using Compacted Clay Liner and Grain Filter

The first case represents the classical case of using natural materials for both the sealing and the drainage layers. The thickness of clay liner is proposed 0.6 m, compacted in three layers each having 0.2 m. The thickness of the other layers is as follows – sub base 0.3 m, drainage layer 0.3 m, mantle 0.5 m and top soil 0.15 m – see Fig. 8.44b and Table 8.2. In this table the radon emission is calculated for the extreme case that the radon emission on the top of the deposited uranium

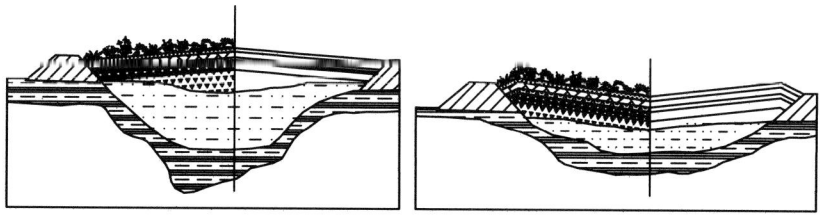

Fig. 8.43 Recommended arrangement of the surface, roof – shaped dewatering and valley dewatering, not in scale

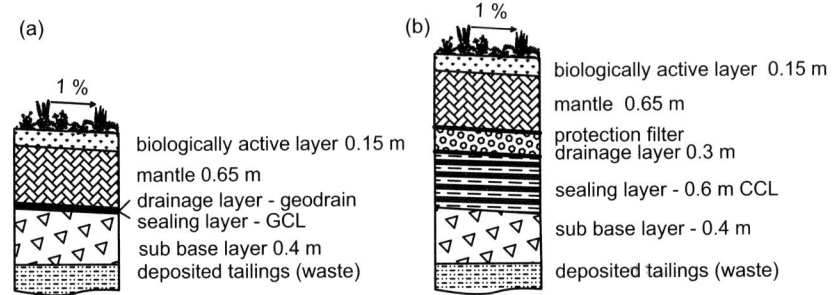

Fig. 8.44 Two compositions of the capping system. (**a**) with geosynthetic clay liner GCL and geodrains. (**b**) with compacted clay liner and natural drain

tailings is $24\,\text{Bq.m}^{-2}.\text{s}^{-1}$. Calculated radon emission on the remediated surface is $0.11\,\text{Bq.m}^{-2}.\text{s}^{-1}$ fulfilling the demanded value.

Cross Section Using Geosynthetic Clay Liner and Geodrain

The second case Table 8.3 and Fig. 8.44a, represents utilisation of geosynthetics materials. For sealing layer geosynthetic clay liner GCL – known also as bentonite mattresses – is proposed and for drainage layer geodrain – e.g. type Colbonddrain. The thickness of each layer is roughly 10 mm. The thickness of the sub base layer (0.7 m) and mantle (0.65 m) is bigger than in the previous case. The final value of radon emission is also lower than the required value.

For the other tailing dams with radon emission on the surface $9\,\text{Bq.m}^{-2}.\text{s}^{-1}$ the final emission on the surface of the same cross section is lower than $0.04\,\text{Bq.m}^{-2}.\text{s}^{-1}$. For this case the following composition is sufficient to protect an environment against radon emission: sub base layer 0.4 m, GCL 0.005 m, geodrain 0.01 m, mantle 0.65 m and top soil 0.15 m. To summarize the differences between calculated two cross sections we can state:

– Total thickness of the capping system can be decreased because roughly 20 mm of geosynthetics can substitute 0.9 m of the classical sealing and draining layers

Table 8.2 Calculation of radon emission on the top of capping system with CCL

Layer	Depth of layer [m]	Radon emission on the surface of layer [$\text{Bq.m}^{-2}.\text{s}^{-1}$]
Deposited tailings – initial level		24
Sub base layer	0.3	5.17
Compacted clay liner CCL	0.6	0.24
Drainage layer	0.3	0.21
Mantle	0.5	0.13
Top soil	0.15	0.11

Table 8.3 Calculation of radon emission on the top of capping system with GCL

Layer	Depth of layer [m]	Radon emission on the surface of layer [$Bq.m^{-2}.s^{-1}$]
Deposited tailings – initial level		24
Sub base layer	0.7	1.37
GCL – geosynthetic clay liner/bentonite mattresses (NaBento)	0.009	1.18
Drainage layer – geodrain	0.01	1.18
Mantle	0.65	0.59
Top soil	0.15	0.56

(but not fully to guarantee the required values for gamma radiation – minimum 1.0 m – and for radon emission) and so to decrease the total settlement of the deposited tailings.
- Strict requirements for sealing and drainage layers are substituted by unsorted soil, which is cheaper and can be found in near surroundings.
- The geosynthetic clay liner is more flexible than the classical compacted clay where tensile cracks as the result of differential settlement can be a significant problem – Vaníček (2002b).

Field Investigation and Laboratory Tests

To be able to predict the total settlement and consolidation in time t, the first phase of site investigation started. For the first phase of the investigation the light penetrometer with height of fall 0.5 m, drive head diameter 25.2 mm and falling weight 10 kg was used. The reason was a limited access on the tailing ponds. The tests were performed only very closely to the embankment, because the rest of the pond was flooded by water. Nevertheless this penetrometer device proved to be a very useful tool for this type of material, the resistance was not so high to protect penetration of the tip even for the coarser material which is usually around periphery of the pond. In the centre of the pond the deposited material is generally finer. Therefore the results are mostly on the optimistic side. Briefly summarized, this penetrometer, which is easily mobile, can recognise the total depth of deposited material and give us the first information about anisotropy, change of properties as a function of position and depth and also values of modulus of deformation as a function of observed profile. Grain size distribution tests were performed on samples collected for oedometric tests; results are showing that the material has a character of fine sand. As mentioned above, the grain size composition in the centre of the tailing pond is much finer in most cases; it contains higher percentage of particles lower than 0.06 mm.

Note: Newson and Fahey (2001) describe a modified small hovercraft, fitted with equipment to enable shear vane tests, piston samples, BAT water samples, CPT and spherical penetrometer tests to be conducted in locations that would otherwise be inaccessible.

For the calculation of total settlement the deformation characteristics have to be determined and the following methods were used for a rough estimation:

- With the help of classical oedometric test on samples with dia 120 mm which were collected close to the place where penetrometer tests were performed. For a lower range of stresses the oedometric modulus was lower than 10 MPa, for a higher range of stresses between 10 and 20 MPa.
- From the penetrometer test the result of which is shown on Fig. 8.45. It is evident that some layers are very soft; the modulus of deformation is close or even less than 1 MPa, even at the place situated very close to the pond periphery.
- From the back analysis. For the first tailing pond K I, covered by protection layers with thickness roughly 2 m, the maximum settlement measured on the surface was up to 750 mm after 3.5 years. From the consolidation curve it is obvious that the further settlement can be expected. For the maximum depth of the deposited material 23 m there, the expected oedometric modulus of deformation E_{oed} (for 1D) is close to 1 MPa.

8.2.3.3 Estimation of the Total Settlement

The first estimation was calculated for maximum depth (10 m) of the deposited uranium tailings in the middle of the tailing pond and for minimum one (1 m)

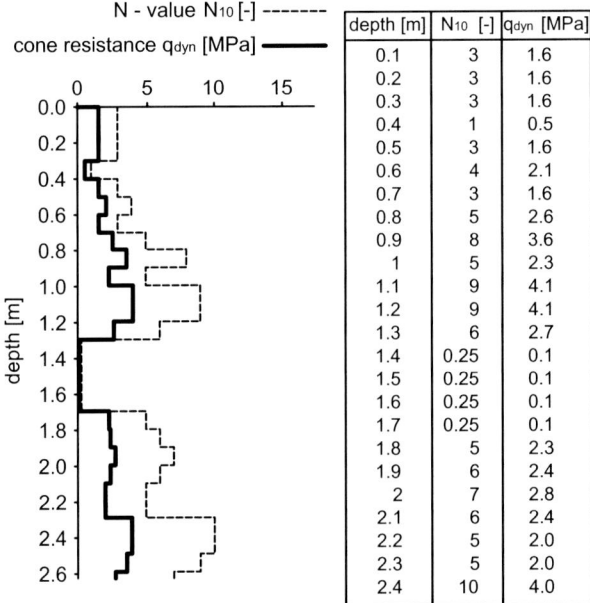

depth [m]	N_{10} [-]	q_{dyn} [MPa]
0.1	3	1.6
0.2	3	1.6
0.3	3	1.6
0.4	1	0.5
0.5	3	1.6
0.6	4	2.1
0.7	3	1.6
0.8	5	2.6
0.9	8	3.6
1	5	2.3
1.1	9	4.1
1.2	9	4.1
1.3	6	2.7
1.4	0.25	0.1
1.5	0.25	0.1
1.6	0.25	0.1
1.7	0.25	0.1
1.8	5	2.3
1.9	6	2.4
2	7	2.8
2.1	6	2.4
2.2	5	2.0
2.3	5	2.0
2.4	10	4.0

Fig. 8.45 Mydlovary tailings dam – the result of penetration test

around periphery of the pond and for the pond width of 400 m. For the roof-shaped dewatering and demanded inclination of the surface only 1%, the thickness of the sub base layer will be changed from minimally assumed 0.4 up to 2.4 m only for the dewatering, plus additional part due to the calculated settlement. The minimum total thickness of the capping system was previously calculated for 1.2 m assuming geosynthetics in cross section and radon emission on the top of deposited uranium tailings 9 Bq.m^{-2}.s^{-1}. The thickness of deposited material 1.0 m is assumed to be around the periphery of the tailing ponds. For expected modulus of deformation $E_{oed} = 1$ MPa, calculated settlement of deposited material is roughly 24 mm there. In the centre of the pond the total thickness of the capping system (without additional settlement) is supposed to be 3.2 m and therefore the expected settlement of the deposit there is close to 640 mm. Therefore we must add another 640 mm to the sub base layer which will increase the total settlement to the value 728 mm. By this step by step method roughly 800 mm of additional sub base material have to be added in the centre of the tailing pond to eliminate the expected settlement and to guarantee the surface of capping system inclination on average 1%, see Fig. 8.46. The advantage is that the consolidation close to the periphery will be quicker due to more coarse (and permeable) character of deposited material, so the inclination of the surface during 1st years will be a little bit steeper than 1%.

The presented case of the remediation of tailings ponds with uranium tailings in Mydlovary shows that the solution is a very complicated problem combining classical geotechnical solution with environmental one.

8.2.4 Tailings Dams with Other Sediments

Wide variability of other sediments stored in tailings dams varies from relatively environmental friendly as paper mill sludge to very dangerous as some chemical sludges.

Paper mill sludges are composed of kaolinite clay, organic fibres and tissue, and water. Initial water content is very high and very often in the range of 100–200%. Therefore the tailings dams are very similar to the classical dams, only that the deposited sludge has relatively low permeability. Therefore the consolidation is going slowly compared to the other tailings dams. Moo-Young and Zimmie (1996) describe the paper mill sludges as sealing layer for landfill covers.

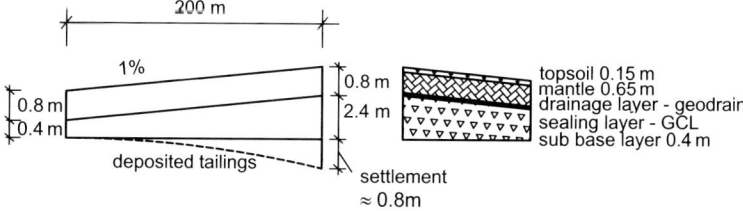

Fig. 8.46 Expected capping profile taking into account the settlement

Chemical sludges are often composed from semi-liquid material with high viscosity, where tar is a main compound. These sludges represent a potential risk to our environment and have to be remediated. Very often the remediation is done by off – site incineration method. But when encapsulation method is used the biggest problems are connected with capping system. Local bearing capacity is extremely low, but when plane load is applied the settlement is theoretically zero. This is due to the fact that undrained 1D compression is practically negligible. Practical application of plane loading is very complicated, material is covered by geomembrane and spilled step by step. Therefore there is a tendency to infill the sludge by coarser material, e.g. by slag, to create skeleton the pores of which are filled by sludge. But in this case the volume of deposited material which has to be protected by capping system is going up.

Morgenstern (1994) describes the problems of tailings dams composed from oil sands in Alberta, Canada. Above all he is emphasizing the application of observational method of design on this case. Details on the geotechnical aspects of oil sand recovery are presented by Morgenstern et al. (1988). Approximately 475×10^6 m^3 of sand, 400×10^6 of sludge and 50×10^6 m^3 of free water will require storage for several decades. To accommodate these volumes approximately 18 km of dyke ranging from 32 to 90 m in final height have been constructed. The compacted shell is constructed by utilizing hydraulic construction techniques employing dozer compaction. During the winter months when this is not feasible, the tailings stream is discharged upstream of the compacted shell to form a beach. The overall stability is influenced by highly plastic clay-shale which is beneath much of the dykes. Initial design – the inclination of downstream slopes 1:4 – was based on approximately average properties. The design was supported by large field instrumentation, above all by inclinometers and piezometers. When shear movements in subsoil were detected the dyke slope was changed to 1:8.5. Movements continued in the foundation with the addition of each 3 m lift of the dyke. Morgenstern emphasizes that the traditional limit equilibrium analysis proved powerless to deal with the assessment of these movements and the forecast of ultimate stability. Construction proceeded cautiously. Under the support of advanced stress analysis the dyke was completed on the basis of strain monitoring. In this way in excess of 0.4 m of slip was accommodated in the foundation of the dyke in a safe manner. Morgenstern resumes that there was always the additional option of more slopes flattening but at substantial cost.

8.3 Landfills

Deposition of waste on landfills is still in many countries the most often used manner of waste liquidation. Stiff municipal waste materials create the most often used waste stored in landfills (sanitary or municipal landfills). However even in countries where this manner of municipal waste liquidation is limited to zero, they have to construct landfills because even municipal waste which is burned in refuse incinerating plants creates ash (residues after burning this waste) which have to be stored somewhere. Landfills are also used for many other different waste materials starting with practically inert materials and ending with dangerous waste materials coming for example from chemical factories.

The principle of land filling consists of creating a protective envelope around deposited waste.

Schematic cross section is shown in Fig. 8.47.

Depositing of waste starts after the construction of bottom sealing system and subsequently is covered by capping sealing system. Overall design and performance have to guarantee safe protection of surrounding environments against potential contamination spreading from deposited material.

8.3.1 Landfill Classification

Landfill classification can be performed under the different views according to the type of deposited waste, landfill positioning, cross sections, according to volume of landfills and so on. Next only some basic aspects will be briefly specified.

8.3.1.1 Classification According to the Type of Deposited Waste

Potential risk of the deposited waste is the main view point of landfills classification. This potential risk is judged on the base of chemical composition of leachate. Leachate tests are defined by national standards as well as recommended division into different groups. In the simplest version the waste and thus type of landfills are divided into three categories:

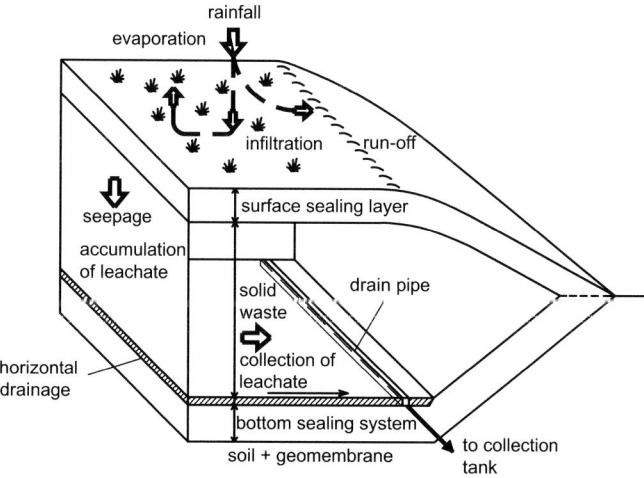

Fig. 8.47 Schematic cross section through landfill

- Inert waste – materials with very low potential risk fall into this category for example building waste, soil excavated during foundation of new structures and others,
- Stiff municipal waste – waste with no strictly defined content, potential risk to the environment is changing with time in dependence on degradation processes which proceeds here,
- Dangerous waste material – for example chemical waste material which can be on one side very dangerous, but on the other one it is very often waste with well defined properties (one-component) the protection of which can be better defined. This waste can be also deposited into the landfills space in containers and space between containers is filled by low permeable material with high sorption potential. It can be clay but also a mixture of soil, ash and bentonite with some other additives as e.g. zeolite.

The amount of stiff municipal waste produced by one inhabitant per day is different but more often in the range of 0.5 kg–2.0 kg. The final result is different for individual countries; in general wealthier countries produce more waste. Differences are also between big cities and small towns, villages, not only with respect to the amount but also to the composition of the municipal waste. The ratio between individual components is changing; basically waste is composed from paper, plastics, rest of foods, textile, glass, metals and garden garbage. However municipal waste can also contain different diluents, rest of medicaments, drugs, batteries and others with potentially a higher negative impact on environment. In individual countries extensive programs are elaborated for separate collection of waste as for paper, glass, plastics in the close vicinity to inhabitants, or drugs in chemist's shops (drugstores), batteries in salesrooms of electric products. The general policy is valid – to decrease the amount of deposited materials, to separate it, to reuse it, Vaníček (2001a).

Principles of the well governed municipal landfill consist of the following four steps.

i) Collection and transport of the municipal waste on selected place, which is technically arranged in such a way to be able to guarantee highest demands on the protection of underground water and environment generally against negative impact of the leachate and landfill gas,
ii) Collected waste is deposited in agreement with recommended technology of landfilling,
iii) Technical measures ensure the safe collection of by-products developing during the degradation of deposited waste – collection of leachate and of landfill gas,
iv) After the end of landfilling the landfill is overlaid, surface sealed and recultivated and long term monitoring is guaranteed.

8.3.1.2 Processes Proceeding in the Landfill of Municipal Waste

Physical, chemical and biological processes, which are changing the character of deposited waste, start just after the waste deposition, Vaníček and Schröfel (1991,

8.3 Landfills

2000). Initial physical changes are connected with waste porosity (macro porosity) decrease as a result of crushing, compaction and by additional loading from subsequent deposited layers. Deformation is time depending namely as in the form of primarily consolidation so in particularly in the form of secondary consolidation as a result of deformation on the contacts of individual stiff particles. Biological processes which reduce volume of waste also participate on deformation. All these forms of deformation can reduce the total height of deposited waste by 30%.

Water which is part of the deposited waste together with water which leak in create the environment in which the soluble substances are easily washed out and are farther floated to the bottom of landfill together with fine particles, which do not participate on degradation processes. The amount of leachate depends on the amount of water which leak in. Degradation processes in the landfill can be supported by artificial humidification, mostly by collected leachate. Chemical reactions going on in landfill include, but shall not be limited to, oxygen, organic acids, carbon dioxide, in some cases produced during degradation process. Carbon dioxide together with water dissolves calcium and magnesium and increases the turbidity of leachate.

Biological reactions in landfill are checking not only composition of landfill gas (bio-gas) but also composition of leachate. Generally speaking the landfill is passing through three stages of the degradation process with different bacteria which dominate to individual stages:

- First phase – aerobic phase – waste degrade at the present of oxygen together with carbon dioxide, water and nitrite; as the oxygen is consumed the second phase can start,
- Second phase – first period of anaerobic process – when anaerobic bacteria prevail, organic acids reduce pH to the of 4–5, importantly for dissolving of some inorganic materials – this phase is characterized by high conductivity and low production of methane,
- Third phase – second period of anaerobic process – when bacteria which produce methane prevail. These bacteria degrade volatile acids to methane and carbon dioxide, result of this degradation process is increase of pH value nearly to the neutral value. Degradation of inorganic salts is going down leading to the decrease of conductivity.

The amount of produced bio gas depends on the scale and velocity of degradation process. Theoretical maximal value gainable by total degradation is roughly $400 \, m^3 \cdot t^{-1}$. Energetic valuation of bio gas is a little bit lower than half valid for natural gas. Production of methane is decreasing as process of degradation of organic material progresses. However the production of methane can continue for some years, even decades. Velocity of degradation process strongly depends on moisture content, on density of deposited material. Moisture content is the main factor of bacteria survival, biological reaction is sharply going down when moisture content is lower than 55–60%.

Good compaction of deposited waste is one of the main factors of well governed municipal landfill. Under the good compaction the amount of stored waste in 1 cubic meter is increasing, reaching in average values as 600–800 kg.m^{-3}. Good compaction also supports biodegradation, because water is easily retained in. For the compaction heavy vibratory rollers are recommended, often called compactors, when roller has character of sheep foot roller, where prickles are about 0.2 m high. Good compaction immediately after deposition can also reduce windy erosion of the deposited material, primarily of plastics. At the end of each day, deposited waste is covered by inert material and this layer is compacted as well. Not only windy erosion is limited but also the access of birds to the deposited material.

In a short resume it can be stated that the production of leachate and bio gas represents the highest risk for the environment. Bio gas also contains a very small amount of gasses which smell badly. Methane with oxygen has high explosive potential.

Part of each well governed landfill is therefore collective tank, where leachate is collected, controlled and eventually directly cleaned. The amount of leachate is time dependant, during the landfilling time is much higher than after the closing of landfill.

Also landfill gas should be collected, not only due to the mentioned smell and risk to the explosion but also as a significant source of energy. During active landfill gas collection the collection system is constructed already during the waste deposition. After the surface closing the gas is sucked from this gas collection system and subsequently used.

8.3.1.3 Site Selection for Landfills

Selection of appropriate site for the landfills (municipal waste disposal) is rather a complicated process. This process is not only a technical problem, but also an ecological, economical and sociological one. Theoretically the proper process starts with the selection of wider territory, where the impact on the environment is acceptable. Protection of underground water, water resources, and healing springs is playing the most important role. But it is necessary to take into account also the other aspects as is the protection of flora and fauna; it is necessary exclude the unstable territory (slide prone area, area with solution cavities...), flooded area etc. To the last aspect it is possible to make a short affirmative note; during extreme floods in the Czech Republic in 2002, when a high percentage of the country was flooded, no one from the more than 150 operated landfills was in contact with free water.

After the selection of wider territory the selection continues taking into account all the above mentioned problems. Sociological views requiring the approval of habitants in the selected locality is sometimes very complicated. Very often the syndrome NYMBY – not in my back yard – is mentioned – we recognize the necessity to store waste somewhere, but not in our vicinity (not in my backyard).

8.3.1.4 Incorporation of Landfills into Surrounding Terrain

According to the incorporation of landfills into surrounding terrain the following types of landfills can be distinguished:

- Pile landfills, situated roughly on horizontal terrain; either

  a) Directly on the surface – Fig. 8.48a or
  b) Partly or fully embedded into natural terrain – Fig. 8.48b

- Inclined landfills, situated on slope and again with two alternatives – Fig. 8.48c, d.

Maximum underground water table should be at least 1 m below the bottom of landfill embedded into natural terrain.

8.3.2 Approaches to the Landfill Design

Fundamental approach to the landfill design abides by a mutual interrelationship between deposited waste, natural (geological) protection barrier in the landfill vicinity and by artificial – engineered barrier (demands on structural system of landfill bottom, side wall and on capping system, where the soil is playing a most important role). This mutual interrelationship can be expressed as follows, (Vaníček 1992).

Natural (geological) barrier × Artificial (engineered) barrier × Category of deposited waste.

If the total demanded protection towards a certain type of waste is expressed by defined vector, hereafter optimal design results from a condition that the extent of protection from natural barrier plus the extent of protection from artificial barrier is equal to this total demanded protection, Fig. 8.49.

It is obvious that the extent of the total demanded protection is changing with time – usually going up – and must take into account the possibilities of given society at given time – it means from the degree of the problem understanding and from degree of financial possibilities. Another view is also obvious, for more significant natural protection barrier, the demands for artificial one are lower. This

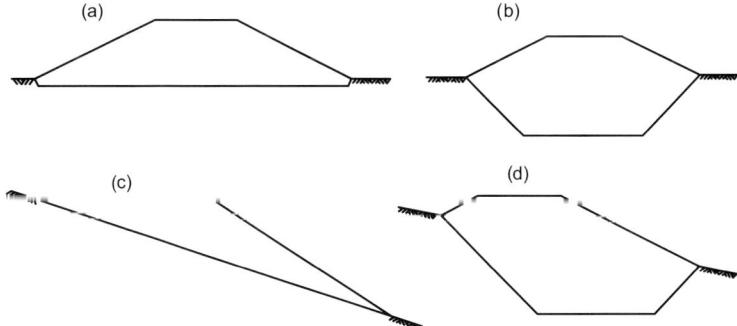

Fig. 8.48 Different cross sections of landfills for horizontal and inclined surface. (**a**) landfills situated on surface. (**b**) landfills partly embedded to the natural ground. (**c**) landfills situated on surface. (**d**) landfills partly embedded to the natural ground

Fig. 8.49 Principle of the optimum landfill design composed from natural (geological) barrier and from artificial (engineered) barrier

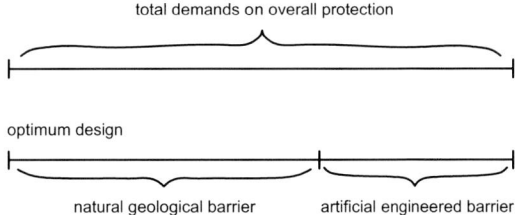

principle was accepted in standard of the Czech Republic, even if in reduced form. These standards are taking into account the thickness of geological barrier below the landfill and its coefficient of conductivity. But some other factors of this geological barrier such as sorption and diffusion coefficients, depth of underground water below the landfill and its direction and velocity, and others are influencing the quality of geological barrier.

In principle the total demand can be therefore defined by two different methods:

– By specification of the composition and properties of the artificial protection barrier,
– By specification of maximum allowed contaminant concentration when reaching the outer periphery of the artificial barrier of the landfill or when reaching recipient (e.g. underground water, water resources).

Here it is necessary to accept the fact that 100% protection against contaminant spreading is practically impossible. Always it is necessary to count with this fact, even if the degree of contamination is lower than the differentiation ability of its detection or lower than is the concentration of a certain chemical compounds in the natural backgrounds or than is the background radiation (for radioactive waste material).

For municipal landfills the first approach is preferred in most countries. The second approach is very close to the approach which was used when studying the contaminant spreading from the waste (e.g. ash) used in the earth structure of transport engineering. Hereafter both approaches will be specified in more detail.

When speaking about approaches to the landfill design there is also the discussion about different demands on the base and capping sealing systems. In general stricter demands are defined for base sealing system. However for the theoretical case of decrease or even loss of functionality of the drainage system carrying leachate out of landfill, thereafter for the higher hydraulic conductivity of the capping system the inflow can be higher than the outflow.

8.3.2.1 Demands on the Composition and Properties of the Engineered Sealing System

The fundamental types of sealing systems are:

- Compacted clay liner – CCL,
- Geomembrane liner – GL,
- Geosynthetics clay liner – GCL.

The sealing systems based on low hydraulic conductivity of stabilized ash or asphalt is usually non-standard solution (e.g. Scapozza and Amann 1998). Sarsby and Finch (1995) aside from stabilized ash are mentioning also paper sludge or this sludge with smaller amount of bentonite. Zimmie et al. (1995) states that even paper sludge without any additives, composed roughly from 50% of kaolinite and 50% of organic compound, can reach low permeability ($k \leq 1.10^{-9}$ m.s^{-1}). Also De Mello (1998) mention the application of cement-stabilized fly ash as sealing system. Low permeability of this sealing system during practical application is partly explained by the effect of capillary barrier.

Compacted Clay Liner – CCL

Minimum demands on the compacted clay liner are most often defined by minimum thickness of 0.6 m, compacted in three layers each 0.2 m and with coefficient of conductivity $k \leq 1.10^{-9}$ m.s^{-1}. As was said before it is possible according to the deposited waste to define stricter demands, e.g. minimum thickness of 1.0 m compacted in five layers with coefficient of conductivity $k \leq 1.10^{-10}$ m.s^{-1}.

Geomembrane Liner – GL

Different types of geomembranes were described in Chapter 3. For landfills most often the HDPE (high density PE) membranes are applied together with VLDPE (very low density PE) or with PVC ones. Last type is used only in some countries however only for capping system, because this type of membrane is not as chemically resistant as are membranes based on polyethylene (PE). Generally demanded thickness of geomembranes is in the range of 1.0–2.5 mm. Laid down geomembrane strips have to be welded and tested. For sealing of side walls the other properties are defined as demanded tensile strength, elongation at failure as well the specification if smooth or structural geomembrane is preferred.

Geosynthetic Clay Liners – GCL

Different types were also described in Chapter 3. For the application the following specification is described as permeability – coefficient of conductivity is usually 100× lower than is compacted clay liner, manner of laying, overlapping and binding of individual strips, tensile strength and elongation at failure.

Liner Composition

Elementary sealing system, composed by the above specified types, is applied only in the case of very good quality of natural protection barrier. For example the Czech standards require natural barrier composed from clay with coefficient of conductivity $k \leq 1.10^{-8} \text{m.s}^{-1}$ and with thickness at least 3 m, or $k \leq 1.10^{-7} \text{m.s}^{-1}$ and thickness at least 30m.

Therefore most often the recommended sealing system is defined as *combined sealing system* composed from compacted clay liner supplemented in most cases by geomembrane liner. The fundamental cases of combined construction systems for capping and bottom sealing are shown in Fig. 8.50. The comparison of the different combined systems used in different countries for surface and bottom liners are presented e.g. by Manassero et al. (1998).

In the exceptional cases doubled combined systems are proposed. Comparison of different systems from the view of the amount of leakage are presented e.g. by Giroud et al. (1994), see also Manassero et al. (1998) – Table 8.4. Significant influence of the combined system compared with the elementary system is easily visible. Instead of leakage roughly around 1m³ per hectare per day is leakage decreased to roughly 0.0001 m³ per hectare per day, it means 10 000 times.

But these values are valid for very good contact of geomembrane with compacted clay liner. A much worse situation is for bad contact Fig. 8.51. Therefore great care should be devoted to this contact, so called "waves" have to be eliminating as much as possible. Better contact can be reached for more flexible geomembrane

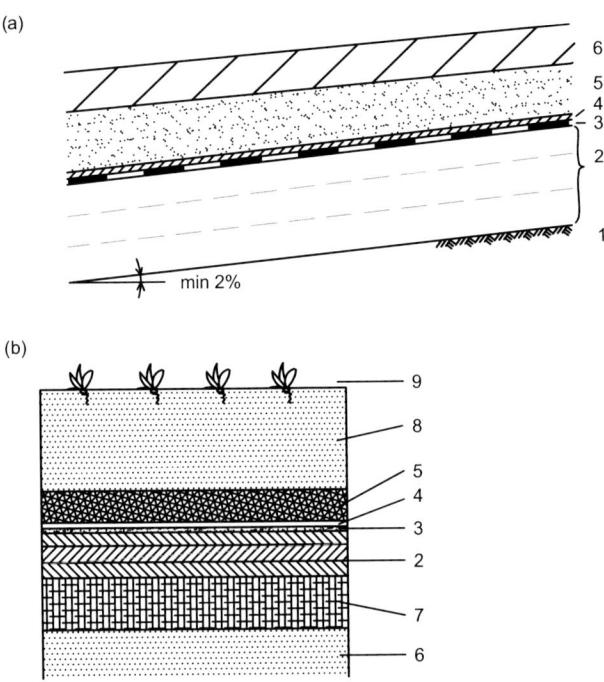

Fig. 8.50 Composite barriers. (**a**) bottom liner; 1 – compacted subsoil; 2 – clay liner (3 × 0.2 m); 3 – geomembrane; 4 – protection geotextile; 5 – granular drainage layer; 6 – deposited waste. (**b**) capping system; 7 – regulating or gas collection layer; 8 – topsoil; 9 – vegetation

8.3 Landfills

Table 8.4 Rate of leachate migration through various types of liners (according to Giroud et al.)

	Volume of seepage in [m³/ha/d]	
Compacted clay liner	-100 m³/ha/d	for $k \sim 10^{-7}$ m.s^{-1}
	-10 m³/ha/d	for $k \sim 10^{-8}$ m.s^{-1}
	-1 m³/ha/d	for $k \sim 10^{-9}$ m.s^{-1}
Geomembrane liner	-0.1 m³/ha/d	
Simple composite liner	-0.01 m³/ha/d	for clay liner $k \sim 10^{-7} - 10^{-8}$ m.s^{-1}
	-0.001 m³/ha/d	for clay liner $k \sim 10^{-9}$ m.s^{-1}
Double composite liner	-0.0001 m³ l/ha/d	for clay liner $k \sim 10^{-9}$ m.s^{-1}

as for VLDPE membrane. Waves on slopes just after laying the geomembrane are very dangerous. Usually black geomembranes are sensitive to temperature increase, especially during sunshine. Temperature below black membrane sharply increases. Compacted clay liner is dried, evaporated water condensates on the inner surface of geomembrane and subsequently runs down. In the upper part shrinkage cracks can be created and in the lower part the zones with very high moisture content.

A non-standard situation is described by Vaníček and Kazda (1995), when sanitary landfill was situated inside of inner spoil heap body – see also Chapter 5 – numerical modelling of contaminant spreading.

When using old pits after open pit mining, a very difficult problem is connected with underground water. To eliminate direct contact of underground water with landfill, three compacted clay liners were proposed by the authors, Fig. 8.52. First drain is supposed to collect leachate water, second drain is a control drain for leachate passing through first compacted clay liner and the third drain will collect mostly underground water passing through third compacted clay liner.

8.3.2.2 Contaminant Spreading Through the Compacted Clay Liner

The second possibility how to define the demands on the sealing systems is to apply the contaminant spreading modelling. The design of the protection barrier is after that realized in such a way that all the demands on the concentration of a selected contaminant on the outer side of the protection barrier (or when these contaminants are reaching water sources) as a function of time and position are fulfilled.

To show the basic problem the analytical solution of the contaminant transport in 1D form according to Ogata-Banks (e.g. Yeung and Jiang 1995) can be used:

$$\frac{c(x = L, t)}{c_0} = 0.5 \cdot (\text{erfc}(\xi_1) + \exp(\xi_2) \cdot \text{erfc}(\xi_3)) \tag{8.10}$$

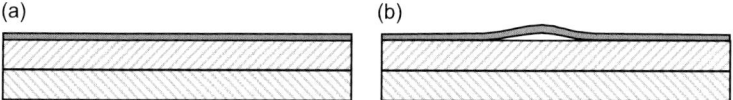

Fig. 8.51 Composite barrier – problem of contacts of individual layers. (**a**) good contact of geomembrane with clay liner. (**b**) insufficient contact

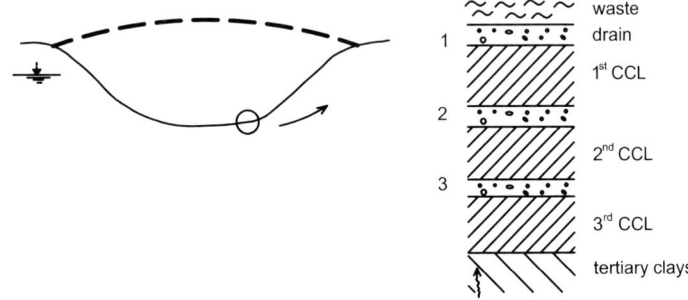

Fig. 8.52 Non standard proposal of landfill in old pits after open pit mining. 1 – leachate drain; 2 – control drain for leachate; 3 – control drain for underground water

where

$$\text{border conditions}: c(0,t) = c_0; c(\infty, t) = 0 \tag{8.10a}$$

$$\text{initial condition}: c(x,0) = 0 \tag{8.10b}$$

$$\xi_1 = \frac{R_d\, L - vt}{2\sqrt{R_d\, Dt}} \tag{8.10c}$$

$$\xi_2 = \frac{vL}{D} \tag{8.10d}$$

$$\xi_3 = \frac{R_d\, L + vt}{2\sqrt{R_d\, Dt}} \tag{8.10e}$$

v... Darcy's velocity
D... diffusion coefficient
R_d... retardation factor
L... distance from the surface
t... time of the solution
c_0... constant concentration on the surface

For this equation a software using spreadsheet was prepared, Vaníček M et al. (1998), where it is possible to change individual input data which have influence on contaminant spreading. After that it was possible to judge their importance for a simple example of the landfill liner consisting only from 0.6 m of clay, see Fig. 8.53.

Further we assumed 0.05 m of leachate in drainage layer. Coefficient of conductivity (filtration, permeability) k was alternated in the range $1 \cdot 10^{-9} - 1 \cdot 10^{-10}$ m.s^{-1}.

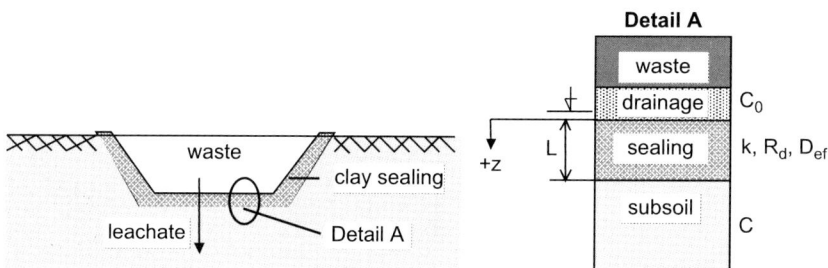

Fig. 8.53 Schematic cross section of landfill for numerical modelling of contaminant spreading

Dispersion and diffusion generally increase contaminant spreading. For small seepage velocity the coefficient of mechanical dispersion was neglected and therefore hydrodynamic dispersion is defined by the effective diffusion coefficient D_{ef}. According to the recommendation of Mitchell (1993) the range of this coefficient was changed in the range $2 \cdot 10^{-10} - 2 \cdot 10^{-9}\,\mathrm{m^2.s^{-1}}$.

Retardation factor R_d covers first of all sorption properties, slowing down the contaminant spreading. Values of R_d are given for example by Suthersan (1997) in the range 1.0–100.0. Value 1.0 in principle represents soil minerals without sorption capacity. Therefore the selected value will depend on the used clay mineral and on the type of contaminant in the leachate.

Some calculated results are presented in Figs. 8.54, 8.55, 8.56, where the influence of different input factors are studied, Vaníček (2002b).

This approach can improve the understanding of the different factors influencing contaminant spreading. From the obtained results it is obvious that not only permeability but also effective diffusion characteristics as well sorption capacity have very significant influence on the final results. For ordinarily demanded coefficient of filtration and $D_{ef} = 2.10^{-10}\mathrm{m^2.s^{-1}}$ and $R_d = 14.5$ contaminants are reaching middle of the clay barrier in its initial concentration after 50 years. But on the outer face of the clay barrier the concentration is still close to zero. These results at the same time make possible the new insight on the behaviour of landfills after some decades. The solution for the practical case is more complicated because the input data have to be tested for each chemical component. There are some problems if the type of contaminant is not specified as it is for municipal waste material, but for dangerous chemical landfill (with much less contaminant spectrum) there is more chance to obtain correct input values.

8.3.3 Demands on Sealing Systems for Landfills

Up to now the individual demands were specified only in very broad level. Because the GL and GCL systems are manufactured products and on place only completed, main attention is devoted to the CCL, which is realized on place and finally its quality is checked.

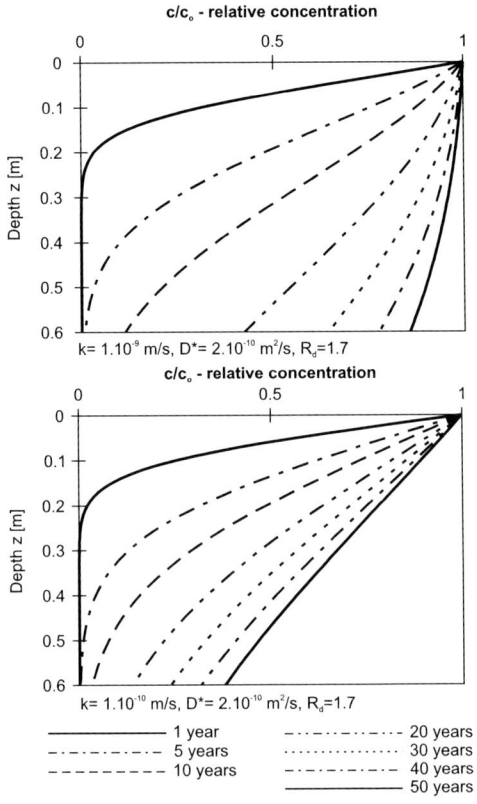

Fig. 8.54 Contaminant spreading for different coefficients of filtration

8.3.3.1 Demands on Compacted Clay Liners – CCL

Selection of Appropriate Soil

When using clay soil for the sealing system the following properties are considered:

- Ability to fulfil the criterion of low permeability (usually $k \leq 1.10^{-9}$ m.s^{-1}),
- Ability of good workability – compaction on site,
- Initial moisture content of the soil in the borrow pit has to be compared with the recommended one.

Local soils are preferred if fulfilling all demands defined for the compacted clay liner. As far as the local soils are not fulfilling some criteria, above all the criterion of low permeability, two possibilities are compared – import of more appropriate soils or the improvement of local ones. Criterion of low permeability are fulfilled by majority of fine –F– soils, according to the classification system presented in Chapter 2, firstly by soils marked by symbols CH, CV, CE – clays with high, very high and extremely high plasticity but also by "M" soils with high plasticity as MH, MV, ME. However also soils composed by mixture of fine soils with sand and gravel

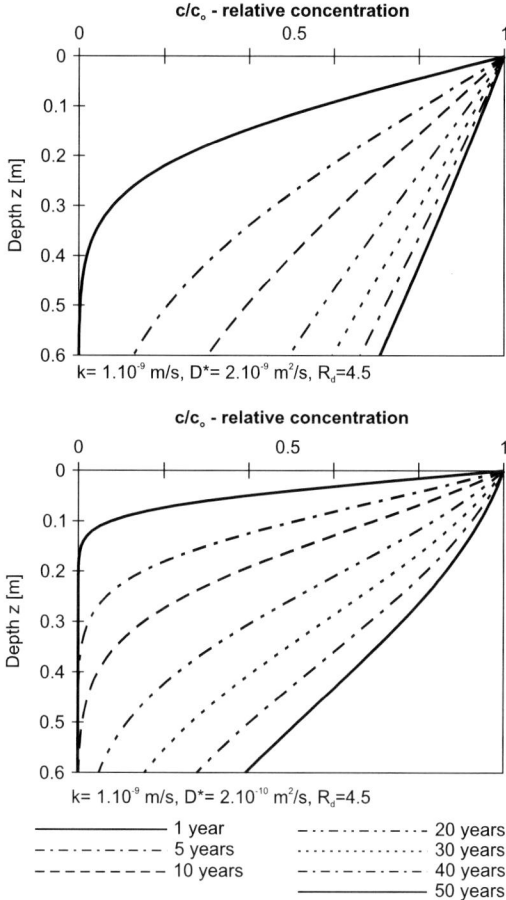

Fig. 8.55 Contaminant spreading for different diffusion coefficients

can fulfil the demanded permeability if the fine fraction contains more active clay minerals and the ratio of fine soils is at least close to 35%.

But on the other side the workability is more difficult if fine soils have high plasticity. Therefore a certain compromise has to be chosen, and more often soil with good workability, which is still able to fulfil demanded permeability criteria is selected for.

For example tertiary clays, which were specified in Section 8.1, are on one side very good sealing material (Madsen et al. 1995), of which a large amount is at our disposal. However due to their initial state – great clods and low moisture content – are practically unworkable even with the help of heavy sheep foot rollers. To be able to use this material it is helpful to leave it on interim dumps, where great clods degrade and moisture content is increased, Vaníček (1995a). The experiences with these tertiary clays when applied for compacted clay liner are described e.g. by Herštus et al. (1997).

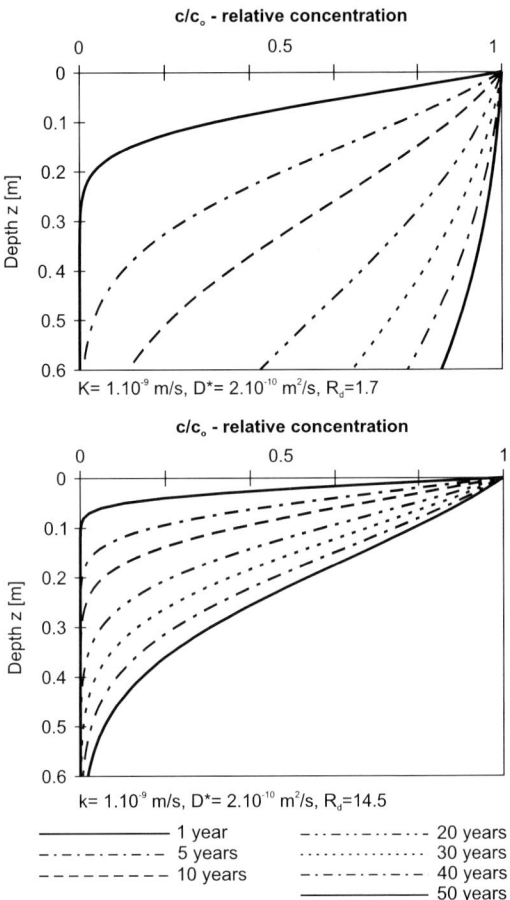

Fig. 8.56 Contaminant spreading for different retardation factors

Most often recommended limits for the optimal composition of the compacted clay liner are:

i) Ratio of clay fraction (< 0.002 mm) $\geq 20-30\%$
ii) Index plasticity $I_p \geq 20-30\%$
iii) Ratio of gravel fraction (2–60 mm) $< 30-40\%$
iv) Maximal size of grains 60 mm

If the soil which is at disposal cannot fulfil the condition of permeability, it can be treated by:

– Screening out of coarser particles,
– Moisture content increase,
– Adding some admixture, e.g. bentonite, of which even low percentage (3–5%) can significantly decrease the coefficient of conductivity. It is possible

to add the bentonite before the application in mixer or on the site with the help of soil stabilizer. In both cases it should be a quick process. The admixture should be compacted on place during 0.5–3.0 h after mixing depending on bentonite and water amount. The lower boundary guarantee minimal time for chemical reaction and the upper boundary should ensure the swelling of bentonite into pores of soils. The results of laboratory tests show that even admixtures of sand with bentonite can fulfil demanded permeability criteria, see e.g. Kenney et al. (1992).

Technology of Compaction of Clay Liner

Proper compaction of widespread layers can take place in standard technique with the help of compaction rollers in such a way that the thickness of individual layers after the compaction is roughly equal to the demanded value of 0.2 m. The number of passes is better to determine on the test site, as well as velocity of roller in agreement with demanded compaction degree. Compaction during freezing, raining should be excluded. Compacted surface should be protected against splash creation, drying, freezing, and mechanical disturbance. The compaction control on realized landfills show the better results for horizontal liner than for inclined liner. It is given by the difficulty of compaction on slope. On slopes the compaction can be performed either in a horizontal direction or in a direction parallel to the slope. For slopes steeper than 1:2.5 the compaction should be realized in horizontal layers. The problem is that the local lower compaction degree has more negative impact for horizontal compaction than for parallel one, see Fig. 8.57.

Sheep foot rollers, compactors, where prickles are about 0.2 m high are usually preferred. They are more easily able to crush down and better interconnect the individual particles, clods. Also better interconnections of individual layers are guaranteed. However the surface of the last layer should be compacted and smoothed by smooth roller to eliminate the infiltration and to guarantee good contact with geomembrane during the application of combined liner. The last tendency is to increase the number of layers. For example to apply four layers when total thickness is 0.6 m or six layers when total thickness is 0.9 m.

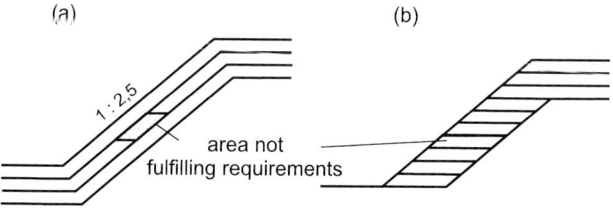

Fig. 8.57 Compaction of clay liner on slope – the influence of compaction imperfection. (**a**) compacted layers are parallel with slope. (**b**) compacted layers are horizontal

Factors Affecting Final Permeability

The final permeability is strongly affected by compaction technology. The attention should be dedicated to the following factors, Vaníček (1995b):

– Initial moisture content,
– Size of individual clods,
– Used compaction equipment.

All these factors have some influence on the soil structure, both micro structure and macro structure. The samples compacted under moisture content lower than the optimal have much larger macro pores between individual clods and therefore the permeability is higher. The compaction for moisture contents higher than the optimal leads to the closure of macro pores due to shear deformations and to nearly parallel arrangement of individual particles and finally to lower permeability. The influence of initial moisture content on permeability was also described in Chapter 2.

Differences between individual demands for compaction of sealing zones in earth dam engineering and for compaction of sealing layers in landfills are shown in Fig. 8.58. For earth and rock fill dams more attention is devoted to the shearing resistance and pore pressure development. For sealing layers of landfills more attention is devoted to the permeability and therefore the area recommended for compaction is much more on the wet side of the Proctor curve. Most frequently used recommendation for compaction of clay liner states:

$$w = w_{opt} + 1.5\% \tag{8.11}$$

$$\rho_d \geq 0.9(0.95) \cdot \rho_{dmax} \tag{8.12}$$

The influence of particle size was mentioned for the case of tertiary clays from Northern Bohemia. The lower clods, the macro pores are more easily closed. This factor was laboratory investigated e.g. by Benson and Daniel (1990).

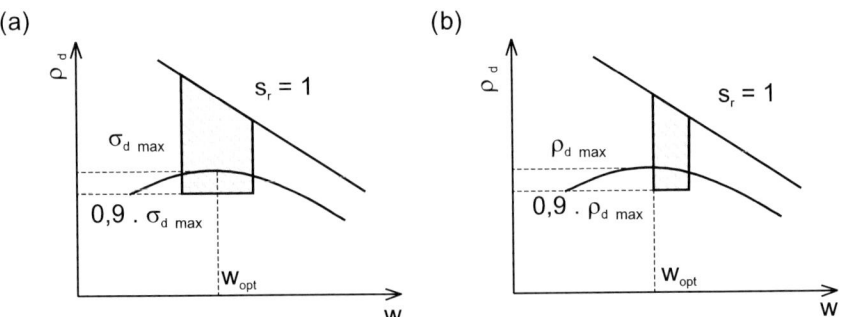

Fig. 8.58 Classical demands for clay sealing layers compaction. (**a**) clay sealing in fill dam. (**b**) clay sealing in landfill

Higher homogeneity of compacted soil can be reached also by higher compaction energy. Therefore the utilization of compaction rollers with higher specific loading per one meter is advisable. From this follows that the recommended values of moisture content w and dry density ρ_d should be closer to the line of 100% of saturation ($S_r = 1$). Nevertheless some authors as e.g. Daniel (1993), Manassero et al. (1998) highlight that in some cases not only permeability but also shear strength have a great influence on overall landfill behaviour, especially in places where shear failure can be expected as well as deformation, susceptibility to shrinkage and others. After that the recommended zone for compacted clay liner can be limited to the zone of interaction of different demands, see Fig. 8.59.

Permeability Verification

In agreement with defined demands on performance of compacted clay liner it is necessary to perform permeability tests. Firstly the tests can be performed in the laboratory on samples collected from the realized sealing layer; secondly the tests can be performed in situ directly on the realized liner.

In this phase it is necessary to mention the influence of tested sample size on the variability of the obtained results. Liner will never be homogenous no matter if the result of better or worse interconnection of individual clods, or if the result of small cracks is sometimes visible even on the surface of well compacted liner. Therefore the larger sample will with higher probability contain the places of weakening, whereas the smaller samples do not have to contain these places of weakening (but they can). Therefore generally speaking better results are obtained from smaller samples. For this reason the laboratory tests are giving better results than in situ tests.

When performing laboratory tests these two views are important:

- Representability of tested sample,
- Possibilities of laboratory devices.

The second view will depend on ability:

- To measure with high preciseness a very small amount of water passing through the sample,
- To measure permeability for different hydraulic gradients,

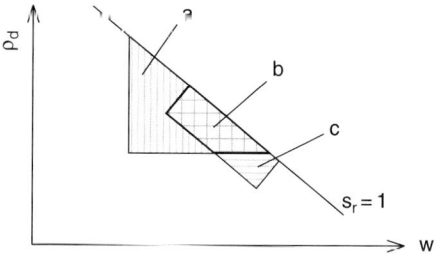

Fig. 8.59 Overall acceptable zone for compacted clay liner; (**a**) acceptable zone based on shear strength. (**b**) overall acceptable zone. (**c**) acceptable zone based on hydraulic conductivity

– To apply back pressure to guarantee gradual dissolving of air bubbles,
– To apply outer loading on sample.

From the different laboratory permeameters mentioned in the Czech standard for permeability tests the permeameter of type G – permeameter on the principle of triaxial chamber – is recommended, see Fig. 8.60. With back pressure the full saturation of the sample can be reached and so to eliminate the influence of the air bubbles on final permeability. The description of different permeameters for testing the clay liner quality was published e.g. by Benson and Daniel (1990).

Field permeability tests have lower chance to affect the sample saturation or to change the hydraulic gradient. Differences in results can be affected also by hydraulic anisotropy. In the laboratory tested samples are usually sampled in vertical direction and also such tested. Very often the permeability in the horizontal direction is higher than in the vertical direction. During field test this fact can have some influence on the final result; therefore there is a tendency to eliminate this effect. From the different field permeability tests the four fundamental principles are mentioned, see Fig. 8.61:

– Test in the shallow borehole,
– Test with porous probe,
– Different types of infiltrometers,
– Method of checking pan – lysimeter.

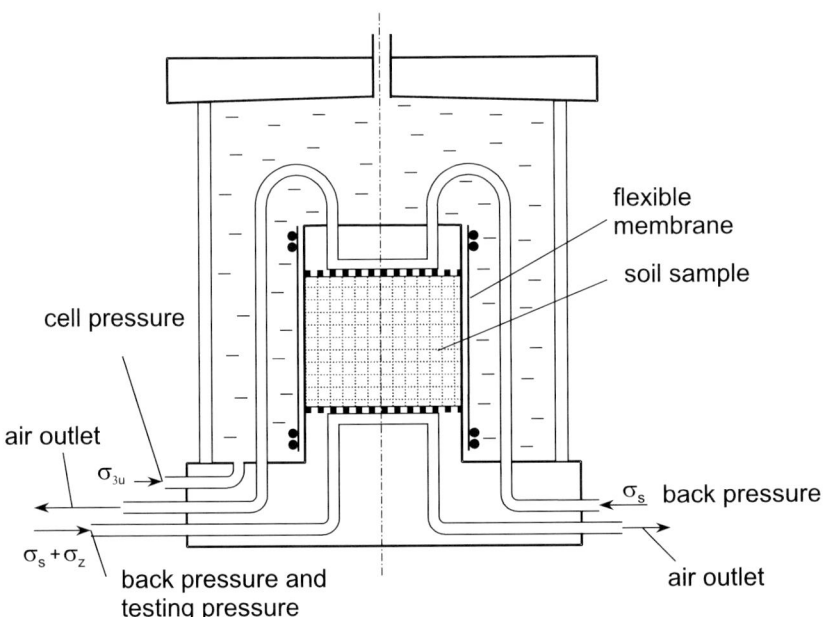

Fig. 8.60 Scheme of the permeameter on the principle of triaxial cell – methods G in ČSN 721020

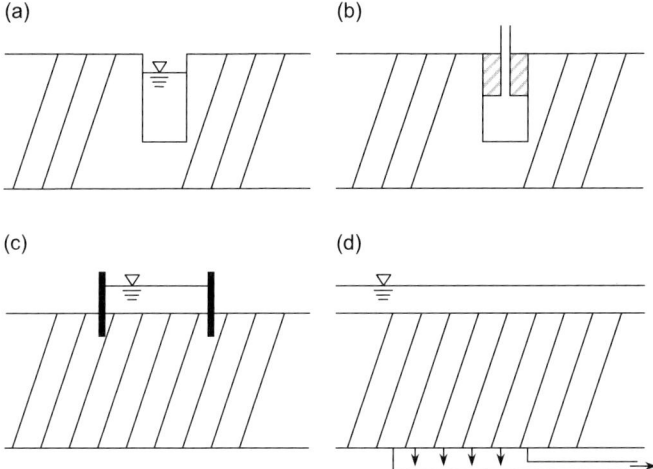

Fig. 8.61 Different field permeability tests. (**a**) shallow borehole. (**b**) porous probe. (**c**) infiltrometer. (**d**) lysimeter

Detailed evaluation of the individual tests are specified e.g. by Trautwein and Boutwell (1994). From the different infiltrometers best evaluation obtained the covered double-cylinder infiltrometer, where second cylinder is ensuring vertical infiltration and by covering the losses of water (e.g. by evaporation) are eliminated.

The authors obtained good experiences with large scale laboratory permeameter, which can fall into the category of lysimeters, Vaníček et al. (1995). Section plan is 2×2 m and soil is compacted by compaction devices as in reality (but preferably by compaction rammers). Firstly it is possible to simulate different modification of clay liner as is the initial moisture content, size of particles on final permeability and others. Secondly this permeameter is used also for testing of different field permeameters as the obtained result can be directly compared with the result obtained for the area of $4\,\mathrm{m}^2$.

Clay Liners Durability Against Pollutants Attack

The concern over the potential detrimental effects of exposure of clay liners to liquid chemical wastes and landfill leachates has sprawled a large number of studies of compacted clay-chemical compatibility. Mitchell and Madsen (1987) presented a summary and a review of chemical effects on clay hydraulic conductivity. Among the main conclusions were: The consolidometer permeameter system may offer the best means for quantitative assessment of chemical interactions, induced volume changes, and for testing clays under confining stress-states that are most representative of those in the field. The influence of chemicals on the hydraulic conductivity of high water content clays may be much greater than on lower content compacted clays owing to the greater particle mobility and easier opportunity for fabric changes in the high water content system. But the main finding was that serious problems are likely only when concentrated hydrocarbons come in contact with a clay liner under no, or very small, overburden, or if the soil minerals are soluble in the type of leachate generated. Madsen (1994) therefore stated that the question about clay liners is normally not about the life-time of the liner material itself, but how long the liner is able to prevent a certain amount of pollutants to penetrate through the liner – the problem which was discussed in the previous Section 8.3.2.2.

General Control of Clay Liners

The significance of permeability control is very high and predominant from the view of fulfilment of fundamental demands. However the need to quickly react leads to the other tests which are more flexible, above all to the index tests. Before the compaction, the tests as the initial moisture content, consistency limits, sometimes grain size distribution curve are performed for the verification that the used soil is falling into the acceptable range for the given case.

The tests performed on compacted liner are in particular verifying the compaction degree as is moisture content and dry density. But some other tests are very useful such as the thickness of individual layers and content of organic matters.

European technical committee No. 8 (ETC 8 – 1993) of the ISSMGE recommended following frequency of control:

- Characteristics of the used soil – grain size distribution curve, consistency limits, moisture content – each $1\,000\,m^2$,
- Moisture content during deposition, homogeneity of deposited material, number of roller passes, volume of added water (if needed) – each $1\,000\,m^2$,
- Maximal size of clods, weight of additives (e.g. bentonite) if they are applied together with a manner of mixing – each $1\,000\,m^2$,
- Thickness of compacted layers, their levelling and adhesion – each $500\,m^2$,
- Degree of compaction – moisture content w, dry density ρ_d plus grain size distribution curve and plasticity if appropriate, and the homogeneity of the results – each $1\,000\,m^2$,
- Determination of hydraulic conductivity – each $2\,000\,m^2$.

8.3.3.2 Demands on Geomembrane Liners – GL

Geomembrane is generally regarded as impermeable. Therefore the most attention is devoted to the connection and testing of individual strips to be sure that the quality of this connection is practically the same as for individual strip. Resistance to the mechanical failure is also playing very significant role because the protection of the geomembrane surface on a very large area is problematic. From the practical point of view it is almost impossible to prevent even very small local failure. The attention is afterwards devoted to identification and repair of such local failures.

8.3 Landfills

Last but not least the chemical resistivity of geomembrane with direct contact with chemical compounds has to be proved. The effect of chemical compounds is judged after 30, 60, 90 days of direct contact at temperatures 23°C and 50°C respectively. Afterwards tests of mechanical resistivity are performed and compared with initial ones. As was said before the chemical resistivity of HDPE geomembrane is very high and therefore this type is usually preferred. Koerner, e.g. Koerner et al. (1990, 1992) devoted to the chemical degradation of geomembranes a great attention. And finally the biological resistivity, especially against the effect of bacteria, mildew and also the resistivity against rodents should be considered.

Geomembrane Laying and Welding

From the view of possibility of mechanical failure strict demands are defined for geomembrane roll handling, for installation, subsoil adjustment, transport limitation and others. Welding guarantees not only a good connection of the individual strips but also a good sealing when different tubes (for drainage, degassing) are passing through the geomembrane.

Between the fundamental methods of geomembrane welding, belongs extrusion welding and thermal fusion or melt bonding. The last method is appropriate for good connection of all types of thermoplastic geomembranes when two strips are partly melted on the contact and subsequently left to cool down under pressure. Here the most often used method is welding by hot wedge, see Fig. 8.62a. Extrusion welding is conjoining two parts with the help of added melted plastics of the same origin as are two bonded parts, see Fig. 8.62b. To get good interconnection between membrane sheets and different pipes during extrusion welding, the pipes are from the same material as sheet is, which means also from HDPE. Examples are shown in Fig. 8.63.

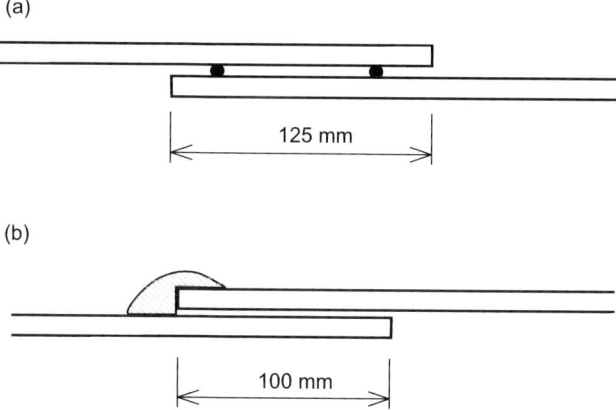

Fig. 8.62 Geomembrane welding. (**a**) thermal fusion – hot wedge. (**b**) extrusion welding

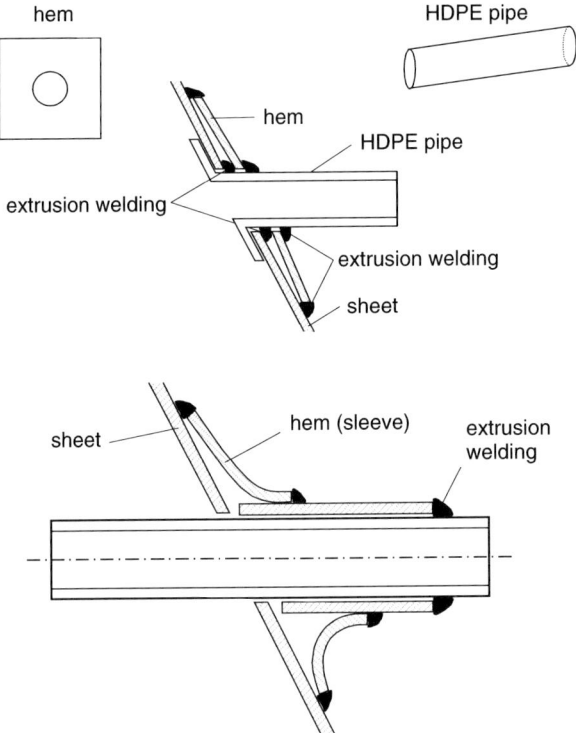

Fig. 8.63 HDPE sheet and pipe connections – with the help of extrusion welding

Quality Control

Welding quality control can have different possibilities as:

– Visual control,
– Non-destructive tests on site,
– Destructive tests in laboratory.

The advantage of welding with hot wedge with double track weld, Fig. 8.62a is the existence of airspace which can be non-destructively tested for overpressure or underpressure, vacuum. During the overpressure test this airspace is closed on one side and pressurized from the second one. Bad connection is indicated by pressure lost during the test duration of 5 minutes. After the test is finished the opposite end is opened and the pressure decrease indicates that the weld was tested all over length.

Laboratory tests are performed on pieces of welds sampled from the connected sheets. The results of tensile, shear and other tests on welds are compared with results obtained on samples from undisturbed sheet. The results obtained on welds should be relevant for the tests on sheets.

Visual control is done just before geomembrane is covered by protective geotextile, geodrain. For the mechanical failure control the geophysical methods can be used as well. Nosko et al. (1996) describe the method based on installation of electric current sensors. In this case the contaminant spreading through small hole in geomembrane is simulated by an electric field. The electrical field is measured with the help of sensors installed above and below geomembrane.

Nosko also states an interesting view with respect to the number of failures and reasons of their development. For example in the phase of covering he states that 68% of failures is caused by coarse grains and 16% are cause by heavy machineries used there. Therefore great care should be devoted to the geomembrane protection. Subsoil is settled, sharp-edged grains removed, surface well compacted. Analogically the surface of laid geomembrane is protected, especially on contact with drainage layers from gravel particles. When the geodrains are used the potential risk of mechanical failure is very small. In the case of coarse grain drains (usually fraction 16–32 mm is recommended, when for smaller grains the potential risk of drain choking is higher) the potential risk of failure is much higher and therefore the protective geotextile is applied. Geotextiles with minimum unit weight of 600 g.m^{-2} are recommended. Reinforced geocomposite, consisting of woven reinforcing geotextile covered from both sides by non-woven geotextile are prioritized during the last period. The authors in a contested case recommend to verify the design in the laboratory in large scale oedometer, where there is a possibility to change the grain-size distribution of drain, applied loading, the influence of individual protective geotextile and others.

8.3.4 Drainage Systems for Landfills

Drainage layer situated above sealing layer is firstly fulfilling these two functions:

- Lowering of pore pressures with positive impact on stability,
- Lowering of infiltration into sealing layer as a result of lower hydraulic gradient.

Nevertheless the demands on the drainage layer of the capping construction system differ from the demands on the drainage layer of the bottom construction system. Drain of the capping system is not in contact with contaminated water, however it can be clogged by roots of surface vegetation. Drain of the bottom system is in direct contact with leachate which can carry also very fine particles, again with a high risk of clogging. Therefore the protection of drains with filters is desirable, but the fine rootlets can pass through this filter. The fundamental part of each drainage system is planar drain which can be either constructed from sandy gravel grains or from geosynthetics.

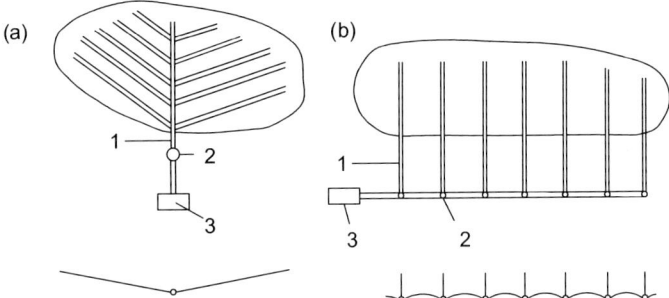

Fig. 8.64 Layout of the collecting drains. (**a**) classical arrangement – main drain in the lowest place. (**b**) solution enabling to control and clean all drains. 1 – collecting drain; 2 – inspection chamber; 3 – collecting tank

Thickness of the planar grain drain is usually 0.3–0.5 m with minimum inclination 2%. Water from the planar drain is collected by collecting drain with inner diameter at least 200 mm. Perforated HDPE pipes with sufficient stiffness are recommended, to guarantee profile stability. Arrangement of collecting drains is shown in Fig. 8.64. Classical arrangement is presented on Fig. 8.64a where the main drain is situated in the lowest place and is collecting leachate from the lateral collecting drain. With increasing demands on continuous long term control, e.g. with the help of television cameras and with a possibility to clean the drains the arrangement which is able to fulfill these demands is shown in Fig. 8.64b. Here the individual collecting drains are flowed into main drain which is situated out of landfill plan. Inspection chambers situated in place of this interconnection make it possible to control the quality of drains and their maintenance. To protect the biogas outflow water-seals (siphon traps) are situated front of the inspection chambers. Details of the situation of collecting drains in relation to the cross-section of the whole bottom construction system are shown in Fig. 8.65. Capping drainage system is not using collecting drains so often. The detail of the outfall of the drainage system out of plan of landfill is shown on Fig. 8.66a. For larger landfills a small sealing hedge is proposed on the slope which helps to divert water into collecting drains, see Fig. 8.66b. This solution was proposed for example by Hoekstra (1995), which states good experiences with geodrain Enkadrain used for landfill Limburg in the Netherlands.

At the same time there is also the composition of the whole capping construction system, which is at least 1 m thick when upper part of ca 0.3 m is created by top soil. The soil layer is able to accumulate rainfall water and at the same time protect the sealing layer against freezing and drying. The type of vegetation is selected with shallow roots to protect root penetration into planar drainage layer. But the main aim of vegetation is to consume significant percentage of the rain water and to guarantee the surface erosion protection. Vegetation is combined with anti-erosion mattresses on steeper slopes.

8.3 Landfills

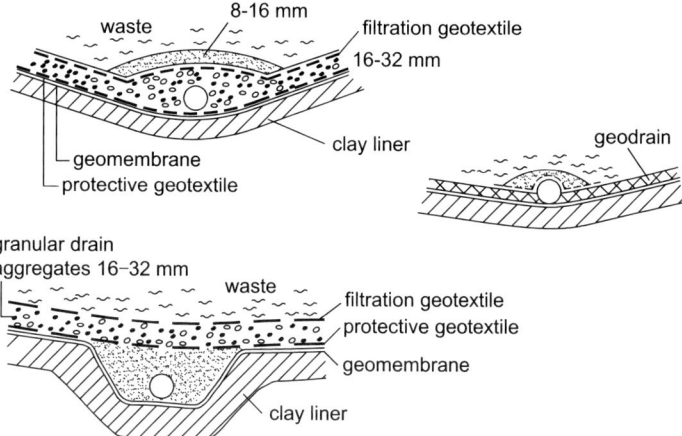

Fig. 8.65 Details of the collecting drains positioning

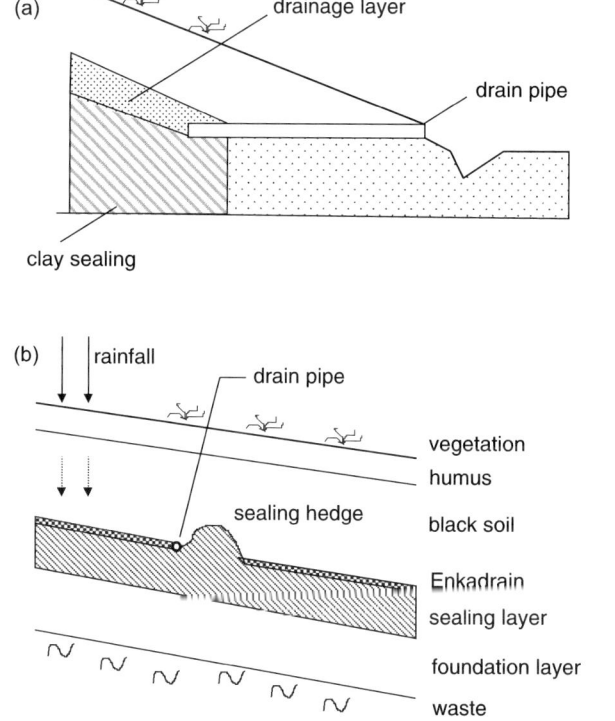

Fig. 8.66 Drainage of the landfill cover

8.3.4.1 Gas Collection/Planishing Layer

Layers situated just below sealing layers of the capping system are levelling of the landfill surface into demanded shape. Therefore the thickness can be unequal. This layer can be composed also from inert material as excavated soil, building waste, ash. If this material is permeable enough, after that it can be used as a gas collection layer from which the gas is exhausted. HDPE gas collection pipes passing through the geomembrane are sealed with the help of extrusive welding. When passing through clay liner or geosynthetic clay liner the bentonite powder is used for sealing of this contact.

8.3.5 Properties of Stored Waste in Municipal Landfill

For the design of landfill as a special type of the earth structure, it means for the calculation of different limit states as stability, deformation we have to know not only the properties of soil but also deposited waste material. This last requirement is rather complicated not only due to high heterogeneity of the deposited waste (problem of representative sample) but also with respect to the change of waste properties with time. Nevertheless a certain range of expected properties are determined on the base of increasing number of practical findings. A very comprehensive survey in this direction is presented by Manassero et al. (1997).

8.3.5.1 Index Properties

Between index properties, belong above all moisture content, grain size distribution and volume density. Information about temperature is also important.

Moisture content – strongly depends on many different factors as initial composition of waste, local climatic conditions, technology of deposition, whether the landfill is partly covered or not and others. As was said before moisture content is sometimes artificially increased just as support of the degradation processes in the landfill. Recommended value is about 50–60%.

The determination of moisture content of waste is different than for soils, and is determined from the water lost for temperature $60°C$ to avoid combustion of volatile material. Obtained values are in a large range of 15–130%, strongly depending on climatic conditions. Variability of moisture content is multivalent; in some cases moisture content is increasing with depth in other ones the opposite is valid.

Unit weight – also very variable value; is depending on waste composition, technology of deposition, compaction effort, local moisture content. Very often the unit weight is in the range of $600–1\,000\,kg.m^{-3}$ for well compacted waste. Unit weight is changing with landfill age. At the end of degradation process unit weight can reach values as $1\,200–1300\,kg.m^{-3}$. For elder landfills with high content of mineral components this value can be even higher. For the above mentioned values of unit weight the typical void ratio is between 15 for uncompacted waste and two for well compacted waste.

Particle size distribution – the determination of this value is very problematic. For the waste with higher content of mineral components classical soil mechanics methods are used, but for other parts of the deposited waste this way is problematic with respect to the high content of paper, plastics, textile and others. For the first case the content of fine particles is up to 20%, and the biggest particle depends on the waste character, whether the waste was crushed down before the deposition or not. The above mentioned content of plastics, textile can have some reinforcing effect when determining mechanical properties of waste.

Temperature – monitored results on monitored landfill depend on landfill age, depth of measurement. Most often mentioned values are in the range of 40–60°C.

8.3.5.2 Physical-Mechanical Properties

Permeability – also the results of permeability of the waste material depend on the above mentioned factors, monitored results of the filtration coefficient are in the range of $10^{-4} - 10^{-8}$ m.s^{-1}.

Deformation characteristics – Settlement of the landfill surface as a function of time is very important from the view of safe remediation and recultivation processes as well from the view of potential utilization of landfill surface for new construction. But this time dependent characteristic is influenced by all physical, chemical and biological processes described earlier.

First practical results were published by Sowers (1973). He concluded that:

- The mechanical settlement occurs rapidly with little or no pore pressure build-up. The initial and primary phase is complete in less than 1 month. The total settlement for the 1st month can be found from a compression index C_c that is related to the void ratio e: $C_c = 0.15.e_0$ for fills low in organic matter to $C_c = 0.55.e_0$ for fills high in organic matter. $\Delta e = -Cc \log((\sigma_0' + \Delta \sigma)/\sigma_0')$
- Continuing settlement is analogous to secondary compression and is also related to the environment for bio-chemical decay. The rate can be expressed by $\alpha = 0.03e_0$ for conditions unfavourable to decay to $\alpha = 0.09e_0$ for conditions favourable to decay. $\Delta e = -\alpha \log(t_2/t_1)$.

The secondary compression supposes that the settlement – log time relationship is more or less linear. Nevertheless for waste material containing higher percentage of organic compound the material is more compressible after some time period. Manassero et al. (1997) recommend using the expression for secondary settlement of the surface of landfill in the more ordinary form:

$$s_s = C_\alpha . H . \log . t_1/t_2 \qquad (8.13)$$

Where H is landfill thickness and C_α secondary compression index.

From the surface settlement measurement index of secondary compression for given H can be calculated as well further settlement of the surface in the future.

Shear Parameters

Although we are not working with classical soil some principles of soil mechanics are applied even on shear characteristics; with the help of angle of internal friction φ and cohesion c. Due to a great heterogeneity of the deposited waste the maximum shear strength is reached for much larger shear deformation than is valid for soils. Therefore sometimes it is better to define shear strength for defined acceptable strains. Especially for cases where shear plane is passing also through soil, where maximum shear strength is reached for much lower shear strain.

Interesting results of the determination of strength parameters of a fresh, 1–3 years old mixed waste sampled in a pit at a municipal solid waste landfill are described by Kockel and Jessberger (1995). They used triaxial tests with dia 100 mm and large uniaxial compression tests with dia 550 mm. For the waste tested in large uniaxial compression almost no processing is necessary; here only bulky components were removed (composite matrix−2). For triaxial tests maximum particle size of 15 mm were left (basic matrix−1). Triaxial tests were also performed with addition of previously removed plastic foils ($l_{max} \leq 32$ mm) as plastic can be regarded to be one of the dominant components of the reinforcing matrix (3) of mixed waste. The results of the triaxial tests are presented in Fig. 8.67. With large strains a maximum friction angle $\varphi(42° - 49°)$ is activated which is only slightly influenced by

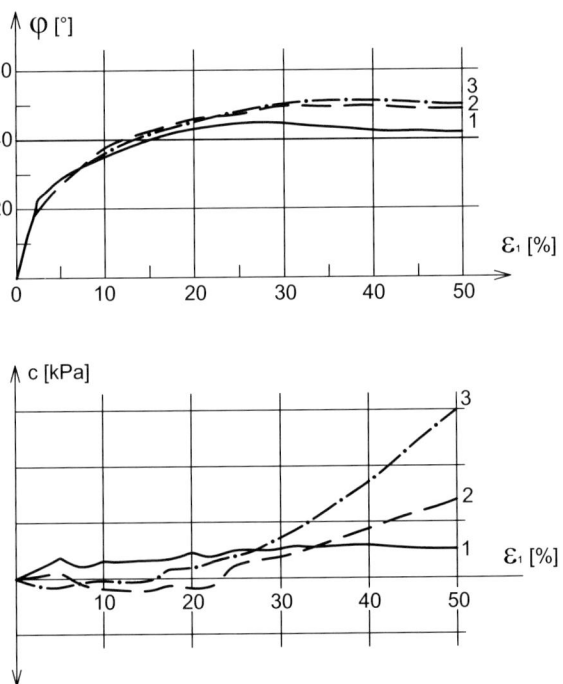

Fig. 8.67 Results of the municipal waste triaxial tests containing different amount of plastics, (according to Kockel and Jessberger)

the reinforcing plastics. A reinforcing matrix provides an additional cohesion-like strength; its activation requires large strains and starts at about axial strain about 20% when φ is almost completely activated. For large uniaxial compression tests the following uniaxial compressive strength were obtained:

- $q_u = 215\,\text{kN.m}^{-2}$ for composite matrix,
- $q_u = 75\,\text{kN.m}^{-2}$ for basic matrix (but with maximum particle size of 120 mm)

From these results Kockel and Jessberger concluded that the maximum shear strength of the 1–3 years old mixed waste is given by $\varphi = 42°\ldots 49°$ and $c = 51\ldots 41$ kPa.

Very high cohesion of deposited waste was mentioned also by Mitchell (1997) when showing ca 36 m high nearly vertical wall of waste material, which stayed stable after the rest of the landfill, was shifted out as a result of the slope instability.

8.3.6 Limit States of Landfills

Also during the landfill design it is necessary to look on it as a very complicated geotechnical structure where all limit states have to be controlled. Especially the following three limit states have to be checked:

- Limit state of stability,
- Limit state of deformation,
- Limits state of surface erosion.

Surface erosion is important in the phase of landfill closing, and the fundamental question is connected with control whether vegetation on the surface is able to protect surface erosion or have to be strengthened by anti erosion geosynthetic mattresses.

8.3.6.1 Landfill Stability

When checking landfill stability, e.g. Brandl (2000, 2001d), it is firstly needed to check for *overall stability*, when potential slip surface is passing not only through landfill basement (above all when landfill is founded on soft subsoil) but also through landfill body, see Fig. 8.68a. The discussion about development of shear resistance with shearing mentioned just before have to be considered as well. An interesting practical example of this problem was mentioned by Mitchell (1996), when performing back-analysis of municipal waste landfill slide. The failure occurs in the municipal waste at large shearing strains when in subsoil peak strength was exceeded and only residual shear strength was active there.

Local stability should be checked above all on contacts in individual layers, for example along contact of geomembrane with compacted clay liner or along the contact of protecting geotextile on contact with geomembrane, see Fig. 8.68b. With

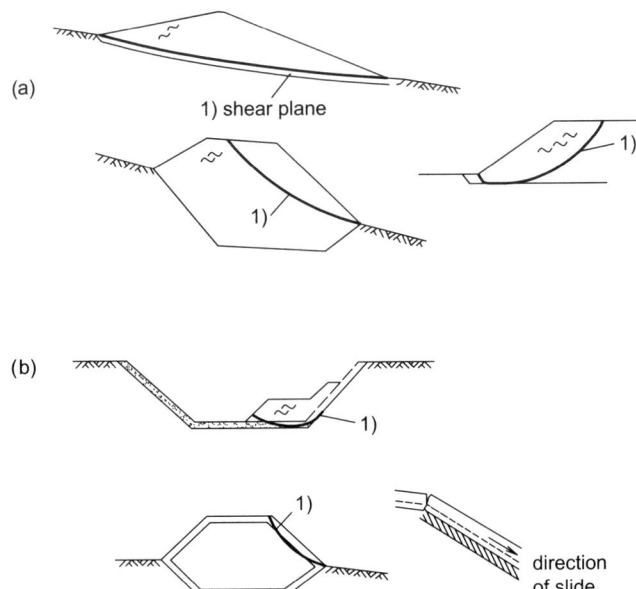

Fig. 8.68 Landfill stability. (**a**) overall stability. (**b**) local stability

respect to the very low value of angle of friction along these contacts it is necessary to anchor geomembrane and protective geotextile in the upper part of slope, see Fig. 8.69.

Furthermore it is necessary to check two other possibilities of failure:

– Tensile strength both of these materials is not sufficient for the transfer of shearing forces,
– Angle of friction along contact of protective geotextile with grain drain is lower than is an angle of the landfill slope.

In this case it is appropriate to cover protective geotextile with reinforcing geogrid which is able to transfer shear forces and also has higher angle of friction along the contact with grain drain. But also geogrid should be anchored in the upper part of

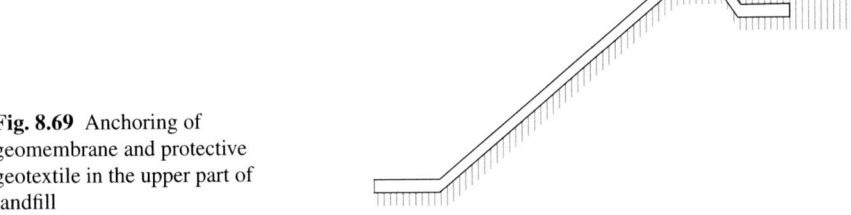

Fig. 8.69 Anchoring of geomembrane and protective geotextile in the upper part of landfill

slope. Hall and Gilchrist (1995) recommend placing reinforcing geogrid on filtration geotextile, however they describe the case when drainage was ensured by geodrain.

When checking the stability of the grain drain of the capping system the case of heavy rainfall must not be neglected, when the drain can be fully saturated thereby affecting grains by considerable seepage force. This problem is solved similarly as stability of slope from granular soils when seeping water is parallel with slope.

8.3.6.2 Landfill Deformation

Due to the different physical, chemical and biological processes inside deposited material, settlement of the landfill surface is sometimes very high. In previous times it has very often been pointed out, that surface clay sealing systems can embody differential settlement causing tensile cracks – Jessberger and Stone (1991), Daniel (1995). As an example a crater with the dia $\Delta L = 5\,m$ with maximum depression $\Delta s = 0.25$–$0.5\,m$ is mentioned as a typical case observed on the landfill surface, Fig. 8.70, Vaníček (2002b).

This differential settlement corresponds to the elongation of 0.1–1.0%, which can cause the tensile crack development with preferential infiltration into landfill body. Daniel for example indicates that many, if not all, covers for municipal solid waste landfills have areas with distortion of this magnitude or larger. This is also one of Daniel's arguments for giving a preference to the GCL – geosynthetic clay liner in the capping system. But on the basis of own experience with tensile tests performed for the checking of the potential of the tensile crack development in the core of the earth and rock fill dams, see Chapter 7, the authors are not so sceptical towards the utilization of clay liners in landfills, mainly for the following reasons:

- Maximum tensile elongation for cohesive soils compacted for optimum moisture content according to the Proctor standard test is really in the range of 0.1–1.0%, but grows with moisture content increase. This significant increase is seen from the Fig. 7.19, results of which were obtained from the bending test – Vaníček (1977a, b). Because the compaction of the clay liners is usually performed for moisture content higher than optimum (due to decrease of permeability), this aspect is on the positive side.
- Up to now mentioned results were obtained for the tensile tests which can be labelled as undrained tests. For the real conditions the crater development on

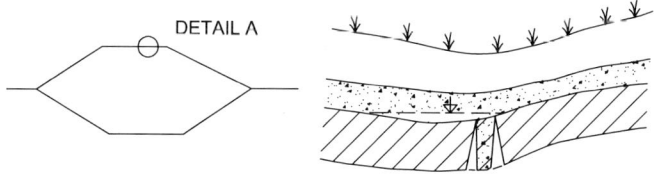

Fig. 8.70 Deformation of the landfill surface – local differential settlements affecting functionality of sealing and draining layers

the landfill surface is not such a quick process, and is time dependant. The results obtained under drained conditions in the triaxial hydraulic apparatus under the scheme shown in Fig. 7.20 – are rather different from the results obtained for undrained conditions. While the undrained tensile strength is in the range 30–80 kPa, the drained tensile strength is much lower, roughly 10 times – 3–8 kPa. However drained tensile elongation at failure is much higher than the undrained one, usually 2–5%. For the capping clay liners partly drained conditions can be expected and therefore also the lower probability of tensile cracks development.
– Last chance for improvement is the swelling potential – after the potential opening of the tensile crack and first water infiltration through it the crack can be closed by the swelling potential of clay minerals. The authors have also previously discussed some aspects of this process, namely in relation to the earth and rockfill dams in Chapter 7.

Therefore we can conclude that the potential risk of the tensile crack development exists but is not as high as was believed at the beginning of the last decade.

Nevertheless Daniel (1995) also mentioned the potential risk of crack development due to desiccation of capping clay liner from below. But he stated that the issue of desiccation from below has not been studied in landfills.

The potential risk of crack development can lead to the:

- Higher compaction effort for gas collection/foundation layer as well for clay liner,
- Increase of the thickness of the gas collection/foundation layer,
- Reinforcement of this layer, because the reinforcing element can significantly reduce the potential to crack development.

In some cases this scepticism can lead to the preference of sealing systems which are more flexible as geosynthetic clay liners or geomembranes liners or to the combination of compacted clay liners with GCL or GL.

8.3.7 Application of Geosynthetics in Landfill Construction

The use of geosynthetics for waste containment application is widespread and is still growing. One of the advantages is that they are made in factories where the quality control can guarantee not only specific properties but also low variability in these properties. When compared to classical soil above all for sealing and drainage layers, total thickness of geosynthetics is much lower, so the space of landfill for waste containment can be used more effectively. It is not our intention to go into more details because many aspects of the geosynthetics application were mentioned in this chapter and in Chapter 3 as well. Just to briefly summarize the most important

8.3 Landfills

technical issues related to geosynthetics, which are also mentioned by Daniel and Bowders (1996):

- Service life,
- Construction quality assurance,
- Performance in the field,
- Interface shear strength.

Our intention is just to show how big an influence geosynthetics have in construction of earth structures, especially in the field of environmental engineering. To stress this Daniel and Bowders presented the picture of an extreme application shown in Fig. 8.71. This figure illustrates a double liner system involving heavy use of geosynthetics. Authors of this picture emphasized that not all of the components shown there would normally be needed at any one landfill facility, but all the geosynthetics shown there have been used at landfills.

8.3.8 Landfill Monitoring

Landfill as a very sensitive structure in our environment needs to be carefully monitored. Demands are defined by National standards, as by the Czech standard ČSN 838036 Waste landfilling – monitoring of landfills:

a) Permanent monitoring of landfill ability to fulfil safely and reliably all functions for which was constructed,

b) Permanent monitoring of all environmental impacts and so timely locate phenomena which can be a sign of defects,

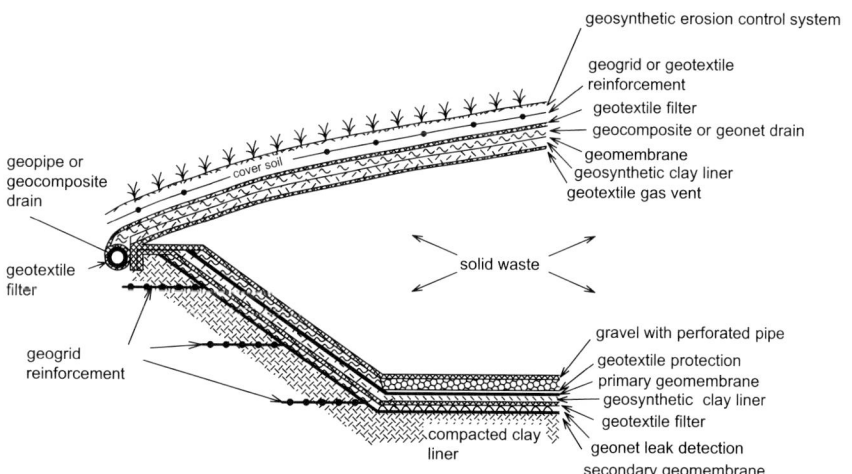

Fig. 8.71 Landfill cell with intensive use of geosynthetics, (according to Daniel and Bowders)

c) Monitoring of technical quality of landfill facilities and of project assumption,
d) Evaluation and refinement of realized site investigation for design of supplemental measures leading to the improvement of individual facilities.

Greatest attention is devoted to the quality of underground water. Before construction an initial quality has to be documented to be able to compare measured values with it. In control points monitoring wells, piezometers are installed which are able to determine water pressure in different levels and at the same time to collect water samples from these depth for chemical analysis. Recommended minimum is 2 monitoring wells "above" landfill and 3 "below" landfill. The situation is schematically demonstrated in Fig. 8.72 according to Brandl and Robertson (1996).

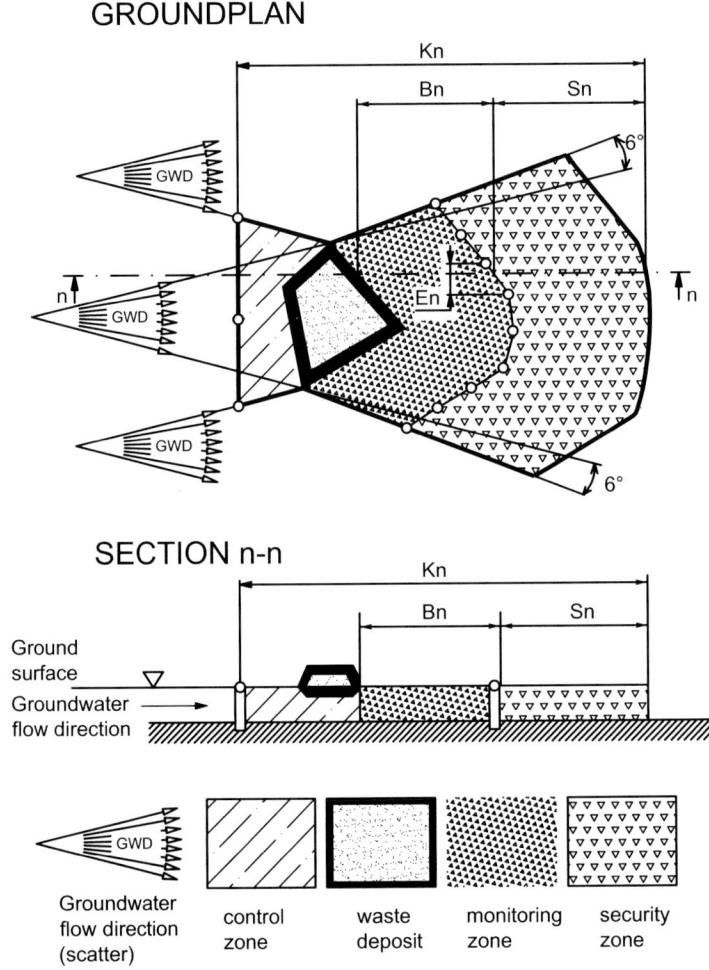

Fig. 8.72 Recommended distribution of monitoring wells, (according to Brandl and Robertson)

Besides the quality of underground water, the quality of leachate collected at the collecting reservoir is observed as well as the quality of surface water course collecting water from outer drainage system of the landfill. The attention is also devoted to the landfill gas, its composition and quantity.

Separate attention is devoted to the landfill surface settlement to be able to predict settlement in the future. Especially in the case when the surface of landfill is proposed for further utilization.

8.3.9 Remediation of Old Landfills

In many countries the old landfills were constructed without any artificial sealing systems or only with systems which are now recognised as insufficient. When checking the potential present-day negative impact on environment, the distinction between municipal and chemical waste landfills has to be done. Historical municipal landfills were usually smaller than those today, and because the degradation processes are over, the potential risk is in most cases, rather a small one. Therefore very often the capping system is applied just to protect direct contact with deposited waste and to decrease the amount of water passing through it. Methods of construction as described in the previous part can be used.

Old chemical landfills with large volume are more problematic. Usually method of encapsulation is selected. To the capping system the vertical sealing walls are added. But because there was a tendency to store as much waste on one place as possible, some problems with slope stability can arise when reshaping the final profile of the landfill before construction of capping system. Landfill in Chabařovice is one such example, Vaníček et al. (2003, 2005).

Landfill Chabařovice is situated close to the town Ústí nad Labem along the road connecting Ústí nad Labem with small town Chabařovice. Storing of the waste material started roughly 100 years ago. Originally landfill was used for storage of slag and ash from chemical, glass and textile industries. Later on different chemical waste materials were stored there as different sludge from the production of organic pigments, from production of caustic lye by mercury electrolysis, from production of potash, arsenic sludge and many others. In the southwest part barrels with hexachlorbenzene were stored. Barrels were covered by acid gypsum containing 2–3% of sulphur acid. Landfill was also used for storing different waste material from industrial accidents. Detailed information is unknown about all the stored materials during the lifetime of this landfill.

A basic assumption of the landfill remediation is based on the principle that no material will be transported out of the landfill area. Therefore the individual steps of the remediation process are as follow:

– Ground reshaping,
– Measures for drainage of surface water as well as for internal contaminated water,
– Vertical sealing system all around landfill,
– Measures against dusting and combustion processes inside of landfill body,

- Measures to increase slope stability of the landfill body,
- To protect infiltration into landfill body by horizontal sealing system,
- Recultivation of remediated surface and surrounding landscape,
- Monitoring remediation actions.

All protected area inside fencing is roughly 40.8 ha with 26.8 ha surrounded by vertical sealing wall. Top of the landfill is the highest place with slopes all around. Close to this peak was small lagoon where liquid waste material was stored, Fig. 8.73.

The body of the landfill was not stable in the history, different movements were observed, very often connected with small – creep – movement, and it was very sensitive to the rainfall. Landfill reshaping has to guarantee not only that the surface sealing system will be gravitationally drained but also that the lower general slope inclination will increase slope stability. Additionally the loading berm was proposed and constructed in the north-east part of landfill, where previous instability was observed. But in fact first problems started during the excavation for loading berm when local slips occurred. In summer 2001 the sealing wall was finished as well as the capping system in the central part of the landfill. After heavy rainfalls in September 2001 first cracks were observed in the north-east part of the landfill, roughly in the middle of the slope, where ground shaping was not fully finished yet.

The reason why the cracks were observed in the middle of the slope was not so difficult to explain due to an injection of the upper part where barrels with liquid

Fig. 8.73 Chabařovice landfill. Aerial view during remediation

waste material were expected and also due to higher infiltration of the surface water from the upper part, where capping system was finished. Visual and geodetic observation proved lower speed of deformation in the autumn and winter time, but after another speed increase seven inclinometers were installed in March 2002. They proved shear movement in depth of about 20 m in inclinometers just below observed cracks, about 5–7 m just front of the sealing wall and no deformation behind sealing wall. Cross section is shown in Fig. 8.74. From the beginning the main discussion was associated with character of slope instability, because especially in IJ5 shear deformation was only in the order of mm per month. Results from inclinometers IJ4 and IJ5 (and similar ones in the second cross section) were indicating slip surface, but on the other hand in the lower part no indication of surface uplift was observed to balance the mass lost in the upper part. To check the influence of creep deformation on the slurry wall, the surface was cleaned (protected during winter time) and only small cracks (1–3 mm) were observed, perpendicular to the longitudinal axis of the wall and guiding concrete strips. Shearing along these cracks reached max. 10 mm. Therefore the outer face of the vertical sealing wall was exposed and cracks were observed there, partly filled by clay. Width of these cracks was higher than on the surface, therefore the discussion of the width of the pit to allow bending loading of the exposed wall started. After that the discussion if the slope deformations are due to slip shearing, or as the results of continuous deformation of the deposited clay, or as the result of collapse of spaces after old deep mining activity (roughly 20 m under surface of tertiary clays), was not so important.

At this stage it was doubtless that some additional steps should be taken and that the price of the remediation will go up. The problem was that so many members started with some initiative – designer, investor, contractor, supervisor etc. Field investigation included another inclinometers (result of which was that the slip surface can be even deeper in the lower part, but shear deformations were again very low there), and pore pressure measurement. Pore water pressure was very high, exceptionally even higher than the surface at this time, probably as the result of the additional loading in this part, indicating that the pore pressure parameter B is close to 1.

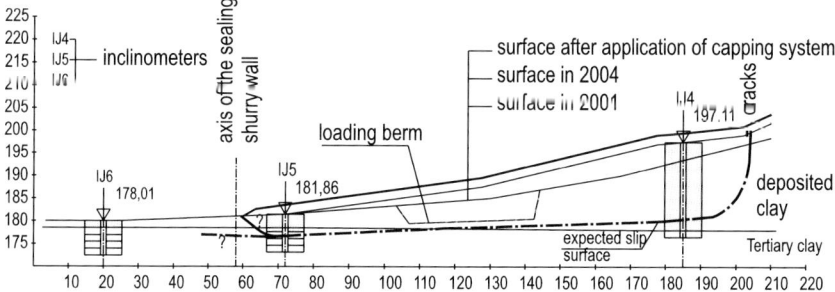

Fig. 8.74 Section through reclaimed Chabařovice landfill

Many different slope stability calculations were executed, as methods of limit equilibrium so FEM. For average inclination between inclinometers IJ5 and IJ4 roughly 8° and residual angle of the internal friction roughly 11.5° it is obvious that pore pressure plays most significant role. Potential slip surface is situated close to the boundary of origin tertiary clays with new deposited clays, where lowest shear strength can be expected as the result of water inflow and kneading in newly deposited clay and due to upheaval and swelling at the surface of origin tertiary clay. When using wedge method of calculation it was concluded, that the potential slip surface will go up in the lower part close to the axis of the sealing wall. Therefore when discussion what method of stabilization of slope deformation have to be used the method of lowering pore pressure by vertical drains supplemented with small additional loading berm was preferred front of the methods proposing retaining wall from deep piles.

Still the investor preferred to prove the proposed method on limited area. Therefore at the end of 2003 vertical geodrains were installed on selected area – roughly 30 000 m with average depth about 12 m (between 7 and 17 m). Proposed depth 17 m was reached only in the exceptional case because the resistance was bigger than expected from the penetration tests. Results of the pore pressure measurement proved quick pore pressure dissipation and the same result was obtained when additional loading berm was applied above area with vertical drains.

Final recommendations therefore include the following steps:

- To realize another vertical drains in strip above sealing wall (roughly 80 000 m with average depth 15 m),
- To add loading berm above this part,
- To add additional thin vertical sealing wall below existing one,
- To substitute proposed compacted clay liner (CCL) on the surface by bentonite mattress (GCL) to decrease additional loading and to improve the flexibility of the surface sealing.

8.3.10 Utilization of Landfill Surface for New Construction

With the up to date philosophy of the sustainable construction to use brownfields as much as possible for new construction to safeguard greenfields, requirements for utilization of landfill surface will increase. An interesting summary was presented by Onitsuka et al. (1996) from Japan, where this problem is most sensitive. Some landfills are constructed at the coastal disposal site where clay formations at the bottom are utilized. Degree of landfill use is classified into three categories, namely low degree such as farm and park, intermediate degree such as low-rise buildings and finally high-rise degree such as intermediate and high buildings. From 657 municipal waste disposal sites 70% of sites are used, 52% as agricultural land or wood land, 14% as park, open space, disaster prevention land and 34% for construction

8.3 Landfills

of housing, schools, and factories. Same percentage of total utilization is valid for industrial waste disposal sites, coastal disposal sites are often used as harbour facilities. As a practical example one school building is described in more details. Building was constructed on 10 m high deposit of municipal waste. The facilities were a three storey school building and a gymnasium. Foundation of the building have box section, the inside space used as a gas collection system.

8.3.10.1 Neratovice Landfill

The example of combination of remediation of old chemical landfill with a construction of a new one on its surface is described by authors – Vaníček et al. (1997, 2003).

The Neratovice landfill is situated north of Prague a few hundred metres from the Labe river. Geology of the site is simple – gravely sand of the depth of about 10 m is underlain by cretaceous marlstones that are practically impervious. The underground water table corresponded originally with the stable water table of the river, being about two metres under the original surface of the site. However due to activities in the nearby fly ash lagoons the groundwater level rose by about 1 m, reaching the bottom layers of the present dump. Monitoring of ground water contamination showed a wide range of pollution, in many cases exceeding the standardized limits. This fact was the initial point for remediation, which was performed between 1996 and 1997.

Dumping in the site started in the early 1950s. Sand was excavated just above the original GWL and embankments were built around the area of about 450×300 m. In this lagoon fly ash was deposited. In the mid 1960s the ash was partly excavated and used to build new neighbouring lagoons. Slurries from a water treatment plant were deposited in one part of the site. Later, depositing of waste from the chemical plant took place and continued until the late 1980s. The waste material was dumped randomly; no reliable information about the placement of individual materials and about the techniques of dumping is available. Some chemicals were deposited in barrels. The estimated volume of the industrial chemical waste from the chemical plant is $200\,000\,\text{m}^3$, two thirds of it being chemical sludge and sediments and about one third consisting of construction rubble with admixtures of plastics, wood, fabric etc. Further to these wastes a few thousand m^3 of sewage sludge was deposited on the top of the fill. Chemical plant Spolana Neratovice is one of the biggest chemical plants in the Czech Republic and is located on the opposite side of the river Labe than the described chemical landfill.

Basic Principles of Landfill Remediation

Because the chemical factory also in the future has to store some waste material in this area it was decided that the remediation should be based on the following principles:

– To stop the contaminant spreading from the waste dump,
– To partly use the locality for storage of a new chemical waste material.

The classical remediation consists from the construction of underground sealing wall in combination with hydraulic barrier, with continuous pumping inside of the protected area to decrease underground water level there, Fig. 8.75.

The horizontal sealing system above old stored chemical material is constructed by such a manner that can be used as a bottom sealing system for new chemical landfill and can also fulfil all higher demand on it. Final capping sealing system will be constructed only after storage of the additional chemical waste material. Because the old waste material was expected to be significantly compressible and to get more space for new deposited material, the inclination of the constructed capping sealing system was small, and allowing only the differential settlement up to 1/400, it was necessary to compact deposited waste to be sure that the new drainage system will work perfectly and that the clay sealing layer will not be damaged by tensile cracks.

From the individual construction steps, the four main ones will be described in more detail:

– Compaction of the old chemical waste material in such a way, that long-term safety of the sealing and drainage systems on the top of it can be guaranteed.
– Construction and control of the underground sealing wall.
– Construction and control of the horizontal sealing system.
– Monitoring system.

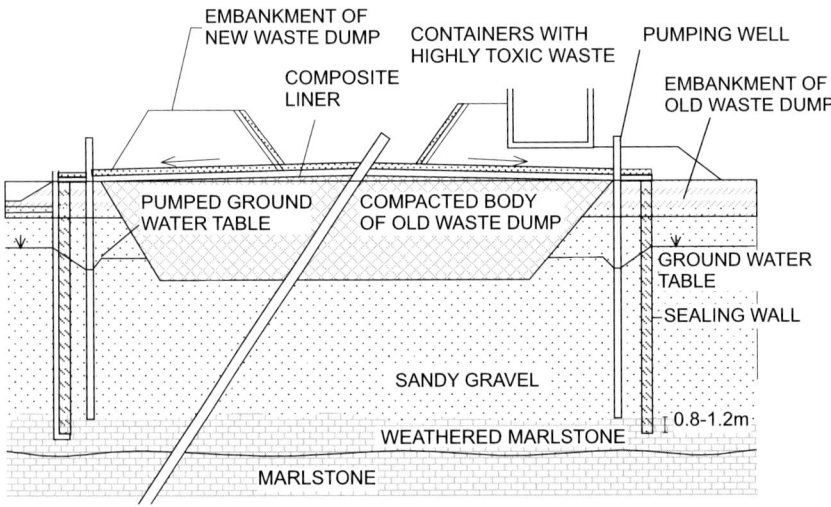

Fig. 8.75 Cross section of the Neratovice chemical landfill remediation with sealing system on the top for new landfill situated on the old one

8.3 Landfills

Dynamic Consolidation

Preparation

The following requirements for the old chemical waste dump treatment were defined in the design:

- Total settlement of the surface should be lower than 100 mm
- Differential settlement of the compacted surface, i.e. also of the horizontal bottom liner under the surcharge by new chemical waste material stored on it, should be lower than 1:400
- Bearing capacity in the place where special containers with most dangerous material will be stored should be higher than 0.2 MPa.

The first two conditions are connected with a good function of the horizontal sealing layer. The first one suggests there is a possibility that the clay liner system can be fractured due to tensile cracks. Some previous results concerning the bending tensile tests were used to check what tensile elongation such clay liner can withstand. The other condition is connected with a good function of the drainage system above the horizontal sealing system because the surface inclination is only 1%.

In order to fulfil all these conditions it was necessary to:

- Check the quality of the old chemical waste material
- Improve the quality of this material to the depth up to 10 m.

The dynamic consolidation method was chosen not only in order to improve the quality of the material to a much greater depth than using other methods, but also because the observation method can be used for a step by step improvement by the monitored response of the soil.

Additional Geo-Environmental Investigation

Time factor played a very important role at this stage, because there were only several weeks allowed for the additional investigation and the execution design, and both physical and chemical properties of the waste dump remained a mystery.

The only possibility was the retrospective analysis of the dump genesis from existing scarce records on industrial production, interview with witnesses, archive military air photos and supplementary field-testing. Synthesis of the recovered data created an overall scheme about the composition and the characteristics of materials dumped there during 30 years as well as the work's valid hypothesis including qualitative features:

- The dump body was created from different layers (from the bottom)
  1–1.5 m sluiced ashes and clinkers, with glue residues
  4–6 m chemical waste in mostly granular, pseudo-layered shape, mega-structured in lenses 0.3–1 m of ashes
- Character of the pseudo-soil is mostly fine-grained, mainly in loose state of porosity

- Ground water level is at 4–6 m, locally closed in lenses, dangerous for dissolution or liquefaction
- The pseudo-soils could not be classified as normal soils. Their behaviour had to be assessed by empirical correlations in relative values, based on phenomenological in situ testing, like penetration and loading tests.

Penetration testing (CPT) proved to be very useful for checking the quality of the chemical waste material. It showed that the material is really very heterogeneous, at some places with nearly zero strength.

A special problem was connected with the protection against an unexpected influence of the compaction on the possible chemical processes inside the dump body. During the investigation and the compaction works a special group of chemical engineers was present and followed a special safety programme.

Execution of the Dynamic Consolidation Method

The method is based on the simple principle of compaction by the fall of the pounder weighing 15 tonnes from the height of 20 m – Fig. 8.76a. This method was chosen in order to improve the foundation property of the dump site due to its undisputed advantages. The compaction programme of several series started in ground points which create rectangular net with side of 10 m. During the compaction process the settlement for each compaction print was measured as well as the movement of the surrounding surface area and evaluated by on-site computer – examples see Fig. 8.77a, b, c. After evaluating these results, the process followed on the ground in the distance between individual points up to 5 m and on places with worst conditions up to 2.5 m. By this way much more information about heterogeneous conditions inside the landfill body were gained than from the previous site investigation. Where the weak spots were discovered they were then reinforced with the creation of the gravel pillars. The final "ironing" impacts were carried out at the net of 2.5 m – Fig. 8.76b. High productivity rates were achieved during the works thanks to the employment of a semi-automatic equipment, the laser-beam measurement and the above mentioned computerized evaluation. The consolidation energy varying at 227–345 tm per m^2 was impacted into the foundation ground within 2 months.

The effectiveness of this method was demonstrated not only by a considerable immediate compression of the surface -0.567 m on average – but also with the consolidation reaching a considerable depth -10 m. At the same time, the internal body of the dump was not penetrated and material was not carried to the surface but covered with a spreading gravel layer so that the danger of toxic emissions was kept at a minimum level as confirmed in the course of works.

Testing of Results

The improvement of the pseudo-soil achieved by the dynamic compaction has been quantified by two series of the Dutch cone penetration tests (CPT). The first series of CPTs was carried out before the dynamic compaction started. After the dynamic

8.3 Landfills

(a)

(b)

Fig. 8.76 Application of dynamic consolidation. (**a**) Compaction performance. (**b**) "Ironing" stage of compaction

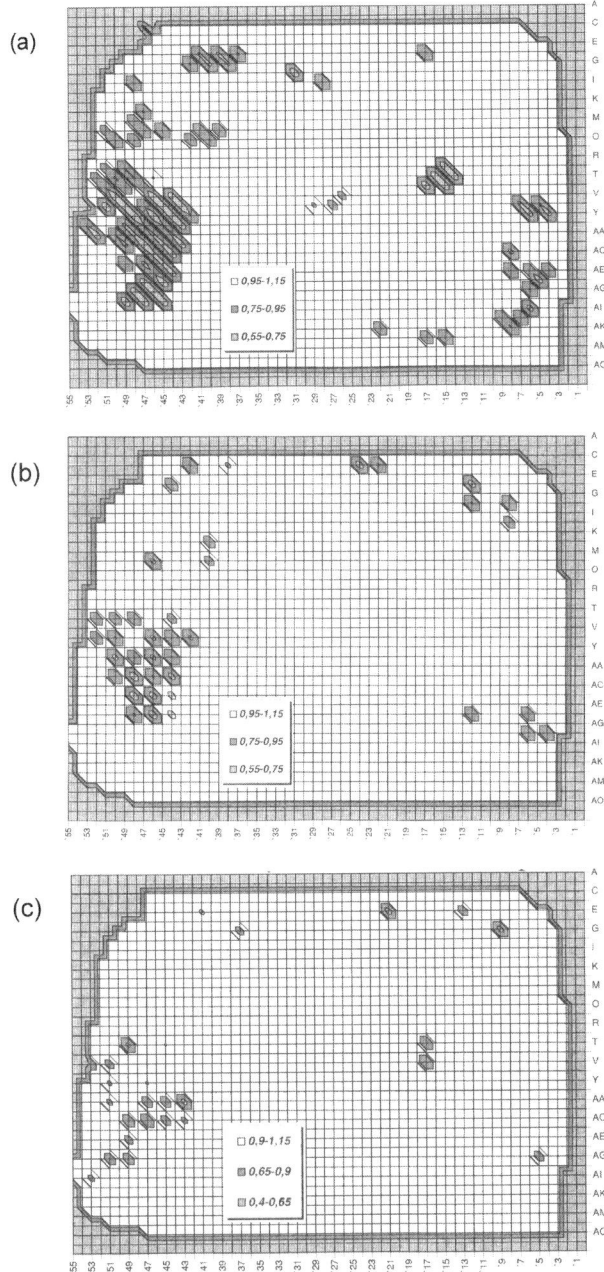

Fig. 8.77 Evaluation of the compaction. (**a**) after 1st series. (**b**) after 2nd series. (**c**) after 3rd series

8.3 Landfills

compaction was completed, CPTs were repeated at the corresponding points. The effectiveness of the dynamic compaction was evaluated comparing the cone resistance q_c of the first and second series, i.e. prior and after the treatment.

In analyzing the individual CPT soundings the raw data were smoothed by averaging within every set of neighbouring readings. This procedure yielded relatively smooth CPT profiles that characterized the mechanical properties of the ground. An example of such data is shown in Fig. 8.78a. To quantify the improvement achieved by dynamic compaction, an improvement ratio has been defined as the ratio of the cone resistance after and before compaction. This way the improvement at every individual sounding can be computed as well as the overall improvement of the site. To characterize the overall improvement an idealized profile has been constructed in which the values at individual depths were computed as averages from all raw readings at a particular depth. Two sets of the averages were computed, before and after the compaction. Comparison of the two states can be seen in Fig. 8.78b. From Fig. 8.78b one can conclude that the dynamic compaction was effective mostly in the zone from 1 to 7 m below the surface. The improvement achieved in this zone can be expressed by the average value or the improvement ratio of 1.9. Deeper below the surface, from 7 m down to 10m the influence of the dynamic compaction was not so high. However, even in the deeper layers a positive effect of the treatment has been verified, as documented for example in Fig. 8.78a.

The average cone resistance before the compaction computed from all the raw data from all soundings was $q_{c,before} = 8.8$ MPa. The corresponding value after the compaction $q_{c,after} = 12.3$ MPa shows that a 50% increase in cone resistance was achieved. Moreover, the number or extremely low values of $q_c = 1$ MPa (or even lower) that were found with some soundings under the original state, was substantially limited and such values were rather exceptional, with the soundings carried out after the compaction. The average value $q_c = 12$ MPa corresponds roughly to the required design bearing capacity 0.2 MPa.

In order to measure the stiffness of the material of the dump after the treatment by the dynamic compaction, a trial embankment for load testing was built on its top. The embankment was 3 m high, 6 m wide at the top and 20 m long. Surveying a number of reference points shown in Fig. 8.79a monitored the settlement. The measured values are in Fig. 8.79b. The following conclusions were drawn from the loading test:

- The maximum settlement measured was 38 mm.
- Figure 8.79b shows that the settlement of the dump below the embankment was relatively quick, probably due to only a partial saturation of its material. However, occurrence of saturated material of low permeability with a slow dissipation of pore pressures and slow consolidation is possible in the heterogeneous dump.
- The differential settlement of the trial embankment was about 1/1000 with the acting contact pressure of about 60 kPa. The design surcharge is 130 kPa. Therefore, differential settlement of about 1/500 should be anticipated. This value complies with the project requiring allowable differential settlement of up to 1/400.

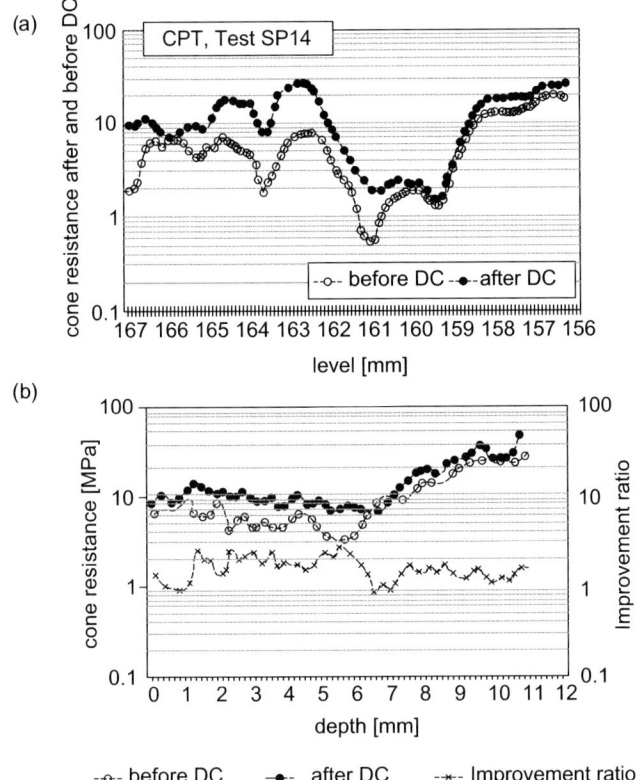

Fig. 8.78 Evaluation of the compaction effort by cone penetration tests – CPT. (**a**) results of the test SP14 before and after dynamic consolidation (DC). (**b**) idealized sounding characterizing the whole site

- From the settlement data a constraint modulus of 9–12 MPa was derived. Using these values the final settlement of 60–95 mm after loading the surface by the new surcharge has been computed. The allowable settlement according to the project is 100 mm.

The trial embankment proved to be a successful tool in investigating the deformation behaviour of the dump, namely in quantifying the stiffness of the compacted heterogeneous fill of the dump and in estimating the maximum and differential final settlements under the planned surcharge.

Cut-Off Diaphragm Sealing Wall

An important component of the geo-containment of the encapsulated old landfill was the 600 mm thick cut-off diaphragm wall of the depth of 11–20 m. In this given case, the technical specifications for the underground sealing wall were as follows:

8.3 Landfills

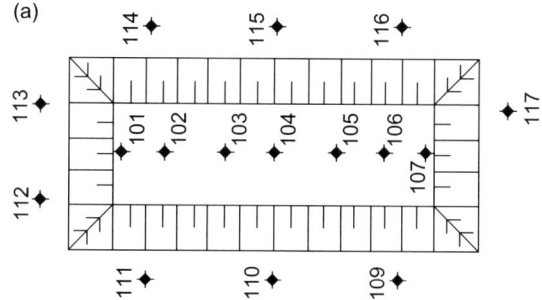

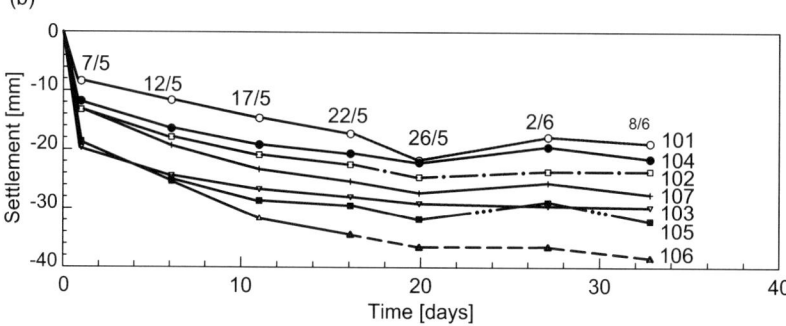

Fig. 8.79 Evaluation of the compaction effort by trial loading test embankment. (**a**) Scheme of the trial embankment. (**b**) Measured settlement of the trial embankment

- Minimum permeability, characterized by the filtration coefficient $k_f \leq n.10^{-10} m.s^{-1}$
- Long-term life expectancy in the landfill's contaminated environment
- Ability to limit the leakage of contaminated material to the maximum degree.

To meet the above demands 24-h supervision by an independent geotechnical consulting firm was carried out on the site. This supervision concentrated on the following points:

- Checking the quality of the used self-hardening slurry before its use – viscosity, volume density and velocity of sedimentation
- Checking the depth for each excavation bit. The main condition was that the underground sealing wall should reach at least 0.8 m of the weathered marlstone in order to fulfil the condition that the sealing wall would have a good embedment into the competent impermeable subsoil
- Checking the permeability and compression strength in the laboratory. Tests on the samples extracted by a special procedure were conducted after 90 days
- Checking the permeability of the constructed cut-off diaphragm wall. In order to prevent its damage, it was finally decided to perform this test only on parts, which are in fact out of the perimeter. After demanded time for the hardening of the wall the core-boring started and the permeability tests were made not only on

the cores obtained from this boring but also directly inside the boreholes in the wall body.

None of the tested samples (nor the test bodies) did overlap the lower demanded boundary. Maximum value: $k_f = 9.5.10^{-11} \mathrm{m.s}^{-1}$, minimum value: $k_f = 1.7.10^{-12} \mathrm{m.s}^{-1}$.

Horizontal Sealing System

Horizontal sealing system was designed and performed according to existing Czech standard, which is very similar to standards in the other countries for this type of landfill. This system consists from clay liner with thickness 1.0 m compacted in five layers and covered by geomembrane HDPE 2.0 mm. On compacted old toxic waste surface drainage layer – sandy gravel – was used – then described sealing system, where coefficient of permeability for clay liner was defined by minimum value $1.10^{-9} \mathrm{m.s}^{-1}$ – geomembrane was protected by unwoven geotextile and finally drainage layer from sand was used with minimum coefficient permeability $1.10^{-4} \mathrm{m.s}^{-1}$.

Clay liner was controlled by three basic ways

- Index properties of the material used for clay liner (grain size distribution, moisture content, plastic and liquid limits, content of organic matter),
- Degree of compaction (volume density and moisture content) performed not only in the laboratory but also in the field by TROXLER radiometer,
- Coefficient of permeability – as the main criteria – mostly in the laboratory on small samples.

Quality of double weld of geomembrane was checked by pressure test. For the overall control of geomembrane impermeability a special monitoring system was used based on electric conductivity.

Monitoring System

Monitoring system is in principle based on three types of measurement

- Quality of underground water pumped from the installed wells inside and outside of the remediated landfill,
- Settlement of the horizontal sealing system, whether the condition of the allowable differential settlement was reached,
- Electric conductivity measurement under geomembrane to check its long-term functionality.

Vertical deformation (settlement) is monitored by hydrostatic probe. This probe checks differential settlements in horizontal plastic pipe that was inbuilt under the new landfill. Maximum settlement for the last 6 years has been found 70 mm. Also two other measurements proved very good performance of the landfill remediation.

The presented case is the example of the new approach to the problems of old ecological burdens in the Czech Republic. This approach is based on good cooperation between individual partners, where the supervision with the help of control and monitoring systems plays a very significant role. Video has been shot about the landfill remediation process, which is now used as tutorial for students. From the technical point of view unusual case of compaction of chemical waste material was presented and the results show that it is possible to construct new safe horizontal liner on top of such compacted material.

References

Ache W, Holzlohner U, Lehnert J (1983) Improvement of structure-ground systems by preloading. In: Proc. 8th EC SMFE, Helsinki, Balkema, vol 2, pp 559–564.
Adam D (1996) Flachendeckende dynamische Verdichtungskontrolle (FDVK) mit Vibrationswalzen (Continuous compaction control (CCC) with vibratory rollers). Ph.D. thesis, Technical University of Vienna, 268 p.
Adam D, Markiewicz R (2001) Compaction behaviour and depth effect of the Polygon-Drum. In: Correia AG and Brandl H (eds) Geotechnics for Roads, Rail Tracks and Earth Structures, Balkema, pp 27–36.
Adolfsson K, Andreásson P, Bengtsson E, Zackrisson P (1999) High speed X2000 on soft organic clay – measurement in Sweden. In: Barends F et al. (eds) Proc. 12th EC SMGE, Amsterdam, Balkema Rotterdam, vol 3, pp 1713–1718.
Aitchison GD, Wood CC (1965) Some interactions of compaction, permeability and postconstruction deflocculation affecting the probability of piping failure in small earth dams. In: Proc. 6th IC SMFE, Montreal, Toronto University press, vol 2, pp 442–446.
Ajaz A, Parry RHG (1975a) Stress-strain behaviour of two compacted clays in tension and compression. Géotechnique, vol 25, No 3, pp 495–512.
Ajaz A, Parry RHG (1975b) Analysis of bending stresses in soil beams. Géotechnique, vol 25, No 3, pp 586–591.
Akai K, Tanaka Y (1999) Settlement behaviour of an off-shore airport KIA. In: Barends et al. (eds) Proc. 12th EC SMGE, Balkema, vol 2, pp 1041–1046.
Alberro J (1977) Effect of interaction in earth-rockfill dams. In: Proc. 9th IC SMFE, Spec. Sess. 8 – Deformation of Earth-rockfill Dams, Tokyo, JGS, pp 1–18.
Alberro J et al. (1976) Guadalupe dam. In: Behaviour of Dams Built in Mexico. Contr. to 12th ICOLD, Com Mexicano de Grandes Presas, Mexico, 10 p.
Alexiew D (2005) Piled embankments: Overview of methods and significant case studies. In: 16th IC SMGE, Osaka, Millpress, pp 1819–1822.
Alexiew D (2007) Load tests on a geosynthetic reinforced abutment (in German). Geotechnik 2007/2, pp 87–94.
Alexiew D, Elsing A, Ast W (2002) FEM-Analysis and dimensioning of a sinkhole overbridging system for high-speed trains at Gröbers in Germany. In: Delmas, Gourc, and Girard (eds) Geosynthetics – 7th ICG Nice, Lisse, Sweto, Zeitlinger, pp 1167–1172.
An Ninh N (2000) Application of geosynthetic materials in dam construction. Ph.D. thesis Czech Technical University Prague, 184 p.
Andrawes KZ, McGown A, Wilson-Fahmy RF, Mashhour MM (1982) The finite element method of analysis as applied on soil-geotextile systems. In: Proc. 2nd Int. Conf. on Geotextiles, Las Vegas, vol 3, pp 695–700.
Atkinson J (1993) An introduction to the mechanics of soils and foundations. McGraw-Hill International Series in Civil Engineering. London, 337 p.
Baliak F, Malgot J, Kopecký M (1996) Landslides on Slovak transport structures. (In Slovak). In: Proc. 2nd Slovak Geotechnical Conf., Bratislava, Slovak Technical University, pp 12–18.

Barden L (1971) Examples of clay structure and its influence on engineering behaviour. In: Proc. Roscoe Mem. Symp., "Stress-strain behaviour of soils" Cambridge, pp 195–205.

Barron RA (1948) Consolidation of fine grained soils by drain wells. Translation ASCE, vol 113, pp 718–742.

Barvínek R, Havlíček J (1977) Ervěnice corridor. (In Czech). Critical Notice. Stavební geologie Praha, 30 p.

Basset RH (1979) The use of physical models in design. General Report. In: Proc. 7th EC SMFE, Brighton, BGS, vol 5, pp 253–270.

Battelino D (1983) Some experience in reinforced cohesive earth. In: Proc. 8th EC SMFE, Helsinki, Balkema, vol 2, pp 463–468.

Beetstra GW, Stoutjesdijk TP (2005) First approach to rational risk management (RRD) by the Delta Institute, 1 November. GeoDelft, Delft.

Beinbrech G, Hohwiller F (2000) Cushion foundations: Rigid expanded polystyrene foam as a deforming and cushioning layer. Ground Improvement, vol 4, No 1, pp 3–12.

Benson CH, Daniel DE (1990) Influence of clods on hydraulic conductivity of compacted clay. Journal of Geotechnical Engineering ASCE, vol 116, No 8, pp 1231–1248.

Bergado DT, Balasubramanian AS, Fannin RJ, Holtz RD (2002) Prefabricated vertical drains (PVDs) in soft Bangkok clay: A case study of the new Bangkok International Airport project. Canadian Geotechnical Journal, vol 39, No 2, pp 304–315.

Bertram GE (1940) An Experimental Investigation of Protective Filters. Harvard University Soil Mechanics Series, No 7, p 1.

Bhatia SK, Moraille J, Smith JL (1998) Performance of granular vs. geotextile filters in protecting cohesionless soils. In: ASCE Spec. Publ No 78. Filtration and Drainage in Geotechnical/Geoenvironmental Engineering, pp 1–29.

Bishop AW (1955) The use of the slip circle in the stability analysis of earth slope. Géotechnique, vol 5, pp 7–17.

Bishop AW (1957) Some factors controlling pore pressure set up during the construction of earth dams. In: Proc. 4th IC SMFE, London, vol 2, pp 294–300.

Bishop AW (1959) The principle of effective stress. Teknisk Ukeblad, vol 106, No 39, pp 859–863.

Bishop AW (1967) Progressive failure with special reference to the mechanism causing it. In: Proc. EC ISSMFE, Oslo, vol 2, pp 3–10.

Bishop AW (1973) The stability of tips and spoil heaps. Quarterly Journal of Engineering Geology, vol 6, pp 335–376.

Bishop AW, Bjerrum L (1960) The relevance of the triaxial test to the solution of stability problems. NGI Publ. No 34, Oslo.

Bishop AW, Garga VK (1969) Drained tension tests on London clay. Géotechnique, vol 19, No 3, pp 310–313.

Bishop AW, Henkel DJ (1962) The measurement of soil properties in the triaxial test. Edward Arnold Publishers, London, 227 p.

Bishop AW, Wesley LD (1975) A hydraulic triaxial apparatus for controlled stress path testing. Géotechnique, vol 25, pp 657–670.

Bjerrum L (1963) Discussion. Section VI Interaction between structure and soil. In: Proc. 3rd EC ISSMFE, Wiesbaden, vol 2, pp 135–137.

Bjerrum L (1967a) Discussion. In: Proc. 9th ICOLD, Istanbul, vol 6, p 456.

Bjerrum L (1967b) Progressive failure in slopes of over-consolidated plastic clay and clay shales. JSMFD ASCE, vol 93, No SM5, pp 3–49.

Bjerrum L (1973) Problems of soil mechanics and construction on soft clays. In: Proc. 8th IC SMFE, Moscow, 1973, vol 3, pp 109–159.

Bjerrum L, Nash JKTL, Kennard RM, Gibson RE (1972) Hydraulic fracturing in field permeability testing. Géotechnique, vol 22, No 2, pp 223–254.

Blight GE (1994) Environmentally acceptable tailings dams. In: Proc. 1st ICEG, Edmonton, BiTech Publishers, pp 417–426.

Blight GE, Troncoso JH, Fourie AB, Wolski W (2000) Issues in the geotechnics of mining wastes and tailings. GeoEng2000, Melbourne, Technomic Publ. Comp., vol 1, pp 1253–1283.

Blume KL (1996) Long-term measurement on a road embankment reinforced with a high-strength geotextile. In: Proc. EuroGeo 1, Maastricht, Balkema, pp 237–244.

Boden JB, Irwing MJ, Pocock RG (1979) Construction of experimental reinforced earth walls at the transport and road research laboratory. TRRL Supplementary Report, vol 457, pp 162–194.

Bourdeau PL, Pardi L, Recordon E (1991) Observation of soil reinforcement interaction by x-ray radiography. In: McGown, Yeo, and Andrawes (eds) Performance of Reinforced Soil Structures, Thomas Telford, London, pp 347–352.

Brady KC, Masterton GGT (1991) Design and construction of an anchored earth wall at Annan. In: McGown, Yeo, Andrawes, and Thomas (eds) Performance of Reinforced Soil Structures, Thomas Telford, London, pp 127–134.

Brakel J, Coprens M, Maagdenberg AC, Risseeuw P (1982) Stability of slopes constructed with polyester reinforcing fabric, test section at Almere. In: Proc. 2nd Int. Cong. on Geotextiles, Las Vegas, vol 3, pp 727–732.

Brandl H (1999) Mixed-in-place stabilization of pavement structures with cement and additives. In: Barends FBJ et al. (eds) Proc. 12th EC SMGE, Amsterdam, Balkema, Rotterdam, vol 3, pp 1473–1481.

Brandl H (2000) Stability and failures of landfills. In: Proc. IXth Baltic Geotechnics, Pärnu, Estonian G.S. 16 p.

Brandl H (2001a) Compaction of soil and other granular material – interactions. In: Correia AG and Brandl H (eds) Geotechnics for Roads, Rail Tracks and Earth Structures. Balkema, pp 3–12.

Brandl H (2001b) The importance of optimum compaction of soil and other granular material. In: Correia AG and Brandl H (eds) Geotechnics for Roads, Rail Tracks and Earth Structures. Balkema, pp 47–68.

Brandl H (2001c) High embankments instead of bridges and bridge foundations in embankments. In: Correira AG and Brandl H (eds) Geotechnics for Roads, Rail Tracks and Earth Structures. Balkema, pp 13–26.

Brandl H (2001d) Slope stability and spreading of landfills. In: Proc. 2nd Australia-New Zealand Conf. on Environmental Geotechnics, Newcastle-Australia, 6 p.

Brandl H (2001e) Geotechnics of rail track structures In: Correia AG and Brandl H (eds) Geotechnics for Roads, Rail Tracks and Earth Structures. Balkema, pp 271–288.

Brandl H (2004) Geotechnical aspects for high-speed railways. In: Correira AG and Loizos A (eds) Geotechnics in Pavement and Railway Design and Construction. Millpress, pp 117–132.

Brandl H, Adam D (2004) Soil and rock nailing – stability analyse and case studies. In: Proc. Int. Conf. Soil Nailing and Stability of Soil and Rock Engineering, Nanjing, CI-Premier PTE Singapore, pp 1–16.

Brandl H, Blovsky S. (2003) Zustandsbewertung, Überwachung, Sicherung und Sanierung von Deichen bzw. Hochwasserdsämmen. In: Sicherung von Dämmen und Deichen., edt. Hermann and Jensen, Universität Siegen. 34 p.

Brandl H, Robertson PK (1996) Geo-environmental site investigation and characterization. In: Proc. 2nd ICEG, Osaka. Preprint of Special Lectures and State-of-the-Art reports. ISSMGE and JGS, pp 117–140.

Brandl H, Gartung E, Verspohl J, Alexiew D (1997) Performance of a geogrid-reinforced railway embankment on piles. In: 14th IC SMFE, Hamburg, Balkema, pp 1731–1736.

Brandl H, Kopf F, Adam D (2005) Continuous Compaction Control (CCC) with differently excited dynamic rollers. Bundesministerium für Verkehr, Innovation und Technologie, Straßenforschung Heft 553, Wien, p 150.

Brenner RP, Pbebaharan N (1983) Analysis of sandwick performance in soft Bangkok clay. In: Proc. 8th EC SMFE, Helsinki, Balkema, vol 2, pp 579–586.

Broms BB, Anttikoski U (1983) Soil stabilization. In: Proc. 8th EC SMFE, Helsinki, Balkema, vol 3, pp 1289–1301.

Bryhn OR, Loken T, Aas G (1983) Stabilization of sensitive clays with hydro-aluminium compared with unslake lime. In: Proc. 8th EC SMFE, Helsinki, Balkema, vol 2, pp 885–896.

BS5930: 1999 Code of practice for site investigations. BSI, London.
Burd HJ, Brocklehurst CJ (1990) Finite element studies of the mechanics of reinforced unpaved roads. In: Proc. 4th Int. Conf. on Geotextiles, Geomembranes and Related Products, Den Haag, Balkema, Rotterdam, vol 1, pp 217–222.
Burenkova VV, Afinogenova MV (1974) Material analysis of transitional zones from Verchne-Tobolskoj dam. (In Russian) Trudy VNII VODGEO, No 44, Moscow, p 23.
Butterfield R, Harkness RM, Andrawes KZ (1970) A stereophotogrammetric method for measuring displacement fields. Géotechnique, vol 20, No 3, pp 308–314.
Cabrera JG, Woolley GR (1994) Fly ash utilization in civil engineering. In: Goumans, van der Sloot, and Aalbers (eds) Environmental Aspects of Construction with Waste Materials. Elsevier, pp 345–356.
Casagrande A (1950) Notes on the design of earth dams. Journal of the Boston Society of Civil Engineers, vol 37, No 4, pp 405–429.
Casagrande A (1969) Participation in panel discussion on earth and rockfill dams. In: Proc. 7th IC SMFE, Mexico, vol 3, 301 p.
Cazzuffi D, Crippa E (2005) Shear strength behaviour of cohesive soils reinforced with vegetation. In: Proc. 16th IC SMGE, Osaka, Millpress, vol 4, pp 2493–2498.
Cedergren HR (1960) Seepage requirements of filter and pervious bases. JSMFD, ASCE, vol 127, No SM5, pp 15–34
Chabal JP et al. (1983) A novel reinforced fill dam. In: Proc. 8th EC SMFE, Helsinki, Balkema, vol 2, pp 477–480.
Chaddock BCJ (1996) The structural performance of stabilized soil in road foundations. In: Lime Stabilisation. Thomas Telford, London.
Chae YE, Yoon GL, Lee KW, Kim BT (2003) Soft ground improvement at the Incheon International Airport. In: Leung et al. (eds) Proc. 12th AC SMGE, Singapore, Word Scientific, Singapore, vol 1, pp 147–150.
Chandler RJ, Skempton AW (1974) The design of permanent cutting slopes in stiff fissured clays. Géotechnique, vol 24, pp 457–466.
Charles JA, Watts KS (1996) The assessment of the collapse potential of fills and its significance for building on fill. In: Proc. ICE, Geotechnical Engineering, vol 119, January, pp 15–28.
Charles JA, Earle EW, Burford D (1979) Treatment and subsequent performance of cohesive fill left by opencast ironstone mining at Snatchill experimental housing site, Corby. In: Proc. Clay Fills, London, ICE, pp 63–72.
Chen FH (1975) Foundations on Expansive Soils. Developments in Geotechnical Engineering, vol 12, Elsevier Scientific Publ. Co, New York, 280 p.
Christopher BR, Holtz RD, Fischer GR (1993) Research needs in geotextile filter design. In: Brauns, Heibaum, and Schuler (eds) Filters in Geotechnical and Hydraulic Engineering. Balkema, Rotterdam, pp 19–26.
Čištín J (1973) Filtration layers on upstream face of earth-fill dams. (In Czech) Vodní Hospodářství, vol 23, row A, No 2, pp 31–36.
Čištín J (1980) Principles and methods for filter design from technical textiles in hydro engineering. (In Czech). Vodní Hospodářství, vol 30, row A, pp 11–15.
Clark R, Weltman AJ (2003) Man-made deposits – recent and ancient. State of the Art Report. In: Vaníček et al. (eds) Proc. 13th IC SMGE, Prague, Czech Geotechnical Society, vol 3, pp 77–96.
Clough RW, Woodward RJ (1967) Analysis of embankment stresses and deformations. JSMFED, ASCE, vol 93, No SM2.
Collins K, McGown A (1974) The form and function of microfabric features in a variety of natural soils. Géotechnique, vol 24, No 2, pp 223–254.
Consoli NC, Casagrande MDT, Coop MR (2005) Behaviour of a fibre-reinforced sand under large shear strains. In: Proc. 16th IC SMGE, Osaka, Mill press, vol 3, pp 1331–1334.
ČSN 721002: 1993 Soil Classification for Road Constructions, ČNI Prague.
ČSN 721006: 1998 Soil compaction control, ČNI Prague.
ČSN 721018: 1970 Laboratory determination of relative mature of cohesionless soils, ČNI Prague.
ČSN 731001: 1987 Foundation of Structures. Subsoil under Shallow Foundations, ČNI Prague.

ČSN 733050: 1986 Earthworks. General requirements, ČNI Prague.
ČSN 736133: 1998 Road earthwork - design and execution, ČNI Prague.
ČSN 736192: 1996 Impact load tests for road surfaces and subsurfaces, ČNI Prague.
ČSN 736850: 1975 Earth Fill Dams (Bulk Earth Dams), ČNI Prague.
ČSN 838036: 2002 Waste Landfilling – monitoring of Landfills, ČNI Prague.
D'Elia B, Distefano D, Esu F, Federico G (1979) Design criteria for uncompacted clay fills. In: Proc. 7th EC SMFE, Brighton, BGS, vol 1, pp 125–130.
D'Elia B, Distefano D, Esu F, Federico G (1983) Improvement of an uncompacted clay fill. In: Proc. 8th EC SMFE, Helsinki, Balkema, vol 2, pp 745–749.
Daniel DE (1993) Clay liners. Chapter 7 In: Geotechnical Practice for Waste Disposal. Chapman and Hall, London.
Daniel DE (1995) Pollution prevention in landfills using engineered final covers. In: Sarsby (ed) Green 93 Waste Disposal by Landfill. Balkema, Rotterdam, pp 73–92.
Daniel DE, Bowders JJ Jr (1996) Waste containment systems by geosynthetics. In: Proc. 2nd ICEG, Osaka. Preprint of Special Lectures and State-of-the-Art reports. ISSMGE and JGS, pp 49–66.
De Beer EE (1969) Stability problems of slopes in overconsolidated clays and schists. In: Proc. New Advances in Soil Mechanics I, Prague, Czechoslovak scientific and technical society, pp 53–104.
De Mello LGFS (1998) Re-use of by-products. In: Seco e Pinto PS (ed) Proc. 3rd ICEG, Lisbon, Balkema Rotterdam, vol 4, pp 1293–1297.
De Mello VFB (1977) Reflections on design decisions of practical significance to embankment dams. Géotechnique, vol 27, No 3, pp 279–355.
Delmas P, Matichard Y, Gourc JP, Riondy G (1986) Unsurfaced roads reinforced by geotextiles – a seven years experiment. In: Proc. 3rd Int. Conf. on Geotextiles, Vienna, vol 4, pp 1015–1020.
Delmas P, Queyroi D, Quaresma M, De Saint Amand, Puech A (1990) Failure of an experimental embankment on soft soil reinforced with geotextile: Guiche. In: Proc. 4th Int. Conf. on Geotextiles Geomembranes and Related Products, The Hague, Balkema, vol 3, pp 1019–1024.
DIN 1054 (1996) Draft, Soilsand Foundations, Safety in Geotechnical Engineering. DIN, Berlin.
Doleželová M (1970) Effect of steepness of rocky canyons slopes on cracking of clay cores of rock- and- earthfill dams. In: Proc. 10th ICOLD, Montreal, vol 1, pp 215–224.
Duncan JM, Williams GW, Sehn AL, Seed RB (1991) Estimation earth pressures due to compaction. Journal of Geotechnical Engineering, ASCE, vol 117, No 12, December, pp 1833–1847.
Dunlop P, Duncan JM (1970) Development of failure in excavated slopes. JSMFED, ASCE, vol 96, No SM2.
Dykast I (1993) Properties and behaviour of high clayey spoil heaps in the North-Bohemian brown-coal district. (In Czech), Ph.D. thesis, CTU Prague, 126 p.
Dykast I, Pegrimek R, Pichler E, Řehoř M, Havlíček M, Vaníček I (2003) Ervěnice corridor – 130 m high spoil heap from clayey material – with transport infrastructure on its surface. In: Vaníček et al. (eds) Proc. 13th EC SMGE, Prague, Czech Geotechnical Society, vol 4, pp 57–76.
EB GEO (1997) Empfehlungen für Bewehrungen aus Geokunststoffen. (Recommendations for reinforcement with Geosynthetics). (In German). German Society of Soil Mechanics and Geotechnical Engineering (eds), Berlin, Ernst & Sohn.
Eisenstein Z, Krishnayya AVG, Morgenstern NR (1972) An analysis of cracking in earth dams. In: Proc. Appl. FEM in Geot. Eng. US WES Vicksburg, pp 431–456.
Elsing A (2006) HaTelit reinforcement prevents crack propagation at Rosenstrasse (K57), Ochtrup, Germany. HUESKER Job Report, Gescher, 2 p.
Environmental Agency (2001) Contamination Classification Thresholds for Disposal, Version 3. Environmental Agency, UK.
Eriksson L, Ekstrom A (1983) The efficiency of three different types of vertical drain – results from a full scale test. In: Proc. 8th EC SMFE, Helsinki, Balkema, vol 2, pp 605–610.
ETC – 8 (1993) GLR-Recommendations: Geotechnics of Landfill Design and Remedial Works. Ernst and Sohn, Berlin, 2nd edn.
Eurocode 7: Geotechnical Design – Part 1: General Rules (EN 1997-1:2004). CEN, Brussels, p 167.

Fantini P, Roberti R (1996) Highway Frejus Torino: A case history of a green reclamation around a highway viaduct with geogrid reinforcement walls. In: Proc. EuroGeo 1, Maastricht, Balkema, pp 387–392.

Farrar DM (1978) Settlement and pore-presure dissipation within an embankment built of London Clay. Clay Fills. ICE London, pp 101–106.

Faure YH, Imbert B (1996) Use of geotextiles to prevent fine material from polluting railway subgrades. In: De Groot, Den Hoedt, and Termaat (eds) Geosynthetics: Application, Design and Construction. Balkema, Rotterdam, pp 479–486.

Feda J (1977) Mechanics of particulate media. (In Czech), Academia Praha.

Feda J (1982) Mechanics of particulate materials. The principles. Academia Prague, 448 p.

Feda J (2003) Behaviour of double porosity geomaterials. In: Vaníček I et al. (eds) Proc. 8th EC SMCE, Prague, CGtS, vol 1, pp 667–672.

Feda J, Herštus J, Herle I, Šťastný J (1994) Landfills of waste clayey material. In: Proc. 13th IC SMFE, New Delhi, Oxford and IBH Publ. Co, New Delhi, vol 4, pp 1623–1628.

Fell R, Hungr O, Leroueil S, Riemer W (2000) Keynote lecture – Geotechnical engineering of the stability of natural slopes, and cuts and fills in soil. In: GeoEng, Melbourne, Technomic Publishing Co, vol 1, pp 21–120.

Fellenius BH, Wager O (1977) The equivalent sand drain diameter of the band – shaped drain. In: Proc. 9th IC SMFE, Tokyo, JGS, 396 p.

Fiedler J (1990) The settlement of constructions founded on uncompacted fills. In: Glasgow, R (ed) Proc. 3rd Int. Symp. on the Reclamation, Treatment and Utilization of Coal Mining Wastes, Balkema, Rotterdam, pp 283–286.

Fischer GR, Christopher BR, Holtz RD (1990) Filter Criteria Based on Pore Size Distribution. In: Proc. 4th Int. Cong. on Geotextiles, The Hague, Netherlands, Balkema, vol 1, pp 289–294.

Fluet JE Jr (ed) (1987) Geotextile Testing and the Design Engineer. ASTM Publ. 952.

Forssblad L (1965) Investigations of soil compaction by vibration. Acta Polytechnica Scandinavia, Ci-34, Stockholm.

Foster M, Fell R, Spannagle M (2000a) The statistics of embankment dam failures and accidents. Canadian Geotechnical Journal, vol 37, No 5, pp 1000–1024.

Foster M, Fell R, Spannagle M (2000b) A method for assessing the relative likehood of failure of embankment dams by piping. Canadian Geotechnical Journal, vol 37, No 5, pp 1025–1061.

Frydenlund TE, Aaboe R (1994) Expanded polystyrene – a lighter way across soft ground. In: Proc. 13th IC SMFE, New Delhi, Oxford and IBH Publishing Co., New Delhi, vol 3, pp 1287–1292.

Fujii M (1977) Geotechnical aspects of construction of the Shinkansen. In: Proc. 9th IC SMFE, Tokyo, vol 3, pp 237–260.

Furudoi T (2005) Second phase construction project of Kansai International Airport. In: Proc. 16th IC SMGE, Ósaka, Millpress, Rotterdam, vol 1, pp 313–322.

Gässler G (1991) In-situ techniques of reinforced soil. State of the Art report. In: McGown EA, Yeon K, and Andrawes KZ (eds) Proc. Int. Reinforced Soil Conference BGS, Glasgow, London, Thomas Telford, pp 185–196.

Gatti G, Gioda G (1983) Soil improvement by pre-loading: Numerical results vs. in situ measurements. In: Proc. 8th EC SMFE, Helsinki, Balkema, vol 2, pp 617–620.

Gens A (2003) The role of geotechnical engineering in nuclear energy utilisation. Special Lecture. In: Vaníček et al. (eds) Proc. 13th EC SMGE, Prague, Czech Geotechnical Society, vol 3, pp 25–67.

Girijivallanhan CV, Reese LC (1968) Finite element method for problems in soil mechanics. JSMFED, ASCE, vol 94, No SM2.

Giroud JP (1996) Granular filters and geotextile filters. In: Lafleur and Roklin (eds) Proc. Conf. Geofilters 96, Montreal, Bitech Publications, pp 565–679.

Giroud JP, Noiry L (1981) Geotextile-reinforced unpaved road design. JGED, ASCE, vol 107, No GT9, pp 1233–1254.

Giroud JP, Gourc JP, Bally P, Delmas P (1977) Behaviour of a nonwoven fabric in an earth dam. In: Proc. Int. Conf. on the Use of Fabrics in Geotechnics, Paris, pp 213–218.

Giroud JP, Badu-Tweneboah K, Soderman KL (1994) Evaluation of landfill liners. In: Proc. 5th Int. Conf. Geotextile, Geomembranes and Related Products, Singapore.

Gourc JP (1996) Retaining structures with geosynthetics: A mature technique, but with some questions pending. In: Proc. 1st European Conf. EuroGeo 1, Maastricht, Balkema, pp 27–46.

Gourc JP, Perrier H, Riondy G (1983) Unsurfaced roads on soft subgrade: Mechanism of geotextile reinforcement. In: Proc. 8th EC SMFE, Helsinki, Balkema, vol 2, pp 495–498.

Greaves H (1996) An introduction to lime stabilization. In: Proc. Lime Stabilization, London, Loughborough University, Thomas Telford, pp 5–12.

Greenwood JH (1990) The creep of geotextile. In: Proc. 4th Int. Congress on Geotextiles Geomembranes and Related Products, Hague, Balkema, Rotterdam, vol 2, pp 645–650.

Gschwendt I et al. (2001) Roads, Materials and Technology. (In Slovak). Jaga publishing, Bratislava, 207 p.

Hall CD, Gilchrist AJT (1995) Steeply sloping lining systems – Stability considerations using reinforced soils. In: Proc. Green 93 – Waste Disposal by Landfilling, Balkema, Rotterdam, pp 427–432.

Hansbo S (1981) Consolidation of fine-grained soils by prefabricated drains. In: Proc.10th IC SMFE, Stockholm, 1981, vol 3, pp 677–682.

Hansbo S (1983) How to evaluate the properties of prefabricated drains. In: Proc. 8th EC SMFE, Helsinki, vol 2, pp 621–626.

Hansbo S (2001) Consolidation equations valid for both Darcien and non-Darcien flow. Géotechnique, vol 51, No 1, pp 51–54.

Hansbo S, Jamiolkowski M, Kok L (1981) Consolidation by vertical drains. Géotechnique, vol 31.

Havlíček J (1972) Air in pores of soil. (In Czech). Research Report. UTAM Czech Academy of Science. 85 p.

Havlíček J (1978) Foundation settlement. (In Czech). Final Report of the Research Project C 52-347-018, SG Praha.

Havlíček J (1981) Slope stability – Introduction Lecture. (In Czech). National Seminar on Slope Stability. VTS, Karlovy Vary.

Havlíček J, Fiedler J (1969) The Influence of Long Term Rain on Stability of Earth Dams. (In Czech). UTAM CAS, Praha.

Havlíček J, Myslivec A (1965) The influence of saturation and stratification on the shearing properties of certain soils. In: Proc. 6th IC SMFE, Montreal, University of Toronto Press, vol 1, pp 235–239.

Havukainen J (1983) The utilization of compacted coal ash in earth works. In: Proc. 8th EC SMFE, Helsinki, Balkema, vol 2, pp 773–776.

Head M, Lamb M, Reid M, Winter M (2006) The use of waste materials in construction – progress made and challenges for the future. Feature Lecture. In: Thomas H (ed) Proc. 5 ICEG, Cardiff, London, Thomas Telford, vol 1, pp 70–92.

Heerten G (1993) A contribution to the improvement of dimensioning analogies for grain filters and geotextile. In: Brauns, Schuler, and Heibaum (eds) Proc. 1st IC "Geo-filters" Karlsruhe, Filters in Geotechnical and Hydraulic Engineering, Balkema, pp 121–128.

Hegg U, Jamiolkowski M, Lancellotta R, Parvis E (1983) Behaviour of oil tanks on soft cohesive ground improved by vertical drains. In: Proc. 8th EC SMFE, Helsinki, Balkema, vol 2, pp 627–632.

Henkel DJ (1960) The relationships between the strength, pore water content in saturated clays, Géotechnique, vol 10, pp 41–54.

Herle V, Škopek J (2003) Geotechnical problems of shifting of the gothic church in Most. In: Vaníček et al. (eds) Proc. 13th EC SMGE, Prague, CGtS, vol 4, pp 119–130.

Herštus J, Šťastný J (1992) Measurement of spatial stresses and deformations in spoil heap of mine Merkur. (In Czech). AGE Prague.

Herštus et al. (1992) Origin and development of pore pressure in spoil heap body. (In Czech). AGE Prague.

Herštus J, Šťastný J, Málek J, Bílý P (1997) Open-pit mine claystones as material for waste landfill liners. In: Proc. 14th IC SMGE, Hamburg, Balkema, Rotterdam, vol 3, pp 1823–1826.
Hewlet WJ, Randolf MF (1988) Analysis of piled embankments. Ground Engineering, vol 22, No 3, pp 12–18.
Hillis SF, Truscott EG (1983) Magat dams: Design of internal filters and drains. Canadian Geotechnical Journal, No 3, pp 491–501.
Höeg K (2001) Embankment-dam engineering, safety evaluation and upgrading. In: Proc. 15th IC SMGE, Istanbul, Balkema, Lisse, vol 4, pp 2491–2504.
Höeg K (2003) Embankment-dam engineering and safety evaluation. In: Jubilee Volume in Celebration of 75th Ann. of K.Terzaghi's "Erdbaumechanik" and 60th Birthday of Prof. Brandl. Vienna Technical University, 2nd ed., pp 321–341.
Hoekstra SJ (1995) Final capping and finishing of a landfill Limburg, the Netherland. In: Sarsby (ed) Proc. Green '93 Waste Disposal by Landfill, Balkema, Rotterdam, pp 643–648.
Holeyman A, Mitchell JK (1983) Assessment of quicklime pile behaviour. In: Proc. VIII EC SMFE, Helsinki, Balkema, vol 2, pp 897–902.
Hollingworth F, Druyts FHWM (1982) Filter cloth partially replaces and supplement filter material for protection of poor quality core material in rock-fill dam. In: Proc. 14th ICOLD, Rio de Janeiro, vol 4, pp 709–726.
Holm G, Tränk R, Ekström A (1983) Improving lime column strength with gypsum. In: Proc. VIII EC SMFE, Helsinki, Balkema, vol 2, pp 903–907.
Holt CC, Freer-Hewish RJ (1996) Lime treatment of capping layers under the current DoT specification for highway works. In: Proc. Lime Stabilization, London, Loughborough University, Thomas Telford, pp 51–61.
Hoteit N, Faucher B (2003) Overview of geotechnical issues of feasibility study of geological disposal of radioactive waste. In: Vaníček et al. (eds) Proc. 13th EC SMGE, Prague, Czech Geotechnical Society, vol 3, pp 17–23.
Houlsby GT, Jewell RA (1990) Design of reinforced unpaved roads for small rut depths. In: Den Hoedt (ed) Proc. 4th Int. Conf. on Geotextiles Geomembranes and Related Products, Balkema, Rotterdam, vol 1, pp 171–176.
Houlsby GT, Burd HJ (1999) Understanding the behaviour of unpaved roads on soft clay. In: Barends et al. (eds) Proc. XII EC SMGE, Balkema, Rotterdam, pp 31–42.
Hulla J, Kadubcová Z (2000) Erosive filtration in the flood embankments subsoil. In: Wolski and Mlynarek (eds) Filters and drainage in geotechnical and environmental engineering. Balkema, pp 305–312.
Hunter D (1988) Lime-induced heave in sulphate-bearing clay soils. ASCE Journal of Geotechnical Engineering, vol 114, No 2, pp 150–167.
Hutchinson JN (1977) Assessment of the effectiveness of corrective measures in relation to geological conditions and types of slope movement. In: Proc. Symp. IAEG Landslides and Other Mass Movements, Prague, Bull. Int. Assoc. of Engrg. Geol. No 16, pp 131–155.
ICOLD (1986) Bulletin No 56. Quality Control for Fill Dams. Int. Commission on Large Dams. Paris.
ICOLD (1994) Bulletin No 97. Tailings Dams, Design of the Drainage. Review and Recommendations. Int. Commission on Large Dams. Paris.
ICOLD (1996) Bulletin No 106. A Guide to Tailings Dams and Impoundments. Int. Commission on Large Dams. Paris.
ICOLD (2001) Bulletin No 121. Tailings Dams. Risk of Dangerous Occurrences. Lessons Learnt from Practical Experiences. Int. Commission on Large Dams. Paris.
Imbert B, Breul B, Herment R (1996) More than 20 years of experience in using a bituminous geomembrane. In: De Groot, Den Hoedt, and Termaat (eds) Geosynthetics: Application, Design and Construction. Balkema, Rotterdam, pp 283–286.
Indraratna BN, Locke M (2000) Analytical modeling and experimental verification of granular filter behaviour. In: Wolski and Mlynarek (eds) Proc. Conf. Geofilters, Warsaw, Balkema, pp 3–26.

Ingold TS (1983) Some factors in the design of geotextile reinforced embankments. In: Proc. 8th EC SMFE, Helsinki, Balkema, vol 2, pp 503–508.
Ingold TS (ed) (1995) The Practice of Soil Reinforcing in Europe. Thomas Telford, London.
Ingold TS, Miller KS (1988) Geotextiles Handbook. Thomas Telford, London.
Istomina VS, Burenkova VV, Misurova GV (1975) Filtration Stability of Clay Soils. (In Russian). Strojizdat, Moscow, 219 p.
Jamiolkowski M, Lancellotta R, Wolski W (1983) Precompression and speeding up consolidation. In: Proc. 8th EC SMFE, Helsinki, Balkema, vol 3, pp 1201–1226.
Janbu N (1954) Application of composite slip surface for stability analysis. Proceedings of European conference on stability of earth slopes, Stockholm, vol 3, pp 43–49.
Janbu N (1963) Soil compressibility as determined by oedometer and triaxial tests. In: Proc. EC SMFE, Wiesbaden, vol 1, pp 19–25.
Janbu N (1973) Slope Stability Computations. Embankment Dam Engineering. Casagrande Volume. Hirchfeld RC and Poulos SJ (eds). John Wiley and Sons, New York, pp 47–86.
Janbu N, Senneset K (1979) Interpretation procedures for obtaining soil deformation parameters. In: Proc. 7th EC SMFE, Brighton, BGS, vol 1, 185–188.
Jansen HL, Den Hoedt G (1983) Vertical drains: in situ and laboratory performance and design considerations. In: Proc. 8th EC SMFE, Helsinki, Balkema, vol 2, pp 633–636.
Jesenák J (1993) Soil Mechanics. (In Slovak), STU Bratislava, 406 p.
Jessberger HL, Stone K (1991) Subsidence effects on clay barriers. Géotechnique, vol 41, No 2, pp 185–194.
Jewell RA (1990) Strength and deformation in reinforced soil design. In: Proc. 4th IC Geotextiles Geomembranes and Related Products, Den Haag, Balkema, Rotterdam, vol 3, pp 913–946.
Johnson PE, Card GB (1998) The use of soil nailing for the construction and repair of retaining walls. TRL Report 373, 42p.
Jones CJFP (1991) Construction influences on the performance of reinforced soil structures. In: McGown, Yeo, and Andrawes (eds) Performance of Reinforced Soil Structures. Thomas Telford, London, pp 97–116.
Jones CJFP (1996) Earth Reinforcement and Soil Structures. Thomas Telford, London, 379 p.
Kamon M, Hartlén J, Katsumi T (2000) Reuse of waste and its environmental impact. In: GeoEng 2000 vol 1, Invited papers. Technomic Publ. Comp. Lancaster, pp 1095–1123.
Karpoff KP (1955) The use of laboratory tests to develop design criteria for protective filters. In: Proc. Am. Soc. Test. Mat 55, p 1183.
Karpurapu R, Bathurst RJ (1994) Finite element analysis of geotextile reinforced retaining walls. In: Proc. 12th IC SMFE, New Delhi, pp 1381–1384.
Kassif G, Baker R (1971) Ageing effects on swell potential of compacted clays. Journal SMFD ASCE, vol 97, No SM3, pp 529–540.
Kassif G, Henkin EN (1967) Engineering and physico-chemical properties affecting piping failure of low loess dams in Negev. In: Proc. 3rd Asian Conf. SMFE, Haifa, vol 1, p 13.
Kassif G, Zaslavsky D, Zeitlen JG (1965) Analysis of filter requirements for compacted clays. In: Proc. 6th IC SMFE, Montreal, vol 2, p 495.
Kazda I (1990) Finite Element Techniques in Ground Water Flow Studies. Elsevier, Amsterdam.
Kazda I (1997) Underground Hydraulics in Ecological and Engineering Applications. (In Czech). Academia, Praha.
Kempfert HG (1996) Embankment foundation on geotextile coated sand columns in soft ground. In: Proc. EuroGeo 1, Maastricht, Balkema, pp 245–250.
Kempfert HG, Jaup A, Raithel M (1997) Interactive behaviour of a flexible reinforced sand column foundation in soft soils. In: Proc. 14th IC SMGE, Hamburg, Balkema, Rotterdam, vol 3, pp 1757–1760.
Kempfert HG, Göbel C, Alexiew D, Heitz C (2004) German recommendation for reinforced embankments on pile-similar elements. In: EuroGeo 3, Munich DGGT, pp 279–284.
Kenney TC, Van Veen WA, Swallow MA, Sungaila MA (1992) Hydraulic conductivity of compacted bentonite-sand mixtures. Canadian Geotechnical Journal, vol 29, pp 364–374.

Kézdi A (1964) Bodenmechanik I., II., Verlag für Bauwesen, Berlin.
Kjaernsli B, Torblaa I (1968) Leakage Through Horizontal Cracks in the Core of Hyttejuvet Dam. NGI, Publ., Oslo, No 80, 39 p.
Kjellman W (1948) Accelerating consolidation of fine-grained soils by means of card-board wicks. In. Proc. 2nd IC SMFE, Rotterdam, p 302.
Kockel R, Jessberger HL (1995) Stability evaluation of municipal solid waste slopes. In: Proc. 11th EC SMFE, Copenhagen, DGS, vol 2, pp 67–72.
Koerner RM (1990) Designing with Geosynthetics; 2nd edition. Prentice Hall Publ. Co, Englewood Cliffs, NJ.
Koerner RM, Halse YH, Lord AE Jr (1990) Long-term durability and ageing of geomembranes. In: Bonaparte R (ed) Waste containment systems. ASCE, New York, pp 106–134.
Koerner RM, Lord AE Jr, Hsuan YH (1992) Arrhenius modeling to predict geosynthetic degradation. Geotextile and Geomembranes, vol 11, pp 151–183.
Kolářová E, Vaníček I (2006) Basic requirements on the capping system of the uranium mine tailings. In: Thomas HR (ed) Proc. 5th ICEG, Thomas Telford, vol II, pp 917–924.
Kondner RL (1963) Hyperbolic stress-strain response: Cohesive soils. Journal SMFED, ASCE, vol 89, No SM1.
Kopf F, Adam D (2005) Dynamic effects due to moving loads on tracks for high-speed railways. In: Proc. 16th IC SMGE, Osaka, Millpress Rotterdam, vol 3, pp 1735–1740.
Krishnayya AVG, Eisenstein Z (1972) Brazilian tensile test for soils. Canadian Geotechnical Journal, vol 11, No 4, pp 632–642.
Krishnayya AVG, Eisenstein Z, Morgenstern NR (1974) Behaviour of compacted soil in tension. Journal of Geotechnical Engineering Division ASCE, vol 100, No GT9, pp 1051–1061.
Kröber W, Floss R, Wallrath W (2001) Dynamic soil stiffness as quality criterion for soil compaction. In: Correia AG and Brandl H (eds) Geotechnics for Roads, Rail Tracks and Earth Structures. Balkema, pp 189–199.
Kulhawy FH, Gurtowski TM (1976) Load transfer and hydraulic fracturing in zoned dams. JGED, ASCE, vol 102, No GT9, pp 963–974.
Kuntsche K (1992) Soil mechanics settlement prognoses for opencast dumps. (In German), Braunkohle 1-2., pp 17–24.
Lambe TW (1953) The structure of inorganic soils. In: Proc. ASCE, vol 79, Separate No 315.
Lambe TW (1958) The engineering behavior of compacted clay. Journal SMFD ASCE, vol 84 (May).
Lambe TW, Whitman RV (1969) Soil Mechanics. John Willey and Sons, New York.
Lawson CR (1982) Filter criteria for geotextiles: Relevance and use. JGED, ASCE, vol 108, No GT10, pp 1300–1317.
Leflaive E (1985) Soil reinforcement with continuous yarns: The Texsol. In: Proc. 11th IC SMFE, San Francisco, vol 3, pp 1787–1790.
Leitner B, Sobolewski J, Ast W, Hangen H (2002) A Geosynthetic Overbridging System in the Base of a Railway Embankment Located on an Area Prone to Subsidence at Gröbers: Construction Experience. 7th ICG Nice, pp 349–354.
Leonards GA, Altschaeffl AG (1964) Compressibility of clay. Journal of Soil Mechanics and Foundation Division, ASCE, vol 90, No SM5, pp 133–155.
Leonards GA, Narain J (1963) Flexibility of clay and cracking of earth dams. Journal SMFD, ASCE, vol 89, No SM2, pp 47–98.
Leupold D, Oswald KD, Heerten G, Horstmann J (2003) Geotechnical aspects concerning the long-term remediation of tailings management facilities as former uranium ore mining industry legacies. In: Vaníček et al. (eds) Proc. 13th EC SMGE, Prague, CGtS, vol 3, p 277–282.
Lord JA (1999) Railway foundations: Discussion paper. In: Barends et al. (eds) Proc. 12th EC SMGE, Balkema vol 3, pp 1679–1691.
Lukáč M, Bednárová E (2001) Reservoirs and Dams (In Slovak). Slovak Technical University Press, Bratislava, 330 p.
Lunne T, Robertson PK, Powell JJM (1997) Cone Penetration Testing in Geotechnical Practice. Blackie Academic and Professional, London, p 312.

Lushnikov VV, Vulis SP, Litvinov BM (1973) Relationship between compression and tension modulus. (In Russian). Osnov. Fund. Mech. Grunt., vol 15, No 3, pp 18–19.
Madsen FT (1994) Clay and synthetic liners – durability against pollutant attack. In: Proc. 13th IC SMFE, New Delhi, Oxford and IBH publishing co. New Delhi, vol 5, pp 287–288.
Madsen FT, Vaníček I, Amann P, Kahr G (1995) Clay mineralogical investigations of tertiary clay from Northern Bohemia. Geologica Carpathia Clays, Bratislava, vol 4, No 2.
Madshus C (2001) Modelling, monitoring and controlling the behaviour of embankments under high speed train loads. Workshop 6 Embankment of High Speed Trains 15th ICSMGE Istanbul.
Madshus C, Kaynia AM (1999) Dynamic ground interaction; a critical issue for high speed train lines on soft soil. In: Barends F et al. (eds) Proc. 12th EC SMFE, Amsterdam, Balkema, Rotterdam, vol 3, pp 1829–1835.
Manassero M, Van Impe WF, Bouazza A (1997) Waste disposal and containment. In: Proc. 2nd ICEG, Osaka 1996, Balkema, Rotterdam, pp 1425–1474.
Manassero et al. (1998) Controlled landfill design (Geotechnical aspects). In: Seco e Pinto PS (ed) Proc. 3rd ICEG, Lisboa, Balkema Rotterdam, vol 3, pp 1001–1038.
Mannsbart G, Kropik C (1996) Nonwoven geotextiles used for temporary reinforcement of a retaining structure under a railroad track. In: Proc. EuroGeo 1, Maastricht, Balkema, pp 121–124.
Marek J, Bureš V, Rozsypal A (2003) Stability of the Krušné hory hillsides affected by coal mining activity at the foothills. In: Vaníček et al. (eds) Proc. 13th EC SMGE, Prague, Czech Geotechnical Society, vol 4, pp 173–196.
Marsal RJ (1973) Mechanical Properties of Rockfill. Embankment Dam Engineering. Casagrande Volume. Willey, New York.
Marsal RJ (1977) Strain computation in earth-rockfill dams. In: Proc. 9th IC SMFE, Spec. Sess. 8 – Deformation of Earth-rockfill Dams, Tokyo, JGS, pp 5–6.
Marshal RJ (1977) Deformation of earth/rockfill dams. Spec. session 8, In. Proc. 9th IC SMFE Tokyo, JGS, vol 3, pp 525–536.
Marsal RJ, de Arellano R (1967) Performance of El Infernillo dam, 1963–1966. JSMFD, ASCE, vol 93, No SM4, pp 265–298.
Masarovičová M, Slávik I, Jesenák J (2003) Geotechnical aspects of desludging sites. In: Vaníček et al. (eds) Proc. 13th EC SMGE, Prague, CGtS, vol 3, pp 177–181.
Matejčeková M (2005) Design of Soil Nailed Walls. STU, Bratislava, Ph.D. thesis, 164 p.
Matsui T, Oda K, Tabata T (2003) Structures on and within man-made deposits – Kansai Airport. In: Vaníček et al. (eds) Proc. 13th EC SMGE, Prague, ČGtS, vol 3, pp 315–328.
McDaniel TN, Decker RS (1979) Dispersive soil problem at Los Esteros dam. JGED, ASCE, vol 105, No GT9, pp 1017–1030.
McGown A, Andrawes KZ, Hytiris N, Mercer FB (1985) Soil strengthening using randomly distributed mesh elements. In: Proc. 11th ICSMFE, San Francisco, Balkema, vol 3, pp 1735–1738.
McGown A, Murray RT, Andrawes KZ (1987) Influence of wall yielding on lateral stresses in unreinforced and reinforced fills. TRRL Research Report 113, 14p.
McGown A, Yeo K, Andrawes KZ (1991) (eds) Performance of reinforced soil structures. Proc. of the Int. Reinforced Soil Conference Organized by BGS, London, Thomas Telford.
Melvill AL (1980) Discussion to dispersive soil problem at Los Esteros dam. JGED, ASCE, vol 106, No GT11, pp 1282–1283.
Menzies BK, Simons NE (1978) Stability of embankments on soft ground. In: Scott CR (ed) Soil Mechanics Development 1. Applied Science Publ.
Michalowski RL, Čermák J (2003) Triaxial compression of sand reinforced with fibres. Journal of Geotechnical and Geoenvironmental Engineering, ASCE, vol 129, pp 125–136.
Michaud O (2000) European roads for the time frame until 2025. In: Jubilee Volume Vienna Technical University, vol 5, pp 342–345.
Milerski R, Mičín J, Veselý J (2003) Water works. (In Czech). CERM Academic Publ. Brno.
Minář L (2000) Quality of earth plain and its improvement. (In Czech). In: Proc. Substructure, Czech Railways Seminar, Ústí nad Labem, pp 86–91.

Misurova GV (1974) Strength of aggregates created during seepage through crack. (In Russian), Trudy VNII VODGEO No 44, Moscow, 26 p.

Mitarai Y, Fukasawa T, Sakamoto A, Nakamura A (2003) The construction project of Central Japan International Airport – Application of pneumatic flow method for large and rapid work. In: Leung et al. (eds) Proc. 12th AC SMGE, Singapore, Word Scientific, Singapore, vol 2, pp 511–1516.

Mitchell JK (1993) Fundamentals of Soil Behaviour. John Wiley and Sons. New York, sec. edn, first edn 1976.

Mitchell JK (1996) Geotechnics of soil-waste interactions. In: Kamon (ed) Proc. 2nd ICEG, Osaka, Balkema, Rotterdam, vol 3, pp 1311–1328.

Mitchell JK (1997) Understanding soil behavior runs through it – role of soil behavior in continuing evolution of geotechnical engineering. Prague Geotechnical Lecture.

Mitchell JK, Katti RK (1981) Soil improvement. In: Proc. 10th IC SMFE, Stockholm, Balkema, pp 261–317.

Mitchell JK, Madsen FT (1987) Chemical effects on clay hydraulic conductivity. In: Woods RD (ed) Geotechnical Practice for Waste Disposal '87. ASCE, New York, pp 87–116.

Mlynarek J (2000) Geo drains and geo-filters – Retrospective and future trends. In: Wolski and Mlynarek (eds) Proc. Conf. Geofilters, Warsaw, Balkema, pp 27–50.

Moh Z (1962) Soil stabilization with Portland cement and sodium additives. Journal SMFE, ASCE, vol 88, No 6, pp 81–105.

Moo-Young HK, Zimmie TF (1996) Effects of organic decomposition on paper mill sludges used as landfill cover. In: Kamon M (ed) Proc. 2nd ICEG, Osaka, Balkema Rotterdam, vol 2, pp 827–832.

Morgenstern NR (1994) The observational method in environmental geotechnics. In: Proc. 1st ICEG, Edmonton, BiTech Publishers, pp 963–976.

Morgenstern NR (1999) Observations on geotechnics and mine waste management. In: Geotechnics in the New Millenium. An Imperial College perspective. IC London, pp 118–120.

Morgenstern NR, Price VE (1965) The analysis of the stability of general slip surfaces. Géotechnique, vol 15, pp 79–93.

Morgenstern NR, Fair AE, McRoberts EC (1988) Geotechnical engineering beyond soil mechanics – a case study. Canadian Geotechnical Journal, vol 24, pp 637–661.

Moser AP (2001) Buried Pipe Design. McGraw-Hill Professional, 2nd edn, 544 p.

Myslivec A, Eisenstein Z (1965) Utilization of swelling clays in civil engineering. (In Czech) Inženýrské Stavby, vol 13, No 3, pp 307–309.

Myslivec A, Jesenák J, Eichler J (1970) Soil Mechanics. (In Czech). SNTL 1970, Praha, 388 p.

Narain J, Rawat PC (1970) Tensile strength of compacted soils. Journal SMFD, ASCE, vol 96, No SM6, pp 2185–2190.

Newson TA, Fahey M (2001) Site investigation of soft tailings deposits using a hovercraft. Geotechnical Engineering. In: Proc. ICE, April 2001, vol 149, pp 115–126.

Ničiporovič AA (1969) Participation in panel discussion on earth and rockfill dams. In: Proc. 7th IC SMFE, Mexican Soil Mechanics Society, vol 3, p 293.

Nosko V, Andrezal T, Gregor T, Ganier P (1996) SENSOR damage detection system (DDS) – The unique geomembrane testing method. In: De Groot, Den Hoedt, and Termaat (eds) Proc. 1st Europ. Geosynth. Conf. Maastricht. Geosynthetics: Applications, Design and Construction, Balkema, Rotterdam, pp 743–748.

Ohne Y, Narita K (1977) Discussion on cracking and hydraulic fracturing in fill-type dams. In: Proc. 9th IC SMFE, Spec. Sess. 8 – Deformation of Earth-rockfill Dams, Tokyo, JSSMFE, pp 60–63.

Onitsuka K, Kamon M, Katsimu T, Okada J, Nambu M, Ono S (1996) Industrial waste disposal and utilization of waste landfill in Japan. In: Kamon M (ed) Proc. 2nd ICEG, Osaka, Balkema Rotterdam, vol 2, pp 663–668.

O'Riordan N (2004) Channel Tunnel Rail Link – Ground Engineering. Invited Lecture. Czech Geot. Soc. Prague, 11.02.2004

Ovesen NK (1979) Panel discussion. In: Proc. VII EC SMFE, Brighton, vol 4, pp 319–323.
Ovesen NK (1983) Centrifuge tests of embankments reinforced with geotextiles on soft clay. In: Proc. 8th EC SMFE, Helsinki, 1983, Balkema, vol 1, pp 393–398.
Pařízek J (1976) Jirkov Dam. Periodical Report of Technical Security Supervision to the Date 30.6.1976. (In Czech) VRV-TBD Prague, 20 p.
Park et al. (1997) A case study of vacuum preloading with vertical drains. In: Davies MCR and Schlosser F (eds) Ground Improvement Geosystems Densification and Reinforcement. Thomas Telford, London, pp 68–74.
Parry RHG, Nadarajan V (1974) Anisotropy in a natural soft clayey silt. Engineering Geology, vol 8, pp 287–309.
Pašek J, Kudrna Z (1996) Motorway in a landslide area in the České středohoří Mountains. (In Czech). In: Proc. 2nd Slovak Geotechnical Conf. Bratislava, Slovak Technical University, pp 97–102.
Peck RB (1969) Advantages and limitations of the observational method in applied soil mechanics. Géotechnique, vol 19, 1969, pp 171–187.
Pecková L (2005) Monitoring Significance During Deposition of Large-volume Waste Materials. (In Czech). MSc thesis, CTU Prague, 149 p.
Penman ADM (1977) Hydraulic fracture. In: Proc. 9th ICSMFE, Spec. Sess. 8 – Deformability of Earth-rockfill Dams, Tokyo, pp 57–58.
Penman ADM (1986) On the embankment dam. Rankine Lecture. Géotechnique, vol 36, No 3, pp 303–348.
Penman ADM, Charles JA (1976) The quality and suitability of rockfill used in dam construction. In: Proc. 12th ICOLD, Mexico, vol 1, pp 533–566.
Perry J, McNeil DJ, Wilson PE (1996) The uses of lime in ground engineering: A review of work undertaken at the Transport Research Laboratory. In: Rogers CDF, Glendinning S, Dixon N (eds), Lime Stabilization. Thomas Telford, London, pp 27–45.
Peter P (1975) Canal and River Levees. (In Slovak) VEDA, SAV, Bratislava.
Phear A, Dew C, Ozsoy B, Wharmby NJ, Judge J, Barley AD (2005) Soil Nailing: Best Practice Guidance. CIRIA C637, London, CIRIA, 286 p.
Pilarczyk KP (1998) Dikes and Revetments. Design. Maintenance and Safety Assessment. Balkema, Rotterdam, 592 p.
Pinard MI (2001) Developments in Compaction Technology. In: Correia AG and Brandl H (eds) Geotechnics for Roads, Rail Tracks and Earth Structures. Balkema, pp 37–46.
Plant GW (1999) Hong Kong International Airport. In: Barends et al. (eds) Proc. 12th EC SMGE, Balkema, Rotterdam, vol 2, pp 1001–1013.
Plumelle C, Schlosser F (1991) A French national research project on soil nailing: Clouterre. In: McGown A, Yeo K, and Andrawes KW (eds) Performance of Reinforced Soil Structures. Thomas Telford, London, pp 219–223.
Pokharel G, Ochiai T (1997) Design and construction of a new soil nailing. In: Davies MCR and Schlosser F (eds) Ground Improvement Geosystems. Thomas Telford, London, pp 407–413.
Poláček J (1984) Methods of Investigations of Seepage Path in Dams and Subsoil. (In Czech). Research Report VRV Prague, 108 p.
Potts DM, Zdravkovic L (1999) Finite Element Analysis in Geotechnical Engineering: Theory. Thomas Telford, London, 440 p.
Poulos HG, Davis EH (1974) Elastic Solutions for Soil and Rock Mechanics. John Whiley and Sons, New York, 424 p.
Proctor RR (1933) The design and construction of roller earth dams. Engineering News Record III, August 31, September 7, 21, and 28.
Puumalainen N, Vepsalainen P (1997) Vacuum preloading of a vertically drained ground at the Helsinky test field. In: Proc. 14th IC SMGE Hamburg, Balkema Rotterdam vol 3, pp 1769–1772.
Raithel M, Küster V, Lindmark A (2004) Geotextile-Encased Columns-a foundation system for earth structures, illustrated by a dyke project for a works extension in Hamburg. In: 14th Nordic Geotechnical Meeting, Ystad, 10 p.

Reinius E (1948) On the stability of the upstream slope of earth dams. Victor Pettersons Bokindustriaktiebolag, Stockholm.
Rimoldi P, Jaecklin F (1996) Green faced reinforced soil walls and steep slopes: The state-of-the-art in Europe. In: De Groot, Den Hoedt, and Termaat (eds) Geosythetics: Application, Design and Construction. Balkema, Rotterdam, pp 361–380.
Rodriguez-Ortiz JM (2003) Some special geotechnical aspects of recent tailings dams failures. In: Vaníček et al. (eds) Proc. 13th EC SMGE, Prague, CGtS, vol 3, pp 173–176.
Romero US, Rodrígues PJR (1981) Tensile behaviour of compacted clays by field test. In: Proc. 10th IC SMFE, Stockholm, Balkema, vol 3, pp 581–584.
Rosenquist IT (1959) Physico-chemical properties of soils. Soil-water system. Journal SMFD, ASCE, vol 85, No SM2, April.
Rowe PK, Ho SK (1988) Application of finite element techniques to the analysis of reinforced soil walls. In: Jarret PM and McGown A (eds) The Application of Polymeric Reinforcement in Soil Retaining Structures. Kluwer Academic Publishers, pp 541–554.
Saathof LEB, Ketelaars MBG (1999) Re-use of excavated soil from large diameter tunnels in the Netherlands. In: Barends, Lindenberg, Luger, de Quelerij, and Verruijt (eds) Proc. 12th EC SMGE, Balkema, Rotterdam, vol 2, pp 1301–1307.
Sargand SM, Nasada T (2000) Performance of large-diameter honey comb design HDPE pipe under a highway embankment. Canadian Geotechnical Journal, vol 37, No 5, pp 1099–1108.
Sarma SK (1973) Stability analysis of embankments and slopes. Géotechnique 23, No. 3, 423–433.
Sarma SK (1979) Stability analysis of embankment and slopes. JGED, ASCE, vol 105, No GT12, pp 1511–1524.
Sarsby RW, Finch S (1995) The use of industrial by-products to form landfill caps. In: Sarsby (ed) Proc. Green 93, Waste Disposal by Landfill, Balkema, Rotterdam, pp 267–274.
Sawicki A, Swidzinski W (1990) Compaction beneath flexible pavement due to traffic. Journal of GED, ASCE, vol 116, No 11, pp 1738–1743.
Scapozza I, Amann P (1998) Assessment of a new bottom sealing system for landfill sites made of an asphalt sealing layer on a clay foundation layer. In: Seco e Pinto PS (ed) Proc. 3rd ICEG, Lisboa, Balkema, Rotterdam, vol 1, pp 307–311.
Scharle P, Szalatkay I, Vas J (1983) Experiments on plastic grid reinforced sand masses. In: Proc. VIII EC SMFE, Helsinki, Balkema, vol 2, pp 537–538.
Schlosser F (1991) CLOUTERRE – Soil nailing recommendations for designing, calculating and inspecting earth support systems using soil nailing. French National Research Project, ENPC, Paris, English Translation, July 1993, US Department of transportation, Federal highway administration, 302 p.
Schlosser F (1997) Theme lecture: Soil improvement and reinforcement, (In French), In: Proc. 14th IC SMFE, Hamburg, Balkema, Rotterdam, vol 4, pp 2445–2466.
Schlosser F, DeBuhan P (1991) Theory and design related to the performance of reinforced soil structures. In: McGown A, Yeon K, and Andrawes KZ (eds) Proc. Int. Reinforced Soil Conference BGS, Glasgow, London, Thomas Telford, pp 1–14.
Schlosser F, Magnan JP, Holtz RD (1985) Geotechnical engineered construction. Theme Lecture. In: Proc. 11th IC SMFE, San Francisco, Balkema, vol 5, pp 2499–2539.
Schober W (1967) Discussion. In: Proc. 9th ICOLD, Istanbul, vol 6, pp 422–424.
Schober W (1977) Development of longitudinal cracks at Gepatsch dam. In: Proc. 9th IC SMFE, Spec. Sess. 8 – Deformation of Earth-rockfill Dams, Tokyo, JSSMFE, pp 72–74.
Schober W, Teindl H (1979) Filter criteria for geotextiles. In: Proc. 7th EC SMFE, Brighton, BGS, vol 2, pp 121–129.
Schofield AN (1980) Cambridge geotechnical centrifuge operations. Géotechnique, vol 30, No 3, pp 225–268.
Schuler U, Brauns J (1993) Behaviour of coarse and well-graded filters. In: Brauns, Schuler, and Heibaum (eds) Proc. 1st IC "Geo-filters" Karlsruhe, Filters in Geotechnical and Hydraulic Engineering, Balkema 1993, pp 3–18.
Seco e Pinto PS, Das Neves EM (1985) Hydraulic fracturing in zoned earth and rockfill dams. In: Proc. 11th IC SMFE, San Francisco, vol 4, pp 2025–2030.

Seed HB, Duncan JM (1981) The Teton dam failure – a retrospective review. In: Proc. 10th IC SMFE, Stockholm, Balkema, vol 4, pp 219–238.
Seed HB, Woodward RJ, Lundgren R (1962) Prediction of swelling potential for compacted clays. Journal of SMFD, ASCE, vol 88, No SM3, pp 53–87.
Selig ET (1985) Review of Specifications for Buried Corrugated Metal Conduit Installations. In: Transportation Research Record N1008, Culverts: Analysis of Soil-Culvert Interaction and Design. Transport Research Board. Washington, pp 15–21.
Sembenelli P, Ueshita K (1981) Environmental Geotechnics. State of the Art Report. In: Proc. 10th IC SMFE, Stockholm, Balkema, vol 4, pp 335–394.
Serra M, Robinet JC, Mohkam M, Daonh T (1983) Soil improvement of dykes by liming. In: Proc. 8th EC SMFE, Helsinki, Balkema, vol 2, pp 947–950.
Seyček J (1975) Settlement of clay subsoil of earthfill dam measured and predicted on the results of oedometric tests (In Czech). In: Proc. Seminar: Field and Laboratory Tests of Soils. VUHU Most, pp 1–10.
Seyček J (1995) Calculation of 2nd Limit State of Spread Foundations. (In Czech) CTU Prague, PhD thesis.
Sherard JL (1973) Embankment Dam Cracking. Embankment Dam Engineering, Casagrande Volume. Willey, New York, pp 271–353.
Sherard JL, Woodward RJ, Gizienski SF, Clevenger WA (1963) Earth and Earth-rock Dams. Engineering Problems of Design and Construction. John Wiley and Sons, New York, Third ed. 1967, 725 p.
Sherard JL, Decker RS, Ryker NL (1972a) Piping in earth dams of dispersive clay. In: Proc. ASCE Spec. Conf., vol 1, part 1, pp 589–626.
Sherard JL, Decker RS, Ryker NL (1972b) Hydraulic fracturing in low dams of dispersive clay. In: Proc. ASCE Spec. Conf., vol 1, part 1, pp 653–690.
Sherard JL, Dunnigan LP, Decker RS (1976a) Pinhole test for identifying dispersive soils. JGED, ASCE, vol 102, No GT1, pp 69–85.
Sherard JL, Dunnigan LP, Decker RS (1976b) Identification and nature of dispersive soils. JGED, ASCE, vol 102, No GT4, pp 287–301.
Sherard JL, Dunnigan LP, Talbot JR (1984a) Basic properties of sand and gravel filters. JGED, ASCE, vol 110, No 6, pp 684–700.
Sherard JL, Dunnigan LP, Talbot JR (1984b) Filters for silts and clays. JGED, ASCE, vol 110, No 6, pp 701–718.
Sherwood PT (1967) Views of the road research laboratory on soil stabilisation in the United Kingdom. Cement Lime and Gravel, vol 42, No 9, pp 277–280.
Shimobe S, Moroto N (1997) A new engineering classification chart for densification of granular soils. In: Davies MCR and Schlosser F (eds) Ground Improvement Geosystems. Thomas Telford, London, pp 32–37.
Sides G, Barden L (1971) The microstructure of dispersed and flocculated samples of kaolinite, illite and montmorillonite. Canadian Geotechnical Journal, vol 7, No 3, pp 391–399.
Silveira A (1965) An analysis of the problem of washing through in protective filters. In: Proc. 6th IC SMFE, Montreal, University of Toronto Press, vol II, pp 551–555.
Silveira A (1993) A method for determining the void size distribution curve for filter materials. In: Brauns E, Schuler, and Heibaum (eds) Proc. 1st IC "Geo-filters" Karlsruhe, Filters in Geotechnical and Hydraulic Engineering, 1992, Balkema, pp 71–73.
Simek J, Jesenák J, Eichler J, Vaníček I (1990) Soil mechanics. (In Czech). SNTL Prague, p 388.
Šimek M (1976) Methods of failures identification and remediation, Question No 45 ICOLD Congress. Dam days, Žilina vol 2, pp 47–84.
Simon A, Perfetti J (1983) Use of a geotextile on a clayey sea bed. In: Proc. 8th EC SMFE, Helsinki, vol 2, pp 869–872.
Skarzynska KM, Rainbow AM, Zawisza E (1989) Characteristic of ash in storage ponds. In: Proc. 12th IC SMFE, Rio de Janeiro, pp 1915–1918.
Skempton AW (1953) The colloidal activity of clays. In: Proc. 3rd IC SMFE, Zurich, vol 1, pp 57–61.
Skempton AW (1954) The pore-pressure coefficient A and B. Géotechnique, vol 4, pp 143–147.

Skempton AW (1977) Slope stability of cuttings in brown London clay. In: Proc. 9th IC SMFE, Tokyo, JGS, vol 3, pp 261–270.

Skempton AW, Bishop AW (1954) Soils, Chapter X of Building materials, their elasticity and inelasticity. North Holland Pub. Co., Amsterdam, pp 417–482.

Skempton AW, Hutchinson J (1969) Stability of natural slopes and embankment foundations. In: Proc. 7th IC SMFE, Mexico, 1969, Mexican Soil Mechanics Society, State of the Art Volume, pp 291–334.

Škopek J (1968) Uniformly distributed area load in the interior of semi-infinite solid. Acta technica 5, Praha,

Škopek J (1979) Geotechnical problems encountered in moving the church at Most. Canadian Geotechnical Journal, vol 16, No 3, pp 473–480.

Škrabal L (1977) Anisotropy of Most Clay. MSc thesis CTU Prague, 122 p.

Slávik I, Masarovičová M, Jesenák J (1994) Geotechnical properties of ash deposited at Novaky tailings dam. (In Slovak). Inžinierske stavby No 8, pp 265–269.

Smekal A (2000) Utilization of polystyrene in railway engineering. (In Czech) In: Proc. Zeleznicni Spodek 2000. Usti nad Labem CD, pp 156–164.

Smith JH (1996) Construction of lime or lime plus cement stabilized cohesive soils. In: Rogers CDF, Glendinning S, Dixon N (eds), Lime Stabilisation. Thomas Telford, London, pp 13–26.

Smolík R (1996) Measurement and prognosis of pore pressure in soils. (In Czech), Ph.D. thesis, Czech Technical University Prague, 154 p.

Soranzo M (1983) Determination of the horizontal consolidation coefficient of a marine silty clay. In: Proc. 8th EC SMFE, Helsinki, Balkema, vol 2, pp 685–690.

Sowers GF (1973) Settlement of waste disposal fills. In: Proc. 8th IC SMFE, Moscow, vol 2, pp 207–210.

Stewart RA (2000) Dam risk management. GeoEng2000, Technomic Publ. Co, Lancaster, vol 1, pp 721–748.

STN 721191: 1991 Testing of freezing Susceptibility, SNI, Bratislava.

STN 721512: 1991 Dense aggregates for construction - technical requirements, SNI, Bratislava.

STN 736125: 1996 Road construction - Stabilized subgrade, SNI, Bratislava.

STN 736126: 1996 Road construction, SNI, Bratislava.

Stocker MF (1994) 40 year of micropiling, 20 years of soil nailing. In: Proc. 13th IC SMFE, New Delhi, Oxford and IBH Publishing Co., New Delhi, vol 5, pp 167–168.

Stocker MF, Korber GW, Gässler G, Gudehus G (1979) Soil nailing. In: Proc. Int. Conf. Soil Reinforcement, Paris, ENPC, vol 2, pp 469–474.

Štýs S (1998) Metamorphous of lunar landscape. Ecoconsult Pons, Most, 34 p.

Suhonen S, Tupala J (1983) Experiences of vertical drainage in the Lahti motorway. In: Proc. 8th EC SMFE, Helsinki, Balkema, vol 2, pp 691–694.

Šuklje L, Drnovšek J (1965) Investigation of the tensile deformability of soils using hollow cylinders. In: Proc. 6th IC SMFE, Montreal, Toronto University Press, vol 1, pp 368–372.

Sunaga M (2001) Characteristics of embankment vibrations due to high-speed train loading and some aspects of the design standard for high-speed links in Japan. In: Correia AG and Brandl H (eds) Geotechnics for Roads, Rail Tracks and Earth Structures. Balkema, Rotterdam, pp 203–211.

Suthersan SS (1997) Remediation Engineering. Design Concepts. CRC. Lewis publishers, New York, 362 p.

Symes MYPR, Gens A, Hight DW (1984) Undrained anisotropy and principal stress rotation in saturated sand. Géotechnique, vol 34, No 1, pp 11–28.

Tammirinne M, Vepsalainen P (1983) Laboratory testing for design and control of ground improvement. In: Proc. 8th EC SMFE, Helsinki, Balkema, vol 3, pp 1135–1141.

Tateyama M, Murata O (1994) Geosynthetic-reinforced soil retaining walls for abutments. In: Proc. XIII IC SMFE, New Delhi, Oxford and IBH Publishing Co., vol 3, pp 1245–1248.

Tatsuoka F, Tateyama M, Uchimura T, Koseki J (1996) Geosynthetic-reinforced soil retaining walls as important permanent structures. In: De Groot, Den Hoedt, and Termaat (eds) Geosynthetics: Application, Design and Construction. Balkema, Rotterdam, pp 3–46.

Taylor N (1995) Geotechnical Centrifuge Technology. Blackie Academic and Professional, Glasgow, 296 p.

Ter-Martirosjan ZG (1973) Compression-extension as a method of cohesive soil test. In: Proc. 8th ICSMFE, Moscow, vol 4.2, 20 p.

Ter-Martirosjan ZG (1977) Certain results of tests on cohesive soils by the compression-extension method. In: Proc. 5th Danube EC SMFE, Bratislava, CaS NC ISSMGE, vol 4, p 159.

Terzaghi K (1943) Theoretical Soil Mechanics. John Willey and Sons, New York, 510 p.

Terzaghi K (1959) Soil Mechanics in Action. Civil Engineering, vol 69, February, pp 33–34.

Terzaghi K, Peck RB (1948) Soil Mechanics in Engineering Practice. John Wiley and Sons, New York, 566 p.

Thurner HF, Sandström A (2000) Continuous Compaction Control, CCC. In: Proc. European Workshop "Compaction of Soils and Granular Materials", Paris, pp 237–246.

Trautwein SJ, Boutwell GP (1994) In-situ hydraulic conductivity test for compacted soil liners and caps. In: Daniel and Trautwein (eds) Hydraulic Conductivity and Waste Contaminant Transport in Soil. ASTM STP 1142.

Troncoso JH (1996) Geotechnics of tailings dams and sediments. In: Proc. 2nd ICEG, Osaka, JGS, vol 4, pp 173–192.

Troncoso JH (1998) New perspective in management of operating and abandoned deposits in mineral residues. In: Seco e Pinto PS (ed) Proc. 3rd ICEG, Lisbon, Balkema, Rotterdam, vol 4, pp 1221–1228.

Tschebotarioff GP, Ward ER, DePhillipe AA (1953) The tensile strength of disturbed and recompacted soils. In: Proc. 3rd IC SMFE, Zurich, vol 1, pp 207–210.

Tsonis P, Christoulas S, Kolias S (1983) Soil improvement with coal ash in road construction. In: Proc. 8th VIII EC SMFE, Helsinki, Balkema, vol 2, pp 961–964.

Tsuchida T, Kang MS (2003) Case studies of lightweight treated soil method in seaport and airport construction project. In: Proc. 12th AS SMGE, Singapore, Word Scientific, Singapore, vol 1, 249–252.

Turček P, Hulla J (2004) Foundation engineering. (In Slovak). Jaga Group. Bratislava, 360 p.

Turček et al. (2005) Foundation engineering. (In Czech). Jaga Group. Bratislava, 302 p.

Tyc P (2000) Geosynthetic materials in construction layers of substructure. (In Czech). In: Proc. Substructure, Czech Railways Seminar, Ústí nad Labem, pp 61–69.

Tyc P, Vaníček I (1981) Geotextiles in railway track (In Czech) Železniční technika, No 3, pp 226–229.

Tyson SS (1994) Overview of coal ash use in the USA. In: Goumans, van der Sloot, and Aalbers (eds) Environmental Aspects of Construction with Waste Materials. Elsevier, pp 699–707.

U.S. Department of the Interior (1977) Teton dam failure review group. Failure of Teton Dam. A Report of Findings.

Van Impe WF (1989) Soil improvement techniques and their evolution. Taylor and Francis, 131 p.

Van Santvoort GPTM (ed) (1994) Geotextiles and geomembranes in civil engineering. Balkema, Rotterdam, 608 p.

Vaníček I (1971) Results of the Bending Tests for Soils from the Core of the Zarga Project Dam, Jordan. (In Czech), CTU Prague, 21 p.

Vaníček I (1973) Soil mechanics – Solved examples (In Czech). Publishing company of the Czech Technical University, Prague, 127 p.

Vaníček I (1975) The stress-strain behaviour of soils in tension. (In Czech). Ph.D. thesis, Czech Technical University in Prague, 170 p.

Vaníček I (1977a) Discussion to Bending test for compacted clays. JGED, ASCE, vol 103, No GT9, pp 1028–1030.

Vaníček I (1977b) Bending tension test on laboratory prepared and in situ taken sample from the core of rockfill dam. In: Proc. 5th Danube EC SMFE, Bratislava, CaS NC ISSMGE, vol 1, pp 415–419.

Vaníček I (1977c) Some aspects of tension tests. DIC thesis, The University of London, 121 p.
Vaníček I (1977d) Influence of moisture content on failure tensile strain of soils. Czech Technical University Press, Prague series KD1, pp 114–122.
Vaníček I (1978a) Conditions and reasons of tensile cracks development in fill dam body. (In Czech). Vodní Hospodářství, vol 28, row A, No 10, pp 261–267.
Vaníček I (1978b) Control tests of coefficients of permeability and consolidation for clay core of Dalešice dam. (In Czech), Research Report CTU Prague, 18 p.
Vaníček I (1979) Susceptibility of clay core of Dalešice dam to cracking (In Czech). Surveyor's Report. CTU Prague, 12 p.
Vaníček I (1980) A study of the deformation of the slope of reinforced soils by stereophotogrammetric method. In: Proc. 6th Danube European Conf. SMFE, Varna, Sect 3, pp 277–286.
Vaníček I (1981) Categorization of landslides with respects to slope stability calculation. (In Czech). In: National Seminar on Slope Stability. VTS, Karlovy Vary.
Vaníček I (1982a) Soil Mechanics. (In Czech). Publishing company of the Czech Technical University, Prague, 331 p.
Vaníček I (1982b) Simple non-standard laboratory tests before and during construction of Dalešice dam. In: Proc. 14th ICOLD, Rio de Janeiro, vol 1, pp 605–609.
Vaníček I (1983a) Laboratory investigation of the geotextile reinforcement on subsoil stability. In: Proc. VIII EC SMFE, Helsinki, Balkema, vol 1, pp 431–436.
Vaníček I (1983b) Discussion, specialty Session 5, In: Proc. 8th EC SMFE, Helsinki, Balkema, vol 3, pp 1184–1185.
Vaníček I (1985) Development and behaviour of cracks in clay core of fill dams. (In Czech) DSc. thesis, Czech Technical University Prague, 326 p.
Vaníček I (1986) Behaviour of soil of the high clayey spoil heap. In: Proc. 8th EC SMFE, Nürnberg, DGD, pp 235–240.
Vaníček I (1987) Geomechanical problems of spoil heaps. In: Proc. 1st Conf. on Mechanics, Prague, UTAM ČSAV, vol 6, pp 182–185.
Vaníček I (1988) Creation and behaviour of cracks in clay core of earth and rockfill dams. (In Czech). SNTL – Publishing house of technical literature, Praha, 168 p.
Vaníček I (1990) The utilization of clayey spoil heaps for new construction. In: Proc. 11th Polish Nat. Conf. SMFE, Krakow, vol II, pp 455–460.
Vaníček I (1991) Foundation on high spoil heaps. In: Proc. 10th EC SMFE, Florence, pp 629–632.
Vaníček I (1992) Geotechnical aspects of waste deposition. (In Czech). Report Grant G 1125. Ministry of Environment, Prague, 32 p.
Vaníček I (1995a) Ways of utilization of waste clayey material. In: Proc. 10th DEC SMFE, Mamaia, pp 967–974.
Vaníček I (1995b) The conception and solutions of underground waste storage in excavated rock spaces. State Programme – Environmental Protection. Czech Technical University, Prague. 120 p.
Vaníček I (1997) Proposal for TP 97 (Ministry of transport recommendation) – Geotextiles and other geosynthetics in earth structures for road engineering. (In Czech). Geotechconsult Praha, 40 p.
Vaníček I (1998) Tailing dams for flying ash – Some experiences gained in CR. In: Seco e Pinto PS (ed) Proc. 3rd ICEG, Balkema Rotterdam, pp 697–702.
Vaníček I (1999) Underground rock laboratory. (In Czech). Tunel, Praha, pp 18–24.
Vaníček I (2000) Deformation of high spoil heaps from clayey materials. In: Proc. GREEN III, Berlin, Thomas Telford, pp 120–124.
Vaníček I (2001a) Municipal solid waste management. (In Czech). Chapter 3 in Urban engineering 2 Šrytr (ed) Academia Prague, pp 129–181.
Vaníček I (2001b) Reinforced soils – Limit states applied on soil slopes, retaining structures and bridge abutments. (In Czech), Academic publisher CERM Brno, 115 p.
Vaníček I (2002a) Tailing dams – practical problems of ageing. In: de Mello and Almeida (eds) Proc. 4th ICEG, Rio de Janeiro, Brazil, Balkema Publishers, Lisse, vol 1, pp 301–304.

Vaníček I (2002b) Deposition of waste material on the surface. Main Lecture, Session 4: Environmental Protection. In: Proc. 12th Danube-European Conference Geotechnical Engineering. ISSMGE, Passau, Verlag Gluckauf GmbH, Essen, pp 543–554.
Vaníček I (2002c) Limit state approach to the reinforced soils. In: Proc. Geotechnics Through Eurocode 7, Croatian Geotechnical Society, Hvar, pp 59–66.
Vaníček I (2002d) Remediation of Landfills, Old Ecological Burdens. (In Czech). Publishing company of the Czech Technical University, Prague, 247 p.
Vaníček I (2003) Limit state of deformation. Introduction Lecture. In: Vaníček et al. (eds) Proc. 13th EC SMGE, Prague, CGtS, vol 3, pp 353–356.
Vaníček I (2004a) Lessons learned from the failures of ponds on river-basin Lomnice. (In Czech) In: Proc. Dam days. Czech Committee of ICOLD, Ceske Budejovice. 8 p.
Vaníček I (2004b) Dams of ponds and polders. (In Czech) Keynote Lecture. Dam days. Czech Committee of ICOLD, Ceske Budejovice, 13 p.
Vaníček I (2005a) Limit state approach and design of spread foundations. Panel Report. Session 2c: Excavation, Retaining Structures, and Foundations. 16th ICSMGE, Osaka 2005, p 4.
Vaníček I (2005b) Reinforced earth structures (In Czech). In: Turček et al. Foundation Engineering. Jaga, Bratislava, pp 213–224.
Vaníček I (2006) Final Report of the First Year Solution of the Research Project "Sustainable construction", (In Czech), Czech Technical University, 55 p.
Vaníček I, Chamra S (2001) Liapor in earth structures. (In Czech). Internal rep. CTU Prague, 42 p.
Vaníček I, Hasaan AA (2000) Laboratory pull-out tests evaluation for geogrid strip reinforcement. In: Mohamed AMO and Hosani KIAl (eds) Proc. 1st Int. Conf. on Geotechnical, Geoenvironmental Engineering and Management in Arid Lands – GEO2000, United Arab Emirates, Balkema, Rotterdam, pp 193–200.
Vaníček I, Kazda J (1995) Two examples of pollutant transport modelling. In: Proc. 11th EC SMFE, Copenhagen, DGS, vol 2, pp 145–150.
Vaníček I, Kudrnáčová I (1987) Usage of geotextiles in tailings dams design in CR (In Russian). In: Proc. 2nd Bulgarian Conf. on SMFE, Ruse, pp 413–424.
Vaníček I, Pachta V (1976) Problems connected with laboratory permeability determination for clay samples. (In Czech), Vodni Hospodarstvi 12/76, pp 319–323.
Vaníček I, Pauli J (1973) Bending tests for clay core of Rosino dam in Bulgaria. (In Czech). Czech Technical University, 12 p.
Vaníček I, Schröfel J (2000) Environment and civil engineering structures. (In Czech). Czech Technical University Publ Comp., 154 p. (1st edn in 1991).
Vaníček I, Škopek P (1989) Stability calculation of reinforced soil slope. In: Proc. 12th IC SMFE, Rio de Janeiro, 17/28, pp 1321–1324.
Vaníček I, Šťastný J (1981) Oedometric test with continuous loading. (In Czech). In: Proc. National Conf. Foundation Engineering, Brno, 1981, pp 171–174.
Vaníček I, Vacín O (1968) Comparison tests of intergranular equilibrium pressure for Přísečnice dam. (In Czech). Research Report VTS 040, Prague, 21 p.
Vaníček I, Vaníček J (2004) Rehabilitation of old earth dams failed during heavy floods in 2002. In: Wieland, Ren, and Tan (eds) Proc. New Developments in Dam Engineering, Balkema, pp 889–898.
Vaníček I, Vaníček M (1997) Geotechnical design of the bridge abutment from reinforced soil. (In Czech). In: Proc. 25 National Conference Foundation Engineering, CGtS, Brno, 4 p.
Vaníček I, Vaníček M (2001) Geosynthetics reinforcement – limit state approach. In: Proc. 15th IC SMGE, Istanbul 2001, vol 2, pp 1637–1640.
Vaníček I, Vaníček M (2006) Embankment of transport infrastructure and waste or recycled materials. In: Logar J, Gaberc A, and Majes B (eds) Proc. XIII. Danube European Conference on Geotechnical Engineering – Active Geotechnical Design in Infrastructure Development, Ljubljana, Slovenian Geotechnical Society, vol 2, pp 793–79.
Vaníček I, Záleský J (1981) Reasons of high settlements on 5 small dams in South town in Prague. (In Czech), Surveyor's Report. CTU Prague, 33 p.

Vaníček I, Záleský J (1984) Some aspects of oedometric test with continuous loading. In: Proc. 7th Polish Nat. Conf. SMFE, Poznaň 1984, vol 1.
Vaníček I, Škopek P, Herle I (1989) Triaxial tests on ash from electric power station Prunéřov. (In Czech). Research Report, CTU Prague, 32 p.
Vaníček I, Záleský J, Škopek P (1990) Laboratory tests of unsaturated ash. (In Czech). Research Report, CTU Prague, 50 p.
Vaníček I, Záleský J, Lamboj L, Kudrnáčová I (1995) Methods of permeability checking for clay liners. In: Proc. 10th DEC SMFE, Mamaia, pp 975–982.
Vaníček I, Boháč J, Řičica J, Záruba J (1997) Remedy of a chemical waste dump with foundation of a new dump on its surface. In: Proc. 14th IC SMGE, Hamburg, vol 3, pp 1857–1860.
Vaníček I, Řičica J, Záruba J, Šrámek P, Holada J (2003) Remediation of chemical landfills in Neratovice and Chabařovice. In: Vaníček et al. (eds) Proc. 13th EC SMGE, Prague, CGtS, vol 4, pp 279–294.
Vaníček I, Záleský J, Lamboj L, Kurka J (2005) Two examples of clay slope stability in areas affected by previous man-made activity – open pit mines, landfills. In: Proc. 16th IC SMGE, Osaka, Millpress, Rotterdam, vol 4, pp 2602–2606.
Vaníček J (2004) Principles of the reconstruction design for 3 dams on the river-basin Lomnice (In Czech). In: Proc. Dam Days, Czech Committee of ICOLD, České Budějovice, 8 p.
Vaníček J, Vaníček M (1999) Bridge abutment and geosynthetics. (In Czech). Stavitel, No 5/99, pp 48–49.
Vaníček M (1998) Construction of retaining walls from lightweight prefabricates and reinforced soil (In Czech). In: Proc. 26th Int. Conf. Foundation Engineering, Brno, ČGtS.
Vaníček M (2000) Limit design approach of the reinforced soils. Acta Polytechnica 2000, vol 40, No 2, pp 74–77.
Vaníček M (2003) Contaminant transport in the host rock, numerical modelling and laboratory work. Ph.D. thesis, Czech Technical University in Prague, 128 p.
Vaníček M (2006) Risk assessment for the case of waste utilization for embankment construction. In: Logar J, Gaberc A, and Majes B (eds) Proc. XIII. Danube European Conference on Geotechnical Engineering – Active Geotechnical Design in Infrastructure Development, Ljubljana, Slovenian Geotechnical Society, vol 2, pp 799–806.
Vaníček M, Vaníček J (2000) Stability calculation of reinforced slopes using program SVARG. (In Czech). Geotechnika, vol 3, No 2, pp 30–31.
Vaníček M, Košťál J, Vaníček I (1998) Analytical modelling of clay barrier on contaminant spreading. (In Czech). In: Proc. New Demands on Materials and Construction, Prague, The Czech Technical University.
Van Staveren M. (2006) Uncertainty and ground conditions. A Risk Management Approach. Elsevier, Amsterdam, 321 p.
Vaslestad J, Johansen TH, Holm W (1994) Load reduction on rigid culverts beneath high fills. Long term behavior. Directorate of Public Roads, Norwegian Road Research Laboratory, Oslo, Publication 74.
Vaughan PR (1972) Discussion. In: Proc. 5th EC SMFE, Madrid, vol 2, pp 72–75.
Vaughan PR (1976) Embankment Dams. Teaching Scripts. Imperial College, London, 61 p.
Vaughan PR, Soares HF (1982) Design of filters for clay cores of dams. JGED, ASCE, vol 108, No GT1, pp 17–31.
Vaughan PR et al. (1970) Cracking and erosion of Barderhead dam and the remedial works adopted for its repair. In: Proc. 10th ICOLD, Montreal, vol 1, p 73.
Veder CH, Prinzl F (1983) The avoidance of secondary settlements through overconsolidation. In: Proc. 8th EC SMFE, Helsinki, Balkema, vol 2, pp 697–706.
Verfel J (1979) Détermination des causes de défauts et leur reparation sur un barrage en enrochements. In: Proc. 13th ICOLD, New Delhi, vol 2, pp 297–304.
Vestad H (1976) Viddalsvatn dam. A history of leakage and investigations. In: Proc. 12th ICOLD, Mexico, vol 2, pp 369–390.

Větrovský M (2006) Contribution to the Problem of Foundation on Spoil Heaps in North-Bohemian Brown-Coal Basin. (In Czech). Ph.D. thesis. Technical University Ostrava, 169 p.

Vidmar S, Gabrc A, Kardelj E (1981) Successful highway construction on very soft soils. In: Proc. Xth IC SMFE, Stockholm, Balkema, Rotterdam, vol 1, pp 263–268.

Viezee DJ, Voskamp W, den Hoedt G, Troost GH (1990) Designing soil reinforcement with woven geotextiles-The effect of mechanical damage and chemical ageing on the long-term performance of polyester fibres and fabrics. In: Proc. 4th IC Geomembranes and Related Products, The Hague, Balkema, Rotterdam, vol 2, pp 651–656.

Votruba L, Broža V, Kazda I (1978). Dams. (In Czech). Czech Technical University Press, Praha, 330 p.

Vreeken C, van der Berg F, Loxham M (1983) The effect of clay-drain interface erosion on the performance of band-shaped vertical drains. In: Proc. 8th EC SMFE, Helsinki, Balkema, vol 2, pp 713–716.

Watts GRA, Brady KC, Greene MJ (1998) The creep of geosynthetics. TRL Report 319, UK, 40 p.

Whitman RV (1984) Evaluating calculated risk in geotechnical engineering. JGED, ASCE, vol 110, No 2, pp 145–188.

Wilson SD (1973) Deformation of earth and rockfill dams. In: Hirschfeld and Poulos (eds) Embankment-dam engineering, Casagrande volume. John Wiley and Sons, Toronto, pp 365–418.

Wilson SD (1977) Influence of field measurements on the design of embankment dams. In: Proc. 9th IC SMFE, Spec. Sess. 8 – Deformation of Earth-rockfill Dams, Tokyo, JGS, pp 19–36.

Wilson SD, Squier LR (1969) Earth and rockfill dams. In: Proc. 7th IC SMFE, Mexico, Mexican Geotechnical Society, State of the Art vol, pp 137–223.

Wolski W (1965) Model tests on the seepage erosion in silty clay core of an earth dam. In: Proc. 6th IC SMFE, Montreal, Toronto University Press, vol 2, pp 583–587.

Wolski W (2003) Tailings dams – selected question. In: Vaníček et al. (eds) Proc.13th EC SMGE, Prague, CGtS, vol 3, pp 167–172.

Yang KS, Chiam SL (1999) Geotechnical engineering for transportation infrastructures in Singapore. Part B: Airport developments. In: Barends et al. (eds) Proc. 12th EC SMGE, Balkema, vol 1, pp 229–237.

Yashima A (1997) Finite element analysis on earth reinforcement – Current and future. In: Ochiai, Yasufuku, and Omine (eds) Proc. Int. Symp. Earth Reinforcement – IS Kyushu 96, Balkema, vol 2, pp 1111–1116.

Yeung AT, Jiang S (1995) Appraisal of Ogata solution for soluble transport. JGED, ASCE, vol 121, No 2, February, pp 111–118.

Zaretsky JK, Vorontsov EI, Garitselov MJ (1977) Tensile strength of soil. (In Russian) In: Proc. 5th Danube EC SMFE, Bratislava, CaS NC ISSMFE, vol 1, pp 459–467.

Záruba Q (1978) Reasons of the Teton dam failure in USA. (In Czech) Vodní Hospodářství, vol 28, row A, No 12, pp 321–325.

Zavoral J, et al. (1987) Soil Mechanics – Methodology (In Czech). ČGÚ Praha, vol I, 187 p.

Zeman O (1979) Technical Petrography. (In Czech). Publishing company of the Czech Technical University, Prague, 179 p.

Zienkiewicz OC (1967) The finite element method. McGraw Hill, London, 3rd edn 1977.

Zienkiewicz OC, Taylor RL (2000) The Finite Element Method, fifth edn, vol 1: The basis, Butterworth Heinemann, Oxford, 689 p.

Zimmie TF, Moo-Young H, Laplante K (1995) The use of waste paper sludge for landfill cover material. In: Sarsby (ed) Proc. Green 93, Waste Disposal by Landfill, Balkema, Rotterdam, pp 487–496.

Zwek H, Davidenkoff R (1957) Étude expérimentale des filtres de granulométrie uniforme. In: Proc. 4th IC SMFE, London, vol 2, p 410.

Index

A

Aggregates
  recycled, 261
  secondary, 261
Airfield, 361
  artificial island, 362
  pavement, 368
Anchorage length, 271
Angle of shearing resistance
  influence of compaction effort, 66–67
Anti flood protection system
  mobile barriers, 496
Artificial (engineered) protection barrier, 561
Ash liquefaction, 535
Ash properties, 532

B

Bending test
  maximum tensile strength, 417
  tensile strain at failure, 416
  tension modulus, 418
Bentonite
  prefabricated blocks, 498
Borrow pit
  phases of investigation, 27
Bridge abutment, 327
  geosynthetic reinforced soil, 328

C

Cement stabilization, 337
  acceptance tests, 337
  control tests on the construction site, 337
  micro-crack rolling, 337
  mixing-in-place, 337
  mixing-in-place, rotovator, 337
  mixing-in-plant, 336
  suitability tests, 337

Clay
  fissured, 305
  fissured, overconsolidated, 307
  intact, 305
Clayey minerals, 9
  exchangeable cations, 10
  illite, 9
  index of colloidal activity, 10
  kaolinite, 9
  montmorillonite, 9
Coefficient
  pore pressure, 275
Collection
  pond, 377
  surface run-off, 377
Compacted clay liner
  control, 576
  durability, 575
  permeability verification, 573
Compaction, 33
  by free fall, 501
  degree, 34, 52
  degree, required, 39
  dynamic consolidation, 597
  effort, 34
  energy, 34, 38
  field trial, 49, 52
  laboratory methods, 36
  maximum dry density, 34
  mechanism, 34
  method, 33
  minimum dry density required, 39
  moisture content, required range, 39
  number of passes, 52
  optimum moisture content, 34, 36, 39
  quality control, 53
  relative density index, 41
  sensitivity to moisture content changes, 36
  suitable compacting machine, 51

technology, 33
vibration amplitude, 46
vibration effect, 34, 45
vibration frequency, 46
vibration frequency resonance, 46
vibration oscillatory rollers, 47
vibratory plates, 50
vibratory rammers, 50
vibratory roller, 46
Compaction control, 56
  coefficient of machine efficiency, 57
  coefficient of variation, 57–58
  continuous compaction control, 62
  deformation moduli ratio, 61
  direct methods, 59
  dry density determination, 55
  geophysical methods, 55
  modulus of deformation, 60
  moisture content difference, 57
  nuclear methods, 54
  penetration test, 54
  rock fill, 58
  static plate load test, 59
Compaction curve
  dry side, 35
  sets of, 35
  wet side, 34
Compaction effort
  tensile characteristics, 419
Compaction roller, 43
  automatically controlled roller, 47
  compaction efficiency, 47
  high energy impact compactors, 50
  polygon-drum, 49
  rubber-tyre, 43
  sheep-foot, 43
  speed, 47, 52
  vibration, counter-rotating exciter, 47
  with smooth surface, 43
Compaction test
  dynamic, 38
  static, 39
Concentrated leakage, 467
Consolidation
  speeding up, 282
Construction on spoil heaps
  limitation of high settlement, 522
Construction waste, 261
Contaminant spreading
  advection, 233
  diffusion, 233
  dispersion, 233
  dispersion coefficient, 375
  modelling, 371
  retardation factor, 375
  sorption, 233
Contaminated subsoil, 72
Crack
  longitudinal external, 397, 399
  longitudinal internal, 397, 403
  remediation, 470
  transversal external, 397, 398
  transversal internal, 397
Crack development
  desiccation, 398
  differential settlement, 398
  hydraulic fracturing, 398, 405
  seismic effect, 398
Cut widening, 310
Cuttings
  stiff fissured clay, 249

D

Dam construction
  potential risk of tensile cracks, 456
Dam failure
  domino effect, 479
  potential risk, 471
Dam rockfill/earth
  soil suitability, 17
Deformation characteristics
  modulus of deformation, 193
  modulus of elasticity, 190
  oedometric modulus, 190
  Poisson's ratio, 190, 193
  undrained modulus of deformation, 202
Design approaches according to EC 7, 166
Differential settlement, 463
Dikes, 491
Dikes maintenance, 492
Dispersion, 438
Dispersive soils, 441
Dispersive soils identification, 444
  crumb test, 446
  double hydrometer test, 445
  pinhole test, 445
  TDS test, 444
Domino effect, 488, 489
Double porosity, 512
Drainage function
  vertical drain, 93
Drained tensile test
  stress-strain hardening, 428
Dry crack, 392
Dynamic consolidation, 598

Index 631

E
Earth retaining structures, 39
Earth retaining system, 317
  externally stabilized, 317
  hybrid, 317
  internally stabilized, 317
Earth structure design, 153
  based on experience, 8
Earth structures
  environmental engineering, 497
Earth structures limit states
  deformation, 169, 187
  internal erosion, 169
  stability, 169
  surface erosion, 169
Earthwork, 31
Effective tensile strength, 431
EIA process, 26
Embankment
  from ash, 371
  from waste material, 371
  high fill, 268
  sandwich construction, 371
  transport infrastructure, 266
Embankment construction
  in stages, 282
  in winter, 269
Embankment design
  good quality subsoil, 266
  pile foundation, 294
  soft subsoil, 275
  widening, 299
Environmental geotechnics, 72
Erosion protection, 273
Ervenice corridor, 514, 523

F
Factor of safety, 162
Failure by heave, 158
Failure by internal erosion, 158, 161
Failure by piping, 158
Failure localization, 467
Failure remediation, 467
Fill
  reinforcement, 270
Fill dam
  analysis of failure, 385
  fault liability rate, 386
  insufficient slope stability, 386
Fill dam cracking
  dry conditions, 392
  wet conditions, 392
Fill dam deformation
  during construction, 388

during first filling, 387
horizontal displacement, 389
vertical displacement, 389
Fill dam design
  logical way, 475
Fill dam failure
  overflowing, 386
  seepage, 386
Fill lightening, 296
  ash, 297
  expanded clay, 297
  polystyrene, 297
Filter design, 207, 451
Filter material segregation, 209
Filtration criteria
  cohesive undisturbed base soil, 214
  non-cohesive base soil, 207
Filtration function
  blockage (clogging) criteria, 89
  permeability criteria, 89
  retention criteria, 89
Filtration layer thickness, 210
First group of limit states, 157
  EQU, 157
  GEO, 157
  HYD, 157
  STR, 157
  UPL, 157
Flocculated soils, 447
  size of individual floccules, 447
Flocculation, 438
Flood protection
  structures, 491
Flood protection dam
  limit states, 494
  specificity, 493
Flow net, 381
  hydraulic gradient, 381
  pore water pressure, 381
  seepage amount, 381
  seepage pressure, 381

G
Geocell, 100
Geocomposites, 73
Geodrain, 73, 93, 120
  discharge capacity, 121
  drainage core, 93
  equivalent diameter, 121
  filtration properties, 121
  smear zone, 122
Geogrid, 73
  extruded, 73
  prestressing, 293

woven, 73
Geomattress, 73, 100
Geomembrane, 377
Geomembrane sealing control, 579
Geomembrane welding
   extrusion welding, 577
   hot wedge, 577
Geomembranes, 73, 85
   flexibility, 102
   high density polyethylene, 102
   polyvinyl chloride, 102
   properties, 102
   resistivity to chemicals, 102
   very low density polyethylene, 102
Geosynthetic clay liner, 102, 377
   bentonite, 102
   Ca-montmorillonite, 103
   Na-montmorillonite, 103
Geosynthetic properties, 75
   design elongation (strain), 82
   design tensile strength, 82
   abrasion tests, 86
   adhesion between soil and geosynthetics, 83
   anchorage length, 83
   angle of friction between soil and geosynthetics, 83
   characteristic opening size, 77
   creep, 139
   creep coefficient, 141
   elongation (strain) at break, 82
   filtration coefficient, 78
   friction resistance between soil and geosynthetics, 83, 84
   inclined plane test, 83
   interface friction angle, 85
   penetration tests, 86
   permitivity, 79
   pull out test, 83
   shear box test, 83
   stable creep, 140
   tearing tests, 86
   tensile strength, 81, 94
   tensile strength at break, 82
   thickness, 76
   transmissivity, 79
   unit weight, 76
   unstable creep, 140
   UV light resistance, 99
Geosynthetic soil reinforcement
   post-construction elongation, 142
   prestress of reinforcement, 143
Geosynthetics, 71
   filters, 71
   identification on site, 75
   aramid, 72
   bentonite mattress, 101, 103
   design life, 72
   design tensile strength, 182
   drainage function, 92
   drains, 71
   filtration function, 87
   handling on construction site, 105
   polyamide, 72
   polyester, 72
   polyethylene, 72
   polypropylene, 72
   polyvinyl alcohol, 72
   polyvinyl chloride, 72
   protective function, 97
   reinforcing element, 71
   reinforcing function, 94
   sealing function, 101
   separation function, 97
   surface erosion protection function, 98
   tensile strength, 71
Geotechnical categories, 153
Geotechnical design
   associated risk, 475
Geotextile, 71, 73
   high strength, 290
   knitted, 73
   low creep, 290
   natural fibres, 75
   nonwoven, 73
   woven, 73
Geotextile filter, 455
Green reinforced slopes, 273
Groundwater
   lowering, 308
Groundwater level, 32

H

Horizontal sealing system, 604
Hydraulic fracturing, 394

I

Inclinometer, 247
Industrial by-products, 261
Internal erosion, 433, 466–467
   susceptibility, 477
Interparticles forces, 437

L

Laboratory model, 234
   bearing capacity, 236
   centrifugal test, 239

Index   633

deformation, 236
groundwater flow, 244
stereophotogrammetric method, 238
Landfill
   bottom sealing system, 557
   capping sealing system, 557
   design, 561
   drainage system, 579
   gas, 558
   leachate, 558
   limit states, 585
   monitoring, 589, 604
   settlement, 587
   stability, 585
Landfill construction
   geosynthetics application, 588
Landfill remediation
   horizontal sealing system, 604
Landfill surface
   utilization, 594
Landslide prone areas, 262
Leachate
   composition, 371
   testing, 374
Lime stabilization, 129
   light compaction, 132
   lime proportioning, 128
   maturing period, 132
   strength improvement, 128
   sulphate soils, 131
   workability improvement, 128
Limit equilibrium method, 162
Limit state of internal erosion, 204
Limit state of stability, 171
   reinforced soils, 181
Limit state of surface erosion, 217
   dry conditions, 218
   material susceptible to erosion, 218
   wet conditions, 218
Load
   permanent, 155
   variable, 155
Loading states, 156
Logical scheme
   of second degree, 477
Longitudinal outer crack, 461

M
Macro porosity, 501, 512
Methods of limit states verification
   adoption of prescriptive measures, 167
   calculation, 167
   experimental models, 168

   observational method, 168
Modelling
   contaminant spreading, 371, 565
   railway track, 356
Modelling of limit state, 223
   laboratory model, 223
   numerical model, 223
   real-scale model, 223
Municipal landfill
   principles, 558
Municipal waste
   material, 556
   properties, 582
Municipal waste disposal sites, 33, 560

N
Natural (geological) protection barrier, 561
Natural hazards, 491
Neutral line, 265
Noise protection barrier, 378
Numerical model
   boundary conditions, 225, 231–232
   constitutive relationships, 226
   contact elements, 230
   contamination spreading, 233
   failure criteria, 227
   FEM, 224
   reinforced soil, 229
   steady flow, 231

O
Old chemical landfill, 591
   remediation, 591, 595
Old small earth dams failures, 479

P
Parameters
   contaminant transport modelling, 371
Parking lot, 345
Particles orientation, 437
Pavement
   base course, 336
   base course, cement stabilized, 336
   crack, 336
   ruts on asphalt surface, 336
   structure, 334
   surface, 336
Percentage of sodium, 441
Perfect filter, 455
Permeability
   influence of compaction effort, 70
Piezometer, 247
Piping, 433
Pore pressure

dissipation, 283
Pore water pressure
 speed of dissipation, 507
Precompression, 115
Preconsolidation, 115
Predestined seepage path, 432
Preloading
 additional surcharge, 281
 vacuum, 281, 287
Preloading technique, 113, 115
Proctor test, 36
 modified, 36
 standard, 36
Progressive erosion, 432
Progressive failure, 305
 conditions, 306

R

Radioactive waste, 498
Railway system, 346
 high speed, 346, 355
 higher speed, 346
 old lines, 346
Railway track
 allowable differential settlement, 359
 ballast, 347
 earth plain, 347
 high speed, critical speed phenomenon, 358
 modelling, 356
 multi-layered, 356
 reconstruction, 348, 353
Railway track reconstruction
 ballast removal, 353
 recycled ballast, 353
Reinforced earth structure
 facing type, 145
 prestressed, 143
 slope stability, 187
Reinforced soil
 composite model, 229
 discrete model, 229
 pull-out test, 241
Reinforcement
 horizontal, 302
 vertical, 302
Reinforcement of contact, 281, 289, 292
 mining, karstic areas, 292
 sinkhole, 293
Residues after waste burning, 556
Resistance to erosion, 433
Retaining wall, 317
 gabion facing, 321

Retaining wall geosynthetic-reinforced soil
 Retaining Wall (GRS-RW System), 324
Retaining wall L – shaped facing, 321
Retaining wall partial height concrete panel, 319
Retaining wall segmental concrete block, 319
Risk
 analysis, 374
Risk assessment example, 372
Risk principle, 153
Road resurfacing, 344

S

Sand drain, 118
Sandwich embankment, 17
Sealing repair, 409
Sealing system, 562
 GL – geomembrane liner, 563
 bottom, 564
 capping, 564
 CCL – compacted clay liner, 563, 568
 combined, 564
 elementary, 563
 GCL - geosynthetics clay liner, 563
Seepage crack behaviour, 433
 cohesive soils, 434
 frictional soils, 433
Self-sealing by swelling, 458
Serviceability limit state, 155, 316
Settlement
 consolidation settlement, 116, 188
 initial settlement, 116, 188
 secondary consolidation, 116
 total, 200
Slope stability
 combined, 96
 external, 313
 internal, 313
 long-term, 281, 305
 long-term, increasing, 308
 short-term, 276, 305
Small dams failures
 internal erosion, 480
 surface erosion, 480
Small earth dams, 479
Small earth dams failure
 reconstruction, 484
Sodium adsorption ratio, 441
Soil
 alluvial soils, 22
 boulder clay, 22
 boulders, 8

Index   635

clay particles, 8
clayey minerals, 22
cobbles, 8
colluvial soils, 22
eolic soils, 23
fissured tertiary clays, 24
glacial sediments, 23
gravel particles, 8
index tests, 7
loess deposits, 23
man-made sediments, 20
normally consolidated, 24
overconsolidated, 24
quick clays, 24
residual soils, 20
rock-forming minerals, 22
sand particles, 8
sea sediments, 23
secondary minerals, 22
sediments, 20
silt particles, 8
soil classification systems, 7
topsoil, 24
varved clay, 23
weathering process, 21
Soil as structural material
  borrow pit, 20, 26
Soil classification systems
  amount of carbonates, 14
  amount of mica, 15
  Czech classification systems, 12
  organic matter, 12
  soil colour, 15
  USCS, 11
Soil compaction, 7, 42
  compacted layer thickness, 43
  compacting plates, 43
  compacting rammers, 43
  dynamic consolidation, 45
  heavy tamping, 45
  intelligent compaction, 63
  layer thickness, 52
  loading plate compaction controls, 46
  Ménard's method, 45
  rubber-tyre rollers, 44
  specific pressure, 43
  surface-near re-loosening, 45
  thixotropic consolidation, 45
Soil consolidation, 109
  degree of consolidation, 111
  dissipation of pore water pressure, 109
Soil improvement, 17, 71, 109
Soil investigation

data credibility, 29
disturbed sample, 28
geophysical methods, 28
homogeneity of a set, 31
penetration methods, 29
percussion drilling, 28
rotary drilling, 28
undisturbed sample, 28
window sample, 28
Soil nailing
  corrosion protection, 147
  design, 150, 312
  driven nail, 147
  flexible structural facing, 311
  grouted nail, 147
  hard structural facing, 311
  nail spacing, 147
  pull-out resistance, 315
  soft facing, 310
Soil properties
  Atterberg limits, 10
  coefficient of consolidation, 110
  coefficient of pore water pressure, 114
  coefficient of secondary compression, 116
  coefficient of uniformity, 9
  curvature number, 9
  double porosity sample, 69
  dry density, 34, 41
  grain size distribution curve, 19
  liquid limit, 10
  macropores between individual clods, 42
  macrostructure, 69
  maximum density, 41
  maximum dry density, 42
  maximum porosity number, 41
  microstructure, 68
  minimum dry density, 42
  minimum porosity number, 41
  moisture content, 40
  plastic limit, 10
  plasticity index, 10
  relative density index, 67
Soil reinforcement, 17, 71, 133
  anchorage length, 95, 186
  bridge abutment, 136
  contact between embankment and soft
    subsoil, 134
  embankment, 133
  external stability, 96
  geosynthetic soil reinforcement, 133
  internal stability, 96
  macro, 94
  micro, 94

nailing, 145
preloaded and prestressed, 94
retaining wall, 135
soil nailing reinforcement, 133
soil suitability, 138
suitability of geosynthetic reinforcement, 139
Soil sample
technological, 37
Soil stabilization, 123
bonding agents, 125
cement stabilization, 125
frost susceptibility, 124
lime modification, 131
lime stabilization, 125, 131
mechanically stabilized soil, 123
Soil strength anisotropy, 277
geometrical, 277
stress, 277
Speed of filling, 507
Speed of reservoir filling, 475
Spillway
additional, unprotected, 489
Spoil heap
monitoring – Ervěnice corridor, 514
settlement, 517
settlement rate prediction, 519
Spoil heap deformation
collapses phenomena, 513
Spoil heaps, 499
change of properties, 500
deformation characteristics, 511
effective strength, 508
from clayey material, 499
hardening, 502
inner and outer, 500
landslide, 499
pore pressure increase with depth, 511
softening, 502
stability, 506
Stability
long-term, 155
short-term, 155
Static plate load test, 59
plate diameter, 61
time (duration), 62
Stress path
to failure, 308
Stress paths, 195
control, 202
Stress redistribution, 394
Structural strength, 204
Suction potential, 503

Surface erosion, 32
Swelling, 433
capacity, 395
potential, 435, 458

T

Tailing dams, 526
ageing, 539
chemical sludges, 556
construction, 527
drainage system, 530
limit states, 533
location, 527
oil sands, 556
paper mill sludges, 555
remedy measures, 541
slope stability, 534
surface erosion, 536, 545
type, 527
Tailing dams construction
hydraulic filling, 529
hydro cyclones, 529
Tensile crack development
geometrical conditions, 471
material conditions, 472
Tensile cracks
landfill differential settlement, 587
Tensile test
drained conditions, 426
in situ, 422
influence of time, 423
Tensile test of soils
axial test, 412
bending test, 412
hollow cylinder, 412
indirect (Brazilian), 412
triaxial test, 413
undrained conditions, 412
Tensile zone
development, 457
Triaxial tensile test
stress-strain behaviour, 428

U

Ultimate limit state, 155, 313
Underpass, 330
conduit, 330
culvert, 330
high embankment, 331
low embankment, 331
pipe, 330

Undrained shear strength
    influence of compaction effort, 69
Unpaved road, 339
Unsaturated tailings, 535
Uranium tailings, 547
    capping system, 548
    estimation of total settlement, 554
    gamma radiation, 549
    radon emission, 549

V
Vacuum preloading, 122
Vertical drain, 112, 118, 281, 282, 288

W
Waste
    leachate composition, 371
    leachate testing, 374
Waste classification, 557
Wet crack, 393